Vector Bundles

VOLUME 1

Foundations and Stiefel–Whitney Classes

This is a volume in
PURE AND APPLIED MATHEMATICS

A Series of Monographs and Textbooks

Editors: SAMUEL EILENBERG AND HYMAN BASS

A list of recent titles in this series appears at the end of this volume.

Vector Bundles

VOLUME 1

Foundations and Stiefel–Whitney Classes

Howard Osborn

Department of Mathematics
University of Illinois at Urbana–Champaign
Urbana, Illinois

1982

ACADEMIC PRESS
A Subsidiary of Harcourt Brace Jovanovich, Publishers

New York London
Paris San Diego San Francisco São Paulo Sydney Tokyo Toronto

ACADEMIC PRESS, INC.
111 Fifth Avenue, New York, New York 10003

United Kingdom Edition published by
ACADEMIC PRESS, INC. (LONDON) LTD.
24/28 Oval Road, London NW1 7DX

Library of Congress Cataloging in Publication Data

Osborn, Howard.
Vector bundles.

(Pure and applied mathematics;)
Includes bibliographical references and index.
Contents: v. 1. Foundations and Stiefel-Whitney classes.
I. Vector bundles. 2. Characteristic classes.
I. Title. II. Series: Pure and applied mathematics (Academic Press);
QA3.P8 [QA612.63] 510s [514'.224] 82-1639
ISBN 0-12-529301-1 (v. 1)

PRINTED IN THE UNITED STATES OF AMERICA

82 83 84 85 9 8 7 6 5 4 3 2 1

To Jean,
and to our children,
Mark, Stephen, Adrienne, and Emily

Contents

Chapter III **Vector Bundles**

Chapter IV **$\mathbb{Z}/2$ Euler Classes**

Chapter V **Stiefel–Whitney Classes**

Chapter VI **Unoriented Manifolds**

Preface

Vector bundles provide the background for the proper formulation of many classical and modern problems of differential topology, whose solutions via characteristic classes and K-theory are among the mathematical triumphs of the past few decades. These volumes are an introduction to vector bundles, characteristic classes, and K-theory and to some of their applications:

Volume 1: *Foundations and Stiefel–Whitney Classes*
Volume 2: *Euler, Chern, and Pontrjagin Classes*
Volume 3: *K-Theory and Integrality Theorems*

The exposition is based on various courses the author has presented at the University of Illinois, with a one-semester course in singular homology and cohomology as the only prerequisite; no further background is necessary. Appropriate portions of differential topology, Lie groups, and homotopy theory, for example, are explicitly introduced as needed, with complete proofs in most cases; in a few exceptional cases, specific references are substituted for proofs.

Each chapter ends with illustrative remarks and exercises. The remarks constitute a guide to much of the literature on vector bundles, characteristic classes, and K-theory from 1935 to 1981.

This first volume is designed both for self-study and as a classroom text about real vector bundles and their $\mathbb{Z}/2$ characteristic classes. For classroom use, one should perhaps not dwell on the background details of Chapter I, whose results are often taken for granted in any event; only some scattered definitions and the Mayer–Vietoris technique itself need any special emphasis. The remaining chapters can then easily be adapted for presentation either in a one-quarter course or in a one-semester course, with very few omissions in the latter case.

The author is indebted to many mathematical friends for their direct and indirect contributions to this work: to Professor S. S. Chern for an exciting and informative 1953 graduate course based on the newly published Steenrod [4]; to Professor John Milnor, who has been an inspiration since shared undergraduate days at Princeton University, and who provided Milnor [3]; to Professor René Thom, whose pithy and patient explanations turned mathematical abstractions into virtually tactile objects; to Professor Emery Thomas, whose clearly delivered lectures (Thomas [2]) rekindled the author's interest in vector bundles; to Professor Peter Hilton and Professor Raoul Bott, who provided much encouragement and moral support during early stages of the project; to Professor Felix Albrecht and Professor Philippe Tondeur, who read large portions of the completed typescript and provided many constructive suggestions; to Professor Wu Wen-Tsün, who addressed himself to the details of Chapter VI with the astonishing enthusiasm one expects only of someone discovering a new world (which he had in fact helped to create); to Professor Hassler Whitney, who created vector bundles in the first place, and who warmly shared several days of companionship and chamber music; and to many other mathematical friends and colleagues who helped clarify questions as they arose.

The author is equally grateful to Mrs. Doris Bartle for her assistance in proofreading the bibliography and to the staff of Academic Press for their cooperation during the production of the book.

I thank my wife, Jean, and our children, Mark, Stephen, Adrienne, and Emily, for their patience, understanding, and love throughout the entire project.

University of Illinois at Urbana–Champaign HOWARD OSBORN
June, 1982

Introduction

In the first half of this volume both real and complex vector bundles are introduced and classified, and many examples are given which will play a role in later applications. The second half of the volume concentrates exclusively on certain $\mathbb{Z}/2$ cohomology classes assigned to real vector bundles; there are several applications to classical problems in differential topology. (Complex vector bundles reappear in the next volume, where further cohomology classes are assigned to both real and complex vector bundles.)

Let $E \xrightarrow{\pi} X$ be a continuous map from a topological space E (the *total space*) onto a topological space X (the *base space*). Suppose that there is an open covering $\{U_i | i \in I\}$ of X such that each inverse image $\pi^{-1}(U_i) \subset E$ is homeomorphic to the product $U_i \times \mathbb{R}^m$. The homeomorphisms $\pi^{-1}(U_i) \xrightarrow{\approx} U_i \times \mathbb{R}^m$ need not be unique; however, they will belong to a certain family of homeomorphisms whose composition with the first projection $U_i \times \mathbb{R}^m \to U_i$ is the restriction of π to $\pi^{-1}(U_i)$. If an intersection $U_i \cap U_j$ of two sets in the covering $\{U_i | i \in I\}$ is nonempty, then for each $x \in U_i \cap U_j$ one has a composition $\{x\} \times \mathbb{R}^m \xleftarrow{\approx} \pi^{-1}(\{x\}) \xrightarrow{\approx} \{x\} \times \mathbb{R}^m$, hence a homeomorphism $\mathbb{R}^m \to \mathbb{R}^m$. If all the latter homeomorphisms result from the usual action $GL(m, \mathbb{R}) \times \mathbb{R}^m \to \mathbb{R}^m$ of the general linear group $GL(m, \mathbb{R})$, then the *projection* $E \xrightarrow{\pi} X$ is a *coordinate bundle* representing a *real vector bundle* ξ *of rank* m over the base space X. The bundle ξ is also called a *real* m*-plane bundle* over X, its *fiber* being $\mathbb{R}^m$.

For example, any smooth m-dimensional manifold X is covered by open coordinate patches U_i, each of which has coordinate functions $x^1, \ldots, x^m$

used to describe a homeomorphism from U_i to an open set in $\mathbb{R}^m$; the partial differentiations $\partial/\partial x^1, \ldots, \partial/\partial x^m$ at each point of U_i form a basis of a vector space $\mathbb{R}^m$. Let U_j be another open coordinate patch on X such that the intersection $U_i \cap U_j$ is nonempty, and let $y^1, \ldots, y^m$ be coordinate functions on U_j. Then over each point of $U_i \cap U_j$ one has $\partial/\partial y^j = (\partial x^1/\partial y^j)\,\partial/\partial x^1 + \cdots + (\partial x^m/\partial y^j)\,\partial/\partial x^m$ for $j = 1, \ldots, m$, so that the two copies of $\mathbb{R}^m$ over each point of $U_i \cap U_j$ are related by the usual linear action of the jacobian matrix $(\partial x^i/\partial y^j) \in GL(m, \mathbb{R})$. This suggests that X has a well-defined real m-plane bundle associated to its differentiable structure: the *tangent bundle* $\tau(X)$ of X.

Now let $\mathbb{Z}/2$ be the field of integers modulo 2, and let $\mathbb{Z}/2[[t]]$ be the ring of formal power series over $\mathbb{Z}/2$. To each element $f(t) \in \mathbb{Z}/2[[t]]$ with leading term $1 \in \mathbb{Z}/2$, and to each real vector bundle ξ over a base space X, there is a naturally defined cohomology class $u_f(\xi)$ in the direct product $H^{**}(X; \mathbb{Z}/2)$ of the singular cohomology modules $H^p(X; \mathbb{Z}/2)$. In the special case $f(t) = 1 + t \in \mathbb{Z}/2[[t]]$ the class $u_f(\xi)$ is the *total Stiefel–Whitney class* $w(\xi)$ of ξ, which can be used to compute any of the other classes $u_g(\xi)$. For $g(t) = (1 + t)^{-1} = 1 + t + t^2 + \cdots \in \mathbb{Z}/2[[t]]$ the class $u_g(\xi) \in H^{**}(X; \mathbb{Z}/2)$ is the (*total*) *dual Stiefel–Whitney* class $\bar{w}(\xi)$ of ξ.

Here is one of the many classical applications of such $\mathbb{Z}/2$ *characteristic classes*. Let X be a smooth m-dimensional manifold, and let $\bar{w}(\tau(X)) \in H^*(X; \mathbb{Z}/2)$ $[= H^{**}(X; \mathbb{Z}/2)]$ be the dual Stiefel–Whitney class of its tangent bundle $\tau(X)$. Then a necessary condition for the existence of a smooth (proper) embedding of X into the euclidean space $\mathbb{R}^{2m-p}$ is that $\bar{w}(\tau(X))$ vanish in each summand $H^q(X; \mathbb{Z}/2)$ for which $q \geqq m - p$. Using this criterion one easily constructs a smooth closed m-dimensional manifold X with *no* smooth embedding in $\mathbb{R}^{2m-\alpha(m)}$, where $\alpha(m)$ is the number of 1's in the dyadic expansion of m.

One can define vector bundles ξ over arbitrary topological spaces. However, many useful properties of vector bundles depend upon further restrictions; in this book those restrictions are imposed directly on the base spaces themselves. The "category $\mathscr{B}$ of base spaces" is described in Chapter I, along with the category $\mathscr{W}$ of spaces of homotopy types of CW spaces and the category $\mathscr{M}$ of smooth manifolds; since there are inclusions $\mathscr{M} \subset \mathscr{W} \subset \mathscr{B}$ the restriction to vector bundles over base spaces $X \in \mathscr{B}$ is not especially stringent. (For those specialists who prefer to consider *numerable* bundles over arbitrary base spaces, any bundle over a space $X \in \mathscr{B}$ is automatically numerable; numerability is discussed in Remark II.8.4.)

Let $G \times F \to F$ be any effective action of a topological group G on a topological space F. If one substitutes F, G, and $G \times F \to F$ for $\mathbb{R}^m$, $GL(m, R)$, and $GL(m, \mathbb{R}) \times \mathbb{R}^m \to \mathbb{R}^m$ in the earlier sketch of a definition of real m-plane bundles, the result is a *fiber bundle* ξ with *fiber* F and *structure group* G. For example, for each real m-plane bundle ξ there is a corresponding *projective bundle* P_ξ, whose fiber is the real projective space RP^{m-1} and whose structure group is the projective group $PGL(m, \mathbb{R})$. Fiber bundles (over base spaces $X \in \mathscr{B}$) are introduced in Chapter II, which has several results needed in later chapters. One of the main results of Chapter II is that if the structure group of a fiber bundle is a Lie group G, then one can always replace G by any maximal compact subgroup $H \subset G$; for example, the structure group $GL(m, \mathbb{R})$ of any real m-plane bundle can always be replaced by the orthogonal group $O(m) \subset GL(m, \mathbb{R})$. Another major result of Chapter II is the Leray–Hirsch theorem, which asserts for certain coordinate bundles $E \xrightarrow{\pi} X$ that $H^*(E; \mathbb{Z}/2)$ is a free $H^*(X; \mathbb{Z}/2)$-module; this is a crucial step in the later construction of $\mathbb{Z}/2$ characteristic classes.

If $X \xrightarrow{f} Y$ is any map in the category $\mathscr{B}$ of base spaces, and if η is any real m-plane bundle over Y, the *pullback* $f^!\eta$ is a real m-plane bundle over X which is uniquely defined by η and the homotopy class $[f]$ of f. One of the main results of Chapter III is that there is a *universal real m-plane bundle* γ^m over the *real Grassmann manifold* $G^m(\mathbb{R}^\infty)$ such that any real m-plane bundle ξ over any $X \in \mathscr{B}$ is of the form $f^!\gamma^m$ for a unique homotopy class $[f]$ of maps $X \xrightarrow{f} G^m(\mathbb{R}^\infty)$. For this reason $G^m(\mathbb{R}^\infty)$ is also called the *classifying space* for real m-plane bundles.

If $E \xrightarrow{\pi} X$ represents a real m-plane bundle ξ over $X \in \mathscr{B}$, then there is a well-defined *zero-section* $X \xrightarrow{\sigma} E$ carrying each $x \in X$ into that point $\sigma(x) \in E$ whose image under any of the local homeomorphisms $\pi^{-1}(U_i) \to U_i \times \mathbb{R}^m$ with $x \in U_i$ is $(x, 0) \in U_i \times \mathbb{R}^m$. If E^* is the subspace $E - \sigma(X)$ of E, then there is a $\mathbb{Z}/2$ *Thom class* $U_\xi \in H^m(E, E^*; \mathbb{Z}/2)$ which is uniquely characterized in Chapter IV. The composition $X \xrightarrow{\sigma} E \xrightarrow{j} E, E^*$ of the zero-section with the inclusion j then provides the $\mathbb{Z}/2$ *Euler class* $e(\xi) = \sigma^* j^* U_\xi \in H^m(X; \mathbb{Z}/2)$.

Let ξ be a real m-plane bundle over $X \in \mathscr{B}$, and let P_ξ be the corresponding projective bundle. The notation P_ξ is used ambiguously for the total space of any representative coordinate bundle $P_\xi \to X$ of P_ξ, and there is a canonical *real line bundle* (= real 1-plane bundle) λ_ξ with base space P_ξ. One uses the Leray–Hirsch theorem to verify that $H^*(P_\xi; \mathbb{Z}/2)$ is a free $H^*(X; \mathbb{Z}/2)$-module with a basis $\{1, e(\lambda_\xi), \ldots, e(\lambda_\xi)^{m-1}\}$ consisting of cup products of the $\mathbb{Z}/2$ Euler class $e(\lambda_\xi) \in H^1(X; \mathbb{Z}/2)$; the cup product operation $\cup e(\lambda_\xi)$ is itself an endomorphism $H^*(P_\xi; \mathbb{Z}/2) \to H^*(P_\xi; \mathbb{Z}/2)$ over $H^*(X; \mathbb{Z}/2)$. The direct product $H^{**}(P_\xi; \mathbb{Z}/2)$ is similarly a free $H^{**}(X; \mathbb{Z}/2)$-module, so that for any $f(t) \in \mathbb{Z}/2[[t]]$ with leading term $1 \in \mathbb{Z}/2$ one can define the $\mathbb{Z}/2$ characteristic class $u_f(\xi) \in H^{**}(X; \mathbb{Z}/2)$ to be the determinant of the induced free $H^{**}(X; \mathbb{Z}/2)$-module endomorphism $H^{**}(P_\xi; \mathbb{Z}/2) \xrightarrow{\cup f(e(\lambda_\xi))} H^{**}(P_\xi; \mathbb{Z}/2)$; the details of the construction are given in Chapter V.

For any real m-plane bundle ξ over any $X \in \mathscr{B}$ the *(total) Stiefel–Whitney class* $w(\xi)$ is the class $u_f(\xi)$ for $f(t) = 1 + t$. One easily shows that $w(\xi)$ vanishes in all summands $H^p(X; \mathbb{Z}/2) \subset H^{**}(X; \mathbb{Z}/2)$ with $p > m$, so that $w(\xi) = 1 + w_1(\xi) + \cdots + w_m(\xi) \in H^*(X; \mathbb{Z}/2)$, where $w_p(\xi) \in H^p(X; \mathbb{Z}/2)$ for $p = 1, \ldots, m$; furthermore $w_m(\xi)$ is just the $\mathbb{Z}/2$ Euler class $e(\xi) \in H^m(X; \mathbb{Z}/2)$. For any formal power series $g(t) \in \mathbb{Z}/2[[t]]$ whatsoever, with leading term $1 \in \mathbb{Z}/2$, there is an alternative computation of $u_g(\xi) \in H^{**}(X; \mathbb{Z}/2)$ from the classes $w_1(\xi), \ldots, w_m(\xi)$ alone, using the Hirzebruch *multiplicative sequence* associated to $g(t)$; this construction is also given in Chapter V.

According to one of the main results of Chapter III, any real m-plane bundle ξ over any $X \in \mathscr{B}$ is of the form $f^!\gamma^m$ for a unique homotopy class $[f]$ of maps $X \xrightarrow{f} G^m(\mathbb{R}^\infty)$. One easily shows that $w(\xi) = w(f^!\gamma^m) = f^*w(\gamma^m) \in H^*(X; \mathbb{Z}/2)$ for the total Stiefel–Whitney class $w(\gamma^m) \in H^*(G^m(\mathbb{R}^\infty); \mathbb{Z}/2)$ of the universal real m-plane bundle γ^m over $G^m(\mathbb{R}^\infty)$, where $H^*(G^m(\mathbb{R}^\infty); \mathbb{Z}/2) \xrightarrow{f^*} H^*(X; \mathbb{Z}/2)$ is induced by $[f]$. The cohomology ring $H^*(G^m(\mathbb{R}^\infty); \mathbb{Z}/2)$ is therefore of interest: it is the source of all $\mathbb{Z}/2$ characteristic classes. The main result of Chapter V is that $H^*(G^m(\mathbb{R}^\infty); \mathbb{Z}/2)$ is the polynomial ring $\mathbb{Z}/2[w_1(\gamma^m), \ldots, w_m(\gamma_m)]$ with one generator $w_p(\gamma^m) \in H^p(G^m(\mathbb{R}^\infty); \mathbb{Z}/2)$ in each of the degrees $p = 1, \ldots, m$.

Chapter VI contains some of the many applications of $\mathbb{Z}/2$ characteristic classes, including the nonembedding result sketched earlier in this Introduction. The vector bundles of primary interest in Chapter VI are tangent bundles $\tau(X)$ of smooth manifolds X.

The only prerequisite needed for the sequel is a modest background in singular homology and cohomology, as promised in the Preface. Such a background can easily be gleaned from any one or two of the following standard references: Artin and Braun [1, 2], Dold [8], Eilenberg and Steenrod [2], Greenberg [1], Hilton and Wylie [1], Hu [4, 5], Massey [6], Spanier [4, Chapters 4 and 5], Vick [1], and Wallace [6, Chapters 1–4].

Portions of differential topology, Lie groups, and homotopy theory will be introduced in detail as needed. References to other texts containing elementary introductions to topics in fiber bundles in general, vector bundles in particular, and characteristic classes, will also be given in appropriate later chapters.

CHAPTER I

Base Spaces

0. Introduction

A vector bundle over a topological space X consists in part of a projection $E \to X$ from another topological space E onto X; the space X is the *base space* of the given bundle. Vector bundles exist over arbitrary base spaces; however, interesting theorems are accessible only for bundles over appropriately restricted base spaces. Accordingly, in this chapter we introduce a large category $\mathscr{B}$ of topological spaces, and in the remainder of the book we consider only those bundles whose base spaces lie in $\mathscr{B}$.

In §1 the category $\mathscr{B}$ is defined, and it is shown to be closed with respect to finite disjoint unions and finite products. In §§2–7 one learns that $\mathscr{B}$ is indeed a large category, containing all spaces of the homotopy types of CW spaces, for example; reasonable topologists seldom ask for more. The most important feature of $\mathscr{B}$ is that it is a category in which one can prove theorems by the Mayer–Vietoris technique, described and established in §9; several applications of the Mayer–Vietoris technique will appear in later chapters.

1. The Category of Base Spaces

A *map* from a topological space X to a topological space Y is any function $X \xrightarrow{f} Y$ that is continuous in the given topologies. If $[0, 1]$ is the closed unit interval, then any map $X \times [0, 1] \xrightarrow{F} Y$ induces restrictions to $X \times \{0\}$

and $X \times \{1\}$, which can be regarded as maps $X \xrightarrow{f_0} Y$ and $X \xrightarrow{f_1} Y$, respectively; in this case f_0 is *homotopic* to f_1. Two topological spaces X and Y are *homotopy equivalent*, or of the same *homotopy type*, whenever there are maps $X \xrightarrow{f} Y$ and $Y \xrightarrow{g} X$ such that the compositions $g \circ f$ and $f \circ g$ are homotopic to the identity maps $X \to X$ and $Y \to Y$, respectively. A topological space X is *contractible* whenever it is homotopy equivalent to the space $\{*\}$ consisting of a single point $*$.

The category $\mathscr{B}$ will be defined in two stages, the first being an induction on the natural numbers $n \geqq 0$. In describing the inductive step we shall suppose in part that a given topological space X can be covered by q families of open sets, for some natural number $q > 0$, where the sets in each one of the q families will be mutually disjoint; that is, for q unrelated index sets $B_1, \ldots, B_q$ there will be q families $\{U_{\beta_1} | \beta_1 \in B_1\}, \ldots, \{U_{\beta_q} | \beta_q \in B_q\}$ of open sets $U_{\beta_p} \subset X$, which collectively cover X, and for each fixed $p \leqq q$ the sets in the family $\{U_{\beta_p} | \beta_p \in B_p\}$ will be disjoint from one another. For notational convenience we let $\{U_{1,\alpha}\}_\alpha, \ldots, \{U_{q,\gamma}\}_\gamma$ represent the q families of open sets, without explicitly identifying the index sets; for each $p \leqq q$ the notation $\{U_{p,\beta}\}_\beta$ will represent the pth family.

1.1 Definition: A topological space is of 0th *type* if it is a disjoint union of contractible spaces. Suppose that a topological space X can be covered by the union of finitely many families $\{U_{1,\alpha}\}_\alpha, \ldots, \{U_{q,\gamma}\}_\gamma$ of open sets $U_{p,\beta} \subset X$ such that for each $p \leqq q$ the sets $U_{p,\beta}$ in the family $\{U_{p,\beta}\}_\beta$ are disjoint from one another. Then X is of nth *type* whenever there is such a covering with the additional feature that all intersections of the sets in the covering are of $(n-1)$th type. A topological space is of *finite type* if it is of nth type for some $n \geqq 0$.

The sets $U_{p,\beta} \subset X$ in the preceding definition are themselves of $(n-1)$th type since they are singleton intersections. The restriction concerning less trivial intersections is not unduly severe, however, since all intersections of more than q distinct sets in the covering $\{U_{1,\alpha}\}_\alpha, \ldots, \{U_{q,\gamma}\}_\gamma$ are necessarily void.

A *metric* on any set X is any function d from $X \times X$ to the nonnegative real numbers such that for any $(x, y, z) \in X \times X \times X$ one has $d(x, y) = d(y, x)$, $d(x, y) = 0$ if and only if $x = y$, and $d(x, z) \leqq d(x, y) + d(y, z)$. For any $x \in X$ and any $\varepsilon > 0$ there is a subset $U_{x,\varepsilon} \subset X$ consisting of those $y \in X$ with $d(x, y) < \varepsilon$, and the family $\{U_{x,\varepsilon}\}_{(x,\varepsilon)}$ is the basis of the corresponding *metric topology* on X. A given topological space is *metrizable* if its topology is the metric topology of some metric.

We omit the elementary proofs that metrizable spaces have many familiar and desirable features. For example, a metrizable space is *hausdorff* in the usual sense that any two distinct points lie in corresponding disjoint open neighborhoods, and *normal* in the usual sense that any two disjoint closed sets lie in corresponding disjoint open neighborhoods. Some equally elementary but less familiar properties of metrizable spaces will be presented as the need arises.

1.2 Definition: A topological space X is a *base space* if it is homotopy equivalent to a metrizable space of finite type. Morphisms in the *category* $\mathscr{B}$ *of base spaces* are arbitrary maps from one base space to another.

A base space is not necessarily itself metrizable or of finite type; these properties are not preserved under homotopy equivalence.

The principal reason for introducing the category $\mathscr{B}$ will appear in §9. In this section we show only that $\mathscr{B}$ is closed with respect to finite disjoint unions and finite products.

1.3 Lemma: *If X_1 and X_2 are metrizable spaces, then* (i) *the disjoint union $X_1 + X_2$ is metrizable, and* (ii) *the product $X_1 \times X_2$ is metrizable.*

PROOF: (i) For any metric d on any set X there is another metric δ with $\delta(x, y) = d(x, y)/(1 + d(x, y))$ for all $(x, y) \in X \times X$, as one easily verifies, and the topologies induced by d and δ are clearly the same. Hence there are topologies on X_1 and X_2 induced by metrics d_1 and d_2 with values in the half-open interval $[0, 1)$. There is then a metric d on $X_1 + X_2$ given by

$$d(x, y) = \begin{cases} d_1(x, y) & \text{if} \quad (x, y) \in X_1 \times X_1 \\ d_2(x, y) & \text{if} \quad (x, y) \in X_2 \times X_2 \\ 2 & \text{if} \quad (x, y) \in X_1 \times X_2 + X_2 \times X_1. \end{cases}$$

The topology induced on $X_1 + X_2$ by d is the usual topology of $X_1 + X_2$, for the given topologies of X_1 and X_2.

(ii) For any metrics d_1 and d_2 on X_1 and X_2 there is a metric d on $X_1 \times X_2$ given by

$$d((x_1, x_2), (y_1, y_2)) = d_1(x_1, y_1) + d_2(x_2, y_2),$$

and an easy verification shows that the resulting metric topology on $X_1 \times X_2$ is the usual product of the given topologies of X_1 and X_2.

1.4 Proposition: *The category $\mathscr{B}$ of base spaces is closed with respect to finite disjoint unions and finite products.*

PROOF: Since finite sums and finite products of homotopy equivalent spaces are themselves homotopy equivalent, Lemma 1.3 leaves only the task of verifying that finite sums and finite products of spaces of finite type are themselves of finite type. The assertion concerning sums is trivial since spaces of $(n-1)$th type are automatically of nth type. It remains to show (by induction on n) that a product of two spaces of nth type is also of nth type. The case $n = 0$ is clear. Now let X be a space of nth type, covered by q families $\{U_{1,\alpha}\}_\alpha, \ldots, \{U_{q,\gamma}\}_\gamma$ of open sets $U_{p,\beta} \subset X$ as in Definition 1.1, and similarly let Y be a space of nth type, covered by t families $\{V_{1,\delta}\}_\delta, \ldots, \{V_{t,\zeta}\}_\zeta$ of open sets $V_{s,\varepsilon} \subset Y$ as in Definition 1.1. By the inductive hypothesis all intersections of the products $U_{p,\beta} \times V_{s,\varepsilon}$ are of $(n-1)$th type, intersections of more than qt distinct such sets being void, and each of the families $\{U_{p,\beta} \times V_{s,\varepsilon}\}_{(\beta,\varepsilon)}$ in the open covering $\{U_{1,\alpha} \times V_{1,\delta}\}_{(\alpha,\delta)}, \ldots, \{U_{q,\gamma} \times V_{t,\zeta}\}_{(\gamma,\zeta)}$ of $X \times Y$ is mutually disjoint.

2. Some Simplicial Spaces

We now show that many familiar topological spaces are base spaces in the sense of Definition 1.2. Further examples will be given later.

Let K_0 be any set. An *abstract simplicial complex* with vertex set K_0 is a family K of finite subsets of distinct elements $i_0, \ldots, i_p \in K_0$, subject to two conditions: (i) if $\{i_0, \ldots, i_p\} \in K$ and $\{j_0, \ldots, j_q\} \subset \{i_0, \ldots, i_p\}$, then $\{j_0, \ldots, j_q\} \in K$; and (ii) if $i \in K_0$, then $\{i\} \in K$. An element $\{i_0, \ldots, i_p\} \in K$ containing $p+1$ distinct elements of K_0 is a *p-simplex*. Condition (ii) permits one to identify *vertices* $i \in K_0$ with 0-simplexes $\{i\} \in K$.

Let $\tilde{K}_0$ be the family of functions $K_0 \xrightarrow{x} [0,1]$, with value $x_i \in [0,1]$ on $i \in K_0$, such that $x_i = 0$ except for finitely many vertices $i \in K_0$, and such that $\sum_{i \in K_0} x_i = 1$. The value $x_i \in [0,1]$ is the ith *barycentric coordinate* of $x \in \tilde{K}_0$. For each vertex $i \in K_0$ there is a unique element $\delta^i \in \tilde{K}_0$ with value $\delta^i_i = 1$ (no summation), and any $x \in \tilde{K}_0$ is uniquely of the form $\sum_{i \in K_0} x_i \delta^i$ with $\sum_{i \in K_0} x_i = 1$. For each simplex $\{i_0, \ldots, i_p\} \in K$ the corresponding *geometric simplex* $|i_0, \ldots, i_p| \subset \tilde{K}_0$ consists of those $x \in \tilde{K}_0$ such that $x_i = 0$ for $i \notin \{i_0, \ldots, i_p\}$. The *simplicial space* $|K| \subset \tilde{K}_0$ is the union over all simplexes $\{i_0, \ldots, i_p\} \in K$ of the geometric simplexes $|i_0, \ldots, i_p| \subset \tilde{K}_0$. The family $\tilde{K}_0$ and the simplicial space $|K| \subset \tilde{K}_0$ can both be regarded as sub*sets* of the *direct sum* $\coprod_{K_0} \mathbb{R}$, which consists of all functions $K_0 \xrightarrow{x} \mathbb{R}$ such that $x_i = 0$ except for finitely many vertices $i \in K_0$; that is, $\coprod_{K_0} \mathbb{R}$ is the real vector space with basis $\{\delta^i \mid i \in K_0\}$.

There are at least two natural topologies on the vector space $\coprod_{K_0} \mathbb{R}$. One such topology arises from the norm $\coprod_{K_0} \mathbb{R} \xrightarrow{\| \, \|} \mathbb{R}$ given by the finite sums $\|x\| = \sum_{i \in K_0} |x_i|$ for each $x = \sum_{i \in K_0} x_i \delta^i$, the resulting metric and metric topology on $\coprod_{K_0} \mathbb{R}$ being given by setting $d(x, y) = \|x - y\|$. The *metric simplicial space* associated to an abstract simplicial complex K is the simplicial complex K is the simplicial space $|K| \subset \coprod_{K_0} \mathbb{R}$ in the relative metric topology it inherits from the preceding metric topology of $\coprod_{K_0} \mathbb{R}$.

An abstract simplicial complex K is *finite-dimensional* whenever there is a natural number $q \geqq 0$ such that K contains no p-simplexes with $p > q$; K is *q-dimensional* if q is the least such number. The vertex set K_0 of a finite-dimensional simplicial complex K need not be finite; however, if K_0 *is* finite, then K is a *finite simplicial complex*, trivially finite-dimensional. A simplicial space $|K|$ is *finite*, *q-dimensional*, or *finite-dimensional* whenever the underlying simplicial complex K has these properties. In this section we show that any finite-dimensional metric simplicial space is a base space in the sense of Definition 1.2. We shall later obtain the same result for any metric simplicial space whatsoever.

Let K be an abstract simplicial complex with vertex set K_0 as before. The *first barycentric subdivision* of K is an abstract simplicial complex K' with vertex set K, constructed as follows: for any distinct simplexes $I_0, \ldots, I_q$ in K one has $\{I_0, \ldots, I_q\} \in K'$ if and only if one can relabel $I_0, \ldots, I_q$ in such a way that as subsets of K_0 they satisfy $I_0 \subset I_1 \subset \cdots \subset I_q$. One easily verifies that K' is indeed an abstract simplicial complex, and that K' is q-dimensional if and only if K is q-dimensional.

Let $\tilde{K}$ be the family of functions $K \xrightarrow{X} [0, 1]$, with value $X_I \in [0, 1]$ on $I \in K$, such that $X_I = 0$ except for finitely many vertices $I \in K$, and such that $\sum_{I \in K} X_I = 1$. For each vertex $I \in K$ there is a unique element $\delta^I \in \tilde{K}$ with value $\delta^I_I = 1$ (no summation), and any $X \in \tilde{K}$ is uniquely of the form $\sum_{I \in K} X_I \delta^I$ with $\sum_{I \in K} X_I = 1$. For each simplex $\{I_0, \ldots, I_q\} \in K'$ the corresponding geometric simplex $|I_0, \ldots, I_q| \subset \tilde{K}$ consists of those $X \in \tilde{K}$ such that $X_i = 0$ for $I \notin \{I_0, \ldots, I_q\}$, and the simplicial space $|K'| \subset \tilde{K}$ is the union over all simplexes $\{I_0, \ldots, I_q\} \in K'$ of the geometric simplexes $|I_0, \ldots, I_q| \subset \tilde{K}$. A metric topology is imposed on $|K'| \subset \tilde{K} \subset \coprod_K \mathbb{R}$ in the same way that a metric topology was imposed on $|K| \subset \tilde{K}_0 \subset \coprod_{K_0} \mathbb{R}$.

We now construct a homeomorphism $|K'| \xrightarrow{\Phi} |K|$. For any p-simplex $I = \{i_0, \ldots, i_p\} \in K$ the *barycenter* $B^I \in |K|$ is the point $(p + 1)^{-1} \delta^{i_0} + \cdots + (p + 1)^{-1} \delta^{i_p}$ in the geometric simplex $|i_0, \ldots, i_p| \subset |K|$. For any q-simplex $\{I_0, \ldots, I_q\} \in K'$, and for any point $\sum_{I \in K} X_I \delta^I$ of the corresponding geometric simplex $|I_0, \ldots, I_q| \subset |K'|$ one easily verifies from the conditions $\sum_{I \in K} X_I = 1$ and $X_I = 0$ for $I \notin \{I_0, \ldots, I_q\}$ that $\sum_{I \in K} X_I B^I \in |K|$; in fact, if $I_0 \subset I_1 \subset \cdots \subset I_q \subset \{i_0, \ldots, i_p\}$ for a p-simplex $\{i_0, \ldots, i_p\} \in K$, then $\sum_{I \in K} X_I B^I$ lies in the geometric simplex $|i_0, \ldots, i_p| \subset |K|$. Thus there is a well-defined set-theoretic map $|K'| \xrightarrow{\Phi} |K|$ carrying any $\sum_{I \in K} X_I \delta^I \in |K'|$ into $\sum_{I \in K} X_I B^I \in |K|$. A routine verification shows that Φ is indeed a homeomorphism in the metric topologies of $|K|$ and $|K'|$.

Let $i \in K_0$ be a vertex of an abstract simplicial complex K. The *abstract star* of the 0-simplex $\{i\} \in K$ is the unique smallest subcomplex of K (in the obvious sense) that contains every simplex $\{i_0, \ldots, i_p\} \in K$ with $i \in \{i_0, \ldots, i_p\}$; the corresponding subset of $|K|$ is the *closed star* of the point $\delta^i \in |K|$. In general the abstract star of $\{i\}$ also contains simplexes $\{j_0, \ldots, j_q\}$ that do not contain i, and one obtains the *open star* of $\delta^i \in |K|$ by removing all points of the corresponding geometric simplexes $|j_0, \ldots, j_q|$ in the closed star. Alternatively, the open star of $\delta^i \in |K|$ consists of all points $\sum_{j \in K_0} x_j \delta^j \in |K|$ with $x_i > 0$. Observe that the set of *all* points $\sum_{j \in K_0} x_i \delta^j \in \coprod_{K_0} \mathbb{R}$ with $x_i > 0$ is an *open convex* set in $\coprod_{K_0} \mathbb{R}$ whose intersection with $|K|$ is the open star of δ^i; hence the open star of δ^i is open in the metric topology of $|K|$, and it has a "convexity" property.

If K' is the first barycentric subdivision of K, then for any p-simplex $I \in K$ the open star of the point $\delta^I \in |K'|$ consists of all points $\sum_{I' \in K} x_{I'} \delta^{I'}$ such that $x_I > 0$ which satisfy the following additional condition: if $x_{I'} > 0$, then the subsets $I \subset K_0$ and $I' \subset K_0$ satisfy either $I \subset I'$ or $I' \subset I$ (or both). The open star of $\delta^I \in |K'|$ is denoted $U_{p,I}$ in the proof of the following result, I being a p-simplex of K.

2.1 Proposition: *Any finite-dimensional metric simplicial space $|K|$ is of first type.*

PROOF: Let K' be the first barycentric subdivision of K; since there is a homeomorphism $|K'| \to |K|$, it suffices to show that $|K'|$ is of first type. For any $p \geqq 0$, let I and J be p-simplexes $\{i_0, \ldots, i_p\}$ and $\{j_0, \ldots, j_p\}$ of K, and let $U_{p,I} \subset |K'|$ and $U_{p,J} \subset |K'|$ be the open stars of the points $\delta^I \in |K'|$ and $\delta^J \in |K'|$, respectively. If the intersection $U_{p,I} \cap U_{p,J}$ is nonvoid, any point $x \in U_{p,I} \cap U_{p,J}$ is of the form $\sum_{I' \in K} x_{I'} \delta^{I'}$ with both $x_I > 0$ and $x_J > 0$. It follows that the subsets $I \subset K_0$ and $J \subset K_0$ satisfy either $I \subset J$ or $J \subset I$, and since each of I and J has $p + 1$ elements, either consequence is equivalent to $I = J$. Thus if β denotes an arbitrary p-simplex of K, the family $\{U_{p,\beta}\}_\beta$ of open stars $U_{p,\beta} \subset |K'|$ of points $\delta^\beta \in |K'|$ is mutually disjoint. If K is q dimensional, one thereby obtains a covering $\{U_{0,\alpha}\}_\alpha, \ldots, \{U_{q,\gamma}\}_\gamma$ of $|K'|$ by $q + 1$ such families $\{U_{p,\beta}\}_\beta$ of open sets. All intersections of more than $q + 1$ distinct sets in the covering are void, and to complete the proof it remains only to show that all possible nonvoid intersections are contractible; but this is an immediate consequence of the "convexity" property of the open stars $U_{p,\beta} \subset |K'|$.

2.2 Corollary: *Any finite-dimensional metric simplicial space $|K|$ is a base space.*

PROOF: $|K|$ is metric by definition and of first type by Proposition 2.1.

A *polyhedron* is any metric simplicial space $|K|$ whose vertex set K_0 is finite.

2.3 Corollary: *Any polyhedron $|K|$ is a base space.*

PROOF: If K_0 is finite, then $|K|$ is finite dimensional.

Some elementary introductions to simplicial complexes and metric simplicial spaces are indicated in Remark 10.5.

3. More Simplicial Spaces

Any metric simplicial space $|K|$ whatsoever is a base space, even without the dimensionality condition of Corollary 2.2. However, rather than proving that $|K|$ itself is of finite type, we construct a new metric space $|K|^*$ homotopy equivalent to $|K|$, and we show that $|K|^*$ is of second type; this implies that $|K|$ is a base space as desired.

We first replace the given metric simplicial space $|K|$ by the simplicial space $|K'|$ of the first barycentric subdivision K' of the underlying abstract simplicial complex K; this is harmless since there is a natural homeomorphism $|K'| \xrightarrow{\Phi} |K|$, as noted earlier. For any p-simplex $I = \{i_0, \ldots, i_p\}$ of K the *dimension* dim I is the number p.

3.1 Definition: For any metric simplicial space $|K|$ the *telescope function* $|K'| \xrightarrow{f} [0, 1)$ assigns to each $x = \sum_{I \in K} x_I \delta^I \in |K'|$ the value

$$f(x) = \sum_{I \in K} \frac{x_I \dim I}{2 + \dim I} \in [0, 1).$$

For each $x \in |K'|$ both of the preceding sums are finite, and since $\sum_{I \in K} x_I = 1$, it follows that $f(x) \in [0, 1)$ as indicated.

3.2 Lemma: The *telescope function* $|K'| \xrightarrow{f} [0, 1)$ *is continuous in the metric topology of $|K'|$ and the usual real topology of* $[0, 1)$.

PROOF: The telescope function is the restriction to $|K'| \subset \coprod_K \mathbb{R}$ of a linear functional $\coprod_K \mathbb{R} \xrightarrow{f} \mathbb{R}$, also given by $f(x) = \sum_{I \in K}(x_I \dim I/(2 + \dim I))$. Since $|f(x)| \leqq \|x\| = \sum_{I \in K}|x_I|$ for all $x \in \coprod_K \mathbb{R}$, it follows that the linear functional f is bounded, hence continuous, in the metric topology induced by the norm $\| \;\|$; consequently the restriction of f to $|K'| \subset \coprod_K \mathbb{R}$ is continuous in the relative topology of $|K'|$.

In particular, since the image of the telescope function f lies in the half-open interval $[0, 1)$, for any half-open subinterval $[0, r) \subset [0, 1)$ the inverse image $f^{-1}[0, r) \subset |K'|$ is open in $|K'|$.

3.3 Definition: The *telescope* $|K|^*$ of any metric simplicial space $|K|$ is the subset

$$\bigcup_{q>0} f^{-1}\left[0, \frac{q}{q+2}\right) \times \left(\frac{q-2}{q}, \frac{q}{q+2}\right)$$

of the metric space $|K'| \times (-1, 1)$, in the relative topology.

3.4 Lemma: *Any metric simplicial space $|K|$ is homotopy equivalent to its telescope $|K|^*$.*

PROOF: Since there is a homeomorphism $|K'| \xrightarrow{\Phi} |K|$, it suffices to show that $|K'|$ is homotopy equivalent to $|K|^*$. Let $|K|^* \xrightarrow{g} |K'|$ be the restriction to $|K|^* \subset |K'| \times (-1, 1)$ of the projection $|K'| \times (-1, 1) \to |K'|$, which is trivially continuous. If $|K'| \xrightarrow{\text{id}} |K'|$ is the identity map and $|K'| \xrightarrow{f} [0, 1)$ the telescope function, then the image of $|K'| \xrightarrow{(\text{id}, f)} |K'| \times [0, 1)$ lies in $|K|^*$, and by Lemma 3.2 the restriction $|K'| \xrightarrow{h} |K|^*$ of (id, f) is continuous. Clearly the composition $|K'| \xrightarrow{h} |K|^* \xrightarrow{g} |K'|$ is the identity on $|K'|$, and it remains to show that the composition $|K|^* \xrightarrow{g} |K'| \xrightarrow{h} |K|^*$ is homotopic to the identity on $|K|^*$. For any point $(x, s) \in |K'| \times (-1, 1)$ and any point $t \in [0, 1]$ let $k_t(x, s) = (x, (1-t)s + tf(x)) \in |K'| \times (-1, 1)$. If $(x, s) \in |K|^*$, then $k_t(x, s) \in |K|^*$ for any $t \in [0, 1]$; furthermore, the restriction of k_0 to $|K|^*$ is the identity $|K|^* \to |K|^*$ and the restriction of k_1 to $|K|^*$ is the composition $h \circ g$, which completes the proof.

3.5 Lemma: *For any $q > 0$ the open set $f^{-1}[0, q/(q+2)) \subset |K'|$ is of first type.*

PROOF: For any $p \geqq 0$ and for any p-simplex $\beta \in K$ let $U_{p,\beta} \subset |K'|$ be the open star of $\delta^\beta \in |K'|$ as in the proof of Proposition 2.1, and let $V_{p,\beta} = U_{p,\beta} \cap f^{-1}[0, q/(q+2))$. Since $f(\delta^\beta) = p/(p+2)$ it follows that the $q+1$ families $\{V_{0,\alpha}\}_\alpha, \ldots, \{V_{q,\gamma}\}_\gamma$ of open sets cover $f^{-1}[0, q/(q+2))$. Since each family $\{U_{p,\beta}\}_\beta$ is mutually disjoint, exactly as in Proposition 2.1, it also follows that each family $\{V_{p,\beta}\}_\beta$ is mutually disjoint. It remains to show that nonvoid intersections of (at most $q+1$) of the sets $V_{p,\beta}$ are contractible. Each open star $U_{p,\beta} \subset |K'|$ is the intersection with $|K'| \subset \coprod_K \mathbb{R}$ of the convex set of points $\sum_I x_I \delta^I \in \coprod_K \mathbb{R}$ such that $x_\beta > 0$, and $|K'| \xrightarrow{f} \mathbb{R}$ is the restriction to $|K'| \subset \coprod_K \mathbb{R}$ of a linear functional $\coprod_K \mathbb{R} \xrightarrow{f} \mathbb{R}$. Consequently any nonvoid intersection $V_{p_0,\beta_0} \cap \cdots \cap V_{p_r,\beta_r}$ (for which one necessarily has $0 \leqq r \leqq q$) is the intersection with $|K'|$ of an intersection of convex sets in $\coprod_K \mathbb{R}$. The latter intersection contracts linearly to the barycenter of the geometric simplex $|\beta_0, \ldots, \beta_r| \subset |K'|$, and the contraction restricts to a contraction of $V_{p_0,\beta_0} \cap \cdots \cap V_{p_r,\beta_r}$, as desired.

3.6 Lemma: *The telescope $|K|^*$ of any metric simplicial space $|K|$ is of second type.*

PROOF: For any $q > 0$, let

$$U_q = f^{-1}\left[0, \frac{q}{q+2}\right) \times \left(\frac{q-2}{q}, \frac{q}{q+2}\right) \subset |K|^* \subset |K'| \times (-1, 1)$$

for the telescope function $|K'| \xrightarrow{f} [0, 1)$, so that $|K|^* = \bigcup_{q>0} U_q$ as in Definition 3.3. The two families $\{U_q\}_{q \text{ even}}$ and $\{U_q\}_{q \text{ odd}}$ are each mutually disjoint, and together they cover $|K|^*$. The only nonvoid intersections in the covering are the singleton intersections

$$U_q = f^{-1}\left[0, \frac{q}{q+2}\right) \times \left(\frac{q-2}{q}, \frac{q}{q+2}\right)$$

and the intersections

$$U_q \cap U_{q+1} = f^{-1}\left[0, \frac{q}{q+2}\right) \times \left(\frac{q-1}{q+1}, \frac{q}{q+2}\right);$$

since each $f^{-1}[0, q/(q+2))$ is of first type by Lemma 3.5, all intersections in the entire covering $\{U_q\}_q$ are of first type. Hence $|K|^*$ is of second type, as in Definition 1.1.

3.7 Theorem: *Any metric simplicial space $|K|$ is a base space.*

PROOF: According to Lemma 3.4, $|K|$ is homotopy equivalent to its telescope $|K^*|$, which is of second type by Lemma 3.6. Since $|K|^*$ inherits its topology as a subspace of a product $|K'| \times (-1, 1)$ of metric spaces, $|K|^*$ is metrizable. Thus $|K|$ is homotopy equivalent to a metrizable space $|K|^*$ of finite type, as required by Definition 1.2.

3.8 Definition: Let $\mathscr{W}$ denote the category of spaces of the homotopy types of metric simplicial spaces, morphisms being arbitrary maps.

Another characterization of $\mathscr{W}$ will be given in Corollary 5.3.

3.9 Corollary: *The category $\mathscr{W}$ is a full subcategory of the category $\mathscr{B}$ of bases spaces.*

PROOF: This follows immediately from Theorem 3.7.

The category $\mathscr{W}$ also satisfies an analog of Proposition 1.4.

3.10 Proposition: *The category $\mathscr{W}$ is closed with respect to finite disjoint unions and finite products.*

PROOF: It suffices to show that finite disjoint unions and finite products of metric simplicial spaces are again metric simplicial spaces. The assertion about unions follows immediately from part (i) of Lemma 1.3. As for products, if K and L are abstract simplicial complexes with vertex sets K_0 and L_0, respectively, then one constructs the obvious product complex $K \times L$ with vertex set $K_0 \times L_0$, observing that the resulting simplicial space $|K \times L| \subset \coprod_{K_0 \times L_0} \mathbb{R}$ is the image of $|K| \times |L| \subset \coprod_{K_0} \mathbb{R} \oplus \coprod_{L_0} \mathbb{R}$ under the canonical isomorphism $\coprod_{K_0} \mathbb{R} \oplus \coprod_{L_0} \mathbb{R} \to \coprod_{K_0 \times L_0} \mathbb{R}$ of direct sums. The metric for $|K \times L|$ is precisely the product metric for $|K| \times |L|$ given in part (ii) of Lemma 1.3.

A simplicial space $|K|$ is *countable* whenever the vertex set K_0 of the underlying abstract simplicial complex K is countable in the usual sense.

3.11 Definition: Let $\mathscr{W}_0$ denote the category of spaces of the homotopy types of *countable* metric simplicial spaces, morphisms being arbitrary maps.

3.12 Corollary: *The category $\mathscr{W}_0$ is a full subcategory of the category $\mathscr{B}$ of base spaces.*

PROOF: Clearly $\mathscr{W}_0 \subset \mathscr{W}$, and $\mathscr{W} \subset \mathscr{B}$ by Corollary 3.9.

The category $\mathscr{W}_0$ also satisfies an analog of Proposition 1.4.

3.13 Proposition: *The category $\mathscr{W}_0$ is closed with respect to finite disjoint unions and finite products.*

PROOF: If the abstract simplicial complexes K and L have countable vertex sets K_0 and L_0, respectively, then $K_0 + L_0$ and $K_0 \times L_0$ are both countable, and the remainder of the proof proceeds as in Proposition 3.10.

In summary, we have obtained inclusions $\mathscr{W}_0 \subset \mathscr{W} \subset \mathscr{B}$ of some useful categories of topological spaces, each category being closed with respect to finite disjoint unions and finite products. We shall identify $\mathscr{W}$ in another fashion in Corollary 5.3, and there is an analogous alternate description of $\mathscr{W}_0$. The inclusions $\mathscr{W}_0 \subset \mathscr{W} \subset \mathscr{B}$ guarantee that spaces in the categories $\mathscr{W}_0$ and $\mathscr{W}$ have all the properties of spaces in the category $\mathscr{B}$, many of which are more easily established directly in $\mathscr{B}$, without the extra provisions of $\mathscr{W}_0$ or $\mathscr{W}$.

4. Weak Simplicial Spaces

In this section we give two new descriptions of the metric topology of a simplicial space $|K|$, and we construct a useful locally finite covering of $|K|$. Then we endow $|K|$ with an entirely different topology, in general finer (more open sets) than the metric topology. Although the identity map carrying $|K|$ in the new topology into $|K|$ in the metric topology is not in general a homeomorphism, we shall show that it is always a homotopy equivalence, hence that $|K|$ is a base space in either topology.

Recall that if K_0 is the vertex set of an abstract simplicial complex K, with direct sum $\coprod_{K_0}\mathbb{R}$ over the real numbers $\mathbb{R}$, then $\tilde{K}_0 \subset \coprod_{K_0}\mathbb{R}$ consists of those points $x \in \coprod_{K_0}\mathbb{R}$ with $x_i \in [0, 1]$ and $x_i = 0$ except for finitely many vertices $i \in K_0$, such that $\sum_{i \in K_0} x_i = 1$. The simplicial space $|K|$ is a subset of $\tilde{K}_0$ as before. The metric topologies of $\coprod_{K_0}\mathbb{R}$, $\tilde{K}_0$, and $|K|$ arise from a norm $\coprod_{K_0}\mathbb{R} \xrightarrow{\| \|} \mathbb{R}$, which will be denoted $\| \|_1$ in this section, given by $\|x\|_1 = \sum_{i \in K_0}|x_i|$.

There are other norms on $\coprod_{K_0}\mathbb{R}$, one of which is the norm $\coprod_{K_0}\mathbb{R} \xrightarrow{\| \|_\infty} \mathbb{R}$ given by $\|x\|_\infty = \max_{i \in K_0}|x_i|$. If K_0 is finite, then the metric topology on $\coprod_{K_0}\mathbb{R}$ induced by $\| \|_\infty$ agrees with the metric topology induced by $\| \|_1$. However, in general the only related conclusion one can obtain for $\coprod_{K_0}\mathbb{R}$ itself comes from the obvious inequality $\|x - y\|_\infty \leqq \|x - y\|_1$ for any x and y: the topology induced by $\| \|_\infty$ is finer than the topology induced by $\| \|_1$. The situation is better on the subset $\tilde{K}_0 \subset \coprod_{K_0}\mathbb{R}$.

4.1 Lemma: *For any vertex set K_0 the topologies induced by $\| \|_1$ and $\| \|_\infty$ agree on the subset $\tilde{K}_0 \subset \coprod_{K_0}\mathbb{R}$.*

PROOF: The inequalities $\|x - y\|_\infty \leq \|x - y\|_1$ imply that open sets in the $\| \|_1$ topology of K_0 are open in the $\| \|_\infty$ topology. Conversely, for any $x \in \tilde{K}_0$ and any $\varepsilon > 0$ let $U_{x,\varepsilon} \subset \tilde{K}_0$ be the $\| \|_1$-open neighborhood of x consisting of those $y \in \tilde{K}_0$ such that $\|x - y\|_1 < \varepsilon$. If x lies in a geometric simplex $|I| \subset \tilde{K}_0$ for a p-simplex $I = \{i_0, \ldots, i_p\} \subset K_0$, then $x_i = 0$ except for the $p + 1$ vertices $i \in I$. Let $V_{x,\varepsilon} \subset \tilde{K}_0$ be the $\| \|_\infty$-open neighborhood of x consisting of those $y \in \tilde{K}_0$ such that $\|x - y\|_\infty < \varepsilon/2(p + 1)$. Then for any $y \in V_{x,\varepsilon}$ one has $\sum_{i \in I}|x_i - y_i| \leqq (p + 1)\|x - y\|_\infty < \varepsilon/2$; one also has $\sum_{i \in I} y_i > \sum_{i \in I}[x_i - (\varepsilon/2(p + 1))] = 1 - \varepsilon/2$, which implies $\sum_{i \notin I}|x_i - y_i| = \sum_{i \notin I} y_i = 1 - \sum_{i \in I} y_i < \varepsilon/2$. It follows for any $y \in V_{x,\varepsilon}$ that

$$\|x - y\|_1 = \sum_{i \in I}|x_i - y_i| + \sum_{i \notin I}|x_i - y_i| < \varepsilon/2 + \varepsilon/2 = \varepsilon,$$

hence that $V_{x,\varepsilon} \subset U_{x,\varepsilon}$. Consequently, open sets in the $\| \|_\infty$ topology of $\tilde{K}_0$ are also open in the $\| \|_1$ topology.

4.2 Proposition: *Let K be an abstract simplicial complex with vertex set K_0 and simplicial space $|K| \subset \coprod_{K_0} \mathbb{R}$; then the norms $\| \|_1$ and $\| \|_\infty$ on $\coprod_{K_0} \mathbb{R}$ induce the same metric topology on $|K|$.*

PROOF: Since $|K| \subset \tilde{K}_0$, this is an immediate consequence of Lemma 4.1.

Thus the metric topology of $|K|$ introduced in §2 via the norm $\| \|$ $(= \| \|_1)$ can equally well be described in terms of the norm $\| \|_\infty$; we simply speak of *the* metric topology of $|K|$.

For each vertex $i \in K_0$ the ith *barycentric coordinate function* $|K| \to [0, 1]$ carries each $x = \sum_{i \in K_0} x_i \delta^i \in |K|$ into $x_i \in [0, 1]$.

4.3 Corollary: *The metric topology of a simplicial space $|K|$ is the coarsest topology (fewest open sets) for which each barycentric coordinate function $|K| \to [0, 1]$ is continuous.*

PROOF: Since $\|x - y\|_\infty = \max_{i \in K_0} |x_i - y_i|$, this is an immediate consequence of the fact that $\| \|_\infty$ induces the metric topology.

An open covering $\{U_j\}_j$ of a topological space X is *locally finite* whenever there is an open neighborhood V_x of each $x \in X$ with $U_j \cap V_x$ void except for finitely many indices j. If $X \xrightarrow{h_j} \mathbb{R}$ is a map with value 0 outside a closed subset of U_j, then one can compute $X \xrightarrow{\sum_j h_j} \mathbb{R}$ in terms of finite sums on each V_x. A family $\{h_j\}_j$ of nonnegative such maps h_j is a *partition of unity* subordinate to $\{U_j\}_j$ whenever $\sum_j h_j = 1$. The following specialized partition of unity will be needed later.

4.4 Proposition: *For any metric simplicial space $|K|$ there is a partition of unity $\{h_i | i \in K_0\}$ subordinate to a locally finite cover $\{U_i | i \in K_0\}$ indexed by the vertex set K_0, such that $h_i(\delta^i) = 1$ for each $i \in K_0$; specifically, if $x \in |K|$ lies in the interior of a geometric simplex $|I| \subset |K|$ for $I = \{i_0, \ldots, i_p\} \in K$, then x has an open neighborhood V_x with $U_i \cap V_x$ void for $i \notin I$.*

PROOF: For each $i \in K_0$ let U_i denote the set of those $x \in |K|$ such that $2x_i > \|x\|_\infty$. The family $\{U_i | i \in K_0\}$ clearly covers $|K|$, and since $\| \|_\infty$ and the barycentric coordinate functions are continuous by Proposition 4.2 and Corollary 4.3, each U_i is open. To show that $\{U_i | i \in K_0\}$ is locally finite suppose that $x \in |K|$ lies in the geometric simplex $|I| = |i_0, \ldots, i_p|$. Let V_x be the open neighborhood of x consisting of those $y \in |K|$ with $\|x - y\|_\infty < \frac{1}{3}\|x\|_\infty$; then for any $y \in V_x$ the triangle inequality gives

$$\|y\|_\infty \geqq \|x\|_\infty - \|x - y\|_\infty > \|x\|_\infty - \tfrac{1}{3}\|x\|_\infty = \tfrac{2}{3}\|x\|_\infty.$$

If $i \notin I$, then $x_i = 0$, so that

$$\tfrac{2}{3}\|x\|_\infty > 2\|x - y\|_\infty \geqq 2|x_i - y_i| = 2y_i \qquad \text{for} \quad y \in V_x.$$

Hence $\|y\|_\infty > \frac{2}{3}\|x\|_\infty > 2y_i$ for any $y \in V_x$, except for the finitely many vertices $i \in I$; that is, $U_i \cap V_x$ is void except for $i \in I$, as required.

Now for each vertex $i \in K_0$ and each point $x \in |K|$ let $g_i(x) \in \mathbb{R}$ be the maximum of the values 0 and $3x_i - 2\|x\|_\infty$. Then each g_i is continuous by Proposition 4.2 and Corollary 4.3, vanishing outside an open subset of U_i. Since $\{U_i | i \in K_0\}$ is locally finite, one can form the sum $g = \sum_i g_i$, for which one has $g(x) > 0$ for all $x \in |K|$; the desired partition of unity $\{h_i | i \in K_0\}$ is then obtained by setting $h_i = g_i/g$ for each $i \in K_0$.

If I is any simplex $\{i_0, \ldots, i_p\} \subset K_0$ of an abstract simplicial complex K, then the metric topology induced on the corresponding geometric simplex $|I| = |i_0, \ldots, i_p| \subset |K|$ is precisely the usual real topology of the subset of those points $(x_0, \ldots, x_p) \in \mathbb{R}^{p+1}$ with $x_i \in [0, 1]$ and $x_0 + \cdots + x_p = 1$. Thus the topology of $|I|$ can be independently defined.

4.5 Definition: For any simplicial space $|K|$ the closed sets of the *weak topology* of $|K|$ are those sets $W \subset |K|$ for which $W \cap |I|$ is closed in each geometric simplex $|I| \subset |K|$; in this topology $|K|$ is a *weak simplicial space.*

We shall use the notations $|K|_\mathrm{m}$ and $|K|_\mathrm{w}$ to identify a simplicial space $|K|$ in its metric and weak topologies, respectively. The subscripts m or w will be dropped whenever a topology is not specified or whenever the topology of $|K|$ is clear in context.

Observe that the barycentric coordinate functions $|K| \to [0, 1]$ are automatically continuous in the weak topology, so that by Corollary 4.3 every open set in $|K|_\mathrm{m}$ is also an open set in $|K|_\mathrm{w}$. Hence the identity $|K| \to |K|$ induces a (continuous) map $|K|_\mathrm{w} \xrightarrow{f} |K|_\mathrm{m}$. Although the inverse of f is not necessarily itself continuous, as we shall soon show, there is nevertheless a (continuous) map $|K|_\mathrm{m} \xrightarrow{g} |K|_\mathrm{w}$ such that $f \circ g$ and $g \circ f$ are homotopic to identity maps, so that $|K|_\mathrm{w}$ and $|K|_\mathrm{m}$ are homotopy equivalent; the proof of this result is the business of the remainder of the section.

Caution: In Definition 4.5, and in analogous situations which occur later, we use the terminology of J. H. C. Whitehead [3]: a "weak topology" on X frequently has *more* open sets than some other topology on X. The words "weak topology" are used elsewhere *but not in this book* for the coarsest topology (fewest open sets) on X such that certain maps $X \to Y$ are continuous.

Suppose that K_0 consists of all the natural numbers $n \geqq 0$, and that the simplicial complex K consists of the 0-simplexes $\{n\}$ for all $n \geqq 0$ and 1-simplexes $\{0, n\}$ for all $n > 0$. Then each geometric 1-simplex $|0, n|$ is canonically homeomorphic to the closed unit interval $[0, 1]$. Let $W \subset |K|$ be the set which intersects each $|0, n|$ in the point corresponding to $1/n \in [0, 1]$. Then according to Definition 4.5 W is closed in $|K|_w$; but since W does not contain the point $\delta^0 \in |K|$, the set W is clearly not closed in $|K|_m$. Thus for this simplicial space $|K|$ the continuous bijection $|K|_w \xrightarrow{f} |K|_m$ is *not* a homeomorphism; specifically, the set-theoretic inverse f^{-1} is not continuous.

Despite the preceding example, certain maps $|L|_m \xrightarrow{g} |K|_m$ of metric simplicial spaces *do* remain continuous when one substitutes $|K|_w$ for $|K|_m$. Suppose for each $x \in |L|$ that the image $g(x) \in |K|$ lies in only finitely many geometric simplexes of $|K|$. Then the metric and weak topologies of $|K|$ induce the *same* relative topology on the image $\operatorname{Im} g \subset |K|$, and since the inclusion $\operatorname{Im} g \subset |K|_w$ is continuous the composition $|L|_m \xrightarrow{g} \operatorname{Im} g \subset |K|_w$ is also continuous; that is, the original map g can also be regarded as a (continuous) map $|L|_m \to |K|_w$. This observation is an essential ingredient in the proof of the following result.

44.6 Proposition (Dowker [1]): *For any simplicial complex K the weak simplicial space $|K|_w$ is homotopy equivalent to the metric simplicial space $|K|_m$.*

PROOF: We have already observed that the set-theoretic identity map $|K| \to |K|$ induces a continuous bijection $|K|_w \xrightarrow{f} |K|_m$, which is not necessarily a homeomorphism. Let $\{h_i \mid i \in K_0\}$ be the partition of unity constructed in Proposition 4.4, and suppose that $x \in |K|$ lies in a geometric simplex $|I| \subset |K|$ for $I = \{i_0, \ldots, i_p\} \in K$; then $h_i(x) = 0$ for $i \notin I$, and since $\sum_{i \in K_0} h_i(x) = 1$, it follows that the point $g(x) = \sum_{i \in K_0} h_i(x)\delta^i \in \coprod_{K_0} \mathbb{R}$ also lies in $|I|$. Since the h_i's are continuous in the metric topology, there is an induced (continuous) map $|K|_m \xrightarrow{g} |K|_m$ from $|K|_m$ to itself. By construction, each $g(x)$ lies in only finitely many geometric simplexes in $|K|$, so that g is also continuous as a map $|K|_m \to |K|_w$, as in the preceding paragraph.

To show that $f \circ g$ and $g \circ f$ are homotopic to identity maps let $|K| \times [0, 1] \xrightarrow{F} \coprod_{K_0} \mathbb{R}$ be given by $F_t(x) = (1 - t)x + tg(x)$. For any geometric simplex $|I| \subset |K|$ we already know that if $x \in |I|$, then $g(x) \in |I|$, so that the image of F lies in $|K|$. Since g is continuous as a map from $|K|_m$ to itself F is a homotopy from the identity map $|K|_m \xrightarrow{F_0} |K|_m$ to the composition $F_1 = f \circ g$, and since F restricts to maps $|I| \times [0, 1] \to |I|$, it also follows that F is also a homotopy from the identity map $|K|_w \xrightarrow{F_0} |K|_w$ to the composition $F_1 = g \circ f$.

Some other expositions of Proposition 4.6 are indicated in Remark 10.7.

4.7 Corollary: *The category $\mathscr{W}$ of spaces of the homotopy types of metric simplicial spaces $|K|_m$ is identical to the category of spaces of the homotopy* *types of weak simplicial spaces $|K|_w$; a fortiori every weak simplicial space is a base space.*

PROOF: We showed in Theorem 3.7 that every metric simplicial space is a base space, base spaces being defined in terms of homotopy type.

5. CW Spaces

We now describe a classical generalization of the category of weak simplicial spaces. It appears directly and indirectly throughout much of algebraic and differential topology, and it will appear in later chapters of the present work.

For any natural number $n > 0$ let $\mathbb{R}^n$ be the standard real n-dimensional vector space, in the metric topology arising from the euclidean norm $\mathbb{R}^n \xrightarrow{\|\ \|} \mathbb{R}$, $\|(x_1, \ldots, x_n)\| = \sqrt{x_1^2 + \cdots + x_n^2}$. The closed *n-disk* D^n and $(n-1)$*-sphere* $S^{n-1} \subset D^n$ consist of those points $x \in \mathbb{R}^n$ satisfying $\|x\| \leqq 1$ and $\|x\| = 1$, respectively, in the relative topology of $\mathbb{R}^n$; in particular D^1 is a closed interval and S^0 consists of two points. An (open) *n-cell* is any space homeomorphic to the interior $D^n - S^{n-1}$ of D^n.

Let $\{D_\alpha^n\}_\alpha$ be any family of homeomorphic copies of D^n and $\{S_\alpha^{n-1}\}_\alpha$ the corresponding family of homeomorphic copies of $S^{n-1} \subset D^n$, with disjoint unions $\bigcup_\alpha D_\alpha^n$ and $\bigcup_\alpha S_\alpha^{n-1}$, respectively. For any map $\bigcup_\alpha S_\alpha^{n-1} \xrightarrow{f} Y$ one lets $\sim$ denote the smallest equivalence relation in the disjoint union $Y + \bigcup_\alpha D_\alpha^n$ that identifies each $x \in \bigcup_\alpha S_\alpha^{n-1}$ with $f(x) \in Y$. The *quotient topology* of $Y + \bigcup_\alpha D_\alpha^n/\sim$ is the finest topology (most open sets) such that the projection $Y + \bigcup_\alpha D_\alpha^n \to Y + \bigcup_\alpha D_\alpha^n/\sim$ is continuous, and $Y + \bigcup_\alpha D_\alpha^n/\sim$ is the *adjunction space* of f in this topology. The map f and its restrictions $S_\alpha^{n-1} \xrightarrow{f_\alpha} Y$ are called *attaching maps* of the adjunction space, and the inclusion mapping Y into the adjunction space *attaches n-cells to Y*; specifically, the attached n-cells are the homeomorphic images of members of the family $\{D_\alpha^n - S_\alpha^{n-1}\}_\alpha$.

A *cell complex* is a sequence $X_0 \to X_1 \to X_2 \to \cdots$ of maps such that X_0 is a discrete space and each $X_{n-1} \to X_n$ attaches n-cells to X_{n-1}. The singleton subspaces of X_0 are the 0-*cells of the complex*, and for each $n > 0$ the n-cells attached to X_{n-1} are the *n-cells of the complex*. Each point of X_{n-1} is a member of one and only one q-cell for some $q < n$, and the complex $X_0 \to X_1 \to X_2 \to \cdots$ is *closure-finite* whenever for each $n > 0$ each attaching map $S_\alpha^{n-1} \xrightarrow{f_\alpha} X_{n-1}$ meets only finitely many cells.

For any closure-finite cell complex let $\lim_n X_n$ be the quotient of the disjoint union $\bigcup_n X_n$ by the smallest equivalence relation that identifies each X_{n-1} with its image under the map $X_{n-1} \to X_n$. Each point of $\lim_n X_n$ is a member of one and only one n-cell for some $n \geqq 0$, homeomorphic to the interior $D^n - S^{n-1}$ of D^n in the case $n > 0$, and a set $U \subset \lim_n X_n$ is open in the *weak topology* of $\lim_n X_n$ if and only if the intersection of U with each n-cell is open in the topology of $D^n - S^{n-1}$. A CW *space* is any topological space of the form $\lim_n X_n$ for a **C**losure-finite cell complex $X_0 \to X_1 \to X_2 \to \cdots$ in the **W**eak topology; the cell complex is the CW *structure* of the CW space. Closure-finite cell complexes are also called CW *complexes*. (In much of the literature the word "CW complex" is used both for a closure finite cell complex and for the resulting topological space. We use the terminology of Dold, who calls a space a space; see Dold [8, p. 89].)

The simplest example of a CW space is the n-sphere S^n itself, for any $n > 0$. One starts with a singleton space $X_0 = \{*\}$ and attaches no cells of dimension less than n, giving $X_0 = X_1 = \cdots = X_{n-1}$; finally one attaches a single n-cell $D^n - S^{n-1}$ via the only possible attaching map $S^{n-1} \to X_{n-1} = \{*\}$. (Even the 0-sphere S^0 can be constructed in this fashion as a CW space, with the obvious meaning of attaching a 0-cell.)

For any CW complex $X_0 \to X_1 \to X_2 \to \cdots$ and any $n \geqq 0$ there is an obvious inclusion of each X_n into the CW space $\lim_n X_n$; X_n is the *n-skeleton* of the CW complex, and its image is the *n-skeleton* of the CW space. A CW complex and the corresponding CW space are *finite-dimensional* if the inclusion of the n-skeleton is a homeomorphism for n sufficiently large; in this case the *dimension* is the least $m \geqq 0$ such that the inclusion of X_n is a homeomorphism whenever $n \geqq m$.

5.1 Proposition: *Any weak simplicial space is a CW space.*

PROOF: For any abstract simplicial complex K and any $n \geqq 0$ let K_n be the subcomplex obtained by deleting all p-simplexes with $p > n$. The sequence $K_0 \to K_1 \to K_2 \to \cdots$ of inclusions then induces a corresponding sequence $|K_0| \to |K_1| \to |K_2| \to \cdots$ of inclusions of weak simplicial spaces with $|K| = \lim_n |K_n|$ as a set. As a topological space $|K_0|$ is discrete, and since any geometric n-simplex $|I| \subset \mathbb{R}^{n+1}$ is trivially homeomorphic to the n-disk $D^n \subset \mathbb{R}^n$, it follows that $|K_0| \to |K_1| \to |K_2| \to \cdots$ is a cell complex; closure-finiteness is obvious. The weak topology of $|K|$ was described in terms of closed sets, a set $W \subset |K|$ being closed if and only if $W \cap |I|$ is closed in $|I|$ for each $|I| \subset |K|$. However, one can equally well replace each geometric n-simplex $[i_0, \ldots, i_n| = |I|$ by its interior $|I|^\circ$, consisting of those points $x_0\delta^{i_0} + \cdots + x_n\delta^{i_n}$ with all coefficients $x_0, \ldots, x_n$ positive, which is canonically homeomorphic to the (open) n-cell $D^n - S^{n-1}$; then $U \subset |K|$ is open in the weak topology of $|K|$ if and only if $U \cap |I|^\circ$ is open in $|I|^\circ$ for each $I \in K$.

There is a nontrivial partial converse to Proposition 5.1, which we shall not prove. We formulate the converse primarily to indicate the size of the category $\mathscr{B}$ of base spaces. However, in order to keep the exposition as nearly self-contained as possible, we shall not apply the converse as such.

5.2 Theorem (J. H. C. Whitehead [4]): *Any CW space is homotopy equivalent to a weak simplicial space.*

The proof of Whitehead's theorem involves two steps. One first shows that a map $Y \to X$ of CW spaces is a homotopy equivalence whenever it satisfies an apparently weaker condition; this step can be found in Gray [1, p. 139], Lundell and Weingram [1, p. 125], Maunder [1, pp. 298–300], or Switzer [1, pp. 87–90], for example. The second step consists in assigning to any CW space X a weak simplicial space $|K|$ and a map $|K| \to X$ satisfying the preceding condition; this step can be found in Gray [1, pp. 145–152] and Lundell and Weingram [1, pp. 102–103], for example. The conclusion that the two preceding steps prove Theorem 5.2 is expressed in Gray [1, p. 149] and Lundell and Weingram [1, pp. 126–127].

5.3 Corollary: *The category $\mathscr{W}$ of spaces of the homotopy types of metric simplicial spaces $|K|_m$ is identical to the category of spaces of the homotopy types of CW spaces.*

PROOF: Corollary 4.7 asserts that $\mathscr{W}$ is identical to the category of spaces of the homotopy types of weak simplicial spaces, and Proposition 5.1 and Theorem 5.2 imply that the latter category is identical to the category of spaces of the homotopy types of CW spaces.

(The category $\mathscr{W}$ was initially introduced in Milnor [8] as the category of spaces of the homotopy types of CW spaces. However, the equivalent characterization of Definition 3.8 is implicit in Milnor's paper.)

5.4 Corollary: *Every CW space is a base space.*

PROOF: We showed in Theorem 3.7 that every metric simplicial space is a base space, base spaces being defined in terms of homotopy type.

One can obtain a shorter proof of Corollary 5.4 by using just the material needed for the first step of the proof of Theorem 5.2; a sketch is given in Remark 10.6.

In the remainder of this section we describe CW structures of some other useful CW spaces. We shall later give independent proofs that these spaces are base spaces, without invoking Theorem 5.2 or its corollaries.

We shall describe the CW structures of real and complex projective spaces, which we now define. For any $n \geqq 0$ let $\mathbb{R}^{(n+1)*}$ be the space $\mathbb{R}^{n+1} - \{0\}$ consisting of $\mathbb{R}^{n+1}$ without its origin, in the relative topology. There is an equivalence relation $\sim$ in $\mathbb{R}^{(n+1)*}$ with $x \sim y$ whenever $x = ay$ for some $a \in \mathbb{R}^*$. The *real projective space* RP^n is the quotient $\mathbb{R}^{(n+1)*}/\sim$ in the quotient topology. The space RP^n can equally well be regarded as a quotient of $S^n \subset \mathbb{R}^{n+1}$ by the same equivalence relation.

5.5 Proposition: *RP^n is an n-dimensional CW space, with a cell structure consisting of one cell in each dimension $q = 0, 1, \ldots, n$.*

PROOF: Trivially RP^0 is a singleton space $\{*\}$. Let $RP^0 \to RP^1 \to RP^2 \to \cdots$ be the sequence of inclusions induced by the natural inclusions $\mathbb{R}^1 \to \mathbb{R}^2 \to \mathbb{R}^3 \to \cdots$; that is, for any $q = 1, \ldots, n$ and any $x = (x_0, \ldots, x_{q-1}) \in S^{q-1}$ the map $RP^{q-1} \to RP^q$ carries the point $[x] = [(x_0, \ldots, x_{q-1})] \in RP^{q-1}$ into the point $[x \oplus 0] = [(x_0, \ldots, x_{q-1}, 0)] \in RP^q$. Let $S^{q-1} \xrightarrow{f} RP^{q-1}$ carry $x \in S^{q-1}$ $(\subset D^q \subset \mathbb{R}^q)$ into $[x] \in RP^{q-1}$, and let $D^q \xrightarrow{f} RP^q$ carry $x \in D^q$ into $[x \oplus \sqrt{1 - \|x\|^2}] \in RP^q$; then f and g are continuous, and the inclusions $S^{q-1} \to D^q$ and $RP^{q-1} \to RP^q$ provide a commutative diagram

$$\begin{array}{ccc} S^{q-1} & \xrightarrow{f} & RP^{q-1} \\ \downarrow & & \downarrow \\ D^q & \xrightarrow{g} & RP^q. \end{array}$$

Since $\sqrt{1 - \|x\|^2}$ is positive whenever $\|x\| < 1$, the restriction $g|D^q - S^{q-1}$ is a homeomorphism onto $RP^q - RP^{q-1}$, and it follows that f is an attaching map with adjunction space RP^q, as required.

Projective spaces need not be finite-dimensional. Let $\mathbb{R}^1 \to \mathbb{R}^2 \to \mathbb{R}^3 \to \cdots$ be the sequence of canonical inclusions and set $\mathbb{R}^\infty = \lim_n \mathbb{R}^n$ in the *weak topology*; that is, a set $W \subset \mathbb{R}^\infty$ is closed whenever each $W \cap \mathbb{R}^n$ is closed. Observe that $\mathbb{R}^\infty$ is also a direct sum of real linear spaces $\mathbb{R}^1$, hence itself a real linear space. Specifically, points of $\mathbb{R}^\infty$ are sequences $(x_0, x_1, x_2, \ldots)$ of real numbers, only finitely many of which are nonzero, and addition and multiplication by scalars are defined as in the finite-dimensional spaces $\mathbb{R}^n$. Let $\mathbb{R}^{\infty*} = \mathbb{R}^\infty - \{0\}$ and observe that there is an equivalence relation $\sim$ in $\mathbb{R}^{\infty*}$ with $x \sim y$ whenever $x = ay$ for some $a \in \mathbb{R}^*$. The *real projective space* RP^∞ is the quotient $\mathbb{R}^{\infty*}/\sim$, in the quotient topology.

5.6 Corollary: *RP^∞ is a CW space, with a cell structure consisting of one cell in each dimension.*

PROOF: The inclusions $RP^0 \to RP^1 \to RP^2 \to \cdots$ induced by the inclusions $\mathbb{R}^1 \to \mathbb{R}^2 \to \mathbb{R}^3 \to \cdots$ provide a limit $\lim_n RP^n$, in the weak topology, which is canonically homeomorphic to the definition $\mathbb{R}^{\infty *}/\sim$ of RP^∞.

One can replace the real field $\mathbb{R}$ by the complex field $\mathbb{C}$ throughout the definition of RP^n to obtain the *complex projective space* $CP^n = \mathbb{C}^{(n+1)*}/\sim$, which can equally well be regarded as a quotient $S^{2n+1}/\sim$ of the $(2n+1)$-sphere $S^{2n+1} \subset \mathbb{C}^{n+1}$. [$S^{2n+1}$ consists of those points $z = (z_0, \ldots, z_n) \in \mathbb{C}^{n+1}$ such that $\|z\|^2 = |z_0|^2 + \cdots + |z_n|^2 = 1$.]

5.7 Proposition: *CP^n is a $2n$-dimensional CW space, with a cell structure consisting of one cell in each even dimension $2q = 0, 2, \ldots, 2n$.*

PROOF: By analogy with Proposition 5.5 one uses the map $D^{2q} \xrightarrow{g} CP^q$ carrying $z \in D^{2q}$ ($\subset \mathbb{C}^q$) into $[z \oplus \sqrt{1 - \|z\|^2}] \in CP^q$, where $\| \, \|$ is the usual norm on $\mathbb{C}^q$. Since $\sqrt{1 - \|z\|^2}$ is real and positive for $\|z\| < 1$, the restriction $g|D^{2q} - S^{2q-1}$ is a homeomorphism onto $CP^q - CP^{q-1}$, so that the restriction $g|S^{2q-1}$ is an attaching map $S^{2q-1} \xrightarrow{f} CP^{q-1}$ with adjunction space CP^q.

The definitions of the projective spaces RP^∞ and CP^n suggest the obvious definition of the *complex projective space* CP^∞, and the proofs of the preceding two results suggest the obvious proof of the following result:

5.8 Corollary: *CP^∞ is a CW space, with a cell structure consisting of one cell in each even dimension.*

Many other expositions of the study of CW spaces in general are indicated in Remark 10.8.

6. Smooth Manifolds

In this section we briefly describe the category $\mathcal{M}$ of (smooth) manifolds and we formulate some standard basic results of differential topology. These results quickly imply that any (smooth) manifold is a base space.

A *closed n-dimensional manifold* is a compact hausdorff space X with an open covering $\{U_i | i \in I\}$ and a family $\{\Phi_i | i \in I\}$ of homeomorphisms $U_i \xrightarrow{\Phi_i} \Phi_i(U_i)$ onto open sets $\Phi_i(U_i) \subset \mathbb{R}^n$. The open covering is a *coordinate covering* of X by *coordinate neighborhoods*, and the family $\{\Phi_i | i \in I\}$ of maps is an *atlas* for X.

Suppose that the intersection $U_i \cap U_j \subset X$ of two coordinate neighborhoods is nonempty. Then Φ_i and Φ_j restrict to homeomorphisms of $U_i \cap U_j$ onto nonempty open subsets $\Phi_i(U_i \cap U_j) \subset \mathbb{R}^n$ and $\Phi_j(U_i \cap U_j) \subset \mathbb{R}^n$. A closed n-dimensional manifold X is *smooth* whenever all the induced homeomorphisms $\Phi_i(U_i \cap U_j) \xrightarrow{\Phi_j \circ \Phi_i^{-1}} \Phi_j(U_i \cap U_j)$ are diffeomorphisms of open subsets of $\mathbb{R}^n$; that is, the real-valued functions on $\Phi_i(U_i \cap U_j) \subset \mathbb{R}^n$ and $\Phi_j(U_i \cap U_j) \subset \mathbb{R}^n$ which define $\Phi_j \circ \Phi_i^{-1}$ and $\Phi_i \circ \Phi_j^{-1}$, respectively, all have partial derivatives of all orders. In this case the coordinate covering and atlas form a *smooth structure* on X.

Let X be a smooth closed m-dimensional manifold with coordinate covering $\{U_i | i \in I\}$ and atlas $\{\Phi_i | i \in I\}$, let Y be a smooth closed n-dimensional manifold with coordinate covering $\{V_j | j \in J\}$ and atlas $\{\Psi_j | j \in J\}$, and let $X \xrightarrow{f} Y$ be a map. Then f induces maps $\Psi_j \circ f \circ \Phi_i^{-1}$ of open subsets $\Phi_i(U_i \cap f^{-1}(V_j)) \subset R^m$ into R^n, and f is itself *smooth* whenever all the induced maps $\Psi_j \circ f \circ \Phi_i^{-1}$ are smooth; that is, all partial derivatives exist as continuous functions on $\Phi_i(U_i \cap f^{-1}(V_j))$. The *category of smooth closed manifolds* consists of smooth closed manifolds and smooth maps. If a smooth map $X \xrightarrow{f} Y$ has a smooth inverse $Y \xrightarrow{g} X$, then f is a *diffeomorphism*; in this case we usually do not distinguish between X and Y.

One example of a smooth closed n-dimensional manifold is the real projective space RP^n, whose cell structure was given in Proposition 5.5. To define the smooth structure of RP^n, for each $i = 0, \ldots, n$ let $U_i \subset RP^n$ be the open subset of all points $[(x_0, \ldots, x_n)] \in RP^n$ with $x_i \neq 0$. There is a homeomorphism $U_i \xrightarrow{\Phi_i} \mathbb{R}^n$ with

$$\Phi_i[(x_0, \ldots, x_n)] = x_i^{-1}(x_0, \ldots, x_{i-1}, x_{i+1}, \ldots, x_n),$$

and each composition $\Phi_j \circ \Phi_i^{-1}$ is the diffeomorphism carrying points $(y_0, \ldots, y_{i-1}, y_{i+1}, \ldots, y_n) \in \mathbb{R}^n$ with $y_j \neq 0$ into

$$y_j^{-1}(y_0, \ldots, y_{i-1}, 1, y_{i+1}, \ldots, y_{j-1}, y_{j+1}, \ldots, y_n) \in \mathbb{R}^n.$$

Suppose that one replaces $\mathbb{R}^n$ in the definition of smooth closed n-dimensional manifolds by the closed subset $(\mathbb{R}^n)^+ \subset \mathbb{R}^n$ of points $(x_1, \ldots, x_n)$ with $x_n \geqq 0$; if $x_n > 0$, then $(x_1, \ldots, x_n)$ is an interior point of $(\mathbb{R}^n)^+$, and if $x_n = 0$, then $(x_1, \ldots, x_n)$ is a boundary point of $(\mathbb{R}^n)^+$. The result of this substitution is the *category of smooth compact manifolds*. If $\{U_i | i \in I\}$ and $\{\Phi_i | i \in I\}$ provide the smooth structure of such a manifold, and if $\Phi_i(x)$ is a boundary point of $(\mathbb{R}^n)^+$ for some $i \in I$ and some $x \in U_i$, then necessarily $\Phi_j(x)$ is a boundary point of $(\mathbb{R}^n)^+$ for all $j \in I$ such that $x \in U_j$. The set $\dot{X}$ of such points is the *boundary* of X. One easily verifies that the smooth structure of X induces a smooth structure for which $\dot{X}$ is a smooth *closed* $(n-1)$-dimensional manifold. The boundary $\dot{X}$ of a compact manifold X may be void, in which case X is a closed manifold. We occasionally use the language *compact manifold with boundary* to identify manifolds which are compact but not closed.

An arbitrary hausdorff space X, not necessarily compact, is an *n-dimensional manifold* if it can be covered by denumerably many compact n-dimensional manifolds. Suppose that the manifolds in such a covering are smooth, and let $\{\Phi_i | i \in I\}$ and $\{\Psi_j | j \in J\}$ be the atlases of two such smooth manifolds. Then X is a *smooth n-dimensional manifold* whenever every nontrivial restriction $\Psi_j \circ \Phi_i^{-1}$ is a diffeomorphism of open sets in $(\mathbb{R}^n)^+$. If every point of a manifold X is an interior point of one of the manifolds in its denumerable covering by compact manifolds, then X is an *open manifold*; for example, $\mathbb{R}^n$ is itself a smooth open n-dimensional manifold. *Smooth maps* in the *category* $\mathscr{M}$ *of smooth manifolds* are defined as before in terms of restrictions to coordinate neighborhoods.

Since a smooth manifold is covered by *denumerably* many compact manifolds of the same dimension, its underlying topological space X is necessarily *second countable* in the usual sense that there is a denumerable basis of the open sets of its topology. Conversely, any second countable hausdorff space X with a smooth structure, $\{U_i | i \in I\}$ and $\{\Phi_i | i \in I\}$, is a smooth manifold. Equivalently, one can characterize smooth manifolds as hausdorff spaces with denumerable smooth structures, $\{U_i | i \in \mathbb{N}\}$ and $\{\Phi_i | i \in \mathbb{N}\}$, where $\mathbb{N} = \{0, 1, 2, \ldots\}$. This implies that any smooth n-dimensional manifold is of the form $\lim_i X_i$ for a sequence $X_0 \to X_1 \to X_2 \to \cdots$ of inclusions of compact n-dimensional manifolds $X_0, X_1, X_2, \ldots$.

Given an open covering $\{U_i | i \in I\}$ of a topological space X, another open covering $\{V_j | j \in J\}$ of X *refines* $\{U_i | i \in I\}$ whenever there is a function $J \xrightarrow{\rho} I$ such that $V_j \subset U_{\rho(j)}$ for every $j \in J$. A hausdorff space X is *paracompact* if any open covering of X has a locally finite refinement. The conclusion of the following lemma is somewhat stronger than paracompactness.

6.1 Lemma: *Any open covering $\{U_i | i \in I\}$ of a manifold X has a countable locally finite refinement $\{V_j | j \in \mathbb{N}\}$ such that each closure $\bar{V}_j$ is compact and satisfies $\bar{V}_j \subset U_{\rho(j)}$.*

PROOF: Let $X = \lim_k X_k$ for a sequence $X_0 \to X_1 \to X_2 \to \cdots$ of inclusions of compact manifolds of the same dimension, and let $\mathring{X}_k$ denote the *interior* $X_k - \dot{X}_k$ of X_k. For any open covering $\{U_i | i \in I\}$ of X there is a refinement $\{U_i \cap (\mathring{X}_{k+1} - X_{k-2}) | (i, k) \in I \times \mathbb{N}\}$, where $X_{-2} = X_{-1} = \varnothing$. For each $k \in \mathbb{N}$ the space $X_k - \mathring{X}_{k-1}$ is compact, and it can therefore be covered by finitely many of the open sets $U_i \cap (\mathring{X}_{k+1} - X_{k-2})$. The latter sets, for each $k \in \mathbb{N}$, form the desired refinement $\{V_j | j \in \mathbb{N}\}$.

Suppose that $\{h_j | j \in J\}$ is a partition of unity subordinate to a locally finite cover $\{V_j | j \in J\}$ of a smooth manifold X; then $\{h_j | j \in J\}$ is a *smooth partition of unity* whenever each $X \xrightarrow{h_j} [0, 1]$ is smooth. If $\{V_j | j \in J\}$ happens to refine another open covering $\{U_i | i \in I\}$ of X, then $\{h_j | j \in J\}$ is also regarded as subordinate to $\{U_i | i \in I\}$.

6.2 Lemma: *For any open covering $\{U_i | i \in I\}$ of a smooth manifold X there is a countable smooth partition of unity $\{h_j | j \in \mathbb{N}\}$ subordinate to $\{U_i | i \in I\}$.*

PROOF: Let $\mathbb{R}^n \xrightarrow{f} \mathbb{R}$ be given by

$$f(x) = \begin{cases} e^{-(1-\|x\|^2)^{-1}} & \text{for } \|x\| < 1 \\ 0 & \text{for } \|x\| \geqq 1, \end{cases}$$

where $\mathbb{R}^n \xrightarrow{\|\ \|} \mathbb{R}$ is the usual euclidean norm. Then f is a smooth function which is positive for $\|x\| < 1$ and zero for $\|x\| \geqq 1$. For any $r > 0$ and any point $y \in \mathbb{R}^n$ one easily adjusts f to yield a smooth function $\mathbb{R}^n \xrightarrow{f_{y,r}} [0, 1]$ which is nonzero precisely in the open disk $D_{y,r}$ of radius r about y. Let U be any open subset of R^n, and let V be any open subset with compact closure $\bar{V} \subset U$. Then $\bar{V}$ can be covered by finitely many such open disks $D_{y,r} \subset U$, and the sum of the corresponding finitely many functions $f_{y,r}$ is a nonnegative smooth function $\mathbb{R}^n \to \mathbb{R}$ which is positive on $\bar{V}$ and vanishes outside of U.

One can refine the given open covering $\{U_i | i \in I\}$ of X by taking intersections with coordinate neighborhoods, so that without loss of generality one can assume that $\{U_i | i \in I\}$ is a coordinate covering; furthermore, since Lemma 6.1 provides a locally finite refinement of the latter covering, one may as well assume in addition that $\{U_i | i \in I\}$ is itself a locally finite coordinate covering of X, with atlas $\{\Phi_i | i \in I\}$. A second application of Lemma 6.1 gives a countable locally finite refinement $\{V_j | j \in \mathbb{N}\}$ of $\{U_i | i \in I\}$ such that each closure $\bar{V}_j$ is compact and satisfies $\bar{V}_j \subset U_{\rho(j)}$. Thus there are open sets $\Phi_{\rho(j)}(V_j)$ and $\Phi_{\rho(j)}(U_{\rho(j)})$ in $\mathbb{R}^n$ with $\overline{\Phi_{\rho(j)}(V_j)} = \Phi_{\rho(j)}(\bar{V}_j) \subset \Phi_{\rho(j)}(U_{\rho(j)})$, so that by the result of the preceding paragraph there are smooth nonnegative functions $\mathbb{R}^n \xrightarrow{g_j} \mathbb{R}$ such that each g_j is positive on $\Phi_{\rho(j)}(\bar{V}_j)$ and vanishes outside of $\Phi_{\rho(j)}(U_{\rho(j)})$. Each composition $U_{\rho(j)} \xrightarrow{\Phi_{\rho(j)}} \Phi_{\rho(j)}(U_{\rho(j)}(U_{\rho(j)}) \xrightarrow{g_j} \mathbb{R}$ extends to a smooth nonnegative function $X \xrightarrow{f_j} \mathbb{R}$ that is positive on $\bar{V}_j$ and vanishes outside of $U_{\rho(j)}$. Since $\{U_i | i \in I\}$ and $\{V_j | j \in \mathbb{N}\}$ are locally finite one can compute the sum $f = \sum_{j \in \mathbb{N}} f_j$, and since $\{\bar{V}_j | j \in \mathbb{N}\}$ covers X, it follows that f is everywhere positive. The desired smooth partition of unity $\{h_j | j \in \mathbb{N}\}$ is then obtained by setting $h_j = f_j/f$ for each $j \in \mathbb{N}$.

Suppose that a smooth map $X \xrightarrow{f} Y$ is a homeomorphism; then f need not be a diffeomorphism. For example, if $\mathbb{R} \xrightarrow{f} \mathbb{R}$ is given by $f(x) = x^3$, then f is clearly a smooth homeomorphism; but since the derivative of f vanishes at $x = 0$, the inverse is not smooth. The difficulty can be overcome by imposing the obvious condition on f, which we do in a more general situation.

Let X and Y be smooth manifolds of dimensions m and n with smooth structures $\{U_i \mid i \in I\}$, $\{\Phi_i \mid i \in I\}$ and $\{V_j \mid j \in J\}$, $\{\Psi_j \mid j \in J\}$, respectively, and let $X \xrightarrow{f} Y$ be a smooth map. By definition, the maps $\Psi_j \circ f \circ \Phi_i^{-1}$ carrying open subsets $U_i \cap f^{-1}(V_j) \subset \mathbb{R}^m$ into $\mathbb{R}^n$ are smooth in the sense that all partial derivatives exist as continuous functions on $U_i \cap f^{-1}(V_j)$; in particular, one can compute the $m \times n$ jacobian matrix of $\Psi_j \circ f \circ \Phi_i^{-1}$ at each point of $U_i \cap f^{-1}(V_i)$. The map f is an *immersion* whenever each of these jacobian matrices has rank m.

Next suppose that $X \xrightarrow{f} Y$ is any injective map, not necessarily smooth. Then the image $f(X) \subset Y$, in the relative topology, need not be homeomorphic to X. For example, one easily constructs a (smooth) injective map $\mathbb{R} \xrightarrow{f} \mathbb{R}^2$ with $f(0) = (0, 0)$ and $f(x) = (1/x, \sin(1/x))$ for $x \geqq 1$. In this example the inverse image of every neighborhood of (0, 0) contains arbitrarily large real numbers so that the relative topology of the image $f(\mathbb{R}) \subset \mathbb{R}^2$ is not the usual topology of $\mathbb{R}$; thus f is not a homeomorphism onto its image.

An *embedding* of a smooth manifold X into a smooth manifold Y is a smooth injective immersion $X \xrightarrow{f} Y$ that induces a homeomorphism of X onto $f(X) \subset Y$, in the relative topology.

(A map $X \xrightarrow{f} Y$ of topological spaces is *proper* if and only if the inverse image of every compact set in Y is compact in X. One easily verifies that a smooth injective immersion $X \xrightarrow{f} Y$ is an embedding if and only if it is proper.)

We shall extend the definition of embeddings somewhat. Recall from the preceding section that $\mathbb{R}^\infty$ denotes the limit $\lim_n \mathbb{R}^n$ of the sequence $\mathbb{R}^1 \to \mathbb{R}^2 \to \mathbb{R}^3 \to \cdots$ of canonical inclusions, in the weak topology. For any smooth manifold X a map $X \xrightarrow{f} \mathbb{R}^\infty$ is *smooth* whenever the composition $X \xrightarrow{f} \mathbb{R}^\infty \to \mathbb{R}$ with each projection $\mathbb{R}^\infty \to \mathbb{R}$ is smooth. If one regards points of $\mathbb{R}^\infty$ as row vectors with countably many entries, then for each coordinate neighborhood of X the jacobian matrix of f consists of countably many columns, the number of rows being the dimension of X. A smooth map $X \xrightarrow{f} \mathbb{R}^\infty$ is an *immersion* whenever the rank of each such jacobian matrix is the dimension of X. A smooth injective immersion $X \xrightarrow{f} \mathbb{R}^\infty$ is an *embedding* whenever it induces a homeomorphism of X onto $f(X) \subset \mathbb{R}^\infty$, in the relative topology.

6.3 Proposition: *For any smooth manifold X there is at least one embedding $X \to \mathbb{R}^\infty$.*

Proof: Lemma 6.1 provides a locally finite coordinate covering $\{U_i \mid i \in I\}$ of X such that each closure $\bar{U}_i$ is compact. Let $\{\Phi_i \mid i \in I\}$ be the corresponding atlas; if X is n-dimensional, then each homeomorphism Φ_i from U_i to the open set $\Phi_i(U_i) \subset \mathbb{R}^n$ is given by coordinate functions $U_i \xrightarrow{x_i^1} \mathbb{R}, \ldots,$

$U_i \xrightarrow{x_i^n} \mathbb{R}$. According to Lemma 6.2 there is a countable smooth partition of unity $\{h_j | j \in \mathbb{N}\}$ subordinate to $\{U_i | i \in I\}$; specifically, there is a function $\mathbb{N} \xrightarrow{\rho} J$ such that $X \xrightarrow{h_j} \mathbb{R}$ vanishes outside of some open set $V_j \subset X$, and such that $\bar{V}_j \subset U_{\rho(j)}$, where $\{V_j | j \in \mathbb{N}\}$ is also a locally finite covering of X. For each $j \in \mathbb{N}$ and each $k = 1, \ldots, n$ the product $V_j \xrightarrow{h_j x^k_{\rho(j)}} \mathbb{R}$ therefore extends to a map $X \xrightarrow{y_j^k} \mathbb{R}$ that vanishes outside of V_j, and since $\{V_j | j \in \mathbb{N}\}$ is locally finite, it follows that the denumerably many maps y_j^k induce a map $X \to \mathbb{R}^\infty$. One easily verifies that the latter map is an embedding.

6.4 Corollary: *Any smooth manifold X is metrizable.*

Proof: As in §2 there is a norm $\mathbb{R}^\infty \xrightarrow{\|\,\|} \mathbb{R}$ for which each $\|(x_0, x_1, x_2, \ldots)\|$ is the (finite) sum $\sum_{j \in \mathbb{N}} |x_j|$. For any embedding $X \xrightarrow{f} \mathbb{R}^\infty$ there is then a metric d on X with $d(x, y) = \|f(x) - f(y)\|$.

6.5 Proposition: *For any smooth compact manifold X there is at least one natural number $N \geqq 0$ for which there exists an embedding $X \to \mathbb{R}^N$.*

Proof: If X is compact, then the modifiers "denumerable" or "countable" become "finite" throughout the proofs of Lemmas 6.1, 6.2, and Proposition 6.3.

Propositions 6.3 and 6.5 are definitely not best-possible results. For certain values of n the following result is best possible.

6.6 Theorem (Whitney Embedding Theorem): *Any smooth n-dimensional manifold X has at least one embedding $X \to \mathbb{R}^{2n}$.*

This theorem appears in Whitney [7, p. 236]; its proof is long. A detailed discussion of Theorem 6.6 and related embedding theorems is given in Remark 10.11.

Any n-dimensional sphere S^n has at least one smooth structure, constructed in the obvious way from two coordinate neighborhoods homeomorphic to $\mathbb{R}^n$ itself. Since one can compose an embedding $X \to \mathbb{R}^{2n}$ with the inclusion $\mathbb{R}^{2n} \to S^{2n}$ of either coordinate neighborhood of S^{2n} the following assertion is clearly equivalent to the Whitney embedding theorem: any smooth n-dimensional manifold X has at least one embedding $X \to S^{2n}$.

6.7 Theorem (Cairns–Whitehead Triangulation Theorem): *Any smooth n-dimensional manifold X is homeomorphic to an n-dimensional metric simplicial space $|K|$.*

In fact X has a "smooth triangulation" in the obvious sense. References for proofs of Theorem 6.7 are given in Remark 10.14.

6.8 Corollary: *Any smooth manifold is a base space; hence the category $\mathscr{M}$ of smooth manifolds is a full subcategory of the category $\mathscr{B}$ of base spaces.*

PROOF: According to Corollary 2.2 any finite-dimensional metric simplicial space $|K|$ is a base space.

A strengthened version of Corollary 6.8 is described in Remarks 10.17 and 10.18; any (second countable) *topological manifold* whatsoever is a base space. The proof avoids the Cairns–Whitehead triangulation theorem, which is not available for topological manifolds.

Recall that an embedding is a specialized immersion. Immersions are themselves of interest, however. For certain values of n the following result is best-possible, where S^{2n-1} is the usual smooth sphere of dimension $2n - 1$.

6.9 Theorem (Whitney Immersion Theorem): *Any smooth n-dimensional manifold X has at least one immersion $X \to S^{2n-1}$.*

References to two proofs of Theorem 6.9 can be found in Remark 10.13, along with a detailed discussion of related results.

As in the case of the Whitney embedding theorem there is a trivially equivalent alternative version of Theorem 6.9, but with a dimensional restriction: if $n > 1$, then any smooth n-dimensional manifold X has at least one immersion $X \to \mathbb{R}^{2n-1}$. The reason for the restriction is that S^1 itself clearly does not immerse in $\mathbb{R}^1$.

Suppose that X is a smooth manifold of even dimension $2n$, with smooth structure $\{U_i | i \in I\}$ and $\{\Phi_i | i \in I\}$. The space $\mathbb{R}^{2n}$ is homeomorphic (in many ways) to $\mathbb{C}^n$, so that the maps

$$\Phi_i(U_i \cap U_j) \xrightarrow{\Phi_j \circ \Phi_i^{-1}} \Phi_j(U_i \cap U_j)$$

can be regarded as maps from sets in $\mathbb{C}^n$ to sets in $\mathbb{C}^n$. The manifold X is a *complex manifold* of *complex dimension n* whenever there is a such a homeomorphism for each $i \in I$ for which all the maps $\Phi_j \circ \Phi_i^{-1}$ are holomorphic; that is, the n complex-valued functions on $\Phi_i(U_i \cap U_j) \subset \mathbb{C}^n$ which describe $\Phi_j \circ \Phi_i^{-1}$ have local expansions about each point as Taylor series in n complex variables.

We shall later show that are are many real even-dimensional smooth manifolds which have no complex structure in the preceding sense; for example, the 4-dimensional sphere S^4 has no such structure. However, we have already encountered some topological spaces which *do* have such structures: the complex projective spaces CP^n, whose cell structures were described in Proposition 5.7. To construct a complex structure for CP^n one merely substitutes $\mathbb{C}$ for $\mathbb{R}$ in the construction of a smooth structure for the real projective space RP^n, given at the beginning of this section. Incidentally, as topological spaces the projective spaces RP^1 and CP^1 are the circle S^1 and the 2-sphere S^2, respectively; in particular, S^2 does have the complex structure it receives from its identification with CP^1.

For later convenience we record an obvious consequence of Lemma 6.1. It is in part a specialized version of the *shrinking lemma* of Dieudonné [1], which also appears in Dugundji [2, pp. 152–153].

6.10 Proposition: *Any manifold X has a countable locally finite covering $\{U_n | n \in \mathbb{N}\}$ by open sets $U_n \subset X$ with compact closures $\bar{U}_n$, refining a given open covering of X. Furthermore, for such a covering $\{U_n | n \in \mathbb{N}\}$, there is another countable locally finite covering $\{V_n | n \in \mathbb{N}\}$ by open sets V_n with compact closures $\bar{V}_n \subset U_n$ for each $n \in \mathbb{N}$.*

PROOF: Lemma 6.1 itself guarantees the existence of $\{U_n | n \in \mathbb{N}\}$ as well as a countable open covering $\{W_n | n \in \mathbb{N}\}$ and a map $\mathbb{N} \xrightarrow{\rho} \mathbb{N}$ such that $\bar{W}_n \subset U_{\rho(n)}$ for each $n \in \mathbb{N}$. Let $\rho^{-1}(n)$ denote the set of those $m \in \mathbb{N}$ with $\rho(m) = n$, and set $V_n = \bigcup_{m \in \rho^{-1}(n)} W_m$, for each $n \in \mathbb{N}$. Then $\bar{V}_n \subset U_n$ and $\{V_n | n \in \mathbb{N}\}$ is an open covering of X. Since each $\bar{V}_n$ is thus a closed subset of the compact set $\bar{U}_n$, it is itself compact, and since $\{U_n | n \in \mathbb{N}\}$ is locally finite, so is $\{V_n | n \in \mathbb{N}\}$.

6.11 Corollary: *Let X be a smooth manifold with countable locally finite open coverings $\{U_n | n \in \mathbb{N}\}$ and $\{V_n | n \in \mathbb{N}\}$ as in Proposition 6.10; then there is a smooth partition of unity $\{h_n | n \in \mathbb{N}\}$ such that each h_n is positive on V_n and vanishes on $X - U_n$.*

PROOF: One simply repeats the proof of Lemma 6.2, observing that the identity map $\mathbb{N} \to \mathbb{N}$ now replaces the map $\mathbb{N} \xrightarrow{\rho} I$ used in Lemma 6.2 to describe $\{V_j | j \in \mathbb{N}\}$ as a refinement of $\{U_i | i \in I\}$.

Proposition 6.10 will be used to prove several approximation theorems, whose proofs also depend on the following best-known elementary approximation theorem.

6.12 Theorem (The Stone–Weierstrass Theorem): *Let $A(\bar{U})$ be any real algebra of continuous real-valued functions $\bar{U} \xrightarrow{g} \mathbb{R}$ on a compact hausdorff space $\bar{U}$, and suppose that* (i) *$A(\bar{U})$ contains all real-valued constant functions and that* (ii) *if $x \neq y$ in $\bar{U}$ then there is at least one function $g \in A(\bar{U})$ such that $g(x) \neq g(y)$ in $\mathbb{R}$. Then for any continuous function $\bar{U} \xrightarrow{f} \mathbb{R}$ whatsoever, and for any constant $\varepsilon > 0$, there is a $g \in A(\bar{U})$ such that $|g(x) - f(x)| < \varepsilon$ for every $x \in \bar{U}$.*

PROOF: See Bartle [1, pp. 185–186] or Royden [1, p. 174], for example.

6.13 Corollary: *Let X be any smooth manifold, and let $X \xrightarrow{f} (\mathbb{R}^m)^+$ be any (continuous) map from X to the euclidean half-space $(\mathbb{R}^m)^+$; then, for the usual euclidean norm $(\mathbb{R}^m) \xrightarrow{\|\ \|} \mathbb{R}^+$ and any $\varepsilon > 0$, there is a smooth map $X \xrightarrow{g} (\mathbb{R}^m)^+$ such that $\|g(x) - f(x)\| < \varepsilon$ for any $x \in X$.*

PROOF: The map f consists of m real-valued functions $X \xrightarrow{f_1} \mathbb{R}, \ldots, X \xrightarrow{f_{m-1}} \mathbb{R}$, $X \xrightarrow{f_m} \mathbb{R}^+$. In case X is a *compact* smooth manifold $\bar{U}$, the *smooth* real-valued functions $\bar{U} \xrightarrow{g} \mathbb{R}$ satisfy the conditions required of the algebra $A(\bar{U})$ in the Stone–Weierstrass theorem, so that there are smooth functions $\bar{U} \xrightarrow{g_1} \mathbb{R}, \ldots, \bar{U} \xrightarrow{g_{m-1}} \mathbb{R}$ such that $|g_j(x) - f_j(x)| < \varepsilon/m$ for every $x \in \bar{U}$, where $j = 1, \ldots, m - 1$; a minor modification of the Stone–Weierstrass theorem also provides a smooth function $\bar{U} \xrightarrow{g_m} \mathbb{R}^+$ such that one has $|g_m(x) - f_m(x)| < \varepsilon/m$ for every $x \in \bar{U}$, and it follows that $(g_1, \ldots, g_m)$ is a smooth map $\bar{U} \xrightarrow{g} (\mathbb{R}^m)^+$ such that $\|g(x) - f(x)\| < \varepsilon$ for every $x \in \bar{U}$. Now let X be any smooth manifold whatsoever. By Proposition 6.10 there is a locally finite covering $\{U^n | n \in \mathbb{N}\}$ of X by open sets $U_n \subset X$ with compact closures $\bar{U}_n$, so that the preceding argument provides a smooth map $\bar{U}_n \xrightarrow{g^n} (R^m)^+$ for each $n \in \mathbb{N}$ such that $\|g^n(x) - f(x)\| < \varepsilon$ for each $x \in \bar{U}_n$. However, according to Corollary 6.11 there is a smooth partition of unity $\{h_n | n \in \mathbb{N}\}$ subordinate to $\{U_n | n \in \mathbb{N}\}$, and $\sum_{n \in \mathbb{N}} h_n g^n$ is then a well-defined map $X \xrightarrow{g} (\mathbb{R}^m)^+$ such that $\|g(x) - f(x)\| < \varepsilon$ for every $x \in X$, as desired.

Corollary 6.13 is a special case of a more general approximation theorem that implies that any (continuous) map $X \xrightarrow{f} Y$ of smooth manifolds is homotopic to a smooth map; we need only the latter implication. To start the proof one first uses Proposition 6.10 to find a countable locally finite covering $\{U_n^0 | n \in \mathbb{N}\}$ of Y by open sets U_n^0, each of which has a compact closure $\overline{U_n^0}$ in some coordinate neighborhood of Y. Thus, if Y is m-dimensional, there is an atlas $\{\Phi_n | n \in \mathbb{N}\}$ of diffeomorphisms of open sets in Y onto open sets in $(\mathbb{R}^m)^+$ such that $\Phi_n(\overline{U_n^0}) = \overline{\Phi_n(U_n^0)} \subset (\mathbb{R}^m)^+$ for each $n \in \mathbb{N}$. One then applies Proposition 6.10 repeatedly to find a sequence $\{U_n^0 | n \in \mathbb{N}\}$, $\{U_n^1 | n \in \mathbb{N}\}$, $\{U_n^2 | n \in \mathbb{N}\}$, ... of countable locally finite open coverings of Y such that each U_n^{q+1} has compact closure satisfying $\overline{U_n^{q+1}} \subset U_n^q$. The sets U_n^q and their closures appear throughout the following lemmas.

Two maps $X \xrightarrow{f_p} Y$ and $X \xrightarrow{f_{p+1}} Y$ are *homotopic relative to a subset* $X' \subset X$ if they are the restrictions to $X \times \{0\}$ and $X \times \{1\}$, respectively, of a map $X \times [0, 1] \xrightarrow{F} Y$ such that $F(x, t)$ is independent of $t \in [0, 1]$ whenever $x \in X'$.

6.14 Lemma: *Let $X \xrightarrow{f} Y$ be a (continuous) map of smooth manifolds, and for each $q \in \mathbb{N}$ let $\{U_n^q | n \in \mathbb{N}\}$ be the covering of Y described above. Then there* *is a sequence* $(f_0, f_1, f_2, \ldots)$ *of maps* $X \xrightarrow{f_p} Y$, *beginning with* $f_0 = f$, *such that the following conditions are satisfied for each* $p \in \mathbb{N}$:

(i) f_p *is homotopic to* f_{p+1}, *relative to* $X - f_p^{-1}(U_p^{p+1})$,
(ii) *the restriction of* f_{p+1} *to* $f_0^{-1}(U_0^2) \cup \cdots \cup f_p^{-1}(U_p^{p+2})$ *is smooth,*
(iii) *if* $0 \leqq q \leqq n + 2p$, *then* $f_{p+1}^{-1}(U_n^{q+2}) \subset f_p^{-1}(U_n^{q+1})$ *for all* $n \in \mathbb{N}$, *and*
(iv) *if* $0 \leqq q \leqq n + 2p$, *then* $f_p^{-1}(U_n^{q+2}) \subset f_{p+1}^{-1}(U_n^{q+1})$ *for all* $n \in \mathbb{N}$.

PROOF: We shall show that if a finite sequence $(f_0, \ldots, f_p)$ satisfies all four conditions, then there is an f_{p+1} such that $(f_0, \ldots, f_p, f_{p+1})$ satisfies the same conditions. Since $\overline{U_p^0}$ is compact, it meets only finitely many members of the locally finite covering $\{U_n^0 | n \in \mathbb{N}\}$. Hence there is an $n_p \in \mathbb{N}$ such that $\overline{U_p^0} \cap U_n^q$ is void whenever $n > n_p$, for any $q \in \mathbb{N}$. Let $\{\Phi_n | n \in \mathbb{N}\}$ be the atlas described earlier, and for each $U_n^q \subset Y$ set $V_n^q = \Phi_p(U_p^0 \cap U_n^q) \subset (\mathbb{R}^m)^+$. For each $n \in \mathbb{N}$ and $q \in \mathbb{N}$ the sets $\overline{V_n^{q+1}}$ and $\bar{V}_p^0 - V_n^q$ are disjoint closed subsets of the compact closure $\bar{V}_p^0 = \overline{\Phi^p(U_p^0)} = \Phi_p(\bar{U}_p^0) \subset (\mathbb{R}^m)^+$, and they are therefore separated by a positive euclidean distance $\varepsilon_n^q > 0$, where $\varepsilon_n^q = 1$ if one or both of the subsets is void. One then defines $\varepsilon > 0$ by setting

$$\varepsilon = \min_{(n,q)} \{\varepsilon_n^q | 0 \leqq n \leqq n_p, 0 \leqq q \leqq n_p + 2p + 2\}.$$

Let $f_p^{-1}(\bar{U}_p^0) \xrightarrow{g_p} \bar{V}_p^0 \subset (\mathbb{R}^m)^+$ be the composition $f_p^{-1}(\bar{U}_p^0) \xrightarrow{f_p} \bar{U}_p^0 \xrightarrow{\Phi_p} \bar{V}_p^0$. One applies Corollary 6.13 to find a smooth map $f_p^{-1}(\bar{U}_p^0) \xrightarrow{\tilde{g}_p} (\mathbb{R}^m)^+$ such that $\|\tilde{g}_p(x) - g_p(x)\| < \varepsilon$ for any $x \in f_p^{-1}(\bar{U}_p^0)$. Since $\overline{V_p^{p+2}}$ and $\bar{V}_p^0 - V_p^{p+1}$ are disjoint closed sets in $\bar{V}_p^0$ the method of Lemma 6.2 yields a smooth map $\bar{V}^0 \to [0, 1]$ with restrictions $h | \overline{V_p^{p+2}} = 1$ and $h | \bar{V}_p^0 - V_p^{p+1} = 0$. Since $\overline{V_p^{p+2}}$ and $\bar{V}_p^0 - \bar{V}_p^{p+1}$ are separated by euclidean distance at least $\varepsilon > 0$, the points $(1 - th(g_p(x)))g_p(x) + th(g_p(x))\tilde{g}_p(x)$ of $(\mathbb{R}^m)^+$ all lie in $\overline{V_p^0}$ whenever $(x, t) \in f_p^{-1}(\bar{U}_p^0) \times [0, 1]$, and one thereby obtains a map $f_p^{-1}(\bar{U}_p^0) \times [0, 1] \xrightarrow{G} V_p^0$. By construction, $G(x, t)$ is independent of $t \in [0, 1]$ whenever $x \in f_p^{-1}(\bar{U}_p^0) - f_p^{-1}(U_p^{p+1})$, so that G is a homotopy from g_p to a map $f_p^{-1}(\bar{U}_p^0) \xrightarrow{g_{p+1}} \bar{V}_p^0$, relative to $f_p^{-1}(\bar{U}_p^0) - f_p^{-1}(U_p^{p+1})$. Consequently there is a unique homotopy $X \times [0, 1] \xrightarrow{F} Y$ from f_p to a map $X \xrightarrow{f_{p+1}} Y$, relative to $X - f_p^{-1}(U_p^{p+1})$, such that $F | f_p^{-1}(U_p^0) \times [0, 1] = \Phi_p^{-1} \circ G$. The map f_{p+1} satisfies condition (i) by its very construction. Since g_{p+1} is clearly smooth on the union of $f_p^{-1}(U_p^{p+2})$ with any open set in $\bar{V}_p^0$ on which g_p is smooth, f_{p+1} satisfies condition (ii). Finally, the definition of ε and the property $\|\tilde{g}_p - g_p\| < \varepsilon$ imply both

$$f_{p+1}^{-1}(\bar{U}_p^0 \cap U_n^{q+2}) \subset f_p^{-1}(\bar{U}_p^0 \cap U_n^{q+1})$$

and

$$f_p^{-1}(\bar{U}_p^0 \cap U_n^{q+2}) \subset f_{p+1}^{-1}(\bar{U}_p^0 \cap U_n^{q+1})$$

whenever $0 \leqq n \leqq n_p$ and $0 \leqq q \leqq n_p + 2p + 2$; since $\bar{U}_p^0 \cap U_n^q$ is void for $n > n_p$, the same conclusions apply whenever $n > n_p$, for any $q \in \mathbb{N}$. Since F is a homotopy relative to $X - f_p^{-1}(U_p^{p+1})$, a fortiori relative to $X - f_p^{-1}(\bar{U}_p^0)$, the inclusions $U_n^{q+2} \subset U_n^{q+1}$ themselves imply both

$$f_{p+1}^{-1}((Y - \bar{U}_p^0) \cap U_n^{q+2}) \subset f_p^{-1}((Y - \bar{U}_p^0) \cap U_n^{q+1})$$

and

$$f_p^{-1}((Y - \bar{U}_p^0) \cap U_n^{q+2}) \subset f_{p+1}^{-1}((Y - \bar{U}_p^0) \cap U_n^{q+1}$$

for any n and q whatsoever. Consequently both $f_{p+1}^{-1}(U_n^{q+2}) \subset f_p^{-1}(U_n^{q+1})$ and $f_p^{-1}(U_n^{q+2}) \subset f_{p+1}^{-1}(U_n^{q+1})$ whenever $0 \leqq q \leqq n + 2p + 2$, as required by conditions (iii) and (iv).

6.15 Lemma: *$f_p^{-1}(U_p^{p+1}) \subset f_0^{-1}(U_p^1)$ for any $p \in \mathbb{N}$.*

PROOF: This is a string

$$f_p^{-1}(U_p^{p+1}) \subset f_{p-1}^{-1}(U_p^p) \subset \cdots \subset f_1^{-1}(U_p^2) \subset f_0^{-1}(U_p^1)$$

of applications of condition (iii) of Lemma 6.14.

6.16 Lemma: *$f_0^{-1}(U_p^{2p+2}) \subset f_p^{-1}(U_p^{p+2})$ for any $p \in \mathbb{N}$.*

PROOF: This is a string

$$f_0^{-1}(U_p^{2p+2}) \subset f_1^{-1}(U_p^{2p+1}) \subset \cdots \subset f_{p-1}^{-1}(U_p^{p+3}) \subset f_p^{-1}(U_p^{p+2})$$

of applications of condition (iv) of Lemma 6.14.

6.17 Lemma: *The family $\{U_p^{2p+2} | p \in \mathbb{N}\}$ of open sets U_p^{2p+2} covers Y.*

PROOF: For any $y \in Y$ let $n_y \in \mathbb{N}$ be the largest number such that $y \in U_{n_y}^0$; n_y exists because $\{U_n^0 | n \in \mathbb{N}\}$ is locally finite. Since $\{U_p^{2n_y+2} | p \in \mathbb{N}\}$ also covers Y, and since $U_p^{2n_y+2} \subset U_p^q$ whenever $q \leqq 2n_y + 2$, it follows that

$$y \in U_0^{2n_y+2} \cup U_1^{2n_y+2} \cup \cdots \cup U_{n_y}^{2n_y+2} \subset U_0^2 \cup U_1^4 \cup \cdots \cup U_{n_y}^{2n_y+2}.$$

6.18 Lemma: *The family $\{f_p^{-1}(U_p^{p+2}) | p \in \mathbb{N}\}$ of open sets $f_p^{-1}(U_p^{p+2})$ covers X.*

PROOF: Since $f_0^{-1}(Y) = X$, this is a consequence of Lemmas 6.16 and 6.17:

$$\begin{aligned} f_0^{-1}(U_0^2 \cup \cdots \cup U_p^{2p+2}) &= f_0^{-1}(U_0^2) \cup \cdots \cup f_0^{-1}(U_p^{2p+2}) \\ &\subset f_0^{-1}(U_0^2) \cup \cdots \cup f_p^{-1}(U_p^{p+2}). \end{aligned}$$

6.19 Theorem: *Any (continuous) map $X \xrightarrow{f} Y$ of smooth manifolds X and Y is homotopic to a smooth map $X \to Y$.*

PROOF: Let $(f_0, f_1, f_2, \ldots)$ be the sequence of maps $X \xrightarrow{f_p} Y$ constructed in Lemma 6.14, and regard the homotopy of condition (i) as a map $X \times [p/(p+1), (p+1)/(p+2)] \xrightarrow{F_p} Y$; specifically, F_p is a homotopy relative to $X - f_p^{-1}(U_p^{p+1})$. Since $f_p^{-1}(U_p^{p+1}) \subset f_0^{-1}(U_p^1)$ for any $p \in \mathbb{N}$, by Lemma 6.15, and since $\{f_0^{-1}(U_p^1) | p \in \mathbb{N}\}$ is a locally finite cover of X, it follows for any $x \in X$ that there is a $p_x \in \mathbb{N}$ such that $F_p(x, t)$ is independent of the choice of $t \in [p/(p+1), (p+1)/(p+2)]$ for $p > p_x$; hence F_p is the restriction to $X \times [p/(p+1), (p+1)/(p+2)]$ of a well-defined homotopy $X \times [0, 1] \xrightarrow{F} Y$, the restriction $X \times \{0\} \xrightarrow{F} Y$ being the initial $f = f_0$. Condition (ii) of Lemma 6.14 implies that $X \times \{1\} \xrightarrow{F} Y$ is smooth on any open set $f_p^{-1}(U_p^{p+2}) \subset X$, and since $\{f_p^{-1}(U_p^{p+2}) | p \in \mathbb{N}\}$ covers X, by Lemma 6.18, $X \times \{1\} \xrightarrow{F} Y$ is smooth on X, as required.

Many other expositions of the study of smooth manifolds in general are indicated in Remark 10.10.

7 Grassmann Manifolds

The projective spaces RP^n, RP^∞, CP^n, and CP^∞ of §5 can be regarded as spaces of lines through the origins of the vector spaces $\mathbb{R}^{n+1}$, $\mathbb{R}^\infty$, $\mathbb{C}^{n+1}$, and $\mathbb{C}^\infty$, respectively; that is, their points are the 1-dimensional subspaces of the corresponding vector spaces. More generally, for any natural number $m > 0$ there are spaces whose points are the m-dimensional subspaces of $\mathbb{R}^{m+n}$, $\mathbb{R}^\infty$, $\mathbb{C}^{m+n}$, and $\mathbb{C}^\infty$. These spaces play a crucial role in the theory of vector bundles, and the primary purpose of this section is to develop some of their properties, especially the fact that they are base spaces.

7.1 Definition: For any natural number $m > 0$ let V denote any of the real or complex vector spaces $\mathbb{R}^{m+n}$, $\mathbb{R}^\infty$, $\mathbb{C}^{m+n}$, $\mathbb{C}^\infty$; the usual topology is imposed on $\mathbb{R}^{m+n}$ and $\mathbb{C}^{m+n}$, and the weak topology is imposed on $\mathbb{R}^\infty$ and $\mathbb{C}^\infty$. Let $(V \times \cdots \times V)^*$ denote the set of m-tuples of linearly independent vectors in V, in the relative topology of the m-fold product $V \times \cdots \times V$, and let $\sim$ be the equivalence relation in $(V \times \cdots \times V)^*$ with $(x_1, \ldots, x_m) \sim (y_1, \ldots, y_m)$ whenever the vectors $x_1, \ldots, x_m$ span the same m-dimensional subspace of V as the vectors $y_1, \ldots, y_m$. The *Grassmann manifold* $G^m(V)$ is the quotient $(V \times \cdots \times V)^*/\sim$, in the quotient topology.

For example, the Grassmann manifolds $G^1(\mathbb{R}^{n+1})$, $G^1(\mathbb{R}^\infty)$, $G^1(\mathbb{C}^{n+1})$ and $G^1(\mathbb{C}^\infty)$ are precisely the projective spaces RP^n, RP^∞, CP^n, and CP^∞.

Observe that the sequence $\mathbb{R}^{m+1} \to \mathbb{R}^{m+2} \to \mathbb{R}^{m+3} \to \cdots$ of canonical inclusions induces a corresponding sequence $G^m(\mathbb{R}^{m+1}) \to G^m(\mathbb{R}^{m+2}) \to$

$G^m(\mathbb{R}^{m+3}) \to \cdots$ of inclusions for which $G^m(\mathbb{R}^\infty) = \lim_n G^m(\mathbb{R}^{m+n})$ in the weak topology; similarly $G^m(\mathbb{C}^\infty) = \lim_n G^m(\mathbb{C}^{m+n})$. We shall show that each of $G^m(\mathbb{C}^{m+n})$ and $G^m(\mathbb{C}^{m+n})$ is a smooth manifold, a fortiori a base space. More generally, if V is any of the vector spaces $\mathbb{R}^{m+n}$, $\mathbb{R}^\infty$, $\mathbb{C}^{m+n}$, $\mathbb{C}^\infty$, then $G^m(V)$ is a CW space, hence a base space; however, it will be more convenient to prove directly that $G^m(V)$ is a base space.

We show that if $V = \mathbb{R}^{m+n}$, then $G^m(V)$ has the structure of a smooth closed manifold of dimension mn; the case $V = \mathbb{C}^{m+n}$ is similar. Let $\bigotimes V$ be the tensor algebra generated over $\mathbb{R}$ by V, let $I \subset \bigotimes V$ be the two-sided ideal generated by squares $x \otimes x \in V \otimes V \subset \bigotimes V$, and let $\bigwedge V$ be the *exterior algebra* $\bigotimes V/I$. The tensor algebra is graded by assigning degree m to real linear combinations of elements $x_1 \otimes \cdots \otimes x_m \in \bigotimes V$ for any m vectors $x_1, \ldots, x_m \in V$, and since I is homogeneous, there is a corresponding grading of $\bigwedge V$. The real vector space of elements of degree m in $\bigwedge V$ is the *m-fold exterior product* $\bigwedge^m V$; the image in $\bigwedge^m V$ of $x_1 \otimes \cdots \otimes x_m \in \bigotimes V$ is denoted $x_1 \wedge \cdots \wedge x_m$.

One easily shows that m vectors $x_1, \ldots, x_m$ in V are linearly dependent if and only if $x_1 \wedge \cdots \wedge x_m = 0 \in \bigwedge^m V$. A nonzero element of $\bigwedge^m V$ is *simple* if and only if it is of the form $x_1 \wedge \cdots \wedge x_m$ for m linearly *in*dependent vectors $x_1, \ldots, x_m$ in V. If $\{x_1, \ldots, x_{m+n}\}$ is a basis of V, then the $\binom{m+n}{m}$ simple elements $x_{i_1} \wedge \cdots \wedge x_{i_m} \in \bigwedge^m V$ with $i_1 < \cdots < i_m$ form a basis of $\bigwedge^m V$.

If $y_1, \ldots, y_m$ and $z_1, \ldots, z_m$ span the same m-dimensional subspace of V, then for some $a \neq 0$ one has $y_1 \wedge \cdots \wedge y_m = a z_1 \wedge \cdots \wedge z_m$ so that each point of $G^m(V)$ is identified up to a nonzero factor with a simple element in $\bigwedge^m V$; more specifically, there is an injective map $G^m(V) \xrightarrow{F} G^1(\bigwedge^m V)$ that is a homeomorphism onto the set $\operatorname{Im} F \subset G^1(\bigwedge^m V)$ of 1-dimensional subspaces spanned by simple elements of $\bigwedge^m V$, in the relative topology. Hence it is of interest to develop an algebraic criterion for simplicity of elements of $\bigwedge^m V$.

For $m = 1$ there is a canonical isomorphism $V \to \bigwedge^1 V$, so that every nonzero element of $\bigwedge^1 V$ is simple. In case $m > 1$ let $\bigwedge^{m-1} V \xrightarrow{\Theta} \mathbb{R}$ be any real linear functional on the $(m-1)$-fold exterior product $\bigwedge^{m-1} V$. One easily verifies that there is then a unique linear map $\bigwedge^m V \xrightarrow{\Theta \lrcorner} V$ with value $\Theta \lrcorner (x_1 \wedge \cdots \wedge x_m) = \sum_{i=1}^m (-1)^{i-1} \Theta(x_1 \wedge \cdots \wedge \hat{x}_i \wedge \cdots \wedge x_m) x_i \in V$ on every simple element $x_1 \wedge \cdots \wedge x_m \in \bigwedge^m V$, where $\hat{x}_i$ indicates that x_i is deleted.

7.2 Lemma: *If $m > 1$ and $X \in \bigwedge^m V$ is nonzero, then X is simple if and only if $X \wedge (\Theta \lrcorner X) = 0 \in \bigwedge^{m+1} V$ for every linear functional $\bigwedge^{m-1} V \xrightarrow{\Theta} \mathbb{R}$.*

Proof: If $X = x_1 \wedge \cdots \wedge x_m$, then any $\Theta \lrcorner X$ is a linear combination of $x_1, \ldots, x_m$, and since $(x_1 \wedge \cdots \wedge x_m) \wedge x_i = 0$ for $i = 1, \ldots, m$, it follows that

$X \wedge (\Theta \,\lrcorner\, X) = 0$. Conversely, let $V_X \subset V$ be the subspace of all vectors of the form $\Theta \,\lrcorner\, X$. Since X is nonzero, one has $\dim_{\mathbb{R}} V_X \geqq m$, and if $X \wedge x = 0$ for every $x \in V_X$, then $\dim_{\mathbb{R}} V_X \leqq m$. Hence $\dim_{\mathbb{R}} V_X = m$, so that $X = x_1 \wedge \cdots \wedge x_m$ for some basis $x_1, \ldots, x_m$ of V_X.

For any simple element $X \in \bigwedge^m V$ the identities $X \wedge (\Theta \,\lrcorner\, X) = 0$ are the *Plücker relations.* If $\{x_1, \ldots, x_{m+n}\}$ is a basis of V, then any $X \in \bigwedge^m V$ is uniquely of the form

$$\sum_{i_1 < \cdots < i_m} a(i_1, \ldots, i_m) x_{i_1} \wedge \cdots \wedge x_{i_m}$$

for real numbers $a(i_1, \ldots, i_m)$. For any permutation π of $\{1, \ldots, m\}$ let $a(\pi(i_1), \ldots, \pi(i_m)) = \varepsilon(\pi) a(i_1, \ldots, i_m)$, where $\varepsilon(\pi) = \pm 1$ according as π is even or odd; if two or more of the indices $i_1, \ldots, i_m$ in $\{1, \ldots, m+n\}$ happen to agree let $a(i_1, \ldots, i_m) = 0$. Then for any ordered $(m-1)$-tuple $(i_1, \ldots, i_{m-1})$ and any ordered $(m+1)$-tuple $(j_0, \ldots, j_m)$ of indices in $\{1, \ldots, m+n\}$ the Plücker relations for simple $X \in \bigwedge^m V$ are of the form

$$\sum_{k=0}^{m} (-1)^k a(i_1, \ldots, i_{m-1}, j_k) a(j_0, \ldots, \hat{j}_k, \ldots, j_m) = 0,$$

where $\hat{j}_k$ indicates that j_k is missing. We use the Plücker relations in the latter form for the Plücker *coordinates* $a(i_1, \ldots, i_m)$.

7.3 Proposition: *For any natural numbers $m > 0$ and $n > 0$ the Grassmann manifold $G^m(\mathbb{R}^{m+n})$ has a smooth structure for which it is an mn-dimensional smooth closed manifold.*

PROOF: Let $\{x_1, \ldots, x_{m+n}\}$ be a basis of $\mathbb{R}^{m+n}$, and suppose that

$$\sum_{i_1 < \cdots < i_m} a(i_1, \ldots, i_m) x_{i_1} \wedge \cdots \wedge x_{i_m}$$

is a simple element of $\bigwedge^m \mathbb{R}^{m+n}$, representing a point in the image $\operatorname{Im} F$ of $G^m(\mathbb{R}^{m+n}) \xrightarrow{F} G_1(\bigwedge^m \mathbb{R}^{m+n})$. For m fixed indices $i_1 < \cdots < i_m$ among $\{1, \ldots, m+n\}$ let $U_{i_1, \ldots, i_m} \subset \operatorname{Im} F$ be the open set of those points with $a(i_1, \ldots, i_m) \neq 0$. For any other m indices $j_1 < \cdots < j_m$ among $\{1, \ldots, m+n\}$ suppose that there are $m - p$ indices in the intersection $\{i_1, \ldots, i_m\} \cap \{j_1, \ldots, j_m\}$. The Plücker relations imply that there is a real polynomial $f^{j_1, \ldots, j_m}_{i_1, \ldots, i_m}$ of degree p in mn variables whose value on the mn real numbers

$$\frac{a(i_1, \ldots, \hat{\imath}_k, \ldots, i_m, j)}{a(i_1, \ldots, i_m)}$$

is the quotient

$$\frac{a(j_1, \ldots, j_m)}{a(i_1, \ldots, i_m)};$$

here $\hat{i}_k$ indicates that i_k is missing, and j is any of the n indices distinct from $i_1, \ldots, i_m$. Thus the map $U_{i_1, \ldots, i_m} \xrightarrow{\Phi_{i_1, \ldots, i_m}} \mathbb{R}^{mn}$ carrying the point with projective coordinates $a(j_1, \ldots, j_m) \in \mathbb{R}$ into the point of $\mathbb{R}^{mn}$ with mn coordinates

$$\frac{a(i_1, \ldots, \hat{i}_k, \ldots, i_m, j)}{a(i_1, \ldots, i_m)} \in \mathbb{R}$$

is a homeomorphism. Since the $\binom{m+n}{m}$ open sets $U_{i_1, \ldots, i_m} \subset \operatorname{Im} F$ cover $\operatorname{Im} F$, they form a coordinate covering $\{U_{i_1, \ldots, i_m} \mid i_1 < \cdots < i_m\}$, with corresponding atlas $\{\Phi_{i_1, \ldots, i_m} \mid i_1 < \cdots < i_m\}$. The homeomorphism $\Phi_{j_1, \ldots, j_m} \circ \Phi_{i_1, \ldots, i_m}^{-1}$ from

$$\Phi_{i_1, \ldots, i_m}(U_{i_1, \ldots, i_m} \cap U_{j_1, \ldots, j_m}) \subset \mathbb{R}^{mn}$$

to

$$\Phi_{j_1, \ldots, j_m}(U_{i_1, \ldots, i_m} \cap U_{j_1, \ldots, j_m}) \subset \mathbb{R}^{mn}$$

is given by the mn quotients

$$f_{i_1, \ldots, i_m}^{j_1, \ldots, \hat{j}_l, \ldots, j_m, k} / f_{i_1, \ldots, i_m}^{j_1, \ldots, j_m}$$

of polynomial functions, whose denominators are nonvanishing in

$$\Phi_{i_1, \ldots, i_m}(U_{i_1, \ldots, i_m} \cap U_{j_1, \ldots, j_m}).$$

Since such quotients are trivially smooth this completes the proof.

Alternative proofs of Proposition 7.3 are cited in Remark 10.21.

Each complex Grassmann manifold $G^m(\mathbb{C}^{m+n})$ is similarly a closed complex manifold of complex dimension mn, hence real dimension $2mn$.

7.4 Corollary: *The Grassmann manifolds $G^m(\mathbb{R}^{m+n})$ and $G^m(\mathbb{C}^{m+n})$ are base spaces.*

Proof: According to Corollary 6.8 any smooth manifold is a base space.

There is an alternative way to show that $G^m(\mathbb{R}^{m+n})$ and $G^m(\mathbb{C}^{m+n})$ are base spaces, which applies equally well to $G^m(\mathbb{R}^\infty)$ and $G^m(\mathbb{C}^\infty)$: one shows that they are CW spaces, hence via Corollary 5.4 that they are base spaces. In fact, we have already carried out this program for the case $m = 1$, the Grassmann manifolds $G^1(\mathbb{R}^{n+1})$, $G^1(\mathbb{R}^\infty)$, $G^1(\mathbb{C}^{n+1})$, $G^1(\mathbb{C}^\infty)$ being the projective spaces RP^n, RP^∞, CP^n, CP^∞, respectively; CW structures of these spaces were given in Proposition 5.5, Corollary 5.6, Proposition 5.7, and Corollary 5.8. More generally, CW structures of arbitrary Grassmann manifolds are constructed in Lundell and Weingram [1, pp. 13–15] and (in the real case) Milnor and Stasheff [1, pp. 73–81].

There is also a direct proof that Grassmann manifolds are base spaces.

7.5 Proposition: *All the Grassmann manifolds $G^m(\mathbb{R}^{m+n})$, $G^m(\mathbb{R}^\infty)$, $G^m(\mathbb{C}^{m+n})$, $G^m(\mathbb{C}^\infty)$ are base spaces.*

PROOF: We consider $G^m(\mathbb{R}^\infty)$, the other cases being similar. The space $\mathbb{R}^\infty$ is a direct sum of copies of $\mathbb{R}$, which one endows with the usual inner product. If $\{x_1, x_2, x_3, \ldots\}$ is an orthonormal basis of $\mathbb{R}^\infty$ then the exterior product $\bigwedge^m \mathbb{R}^\infty$ is also a direct sum of copies of $\mathbb{R}$, with an orthonormal basis $\{x_{i_1} \wedge \cdots \wedge x_{i_m} | i_1 < \cdots < i_m\}$. As in Proposition 7.3 $G^m(\mathbb{R}^\infty)$ can be regarded as the submanifold of those points of $G^1(\bigwedge^m \mathbb{R}^\infty)$ whose projective coefficients $a(i_1, \ldots, i_m) \in \mathbb{R}$ satisfy the Plücker relations. In the next two paragraphs we show that $G^m(\mathbb{R}^\infty)$ is of first type and metric, hence a base space.

For each index $(i_1, \ldots, i_m)$ let $U_{i_1, \ldots, i_m}$ denote the open set of those points in $G^m(\mathbb{R}^\infty)$ with $|a(i_1, \ldots, i_m)| > \max|a(j_1, \ldots, j_m)|$, where the maximum is computed over the finitely many indices $(j_1, \ldots, j_m)$ with $a(j_1, \ldots, j_m) \neq 0$ and $(j_1, \ldots, j_m) \neq (i_1, \ldots, i_m)$. Similarly, for any set $\{r_{i_1, \ldots, i_m} | i_1 < \cdots < i_m\}$ of mutually distinct positive numbers, let $V_{i_1, \ldots, i_m}$ denote the open set of those points in $G^m(\mathbb{R}^\infty)$ with $r_{i_1, \ldots, i_m}|a(i_1, \ldots, i_m)| > \max r_{j_1, \ldots, j_m}|a(j_1, \ldots, j_m)|$. Clearly $\{U_{i_1, \ldots, i_m} | i_1 < \cdots < i_m\}$ is a mutually disjoint family, $\{V_{i_1, \ldots, i_m} | i_1 < \cdots < i_m\}$ is a mutually disjoint family, and the two families collectively cover $G^m(\mathbb{R}^\infty)$; it remains to show that each $U_{i_1, \ldots, i_m}$, each $V_{j_1, \ldots, j_m}$, and each nonvoid intersection $U_{i_1, \ldots, i_m} \cap V_{j_1, \ldots, j_m}$ is of 0th type. Each point of $U_{i_1, \ldots, i_m}$ can be regarded as a one-dimensional subspace of $\bigwedge^m \mathbb{R}^\infty$ spanned by an element of the form $(x_{i_1} + y_1) \wedge \cdots \wedge (x_{i_m} + y_m)$, where each of $y_1, \ldots, y_m$ is orthogonal to the subspace of $\mathbb{R}^\infty$ spanned by $x_{i_1}, \ldots, x_{i_m}$. If $t \in [0, 1]$, then $(x_{i_1} + ty_1) \wedge \cdots \wedge (x_{i_m} + ty_m)$ also represents a point of $U_{i_1, \ldots, i_m}$, so that $U_{i_1, \ldots, i_m}$ trivially contracts to the point represented by $x_{i_1} \wedge \cdots \wedge x_{i_m}$; a fortiori $U_{i_1, \ldots, i_m}$ is of 0th type. Similarly each $V_{j_1, \ldots, j_m}$ is of 0th type. An intersection $U_{i_1, \ldots, i_m} \cap V_{j_1, \ldots, j_m}$ is nonvoid if and only if $r_{i_1, \ldots, i_m} < r_{j_1, \ldots, j_m}$ and finitely many other similar inequalities are satisfied. In this case points of $U_{i_1, \ldots, i_m} \cap V_{j_1, \ldots, j_m}$ are represented by points $(x_{i_1} + t_1 x_{j_1} + y_1) \wedge \cdots \wedge (x_{i_m} + t_m x_{j_m} + y_m) \in \bigwedge^m \mathbb{R}^\infty$, where each of $y_1, \ldots, y_m$ is orthogonal to the subspace of $\mathbb{R}^\infty$ spanned by $x_{i_1}, \ldots, x_{i_m}, x_{j_1}, \ldots, x_{j_m}$; the coefficients $t_1, \ldots, t_m$ satisfy $r_{i_1, \ldots, i_m}/r_{j_1, \ldots, j_m} < |t_1 \cdots t_m| < 1$ and finitely many other similar inequalities. If $t \in [0, 1]$, then $(x_{i_1} + t_1 x_{j_1} + ty_1) \cdots (x_{i_m} + t_m x_{j_m} + ty_m)$ also represents a point of $U_{i_1, \ldots, i_m} \cap V_{j_1, \ldots, j_m}$, so that $U_{i_1, \ldots, i_m} \cap V_{i_1, \ldots, i_m}$ contracts to a set represented by a set $W \subset \bigwedge^m \mathbb{R}^\infty$ of points $(x_{i_1} + t_1 x_{j_1}) \wedge \cdots \wedge (x_{i_m} + t_m x_{j_m})$ in which $t_1, \ldots, t_m$ satisfy finitely many inequalities. Furthermore, W is a disjoint union of (finitely many) contractible relatively open sets, so that $U_{i_1, \ldots, i_m} \cap V_{j_1, \ldots, j_m}$ is a corresponding disjoint union of contractible open sets; that is, $U_{i_1,\ldots,i_m} \cap V_{j_1,\ldots,j_m}$ is of 0th type as required.

To show that $G^m(\mathbb{R}^\infty)$ is metric, let $\bigwedge^m \mathbb{R}^\infty \xrightarrow{\|\,\|_2} \mathbb{R}$ be the euclidean norm on the direct sum $\bigwedge^m \mathbb{R}^\infty$, corresponding to the inner product used in the preceding paragraph. Since every element of the projective space $G^1(\bigwedge^m \mathbb{R}^\infty)$ is an equivalence class $[x]$ of nonzero elements $x \in \bigwedge^m \mathbb{R}$, one may as well choose representatives x that lie on the unit sphere: $\|x\|_2 = 1$. There is then a metric $G^1(\bigwedge^m \mathbb{R}^\infty) \times G^1(\bigwedge^m \mathbb{R}^\infty) \xrightarrow{d} \mathbb{R}$ given by

$$d([x],[y]) = \min_{x,y} \|x - y\|_2,$$

where the minimum is computed over representatives x, y of $[x]$, $[y]$ lying on the unit sphere, and the embedding $G^m(\mathbb{R}^\infty) \xrightarrow{F} G^1(\bigwedge^m \mathbb{R}^\infty)$ induces a metric on $G^m(\mathbb{R}^\infty)$, as desired.

8. Some More Coverings

This section contains two unrelated results. The first result asserts that any base space whatsoever is homotopy equivalent to a paracompact space with an especially useful covering. The second result is a more specialized covering theorem.

We start by quoting some classical general topology. Recall that a hausdorff space X is paracompact if and only if any open covering $\{U_i | i \in I\}$ has a locally finite refinement $\{V_j | j \in J\}$, and that any partition of unity $\{h_j | j \in J\}$ subordinate to $\{V_j | j \in J\}$ is also regarded as being subordinate to $\{U_i | i \in I\}$.

8.1 Lemma: *If X is paracompact, then there is a partition of unity subordinate to any open covering of X.*

The proof of Lemma 8.1 can be found in Dugundji [2, pp. 152–153, 170], for example.

8.2 Lemma (Stone [1]): *Any metric space is paracompact.*

Proofs of Lemma 8.2 can be found in Dugundji [2, pp. 167–169, 186], and Kelley [1, pp. 129, 156–160], for example.

8.3 Proposition: *Any base space is homotopy equivalent to a paracompact hausdorff space $X \in \mathcal{B}$ for which there is a countable locally finite open covering $\{U_n | n \in \mathbb{N}\}$ such that each connected component of each U_n is contained in a contractible open set in X.*

PROOF: Definition 1.2 states that any base space is homotopy equivalent to a metrizable space X of finite type. Since metric spaces are trivially hausdorff, Lemma 8.2 implies that X is paracompact. Definition 1.1 states that a space is of 0th type if it is a disjoint union of contractible spaces, of nth type if it has an appropriate covering by spaces of $(n-1)$th type, and of finite type if it is of nth type for some n. It follows that X has a covering by contractible open sets, and since X is paracompact, Lemma 8.1 provides a partition of unity $\{h_j | j \in J\}$ subordinate to that covering. For each $j \in J$ let $V_j \subset X$ be the open set of points $x \in X$ with $h_j(x) > 0$; then $\{V_j | j \in J\}$ is a locally finite covering of X such that each connected component of each V_j is contained in a contractible open set in X.

The covering $\{V_j | j \in J\}$ is not necessarily countable. For any *finite* subset $I \subset J$ let $W_I \subset X$ be the open set of those $x \in X$ such that

$$\min_{i \in I} h_i(x) > \max_{i \notin I} h_i(x).$$

Then W_I is a subset of V_j whenever $j \in I$, so that each connected component of each W_I is contained in a contractible subspace of X; furthermore $\{W_I\}_I$ trivially covers X. Let $|I|$ be the number of indices in I, and suppose for $I' \neq I$ that $|I'| = |I|$; then clearly $W_I \cap W_{I'} = \varnothing$. Hence for each natural number $n > 0$ each connected component of the open set $U_n = \bigcup_{|I|=n} W_I$ is also contained in a contractible open subspace of X.

Trivially $\{U_n | n \in \mathbb{N}\}$ is a countable open covering of X, and it remains only to show that it is locally finite. Since $\{V_j | j \in J\}$ is locally finite, for each $x \in X$ there is an $n > 0$ and an open neighborhood of x which meets at most n of the sets V_j; consequently that neighborhood of x does not intersect U_m for any $m > n$, which completes the proof.

For later use we record the following minor improvement of a special case of Proposition 2.1.

8.4 Proposition: *Let X be a smooth compact manifold with interior $\mathring{X}$. Then there is a finite covering $\{U_1, \ldots, U_q\}$ of $\mathring{X}$ by open cells such that each nonvoid intersection $U_i \cap U_j$ is a cell U_k in the covering, and such that the closures in X satisfy $\overline{U_i \cap U_j} = \bar{U}_i \cap \bar{U}_j$.*

PROOF: By the Cairns–Whitehead triangulation theorem (Theorem 6.7) X is homeomorphic to an n-dimensional metric simplicial space $|K|$, where n is the dimension of X. Since X is compact, one can assume that the vertex set K_0 of the abstract simplicial complex K is finite. For each vertex $i \in K_0$ let $Y_i \subset \coprod_{K_0} \mathbb{R}$ consist of all points $\sum_{j \in K_0} x_j \delta^j \in \coprod_{K_0} \mathbb{R}$ such that $x_i > (n+2)^{-1}$. Then the finite family $\{W_1, \ldots, W_q\}$ of *all* nonvoid intersections of the open convex sets $Y_i \subset \coprod_{K_0} \mathbb{R}$ has the properties that each nonvoid intersection

$W_i \cap W_j$ is of the form W_k for some $k = k(i, j)$, and that $\overline{W_i \cap W_j} = \bar{W}_i \cap \bar{W}_j$. The same two properties are inherited by the family $\{V_1, \ldots, V_q\}$ of intersections $V_i = W_i \cap |K|^\circ$; furthermore $\{V_1, \ldots, V_q\}$ is a covering of $|K|^\circ$ by open n-cells. To complete the proof one chooses $\{U_1, \ldots, U_c\}$ to be the family of inverse images of the sets $V_1, \ldots, V_q$ under the homeomorphism $X \to |K|$.

9. The Mayer–Vietoris Technique

We now develop a technique which will later be applied to prove certain theorems involving an arbitrary space X of finite type (Definition 1.1). Since the theorems will depend only on the homotopy type of X, one can regard the technique as a method for proving corresponding theorems for any space X' homotopy equivalent to a space of finite type; for example, X' might be any base space (Definition 1.2). Specifically, the technique will be used to prove best-possible versions of the Leray–Hirsch theorem (Proposition II.7.2) and the existence of Thom classes (Propositions IV.1.3 and IX.2.1). In a specialized form the technique will be used to prove the existence of fundamental classes (Propositions VI.1.3 and VII.2.3) and the Poincaré–Lefschetz duality theorem (Propositions VI.2.3 and VII.2.7).

For any space X of finite type let $\mathcal{O}(X)$ denote the category of open sets $U \subset X$, morphisms $U \to V$ in $\mathcal{O}(X)$ being inclusions, and let $\mathcal{O}(X) \xrightarrow{h} \mathfrak{M}$ be any contravariant functor from $\mathcal{O}(X)$ to a category $\mathfrak{M}$ of modules over a fixed commutative ring. Such a functor is *additive* whenever for any mutually disjoint family $\{U_\delta\}_\delta$ of open sets $U_\delta \subset X$ the module $h(\bigcup_\delta U_\delta)$ is the direct sum $\coprod_\delta h(U_\delta)$. For any $U \in \mathcal{O}(X)$ and $V \in \mathcal{O}(X)$ there are four morphisms

$$\begin{array}{ccc} U \xrightarrow{i_U} U \cup V & \text{and} & U \cap V \xrightarrow{j_U} U \\ V \xrightarrow{i_V} U \cup V & & U \cap V \xrightarrow{j_V} V \end{array}$$

in $\mathcal{O}(X)$ and four corresponding morphisms

$$\begin{array}{ccc} h(U \cup V) \xrightarrow{i_U^*} h(U) & \text{and} & h(U) \xrightarrow{j_U^*} h(U \cap V) \\ h(U \cup V) \xrightarrow{i_V^*} h(V) & & h(V) \xrightarrow{j_V^*} h(U \cap V) \end{array}$$

in $\mathfrak{M}$. For any $\alpha \in h(U \cup V)$ let $i_{U,V}^*\alpha = i_U^*\alpha \oplus i_V^*\alpha \in h(U) \oplus h(V)$, and for any $\beta \oplus \gamma \in h(U) \oplus h(V)$ let $j_{U,V}^*(\beta \oplus \gamma) = j_U^*\beta - j_V^*\gamma \in h(U \cap V)$. The induced sequence

$$h(U \cup V) \xrightarrow{i_{U,V}^*} h(U) \oplus h(V) \xrightarrow{j_{U,V}^*} h(U \cap V)$$

of homomorphisms is automatically exact at the term $h(U) \oplus h(V)$.

Let $\{h^m | m \in \mathbb{Z}\}$ be any family of additive contravariant functors from $\mathcal{O}(X)$ to $\mathfrak{M}$, indexed by the integers $\mathbb{Z}$, and suppose for each ordered pair $(U, V) \in \mathcal{O}(X) \times \mathcal{O}(X)$ and each $m \in \mathbb{Z}$ that there is a module homomorphism $\delta^m_{U,V}$ for which

$$\cdots \xrightarrow{i^*_{U,V}} h^{m-1}(U) \oplus h^{m-1}(V) \xrightarrow{j^*_{U,V}} h^{m-1}(U \cap V) \xrightarrow{\delta^{m-1}_{U,V}} h^m(U \cup V)$$
$$\xrightarrow{i^*_{U,V}} h^m(U) \oplus h^m(V) \xrightarrow{j^*_{U,V}} h^m(U \cap V) \xrightarrow{\delta^m_{U,V}} \cdots$$

is a long exact sequence. In this case $\{h^m | m \in \mathbb{Z}\}$ is a *Mayer–Vietoris functor on* $\mathcal{O}(X)$.

For example, if $\{h^m | m \in \mathbb{Z}\}$ is the restriction to $\mathcal{O}(X)$ of singular cohomology $\{H^m | m \in \mathbb{Z}\}$ with a given commutative coefficient ring $H^0(\{*\})$, then there are classical connecting homomorphisms $\delta^m_{U,V}$ for which $\{h^m | m \in \mathbb{Z}\}$ is a Mayer–Vietoris functor on $\mathcal{O}(X)$. (The cohomology version of the Mayer–Vietoris sequence can be found in Artin and Braun [2, pp. 137–143], Hu [5, pp. 28–29], and Spanier [4, pp. 186–190, 218, 239], for example. Since U and V are open sets in X, the pair $(U, V) \in \mathcal{O}(X) \times \mathcal{O}(X)$ automatically satisfies the conditions required for the construction of connecting homomorphisms.)

A *natural transformation* θ from a Mayer–Vietoris functor $\{h^m | m \in \mathbb{Z}\}$ to a Mayer–Vietoris functor $\{k^m | m \in \mathbb{Z}\}$ consists of homomorphisms

$$h^m(U) \xrightarrow{\theta_U} k^m(U)$$

in $\mathfrak{M}$ such that the diagram

$$\begin{array}{ccccccc}
h^{m-1}(U \cap V) & \xrightarrow{\delta^{m-1}_{U,V}} & h^m(U \cup V) & \xrightarrow{i^*_{U,V}} & h^m(U) \oplus h^m(V) & \xrightarrow{j^*_{U,V}} & h^m(U \cap V) \\
\downarrow \theta_{U \cap V} & & \downarrow \theta_{U \cup V} & & \downarrow \theta_U \oplus \theta_V & & \downarrow \theta_{U \cap V} \\
k^{m-1}(U \cap V) & \xrightarrow{\delta^{m-1}_{U,V}} & k^m(U \cup V) & \xrightarrow{i^*_{U,V}} & k^m(U) \oplus k^m(V) & \xrightarrow{j^*_{U,V}} & k^m(U \cap V)
\end{array}$$

commutes for each $U \in \mathcal{O}(X)$, $V \in \mathcal{O}(X)$, and $m \in \mathbb{Z}$. It will be assumed that θ respects the additivity of the functors h^m and k^m; specifically, for any mutually disjoint family $\{U_\delta\}_\delta$ of open sets $U_\delta \subset X$ the diagrams

$$\begin{array}{ccc}
h^m\left(\bigcup_\delta U_\delta\right) & \xrightarrow{\approx} & \coprod_\delta h^m(U_\delta) \\
\downarrow \theta_{\cup_\delta U_\delta} & & \downarrow \amalg_\delta \theta_{U_\delta} \\
k^m\left(\bigcup_\delta U_\delta\right) & \xrightarrow{\approx} & \coprod_\delta k^m(U_\delta)
\end{array}$$

commute. The main result of this section is that if X is of finite type and each $h^m(U) \xrightarrow{\theta_U} k^m(U)$ is an isomorphism whenever $U \in \mathcal{O}(X)$ is contractible, then each $h^m(X) \xrightarrow{\theta_X} k^m(X)$ is an isomorphism.

9.1 Lemma (The 5-lemma): *Let*

$$\begin{array}{ccccccccc} A_1 & \longrightarrow & A_2 & \longrightarrow & A_3 & \longrightarrow & A_4 & \longrightarrow & A_5 \\ \downarrow{\scriptstyle \theta_1} & & \downarrow{\scriptstyle \theta_2} & & \downarrow{\scriptstyle \theta_3} & & \downarrow{\scriptstyle \theta_4} & & \downarrow{\scriptstyle \theta_5} \\ B_1 & \longrightarrow & B_2 & \longrightarrow & B_3 & \longrightarrow & B_4 & \longrightarrow & B_5 \end{array}$$

be a commutative diagram in a category $\mathfrak{M}$ of modules over a fixed ring, each row being exact. If $\theta_1, \theta_2, \theta_4, \theta_5$ are isomorphisms, then θ_3 is also an isomorphism.

PROOF: This is an easy and classical diagram chase, which will be left to the reader.

9.2 Lemma: *Let V be a space of nth type, as in Definition* 1.1, *and let θ be a natural transformation of a Mayer–Vietoris functor $\{h^m | m \in \mathbb{Z}\}$ on $\mathcal{O}(V)$ to a Mayer–Vietoris functor $\{k^m | m \in \mathbb{Z}\}$ on $\mathcal{O}(V)$. Suppose for each $U \in \mathcal{O}(V)$ of $(n-1)$th type that each $h^m(U) \xrightarrow{\theta_U} k^m(U)$ is an isomorphism; then each $h^m(V) \xrightarrow{\theta_V} k^m(V)$ is an isomorphism.*

PROOF: By hypothesis there is some finite q for which V is covered by q families $\{U_{1,\alpha}\}_\alpha, \ldots, \{U_{q,\gamma}\}_\gamma$ of open sets $U_{p,\beta} \in \mathcal{O}(V)$, such that each family is mutually disjoint and such that each intersection of any sets in the covering is of $(n-1)$th type. For $p = 1, \ldots, q$ let U_p denote the (disjoint) union $\bigcup_\beta U_{p,\beta} \in \mathcal{O}(V)$, and let $\{\tilde{U}_r\}_r$ denote the finite family consisting of all possible intersections of the sets $U_1, \ldots, U_q$; the sets $U_1, \ldots, U_q$ themselves are in the family $\{\tilde{U}_r\}_r$, which consists of at most $2^q - 1$ distinct open sets, possibly including the void set. Clearly each $\tilde{U}_r$ is a disjoint union of spaces of $(n-1)$th type, although not necessarily itself of $(n-1)$th type. For each $s > 0$ let $\mathcal{Q}_s(V) \subset \mathcal{O}(V)$ consist of those open sets $W \subset V$ which are unions of at most s members of the finite family $\{\tilde{U}_r\}_r$. Then $V = U_1 \cup \cdots \cup U_q \in \mathcal{Q}_q(V)$. We shall show by induction on s that if $W \in \mathcal{Q}_s(V)$, then each $h^m(W) \xrightarrow{\theta_W} k^m(W)$ is an isomorphism.

Since $\mathcal{Q}_1(V)$ is precisely the family $\{\tilde{U}_r\}_r$ itself, it follows that any $W \in \mathcal{Q}_1(V)$ is a disjoint union $\bigcup_\delta U_\delta$ of open sets $U_\delta \subset V$ that are spaces of $(n-1)$th type. The additivity properties of the given functors and of θ itself then provide commutative diagrams

$$\begin{array}{ccc} h^m(W) & \xrightarrow{\approx} & \coprod_\delta h^m(U_\delta) \\ \downarrow{\scriptstyle \theta_W} & & \downarrow{\scriptstyle \coprod_\delta \theta_{U_\delta}} \\ k^m(W) & \xrightarrow{\approx} & \coprod_\delta k^m(U_\delta), \end{array}$$

and since each θ_{U_δ} is an isomorphism by the hypothesis of the lemma, each θ_W is also an isomorphism, for any $W \in \mathcal{Q}_1(V)$.

Now suppose that each $\theta_{W'}$ is an isomorphism for any $W' \in \mathcal{Q}_{s-1}(V)$, and suppose that $W \in \mathcal{Q}_s(V)$. Then W is a union $W' \cup W''$ of some $W' \in \mathcal{Q}_{s-1}(V)$ and some $W'' \in \mathcal{Q}_1(V)$ $(\subset \mathcal{Q}_{s-1}(V))$, and by the de Morgan laws $W' \cap W''$ is also a member $W''' \in \mathcal{Q}_{s-1}(V)$. Consequently the four homomorphisms labeled either $\theta_{W'} \oplus \theta_{W''}$ or $\theta_{W'''}$ in the commutative diagram

$$\begin{array}{ccccc}
h^{m-1}(W') \oplus h^{m-1}(W'') & \xrightarrow{j^*_{W',W''}} & h^{m-1}(W''') & \xrightarrow{\delta^{m-1}_{W',W''}} & h^m(W) \\
\downarrow{\scriptstyle \theta_{W'} \oplus \theta_{W''}} & & \downarrow{\scriptstyle \theta_{W'''}} & & \downarrow{\scriptstyle \theta_W} \\
k^{m-1}(W') \oplus k^{m-1}(W'') & \xrightarrow{j^*_{W',W''}} & k^{m-1}(W''') & \xrightarrow{\delta^{m-1}_{W',W''}} & k^m(W)
\end{array}$$

$$\begin{array}{ccccc}
\xrightarrow{i^*_{W',W''}} & h^m(W') \oplus h^m(W'') & \xrightarrow{j^*_{W',W''}} & h^m(W''') \\
 & \downarrow{\scriptstyle \theta_{W'} \oplus \theta_{W''}} & & \downarrow{\scriptstyle \theta_{W'''}} \\
\xrightarrow{i^*_{W',W''}} & k^m(W') \oplus k^m(W'') & \xrightarrow{j^*_{W',W''}} & k^m(W''')
\end{array}$$

are isomorphisms by the inductive hypothesis. It follows from the 5-lemma that θ_W is also an isomorphism, which completes the inductive step. Since $V = U_1 \cup \cdots \cup U_q \in \mathcal{Q}_q(V)$, each $h^m(V) \xrightarrow{\theta_V} k^m(V)$ is therefore an isomorphism as claimed.

9.3 Theorem (The Mayer–Vietoris Technique): *Let $\mathcal{O}(X)$ be the category of open sets on a space X of finite type, as in Definition 1.1, and let θ be a natural transformation of a Mayer–Vietoris functor $\{h^m \mid m \in Z\}$ on $\mathcal{O}(X)$ to a Mayer–Vietoris functor $\{k^m \mid m \in Z\}$ on $\mathcal{O}(X)$. If each $h^m(U) \xrightarrow{\theta_U} k^m(U)$ is an isomorphism whenever $U \in \mathcal{O}(X)$ is contractible, then each $h^m(X) \xrightarrow{\theta_X} k^m(X)$ is also an isomorphism.*

PROOF: We shall show by induction on n that if $V \in \mathcal{O}(X)$ is of nth type, then θ_V is an isomorphism. If $V \in \mathcal{O}(X)$ is of 0th type, then there is a mutually disjoint family $\{U_\delta\}_\delta$ of contractible open sets $U_\delta \in \mathcal{O}(X)$ such that $V = \bigcup_\delta U_\delta$, so that the additivity properties of the given Mayer-Vietoris functors and θ itself provide commutative diagrams

$$\begin{array}{ccc}
h^m(V) & \xrightarrow{\approx} & \coprod_\delta h^m(U_\delta) \\
\downarrow{\scriptstyle \theta_V} & & \downarrow{\scriptstyle \coprod_\delta \theta_{U_\delta}} \\
k^m(V) & \xrightarrow{\approx} & \coprod_\delta k^m(U_\delta).
\end{array}$$

Since each θ_{U_δ} is an isomorphism by hypothesis, it follows that θ_V is also an isomorphism. Now suppose that $h^m(U) \xrightarrow{\theta_U} k^m(U)$ is an isomorphism for each $U \in \mathcal{O}(X)$ of $(n-1)$th type, and let $V \in \mathcal{O}(X)$ be of nth type. Since V is open in X, it follows that each $h^m(U) \xrightarrow{\theta_U} k^m(U)$ is an isomorphism for each $U \in \mathcal{O}(V)$ of $(n-1)$th type, so that each $h^m(V) \xrightarrow{\theta_U} k^m(V)$ is an isomorphism by Lemma 9.2. This completes the inductive step, and since X is itself of nth type for some $n \geqq 0$, this also completes the proof of the theorem.

As announced at the beginning of this section there will be several later applications of the Mayer–Vietoris technique. The applications will use slightly embellished versions of Theorem 9.3, and for convenience we present those versions as corollaries.

For any space X of finite type one can regard a Mayer–Vietoris functor $\{h^q \mid q \in \mathbb{Z}\}$ as a single functor $\mathcal{O}(X) \xrightarrow{\coprod_q h^q} \mathcal{A}b^\oplus$ carrying each $U \equiv \mathcal{O}(X)$ into a direct sum $\coprod_{q \in \mathbb{Z}} h^q(U)$ of abelian groups, indexed by the integers $\mathbb{Z}$. Let R be a $\mathbb{Z}$-graded ring that is commutative in the usual sense that if $u \in R$ and $v \in R$ are of degrees p and q, then $uv = (-1)^{pq}vu$ is degree $p + q$, and suppose that $\mathfrak{M}_R^\oplus$ is the category of $\mathbb{Z}$-graded R-modules. Then $\mathfrak{M}_R^\oplus \subset \mathcal{A}b^\oplus$ and we shall consider functors $\coprod_q h^q$ from $\mathcal{O}(X)$ to $\mathfrak{M}_R^\oplus$.

9.4 Corollary: *Let $\mathcal{O}(X)$ be the category of open sets on a space X of finite type, let R be a graded commutative ring as above, and let $\coprod_q h^q \xrightarrow{\theta} \coprod_q k^q$ be a degree-preserving natural transformation of Mayer-Vietoris functors from $\mathcal{O}(X)$ to $\mathfrak{M}_R^\oplus$. If θ_U is an R-module isomorphism whenever $U \in \mathcal{O}(X)$ is a contractible open set, then θ is a natural equivalence; in particular, $\coprod_q h^q(X) \xrightarrow{\theta_X} \coprod_q k^q(X)$ is an R-module isomorphism.*

PROOF: Immediate consequence of Theorem 9.3.

The next variant of the Mayer–Vietoris technique is closer to Theorem 9.3 itself.

9.5 Corollary: *Let θ be a natural transformation of Mayer–Vietoris functors $\{h^q \mid q \in \mathbb{Z}\}$ and $\{k^q \mid q \in \mathbb{Z}\}$ as in Theorem 9.3, and let $n \in \mathbb{Z}$ be a fixed integer. If $h^q(U) \xrightarrow{\theta_U} k^q(U)$ is an isomorphism whenever $q \leqq n$ and the open set $U \in \mathcal{O}(X)$ is contractible, then $h^q(X) \xrightarrow{\theta_X} k^q(X)$ is an isomorphism whenever $q \leqq n$.*

PROOF: One proceeds exactly as in the proof of Theorem 9.3, observing that if $p \leqq q \leqq n$ then the 5-lemma produces isomorphisms only when $p < q \leqq n$.

In Chapters VI and VII we shall use a refinement of the Mayer–Vietoris technique, for which we single out particular base spaces. Let X be a smooth compact manifold with interior $\mathring{X}$. According to Proposition 8.4 there is a finite covering $\{U_1, \ldots, U_q\}$ of $\mathring{X}$ by open cells such that each nonvoid intersection $U_i \cap U_j$ is a cell U_k in the covering, and such that the closures in X satisfy $\overline{U_i \cap U_j} = \bar{U}_i \cap \bar{U}_j$. Let $\mathscr{Q}(\mathring{X})$ be the category whose objects are unions of the sets $U_1, \ldots, U_q$. (One does not need "unions of intersections of the sets $U_1, \ldots, U_q$" since each nonvoid intersection is automatically a union of the sets $U_1, \ldots, U_q$.) Morphisms in $\mathscr{Q}(\mathring{X})$ are inclusions. A *Mayer–Vietoris functor on* $\mathscr{Q}(\mathring{X})$ is a family $\{h^q \mid q \in \mathbb{Z}\}$ of contravariant functors from $\mathscr{Q}(\mathring{X})$ to a category $\mathfrak{M}$ of modules for which there is a long exact Mayer–Vietoris sequence for each pair $(U, V) \in \mathscr{Q}(\mathring{X}) \times \mathscr{Q}(\mathring{X})$, as before. The only difference is that we now restrict U and V to $\mathscr{Q}(\mathring{X})$; the category $\mathscr{O}(\mathring{X})$ of all open sets in $\mathring{X}$ is not needed as such.

Here are two examples. Let $H^*(-)$ denote singular cohomology with coefficients $H^0(\{*\})$, and let $H^*(X, -)$ denote relative singular homology with the same coefficients, where X is the given smooth compact manifold.

9.6 Proposition: *Suppose that X is a smooth compact manifold; then there is a Mayer–Vietoris functor $\{h^q \mid q \in \mathbb{Z}\}$ on $\mathscr{Q}(\mathring{X})$ with $h^q(U) = H^q(\bar{U})$ for every $U \in \mathscr{Q}(\mathring{X})$.*

PROOF: It follows from Proposition 8.4 that the closure in X of any element of $\mathscr{Q}(\mathring{X})$ is itself finitely triangulable. Hence one can construct the connecting homomorphisms $\delta^q_{U,V}$ of the classical Mayer–Vietoris cohomology sequence

$$\cdots \to H^{q-1}(\bar{U} \cap \bar{V}) \xrightarrow{\delta^{q-1}_{\bar{U},\bar{V}}} H^q(\bar{U} \cup \bar{V}) \xrightarrow{i^*_{\bar{U},\bar{V}}} H^q(\bar{U}) \oplus H^q(\bar{V}) \xrightarrow{j^*_{\bar{U},\bar{V}}} H^q(\bar{U} \cap \bar{V}) \to \cdots$$

for any $(U, V) \in \mathscr{Q}(\mathring{X})$. Proposition 8.4 also guarantees that $\overline{U \cap V} = \bar{U} \cap \bar{V}$, and since one always has $\overline{U \cup V} = \bar{U} \cup \bar{V}$, this completes the proof.

9.7 Proposition: *Suppose that X is an n-dimensional smooth compact manifold; then there is a Mayer–Vietoris functor $\{h^q \mid q \in \mathbb{Z}\}$ on $\mathscr{Q}(\mathring{X})$ with $h^q(U) = H_{n-q}(X, X - U)$ for every $U \in \mathscr{Q}(\mathring{X})$.*

PROOF: This time each $X - U$ is itself finitely triangulable, so that one can construct the connecting homomorphisms of the classical relative Mayer–Vietoris homology sequence to complete the proof.

We shall construct a third Mayer–Vietoris functor on $\mathscr{Q}(\mathring{X})$ in Chapter VI. Here is the specialized Mayer–Vietoris technique promised earlier.

9.8 Theorem (Mayer–Vietoris Comparison Theorem): *Let X be a smooth compact manifold with interior $\mathring{X}$, and let θ be a natural transformation of Mayer–Vietoris functors $\{h^q \mid q \in \mathbb{Z}\}$ and $\{k^q \mid q \in \mathbb{Z}\}$ on $\mathscr{Q}(\mathring{X})$. If $h^q(U_i) \xrightarrow{\theta_{U_i}} k^q(U_i)$ is an isomorphism for each of the open cells $U_1, \ldots, U_r$ in $\mathscr{Q}(\mathring{X})$ when-*

ever $q \leqq 0$, *it follows that* $h^q(\mathring{X}) \xrightarrow{\theta_{\mathring{X}}} k^q(\mathring{X})$ *is an isomorphism whenever* $q \leqq 0$. *Similarly, if* $h^q(U_i) \xrightarrow{\theta_{U_i}} k^q(U_i)$ *is an isomorphism for each of the open cells* $U_1, \ldots, U_r$ *in* $\mathscr{Q}(\mathring{X})$ *for any* $q \in \mathbb{Z}$, *it follows that* $h^q(\mathring{X}) \xrightarrow{\theta_{\mathring{X}}} k^q(\mathring{X})$ *is an isomorphism for any* $q \in \mathbb{Z}$.

PROOF: The 5-lemma applies exactly as in the proof of Lemma 9.2.

10. Remarks and Exercises

10.1 Remark: The category $\mathscr{B}$ was identified in Osborn [6, pp. 745, 754] as *the* natural category of topological spaces that serve as base spaces, in the sense described in the next chapter. Indeed, the defining properties of Definition 1.2, the closure properties of Proposition 1.4, the inclusion $\mathscr{W} \subset \mathscr{B}$ of Corollaries 3.9 and 5.4, and the Mayer–Vietoris technique of Theorems 9.3 and 9.8 will be used not only in the next chapter, but throughout the entire book. The definition of $\mathscr{B}$, the inclusion $\mathscr{W} \subset \mathscr{B}$, and the Mayer–Vietoris technique were suggested by a construction in Connell [1, pp. 499–501]; Connell attributes the construction to E. H. Brown.

It is possible to replace the category $\mathscr{B}$ by the category $\mathscr{W}$ (Definition 3.8 and Corollary 5.3) or by the category $\mathscr{W}_0$ (Definition 3.11 and the obvious analog of Corollary 5.3). However, many constructions appearing later in the book are most easily carried out in $\mathscr{B}$ itself, rather than in $\mathscr{W}$ or in $\mathscr{W}_0$, and there seems to be little point to frequent appeals to the inclusions $\mathscr{W}_0 \subset \mathscr{W} \subset \mathscr{B}$.

On the other hand, the relative size of $\mathscr{B}$ is not itself a virtue. Although several major existence and uniqueness theorems (Theorems V.5.1, X.4.1, XI.6.1, and XI.7.3) are more easily established for the category $\mathscr{B}$, one is frequently interested in corresponding results (Theorems V.5.2, X.4.2, XI.6.2, and XI.7.7) for the category $\mathscr{M}$ of smooth manifolds. The inclusion $\mathscr{M} \subset \mathscr{B}$ of Corollary 6.8 permits one to use the existence assertions of the former results to obtain corresponding existence assertions in the latter results; however, the uniqueness assertions in the latter results are somewhat more delicate.

10.2 Remark: In Definition 1.1 the terminology "space of finite type" was chosen for convenience; the same phrase undoubtedly appears in many other contexts with different meanings, and no confusion is intended. However, Definition 1.1 itself may recall another definition, and there may indeed be some faint relation between the two concepts: a topological space X is of *Ljusternik–Schnirelmann category* $n \geqq 0$ (according to one of the two most common conventions) if X can be covered by $n + 1$, but not by n, open contractible spaces.

The Ljusternik–Schnirelmann category was introduced in Ljusternik and Schnirelmann [1] as a topological setting for variational problems; it is still used for that purpose, as in Ljusternik [1], Maurin [1], and Palais [1], for example. The homotopy invariance of the Ljusternik–Schnirelmann category suggests its importance in algebraic topology per se, and it has been investigated in that spirit by Fox [1, 2] and many later authors. Ganea [1] provides a 1962 survey of the subject, and there are more recent developments in Berstein [1], Borsuk [2], Coelho [1], Draper [1], Ganea [2, 3], Hardie [1, 2, 3], Hoo [1], Luft [1, 2], Moran [1, 2, 3, 4], Ono [1, 2], Osborne and Stern [1], Singhof [1, 2], and Takens [1, 2] for example; James [3] provides a 1978 survey of the subject.

10.3 Remark: The superficial similarity of spaces of nth type (Definition 1.1) to spaces of Ljusternik–Schnirelmann category n has just been considered. However, the Ljusternik–Schnirelmann category ignores all properties of intersections of the sets in given coverings by open contractible sets, whereas such properties are of paramount interest in Definition 1.1. Intersections in open coverings *are* considered in Nagata [1, pp. 133–137] in the form of "multiplicative refinements." For example, a separable metric space is of dimension $\leqq n$ if and only if for every open covering there is a multiplicative refinement of length $\leqq n + 1$; the appropriate definitions can be found in Nagata [1]. (There is a corresponding result in Hurewicz and Wallman [1, pp. 54, 66, 67], which ignores intersections.) A somewhat more specialized result concerning minimal open coverings of manifolds, with well-behaved intersections, can be found in Osborne and Stern [1].

10.4 Remark: There is an omission in the list $\mathscr{B}$, $\mathscr{W}$, $\mathscr{W}_0$ of categories considered in Remark 10.1. A hausdorff space X is *compactly generated* if a subset $A \subset X$ is closed whenever all intersections $A \cap B \subset X$ with compact subsets $B \subset X$ are closed. The category $\mathscr{K}$ *of compactly generated spaces* (= *Steenrod's convenient category*) was introduced and investigated in Steenrod [6]; the definition and elementary properties can also be found in Cooke and Finney [1, pp. 86–105], Gray [1, pp. 50–61], and G. W. Whitehead [1, pp. 18–20]. For example, any metric space is compactly generated.

Some alternate "convenient categories" are introduced in Vogt [1, 2] and compared to Steenrod's convenient category.

The categories of Steenrod and Vogt are indeed convenient for much of homotopy theory: Steenrod's category suffices for more than 700 pages in G. W. Whitehead [1], for example. However, Steenrod's definition is not itself homotopy invariant, and it would not serve especially well in the very next chapter of this book, which uses frequent homotopy equivalences. One can perhaps replace $\mathscr{B}$ by the category $\mathscr{U}$ of spaces that are homotopy equivalent to compactly generated spaces: $\mathscr{B} \subset \mathscr{U}$ since any metric space is compactly generated, and according to a result in May [1, p. 28], for any $X \in \mathscr{U}$ there is a homotopy class of *weak homotopy equivalences* $X' \to X$ relating some $X' \in \mathscr{W}$ $(\subset \mathscr{B})$ to X. However, $\mathscr{B}$ will suffice for the purposes of the present work.

10.5 Remark: Simplicial complexes and metric simplicial spaces, considered in §2, are also discussed in Cairns [6, pp. 65–89], Hilton and Wylie [1, pp. 14–52], Maunder [1, pp. 31–62], Spanier [4, pp. 107–129], for example.

10.6 Remark: The telescope of Definition 3.3 was suggested to the author by a corresponding construction in Connell [1]. A similar construction is used for a related purpose in Bendersky [2, pp. 16–17]. Incidentally, the inclusion $\mathscr{W} \subset \mathscr{B}$ of Theorem 3.7 (and Corollaries 3.9 and 5.4) was proved more rapidly in Osborn [6], but by a more sophisticated method. Specifically, although Osborn [6] uses a simpler telescope $|K|^*$ than the one constructed in Definition 3.3, there is no direct construction of a homotopy inverse to the obvious projection $|K|^* \to |K|$. Instead, one observes that $|K|^* \to |K|$ is a *weak homotopy equivalence*, and according to a classical result of J. H. C. Whitehead [4] it follows that $|K|^* \to |K|$ is a homotopy equivalence in the usual sense, as in this book. Whitehead's theorem was cited following Theorem 5.2; it can also be found in Gray [1, p. 139], Lundell and Weingram [1, p. 125], Maunder [1, pp. 298–300], and Switzer [1, pp. 87–90], for example.

10.7 Remark: The proof of Proposition 4.6 (Dowker [1]) can also be found in Lundell and Weingram [1, p. 131], Milnor [8, p. 276], and Dold [8, pp. 354–355], for example.

10.8 Remark: CW spaces were first considered in J. H. C. Whitehead [3]. Other expositions can be found in Cooke and Finney [1], Gray [1, pp. 113–121], Hilton [1, pp. 95–113], Hu [3, pp. 111–149], throughout Lundell and Weingram [1], Massey [6, pp. 76–104], Maunder [1, pp. 273–310], throughout Piccinini [1], Rohlin and Fuks [1, Chapter II], Spanier [4, pp. 400–418], Switzer [1, Chapter V], and G. W. Whitehead [1, pp. 46–95].

There are CW spaces which are not simplicial in any sense (other than homotopy equivalence). Specific examples of such spaces can be found in Metzler [1] and in Bognar [1], for example.

10.9 Remark: According to Corollary 5.4 every CW space is a base space, a fortiori homotopy equivalent to a metric space, so that Lemma 8.2 (Stone [1]) guarantees that every CW space is homotopy equivalent to a paracompact space. A stronger statement is available: every CW space is itself paracompact. This was established by Miyazaki [1], following partial results of Bourgin [1] and Dugundji [1]. Proofs of Miyazaki's theorem can be found in Postnikov [1] and Lundell and Weingram [1,pp. 54–55].

10.10 Remark: There are many elementary introductions to the category M of smooth manifolds, which was described very briefly in §6. See Auslander and MacKenzie [2], Boothby [1], Guillemin and Pollack [1], M. W. Hirsch [4], Hu [6], Lang [1], Milnor [15], Rohlin and Fuks [1, Chapter III], or Wallace [5], for example.

10.11 Remark: The Whitney embedding result of Theorem 6.6 is neither the easiest nor the best-possible embedding theorem for smooth manifolds. The "easy" Whitney embedding theorem asserts that any smooth n-dimensional manifold X admits at least one smooth embedding $X \to \mathbb{R}^{2n+1}$. This result was announced in Whitney [1] and proved in Whitney [3]. Other proofs of the "easy" Whitney embedding theorem can be found in T. Y. Thomas [1, 2], in Whitney [9], Auslander and MacKenzie [2, pp. 106–116], de Rham [1, pp. 9–16], Greene and Wu [1], Guillemin and Pollack [1, pp. 39–56], and in the 1966 revised edition of Munkres [1], for example.

The "hard" Whitney embedding result $X \to \mathbb{R}^{2n}$ of Theorem 6.6 appears in Whitney [7, p. 236]; its proof occupies all 27 pages of that paper.

A long-standing "best-possible" embedding conjecture is that any smooth n-dimensional manifold X admits a smooth embedding $X \to \mathbb{R}^{2n-\alpha(n)+1}$, where $\alpha(n)$ is the number of 1's in the dyadic expansion of the dimension n. (See Gitler [1], for example.) One can definitely do no better: in Chapter VI we shall show for each $n > 0$ that there is a smooth closed n-dimensional manifold which *cannot* be embedded in $\mathbb{R}^{2n-\alpha(n)}$; no stronger counterexamples are known.

Although the "best-possible" embedding conjecture remains unproved, there is at least one faint suggestion of its truth. According to a result of R. L. W. Brown [2, 3], every smooth closed n-dimensional manifold is equivalent to one which smoothly embeds in $\mathbb{R}^{2n-\alpha(n)+1}$: the equivalence is "cobordism," which will be discussed in Chapter VI.

In lieu of a proof of the "best-possible" embedding conjecture, there have been many improvements upon the "hard" Whitney embedding theorem. It is known for every $n > 1$, with the possible exception of the case $n = 4$, that every smooth *orientable* n-dimensional manifold embeds in $\mathbb{R}^{2n-1}$; the case $n = 4$ is probably not an exception. It is also known for every natural number $n > 1$ *not* of the form 2^r that every smooth *nonorientable* n-dimensional manifold embeds in $\mathbb{R}^{2n-1}$; the cases $n = 2^r$ definitely *are* exceptions since the real projective spaces RP^n do *not* embed in $\mathbb{R}^{2n-1}$ for these values of n. The fact that RP^n is an exception for $n = 2^r$ will be established in Proposition VI.4.9, and the remainder of the preceding results will be considered in more detail in Remark VI.9.16, with appropriate references.

10.12 Remark: There are several analogs of the Whitney embedding theorems. For example, any n-dimensional separable metric space is homeomorphic to a subset of $\mathbb{R}^{2n+1}$; proofs are given in Hurewicz and Wallman [1, pp. 60–63] and Nagata [1, pp. 101–108].

For any n-dimensional simplicial space $|K|$ there is a simplicial embedding $|K| \to \mathbb{R}^{2n+1}$; this analog of the "easy" Whitney embedding theorem is proved in Seifert and Threlfall [1, German edition, pp. 44–46; English edition, pp. 45–47], and Cairns [6, pp. 78–80]. In case $|K|$ is an n-dimensional triangulated manifold there is a simplicial embedding $|K| \to \mathbb{R}^{2n}$, due to van Kampen [1]. A necessary and sufficient condition for the existence of simplicial embeddings $|K| \to \mathbb{R}^{2n}$ of n-dimensional simplicial spaces $|K|$ in general is given in Wu [10, p. 237], and it is shown to be satisfied by n-dimensional triangulated manifolds in Wu [10, p. 257]; this provides another proof of van Kampen's analog of the "hard" Whitney embedding theorem.

Any two simplicial embeddings $|K| \xrightarrow{h_0} \mathbb{R}^{2n+2}$ and $|K| \xrightarrow{h_1} \mathbb{R}^{2n+2}$ of an n-dimensional simplicial space $|K|$ are always *isotopic* in the sense that there is a simplicial map $|K| \times [0,1] \xrightarrow{h} \mathbb{R}^{2n+2}$ such that each $|K| \xrightarrow{h_t} \mathbb{R}^{2n+2}$ is itself a simplicial embedding; if $|K|$ is a triangulated manifold of dimension $n > 1$, then any two simplicial embeddings $|K| \xrightarrow{h_0} \mathbb{R}^{2n+1}$ and $|K| \xrightarrow{h_1} \mathbb{R}^{2n+1}$ are isotopic. These results, also due to van Kampen [1], are proved in Wu [10, pp. 207, 258].

10.13 Remark: The Whitney immersion result of Theorem 6.9 is neither the easiest nor the best-possible immersion theorem for smooth manifolds. The "easy" Whitney immersion theorem asserts that any smooth n-dimensional manifold X admits at least one smooth immersion $X \to S^{2n}$. This result was announced (along with the "easy" embedding result) in Whitney [1] and proved in Whitney [3]. Another proof of the "easy" Whitney immersion theorem can be found in Auslander and MacKenzie [2, pp. 106–133].

The "hard" Whitney immersion result $X \to S^{2n-1}$ of Theorem 6.9 appears in Whitney [8, p. 270]; its proof occupies all 47 pages of that paper. The same result appears in M. W. Hirsch [1, p. 270], with an alternative proof that has had far-reaching consequences.

A long-standing "best-possible" immersion conjecture has recently been proved by Cohen [1], the proof depending upon results and techniques of M. W. Hirsch [1] and Brown and Peterson [4, 5]: for any $n > 1$ any smooth compact n-dimensional manifold X admits an immersion $X \to \mathbb{R}^{2n-\alpha(n)}$, where $\alpha(n)$ is the number of 1's in the dyadic expansion of the dimension n. This result will be described in more detail in Remark VI.9.14. One can definitely do no better: in Chapter VI, for each $n > 0$, we shall show that there is an easily constructed smooth closed n-dimensional manifold which *cannot* be immersed in $\mathbb{R}^{2n-\alpha(n)-1}$.

As for embeddings, a weaker result of R. L. W. Brown [2, 3] asserts that every smooth closed n-dimensional manifold is equivalent to one which smoothly immerses in $\mathbb{R}^{2n-\alpha(n)}$: the equivalence is "cobordism," which will be discussed in Chapter VI.

10.14 Remark: The Cairns–Whitehead triangulation result of Theorem 6.7 was first proved for closed smooth manifolds in Cairns [1, 2]. The result was extended to arbitrary smooth manifolds in J. H. C. Whitehead [1]. Discussions and simplifications of the combined result can be found in Cairns [3, 4, 5], and other versions of the proof can be found in Whitney [10, pp. 124–135], in J. H. C. Whitehead [5], and in the 1966 edition of Munkres [1], for example.

10.15 Remark: One of the primary goals of differential topology is a reasonable classification (in some sense) of all smooth manifolds. One cannot expect a complete classification, however, even up to homotopy type, for the following reason. At the end of Chapter 7 of Seifert and Threlfall [1] one learns that for *any* prescribed finitely presented group G whatsoever there is a closed oriented 4-dimensional manifold X whose fundamental group is G; the details of the construction can be found in of Massey [4, pp. 143–144], for example. Quite independently of differential topology, Boone [1, 2, 3], Britton [1], and P. S. Novikov [1, 2] showed that the word problem for finitely presented groups is recursively unsolvable; this led Adyan [1] and Rabin [1] to conclude that the isomorphism problem for finitely presented groups is recursively unsolvable. The topological and group-theoretic pieces of the puzzle were juxtaposed by Markov [1, 2, 3] to conclude that no algorithm exists for classifying oriented 4-dimensional manifolds, a fortiori that no algorithm exists for classifying all smooth manifolds. There are some related results in Boone, Haken, and Poénaru [1].

The lack of a universal recipe is no obstacle to useful partial results, however. The classification of compact 2-dimensional manifolds is a classical result, for example, which can be found in Massey [4, pp. 10–18]. Certain 6-dimensional manifolds are classified in Jupp [1]. There is a complete classification (up to homeomorphism) of closed $(n-1)$-connected $(2n+1)$-dimensional smooth n-dimensional manifolds for $n > 7$ due to Wall [2]; the same result is known for certain smaller values of n as a result of Wilkens [1] (and Wall [2]), and related classifications appear in Tamura [1, 2, 3]. Finally, according to Cheeger and Kister [1] there are only countably many closed *topological manifolds*; a fortiori, up to homeomorphism there are only countably many closed smooth manifolds. A 1975 survey of the classification problem appears in Sullivan [2], and a 1978 survey appears in T. M. Price [1].

10.16 Remark: The *topological manifolds* mentioned in the previous Remark are merely locally euclidean hausdorff spaces whose topologies have a countable basis of open sets. For example, the underlying topological space of a smooth manifold is a topological manifold. Before one deals with the inverse question of finding smooth structures on a given topological manifold it is of interest to consider a broader question: can a topological manifold be triangulated? One-dimensional manifolds pose no problem, and the triangulation of surfaces was established by Radó [1], whose proof can be found in Ahlfors and Sario [1, pp. 44–46]. The triangulation of 3-dimensional manifolds was first established by Moise [1], and an alternative method was later supplied by Bing [1], with the same result.

The 4-dimensional case remains a mystery. However, for $n > 4$ the following results were established independently by Kirby and Siebenmann [1] (reproduced in Kirby and Siebenmann [2, pp. 299–306]), and by Lashof and Rothenberg [2]. If X is any closed n-dimensional topological manifold ($n > 4$), or any open n-dimensional topological manifold ($n > 5$), and if $H^4(X; \mathbb{Z}/2) = 0$, then X can be triangulated; furthermore, if $H^3(X; \mathbb{Z}/2) = 0$, then any two triangulations of X are equivalent in an obvious sense.

The $\mathbb{Z}/2$ cohomology conditions are known to be necessary. In fact, in every dimension $n > 4$ there is a closed topological manifold with no triangulation (as a manifold). An elementary discussion of the results described in this remark can be found in Schultz [3].

10.17 Remark: The negative results of the preceding remark are offset by the following homotopy property of arbitrary topological manifolds, which remain locally euclidean hausdorff spaces whose topologies have a countable basis of open sets. *Every topological manifold is homotopy equivalent to a simplicial space* $|K|$.

More specifically, a simplicial space $|K|$ is *locally finite* if and only if each point of $|K|$ has a neighborhood which meets only finitely many geometric simplexes of $|K|$; equivalently, $|K|$ is *locally finite* whenever the Dowker homotopy equivalence $|K|_w \to |K|_m$ of Proposition 4.6 is a homeomorphism. Furthermore, as in §6 a map $X \xrightarrow{f} Y$ is *proper* if and only if the inverse image of any compact set in Y is compact in X; a *proper homotopy equivalence* is a homotopy equivalence in a given category of topological spaces and proper maps. According to a result in Kirby and Siebenmann [2, p. 123] *every topological manifold is proper homotopy equivalent to a locally finite simplicial space* $|K|$.

10.18 Remark: By combining the weaker of the two preceding assertions with Theorem 3.7 one obtains the following generalization of Corollary 6.8: *any topological manifold is a base space; hence the category of topological manifolds is a full subcategory of the category* $\mathscr{B}$ *of base spaces.*

10.19 Remark: The Cairns–Whitehead triangulation theorem (Theorem 6.7) precludes smooth structures for nontriangulable manifolds; in fact, there are even *triangulable* manifolds with no smooth structure, such as the 10-dimensional manifold of Kervaire [3]. However, a topological manifold can also have more than one smooth structure, a result first established for the 7-sphere S^7 in Milnor [1]. Since then, smooth structures on spheres have been thoroughly treated in Milnor [6, 7], Kervaire and Milnor [1], and Eells and Kuiper [3], for example; we shall consider some of this material in Volume 3 of the present work.

Most products $S^p \times S^q$ of spheres have more differentiable structures than the sphere S^{p+q}, and more differentiable structures than the product of the corresponding numbers for the factors S^p and S^q. Some recent results concerning such products can be found in de Sapio [1, 2], Kawakubo [1, 2], and Schultz [1, 2], for example.

10.20 Remark: Here is a procedure that produces new smooth manifolds from old ones. Let X be a given smooth manifold, and let $X \to X$ be a smooth involution that is *free* in the sense that there are no fixed points; this provides an obvious equivalence relation $\sim$ in X for which the quotient $X/\sim$ is a new smooth manifold with X as a double covering. For example, for any $n > 0$ the antipodal map $S^n \to S^n$ of the standard n-sphere S^n is a free involution for which the resulting quotient $S^n/\sim$ is the real projective space RP^n of the same dimension.

In the special case $n = 3$ Livesay [1] shows that the antipodal map is essentially the only free involution of S^3: *any* free involution $S^3 \to S^3$ is smoothly equivalent to the antipodal map. However, according to Milnor [16], Hirsch and Milnor [1], and Fintushel and Stern [1], there are "exotic" free involutions of the standard spheres S^7, S^6, S^5, S^4 whose quotients $S^n/\sim$ are *not* diffeomorphic to the corresponding projective spaces RP^7, RP^6, RP^5,

RP^4. (The "exotic" involution $S^4 \to S^4$ is recent, although the quotient $S^4/\sim$ had been constructed in Cappell and Shaneson [1] earlier by other means. Although $S^4/\sim$ is not diffeomorphic to RP^4, it *is* homotopy equivalent to RP^4.)

More generally, Browder and Livesay [1] provide an invariant, further described in Livesay and Thomas [1], which leads to the following result of Orlick and Rourke [1]: for each $k > 0$ there are infinitely many distinct free involutions of S^{4k+3} (and of certain exotic spheres Σ^{4k+3}). The quotients of S^{4k+3} (and Σ^{4k+3}) provide a large supply of smooth closed manifolds of dimensions $n = 4k + 3$. There is a criterion for the existence of exotic free involutions of S^n (and exotic spheres Σ^n), valid for *any* $n \geqq 5$, in López de Medrano [1, p. 67].

One can also use certain smooth involutions $X \to X$ that are *not* free to create smooth quotient manifolds $X/\sim$; however, the results are not always new. For example, complex conjugation induces an involution $CP^2 \to CP^2$ of the complex projective plane CP^2, for which one easily establishes that $CP^2/\sim$ is a smooth manifold; however, Kuiper [1] and Massey [5] independently established the disappointing result that $CP^2/\sim$ is merely the standard 4-sphere S^4.

Here are some easily constructed smooth manifolds which will play a role in later remarks; their construction is similar to the construction of RP^n from the standard n-sphere S^n. Any odd-dimensional sphere S^{2n+1} can be regarded as the subspace of those points $(z_0, z_1, \ldots, z_n) \in \mathbb{C}^{n+1}$ such that $|z_0|^2 + |z_1|^2 + \cdots + |z_n|^2 = 1$. Let $q_1, \ldots, q_n$ be any integers which are relatively prime to a fixed integer $p > 0$, and let $S^{2n+1} \xrightarrow{h} S^{2n+1}$ be the diffeomorphism given by $h(z_0, z_1, \ldots, z_n) = (e^{2\pi i/p}z_0, e^{2\pi i q_1/p}z_1, \ldots, e^{2\pi i q_n/p}z_n)$. Then h is periodic with period p, with no fixed points, and the quotient of S^{2n+1} by the resulting equivalence relation is a smooth closed $(2n + 1)$-dimensional manifold, the *lens space* $L(p; q_1, \ldots, q_n)$.

The 3-dimensional lens spaces $L(p; q)$ were first constructed in Tietze [1]. Later work of Reidemeister [1] and J. H. C. Whitehead [2] established that $L(7; 1)$ and $L(7; 2)$, for example, are homotopy equivalent but *not* homeomorphic. R. Myers [1] generalizes the result of Livesay [1] concerning S^3 by showing that all involutions of $L(p; q)$ are equivalent to those induced by the action of the orthogonal group; similar results are valid for certain involutions of period greater than 2.

10.21 Remark: In 1844 Grassmann [1] described the Grassmann manifolds $G^m(\mathbb{R}^{m+n})$, the exterior products $\bigwedge^m \mathbb{R}^{m+n}$, the embeddings $G^m(\mathbb{R}^{m+n}) \xrightarrow{F} G^1(\bigwedge^m \mathbb{R}^{m+n})$, and the Plücker relations, all of which are used in Proposition 7.3. Despite the 1862 amplifications in Grassmann [2], Grassmann's work was not fully appreciated until it was reexamined and further developed by

Hermann Cäsar Hannibal Schubert, beginning in 1886; Schubert's work is briefly cited in Remark V.7.3. By 1897 elementary differential geometry was easily presented in Grassmann's setting, as in Burali-Forti [1]. In 1921 Corrado Segre, who had understood and applied Grassmann's work as a student during the 1880's, wrote a much-respected encyclopedia article identifying Grassmann manifolds as the very source of higher-dimensional algebraic geometry; Segre's enthusiasm appears in C. Segre [1, p. 772].

The Plücker relations and Proposition 7.3 can be found in Kleiman and Laksov [1], for example. Alternative proofs of Proposition 7.3 appear in Milnor and Stasheff [1, pp. 57–59] and in Lanteri [1].

10.22 Exercise: Show that any connected base space $X \in \mathscr{B}$ is pathwise connected. *Hint*: First prove the property for metric spaces of finite type, then observe that pathwise connectedness is a homotopy property. Some of the materials for this exercise can be found in Hu [3, pp. 84–90], for example.

10.23 Remark: The fundamental properties of Mayer–Vietoris sequences were discovered long before the general notion of an exact sequence existed, in Mayer [1] and Vietoris [1]. Exact sequences as such were introduced by Eilenberg and Steenrod [1], and later by Kelley and Pitcher [1].

10.24 Remark: Bott and Tu [1, pp. 42–53] prove a generalized version of Proposition 8.4, using Riemannian metrics rather than triangulations, which they apply in their elegant presentation of the Mayer–Vietoris technique for smooth manifolds.

CHAPTER II

Fiber Bundles

0. Introduction

Let $E \xrightarrow{\pi} X$ be a map onto any space X, and suppose for some space F that there is a homeomorphism $E \xrightarrow{\Psi} X \times F$ for which

$$\begin{array}{ccc} E & \xrightarrow{\Psi} & X \times F \\ & {\scriptstyle\pi}\searrow \quad \swarrow{\scriptstyle\pi_1} & \\ & X & \end{array}$$

commutes, where π_1 is the first projection. Then $E \xrightarrow{\pi} X$ represents a *trivial fiber bundle.*

More generally, let $E \xrightarrow{\pi} X$ be *locally trivial* in the following sense, for some space F: there is an open covering $\{U_i \mid i \in I\}$ of X and a corresponding family $\{\Psi_i \mid i \in I\}$ of homeomorphisms $\pi^{-1}(U_i) \xrightarrow{\Psi_i} U_i \times F$ for which each

$$\begin{array}{ccc} \pi^{-1}(U_i) & \xrightarrow{\Psi_i} & U_i \times F \\ & {\scriptstyle\pi}\searrow \quad \swarrow{\scriptstyle\pi_1} & \\ & U_i & \end{array}$$

commutes. Then the *projection* $E \xrightarrow{\pi} X$ represents a *fiber bundle* with *total space* E over the *base space* X, the *fiber* being F.

The preceding description is incomplete. If the intersection $U_i \cap U_j$ of two sets in the open covering $\{U_i | i \in I\}$ of the base space X is nonvoid, then the restrictions of the corresponding homeomorphisms Ψ_i and Ψ_j to $\pi^{-1}(U_i \cap U_j) \subset E$ induce a homeomorphism $\Psi_j \circ \Psi_i^{-1}$ such that

$$\begin{array}{ccc} (U_i \cap U_j) \times F & \xrightarrow{\Psi_j \circ \Psi_i^{-1}} & (U_i \cap U_j) \times F \\ & {\scriptstyle \pi_1} \searrow \quad \swarrow {\scriptstyle \pi_1} & \\ & U_i \cap U_j & \end{array}$$

commutes. Such a homeomorphism is necessarily of the form $\Psi_j \circ \Psi_i^{-1}(x, f) = (x, \psi_i^j(x)(f))$, where ψ_i^j carries each $x \in U_i \cap U_j$ into a homeomorphism $F \xrightarrow{\psi_i^j(x)} F$ of the fiber; in particular, $\psi_j^i(x) \circ \psi_i^j(x)$ is the identity for any $x \in U_i \cap U_j$, so that ψ_j^i and ψ_i^j carry any $x \in U_i \cap U_j$ into inverse homeomorphisms of the fiber. Since any transformations $F \to F$ whatsoever obey the associative law, it follows that each ψ_i^j can be regarded as a map of $U_i \cap U_j$ into a group G of homeomorphisms of the fiber, called the *structure group* of the given fiber bundle; the continuity of the compositions $\Psi_j \circ \Psi_i^{-1}$ imposes a specific topology on G for which the group operation $G \times G \to G$, the group inverse $G \xrightarrow{(\,)^{-1}} G$, and the action $G \times F \to F$ of G on F are continuous. The maps $U_i \cap U_j \xrightarrow{\psi_i^j} G$ are the *transition functions* of the representation $E \xrightarrow{\pi} X$ of the bundle, with respect to the covering $\{U_i | i \in I\}$ and corresponding family $\{\Psi_i | i \in I\}$.

The structure group G is part of the definition of a fiber bundle, and its choice is critical. For example, if one is interested in preserving given properties of the fiber F, one does *not* choose G to consist of *all* homeomorphism $F \to F$; thus if $F = \mathbb{R}^m$ and one wants to preserve vector addition in $\mathbb{R}^m$ and multiplication by scalars, one chooses G to be the general linear group $GL(m, \mathbb{R})$, or one of its subgroups. On the other hand, if G consists only of the identity map $F \to F$, then any projection $E \xrightarrow{\pi} X$ representing the given bundle is necessarily of the form $X \times F \xrightarrow{\pi_1} X$, so that the bundle is trivial.

The classical Möbius band provides the simplest nontrivial fiber bundle. Let $E \xrightarrow{\pi} S^1$ be the projection of the Möbius band E onto the unit circle S^1,

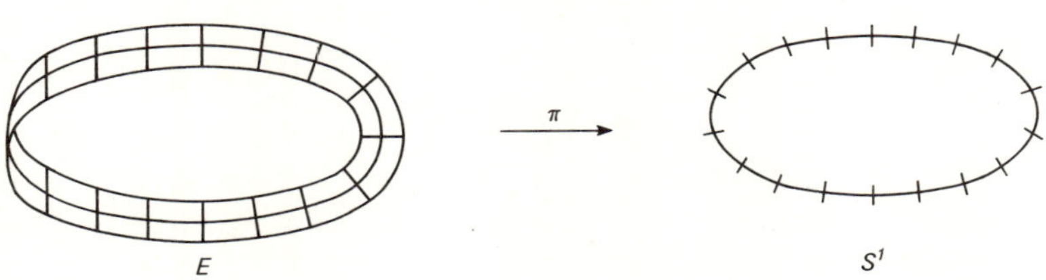

as indicated in the accompanying figure, the fiber F being the closed unit interval $[-1,1]$. One can cover the base space S^1 by two open sets U_0, U_1 homeomorphic to open real intervals, for which there are obvious homeomorphisms $\pi^{-1}(U_i) \xrightarrow{\Psi_i} U_i \times [-1,1]$. The intersection $U_0 \cap U_1$ consists of two disjoint open sets V_0 and V_1, which one can regard as subsets of U_0, for example. The Möbius band E is completely described by requiring the homeomorphism

$$(U_0 \cap U_1) \times [-1,1] \xrightarrow{\Psi_1 \circ \Psi_0^{-1}} (U_0 \cap U_1) \times [-1,1]$$

to carry $(x,f) \in (U_0 \cap U_1) \times [-1,1]$ into (x,f) or $(x,-f)$ according as $x \in V_0$ or $x \in V_1$. In this case one takes the structure group G to be $\mathbb{Z}/2$ in the discrete topology, which acts on $[-1,1]$ via multiplication by $+1$ or -1.

One traditional definition of fiber bundles is given in §1, followed in §2 by an equivalent description which is closer to the preceding sketch. Nothing in these two sections requires any properties of the base spaces X; however, many later results do require some sort of restriction. Accordingly, fib*re* bundles have arbitrary base spaces X, and fib*er* bundles have base spaces X in the category $\mathscr{B}$ of Definition I.1.2. Portions of the rationale for the eventual restriction to fib*er* bundles appear in §§3, 4, 5, and 7; beginning in §7 there are only fib*er* bundles.

Contractible spaces were used in Definition I.1.1 as building blocks for spaces of finite type, leading to the category $\mathscr{B}$. In §3 it will be shown that fibre bundles over contractible spaces are trivial bundles, the building blocks for fibre bundles in general.

In §4 it will be shown for any map $X' \xrightarrow{f} X$ in the category $\mathscr{B}$ and for any fiber bundle over X that the corresponding "pullback" fiber bundle over X' depends only on the homotopy type of f. The proof uses certain properties of the category $\mathscr{B}$, which was defined in terms of homotopy types.

We have already observed that the structure group G is an integral part of the description of a fiber bundle, and that it is desirable to "reduce" G to as small a subgroup $K \subset G$ as possible. In §5 it is shown that if the structure group of a fib*er* bundle (base space in $\mathscr{B}$) is a Lie group with finitely many connected components, then one can "reduce" G to a compact subgroup $K \subset G$. In particular, in §6 it is shown that one can always "reduce" the linear groups $GL(m,\mathbb{R})$, $GL^+(m,\mathbb{R})$, and $GL(n,\mathbb{C})$ to specific compact subgroups $O(m) \subset GL(m,\mathbb{R})$, $O^+(m) \subset GL^+(m,\mathbb{R})$, and $U(n) \subset GL(n,\mathbb{C})$, as structure groups of fib*er* bundles.

In §7 a basic step is provided for assigning cohomology classes in $H^*(X;\Lambda)$ to fiber bundles over base spaces $X \in \mathscr{B}$, where Λ is an appropriate coefficient ring. The provisions describing the category $\mathscr{B}$ are essential for the result, which will be used several times in later chapters.

1. Fibre Bundles and Fiber Bundles

Some of the ingredients of a fiber bundle over a space X are a *fiber* F and a *structure group* G of homeomorphisms $F \xrightarrow{g} F$, as we have just learned. We shall assume that G acts on the *left* of F; that is, $(g_1g_2)f = g_1(g_2f) \in F$ for any $g_1 \in G$, $g_2 \in G$, and $f \in F$. Although the fiber F may be any topological space, the topology of G must be *admissible*: the group operation $G \times G \to G$, the group inverse $G \xrightarrow{(\,)^{-1}} G$, and the action $G \times F \to F$ of G on F must be continuous. The action $G \times F \to F$ must also be *effective*: the identity is the only element $g \in G$ such that $gf = f$ for every $f \in F$. The space X is the *base space* of the bundle; we shall later require X to belong to the category $\mathscr{B}$ of base spaces, as in Definition I.1.2.

For any map $E \xrightarrow{\pi} X$ of a space E onto a space X, and for any $x \in X$, let E_x denote the inverse image $\pi^{-1}(\{x\})$ of $\{x\} \subset X$ under π. A nonvoid set S_x of homeomorphisms $E_x \xrightarrow{h} F$ is *G-related* whenever for any $h \in S_x$, $h' \in S_x$, and $g \in G$ the compositions $E_x \xrightarrow{h} F \xrightarrow{g} F$ and $F \xrightarrow{h^{-1}} E_x \xrightarrow{h'} F$ belong to S_x and G, respectively. Equivalently, for any fixed $h \in S_x$, the set S_x consists of all compositions $E_x \xrightarrow{h} F \xrightarrow{g} F$, where $g \in G$.

Given two maps $E \xrightarrow{\pi} X$ and $E' \xrightarrow{\pi'} X'$ onto spaces X and X', and given points $x \in X$ and $x' \in X'$ and G-related sets S_x and $S_{x'}$ of homeomorphisms $E_x \xrightarrow{h} F$ and $E'_{x'} \xrightarrow{h'} F$, respectively, a *G-related isomorphism* is any homeomorphism $E_x \to E'_{x'}$ such that every composition $F \xrightarrow{h^{-1}} E_x \to E'_{x'} \xrightarrow{h'} F$ belongs to G.

If

$$\begin{array}{ccc} E & \xrightarrow{\mathbf{f}} & E' \\ \Big\downarrow{\scriptstyle \pi} & & \Big\downarrow{\scriptstyle \pi'} \\ X & \xrightarrow{f} & X' \end{array}$$

commutes and $x \in X$, then for $E_x = \pi^{-1}(\{x\}) \subset E$ and $E'_{f(x)} = \pi'^{-1}(\{f(x)\}) \subset E'$ it is clear that $\mathbf{f}$ induces a map $E_x \to E'_{f(x)}$.

1.1 Definition: Given a fiber F and a structure group G of homeomorphisms $F \to F$, a *family of fibers* over a space X is a surjective map $E \xrightarrow{\pi} X$ and an assignment to each $x \in X$ of a G-related set S_x of homeomorphisms $E_x \to F$. If $E \xrightarrow{\pi} X$ and $E' \xrightarrow{\pi'} X'$ are families of fibers with the same fiber F and same structure group G, then a *morphism* from $E \xrightarrow{\pi} X$ to $E' \xrightarrow{\pi} X'$ is a pair of maps

$X \xrightarrow{f} X'$ and $E \xrightarrow{\mathbf{f}} E$ such that

$$\begin{array}{ccc} E & \xrightarrow{\mathbf{f}} & E' \\ {\scriptstyle \pi}\downarrow & & \downarrow{\scriptstyle \pi'} \\ X & \xrightarrow{f} & X' \end{array}$$

commutes, and such that for each $x \in X$ the induced map $E_x \to E'_{f(x)}$ is a G-related isomorphism.

For any family $E \xrightarrow{\pi} X$ of fibers one calls E the *total space*, π the *projection*, and X the *base space*. For the moment the base space X may be any topological space; the restriction $X \in \mathscr{B}$ will be imposed later.

For any given fiber F and structure group G it is clear that families of fibers are the objects of a category $\mathscr{C}(F, G)$ whose morphisms are commutative diagrams. Here is a way to construct new such objects and morphisms.

1.2 Definition: Let $E' \xrightarrow{\pi'} X'$ be a family of fibers in the category $\mathscr{C}(F, G)$, let $X \xrightarrow{f} X'$ be any map, and let $E \subset X \times E'$ consist of those $(x, e') \in X \times E'$ with $f(x) = \pi'(e') \in X'$, in the relative topology. The *pullback of* $E' \xrightarrow{\pi'} X'$ *along* f is the restriction $E \xrightarrow{\pi} X$ to $E \subset X \times E'$ of the first projection $X \times E' \xrightarrow{\pi_1} X$.

1.3 Lemma: *Let $E' \xrightarrow{\pi'} X'$ be a family of fibers in the category $\mathscr{C}(F, G)$, and let $X \xrightarrow{f} X'$ be a map. Then the pullback $E \xrightarrow{\pi} X$ also belongs to $\mathscr{C}(F, G)$, and there is a canonical map* **f** *such that*

$$\begin{array}{ccc} E & \xrightarrow{\mathbf{f}} & E' \\ {\scriptstyle \pi}\downarrow & & \downarrow{\scriptstyle \pi'} \\ X & \xrightarrow{f} & X' \end{array}$$

is a morphism in $\mathscr{C}(F, G)$.

PROOF: This is a direct verification, in which $E \xrightarrow{\mathbf{f}} E'$ is defined as the restriction to $E \subset X \times E'$ of the second projection $X \times E' \xrightarrow{\pi_2} E'$.

Here are the most obvious families of fibers.

1.4 Definition: Given a fiber F and a structure group G, as before, the *product family of fibers* over a space X is the first projection $X \times F \xrightarrow{\pi_1} X$;

For each $x \in X$ the G-related set S_x of homeomorphisms $E_x \to F$ consists of the maps $E_x = \{x\} \times F = F \xrightarrow{g} F$, for all $g \in G$.

1.5 Lemma: *Pullbacks of product families of fibers are product families of fibers.*

PROOF: Let $X' \times F \xrightarrow{\pi_1} X'$ be a product family, and let $X \xrightarrow{g} X'$ be any map. The total space E of the pullback along g consists of those $(x,(x',f)) \in X \times (X' \times F)$ with $g(x) = \pi_1(x',f) = x'$, which is canonically homeomorphic to $X \times F$, and the map $E \xrightarrow{\pi} X$ is the first projection $X \times F \xrightarrow{\pi_1} X$.

The isomorphisms in the category $\mathscr{C}(F,G)$ are clearly those morphisms

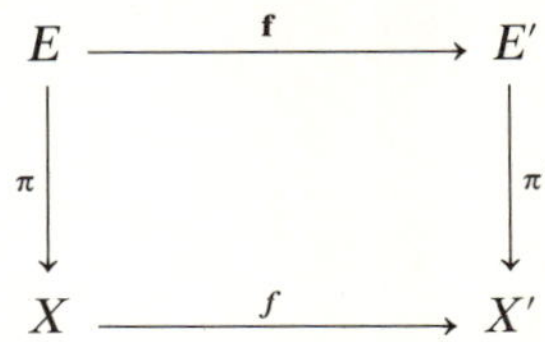

for which both f and $\mathbf{f}$ are homeomorphisms. However, we shall be interested primarily in those isomorphisms in which $X = X'$ and f is the identity; such isomorphisms are of the form

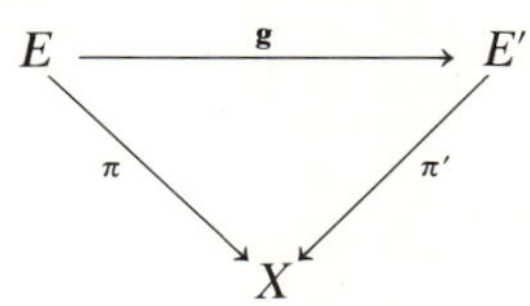

for a homeomorphism $\mathbf{g}$.

1.6 Lemma: *Let $X \xrightarrow{f} X'$ be any map. Then pullbacks along f of isomorphic families of fibers over X' are isomorphic families of fibers over X.*

PROOF: Given a morphism

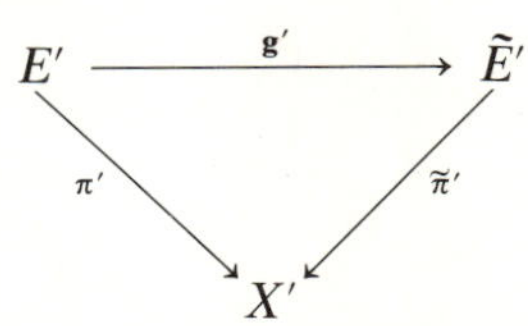

in which $\mathbf{g}'$ is a homeomorphism, the corresponding diagram

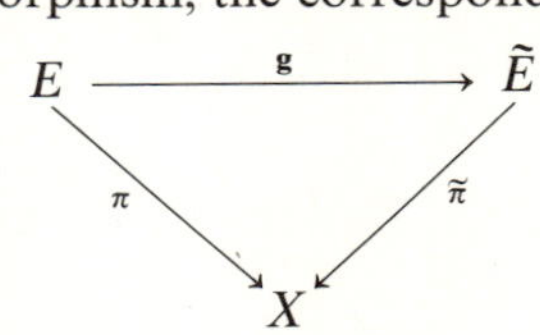

for the pullbacks is just the restriction of

$$\begin{array}{ccc} X \times E' & \xrightarrow{\mathrm{id} \times \mathbf{g}'} & X \times \tilde{E}' \\ & {\scriptstyle \pi_1} \searrow \quad \swarrow {\scriptstyle \pi_1} & \\ & X & \end{array}$$

to $E \subset X \times E'$, which consists of those $(x, e') \in X \times E'$ with $f(x) = \pi'(e')$. The map $E \xrightarrow{\mathbf{g}} \tilde{E}$ is then the restriction to E of a homeomorphism $\mathrm{id} \times \mathbf{g}'$, and hence itself a homeomorphism.

We now pass from the category $\mathscr{C}(F, G)$ of families of fibers, for a given fiber F and structure group G, to the category of isomorphism classes of families of fibers; as before we consider only isomorphisms of the form

$$\begin{array}{ccc} E & \xrightarrow{\quad \mathbf{g} \quad} & E' \\ & {\scriptstyle \pi} \searrow \quad \swarrow {\scriptstyle \pi'} & \\ & X & \end{array}$$

Lemma 1.6 guarantees that if ξ denotes an isomorphism class of families of fibers over a space X', then the pullbacks along any map $X \xrightarrow{f} X'$ also form such an isomorphism class, denoted $f^!\xi$. One simply calls $f^!\xi$ *the pullback of* ξ *along* $X \xrightarrow{f} X'$. Clearly if $X \xrightarrow{f} X' \xrightarrow{g} X''$ is a sequence of maps then for any isomorphism class ξ'' over X'', then one has $f^!g^!\xi'' = (g \circ f)^!\xi''$ as isomorphism classes over X.

1.7 Definition: Let X be an arbitrary topological space. The *trivial bundle* ξ over X with fiber F and structure group G is the isomorphism class of the product family $X \times F \xrightarrow{\pi_1} X$ of fibers.

1.8 Lemma: *Let ξ' be a trivial bundle over a space X', and let $X \xrightarrow{f} X'$ be an arbitary map; then the pullback $f^!\xi'$ is the corresponding trivial bundle over X.*

PROOF: Lemmas 1.5 and 1.6.

Now let ξ be an arbitrary isomorphism class of families of fibers over a space X, and let $U \xrightarrow{i} X$ be an inclusion $U \subset X$. The pullback $i^!\xi$ is the *restriction* $\xi | U$ of ξ to $U \subset X$.

1.9 Definition: For a given fiber F and structure group G, an isomorphism class ξ of families of fibers over a topological space X is a *fibre bundle* over

X whenever there is an open covering $\{U_i | i \in I\}$ of X such that each restriction $\xi | U_i$ is a trivial bundle. A *fiber bundle* is any fibre bundle over a space X in the category $\mathscr{B}$ of base spaces, as in Definition I.1.2.

For the moment we work with fibre bundles in general. However, the major theorems of §4 and §5 will be proved only for fib*er* bundles, and the remainder of the book concerns only certain fib*er* bundles.

1.10 Proposition: *Let $X \xrightarrow{f} X'$ be any map and let ξ' be a fibre bundle over X' for a given fiber and structure group. Then the pullback $f^!\xi'$ is a fibre bundle over X for the same fiber and structure group.*

PROOF: By hypothesis X' has an open covering $\{U'_k | k \in I\}$ such that each restriction $\xi' | U'_k$ is trivial. If $U_k = f^{-1}(U'_k) \subset X$, then $\{U_k | k \in I\}$ is an open covering of X, and for each $k \in I$ the map f restricts to a map $U_k \xrightarrow{g} U'_k$. Let $U_k \xrightarrow{i} X$ and $U'_k \xrightarrow{j} X'$ be the inclusion maps, so that

$$\begin{array}{ccc} U_k & \xrightarrow{\quad g \quad} & U'_k \\ {\scriptstyle i}\downarrow & & \downarrow{\scriptstyle j} \\ X & \xrightarrow{\quad f \quad} & X' \end{array}$$

commutes. Then

$$f^!\xi' | U_k = i^!f^!\xi' = (f \circ i)^!\xi' = (j \circ g)^!\xi' = g^!j^!\xi' = g^!(\xi' | U'_k),$$

and since each $\xi' | U'_k$ is trivial by hypothesis, Lemma 1.8 guarantees that each $f^!\xi' | U_k$ is trivial, as required.

One frequently identifies a fibre bundle over X by choosing a single representative $E \xrightarrow{\pi} X$ of the isomorphism class ξ of families of fibers, arbitrarily calling E *the* total space and π *the* projection of the bundle. This is a harmless abuse of language since $E \xrightarrow{\pi} X$ is related to any other representative $E' \xrightarrow{\pi'} X$ by a commutative diagram

$$\begin{array}{ccc} E & \xrightarrow{\quad \mathbf{f} \quad} & E' \\ & {\scriptstyle \pi}\searrow \quad \swarrow{\scriptstyle \pi'} & \\ & X & \end{array}$$

in which $\mathbf{f}$ is a homeomorphism inducing a G-related isomorphism $E_x \to E'_x$ for each $x \in X$. It will be useful to simplify the preceding criteria for $E \xrightarrow{\pi} X$ and $E' \xrightarrow{\pi'} X$ to represent the same fibre bundle over X.

Recall that the topology of the structure group G is admissible, meaning that the group operation $G \times G \to G$, the group inverse $G \xrightarrow{(\,)^{-1}} G$, and the action $G \times F \to F$ of G on F are continuous. Recall also that G acts effectively on F, meaning that the identity is the only element $g \in G$ such that $gf = f$ for every $f \in F$; that is, a given homeomorphism $F \to F$ is induced by at most one element of G.

1.11 Lemma: *Let* $\mathbf{f}$ *be any map such that*

$$\begin{array}{ccc} X \times F & \xrightarrow{\mathbf{f}} & X \times F \\ & {\scriptstyle \pi_1} \searrow \quad \swarrow {\scriptstyle \pi_1} & \\ & X & \end{array}$$

commutes, and which induces a homeomorphism $F \to F$ *in* G *for each* $x \in X$; *then* $\mathbf{f}$ *is a homeomorphism.*

PROOF: Since G is effective, there is a unique map $X \xrightarrow{\psi} G$ such that $\mathbf{f}(x, f) = (x, \psi(x)(f))$ for every $(x, f) \in X \times F$. Since the topology of G is admissible, the composition $X \xrightarrow{\psi} G \xrightarrow{(\,)^{-1}} G$ is continuous. Hence one can define a continuous inverse $X \times F \xrightarrow{\mathbf{f}^{-1}} X \times F$ by setting $\mathbf{f}^{-1}(x, f) = (x, \psi(x)^{-1}(f))$ for every $(x, f) \in X \times F$.

1.12 Proposition: *Let* $E \xrightarrow{\pi} X$ *and* $E' \xrightarrow{\pi'} X$ *represent fibre bundles* ξ *and* ξ' *with fiber* F *and structure group* G *over the same space* X. *If there is a map* $\mathbf{f}$ *such that*

$$\begin{array}{ccc} E & \xrightarrow{\mathbf{f}} & E' \\ & {\scriptstyle \pi} \searrow \quad \swarrow {\scriptstyle \pi'} & \\ & X & \end{array}$$

commutes and induces G-related isomorphisms $E_x \to E'_x$ *for each* $x \in X$, *then* $\mathbf{f}$ *is necessarily a homeomorphism, and hence* $\xi = \xi'$.

PROOF: There are open coverings $\{U_i | i \in I\}$ and $\{U_j | j \in J\}$ such that the restrictions $\xi | U_i$ and $\xi' | U_j$ are trivial; hence for $K = I \times J$ and $U_k = U_{(i,j)} = U_i \cap U_j$ there is a single open covering $\{U_k | k \in K\}$ of X such that the restrictions $\xi | U_k$ and $\xi' | U_k$ are trivial. The latter restrictions are therefore represented by product families, for which $\mathbf{f}$ induces maps $\mathbf{f}_k$ such that each

$$\begin{array}{ccc} U_k \times F & \xrightarrow{\mathbf{f}_k} & U_k \times F \\ & {\scriptstyle \pi_1} \searrow \quad \swarrow {\scriptstyle \pi_1} & \\ & U_k & \end{array}$$

commutes, inducing a homeomorphism $F \to F$ in G for each $x \in U_k$. By Lemma 1.11 each $\mathbf{f}_k$ is a homeomorphism, so that $\mathbf{f}$ is necessarily bijective. To show that $\mathbf{f}$ is a homeomorphism it remains only to show that it is an open map; but if $V \subset E$ is open, then $\mathbf{f}(V) \subset E'$ is a union of open sets $\mathbf{f}(V \cap \pi^{-1}(U_k)) \subset E'$, also provided by Lemma 1.11.

1.13 Corollary: *Given a fiber F and structure group G, let the families $E \xrightarrow{\pi} X$ and $E' \xrightarrow{\pi'} X'$ of fibers represent fibre bundles ξ over X and ξ' over X', respectively. Then for any morphism*

$$\begin{array}{ccc} E & \xrightarrow{\mathbf{f}} & E' \\ \downarrow{\scriptstyle \pi} & & \downarrow{\scriptstyle \pi'} \\ X & \xrightarrow{f} & X' \end{array}$$

of families of fibers ξ is the pullback $f^!\xi'$ of ξ' along f.

PROOF: The total space E'' of the pullback $E'' \xrightarrow{\pi''} X$ of $E' \xrightarrow{\pi'} X'$ consists of those $(x, e') \in X \times E'$ with $f(x) = \pi'(e') \in X$. Since $f \circ \pi = \pi' \circ \mathbf{f}$, there is consequently a map $E \xrightarrow{\mathbf{g}} E''$ with $\mathbf{g}(e) = (\pi(e), \mathbf{f}(e))$ for each $e \in E$. Trivially

$$\begin{array}{ccc} E & \xrightarrow{\mathbf{g}} & E'' \\ & {\scriptstyle \pi}\searrow \quad \swarrow{\scriptstyle \pi''} & \\ & X & \end{array}$$

commutes, where π'' is induced by the first projection $X \times E' \xrightarrow{\pi_1} X$. Furthermore, for each $x \in X$ the induced map $E_x \to E''_x$ carries $e \in E_x$ into $(x, \mathbf{f}(e)) \in E''_x$; therefore, since $\mathbf{f}$ induces G-related isomorphisms $E_x \to E'_x$, the maps $E_x \to E''_x$ are also G-related isomorphisms. Hence $\mathbf{g}$ is a homeomorphism by Proposition 1.12.

2. Coordinate Bundles

The informal description of fiber bundles given in the Introduction to this chapter began with something more concrete than Definition 1.9. Although the informal description was used primarily as motivation for the structure group G, it too can be molded into a formal definition of fiber bundles, equivalent to Definition 1.9.

Assume that a fiber F and structure group G are given, where G has an admissible topology and acts effectively on F. Let $E \xrightarrow{\pi} X$ be a map onto a space X. Suppose that there is an open covering $\{U_i | i \in I\}$ of X, let $E|U_i = \pi^{-1}(U_i)$ for each U_i, and suppose that there is a corresponding family $\{\Psi_i | i \in I\}$ of homeomorphisms Ψ_i such that each

$$\begin{array}{ccc} E|U_i & \xrightarrow{\Psi_i} & U_i \times F \\ & {\scriptstyle \pi}\searrow \quad \swarrow{\scriptstyle \pi_1} & \\ & U_i & \end{array}$$

commutes. If the intersection $U_i \cap U_j$ of two sets in the covering is nonvoid, then Ψ_i and Ψ_j induce a homeomorphism $\Psi_j \circ \Psi_i^{-1}$ such that

$$\begin{array}{ccc} (U_i \cap U_j) \times F & \xrightarrow{\Psi_j \circ \Psi_i^{-1}} & (U_i \cap U_j) \times F \\ & {\scriptstyle \pi_1}\searrow \quad \swarrow{\scriptstyle \pi_1} & \\ & U_i \cap U_j & \end{array}$$

commutes, and one necessarily has $\Psi_j \circ \Psi_i^{-1}(x, f) = (x, \psi_i^j(x)(f))$ for each $(x, f) \in (U_i \cap U_j) \times F$, where ψ_i^j carries each $x \in U_i \cap U_j$ into a homeomorphism $F \xrightarrow{\psi_i^j(x)} F$.

2.1 Definition: If each of the preceding ψ_i^j's is a (continuous) map from $U_i \cap U_j$ to the structure group G of homeomorphisms $F \to F$, then $E \xrightarrow{\pi} X$ is a *coordinate bundle* with respect to the covering $\{U_i | i \in I\}$ of X.

The maps $E|U_i \xrightarrow{\Psi_i} U_i \times F$ required for Definition 2.1 are the *local trivializations* of the coordinate bundle $E \xrightarrow{\pi} X$, and the induced maps $U_i \cap U_j \xrightarrow{\psi_i^j} G$ upon which the definition is based are the *transition functions.* As before, π is itself the *projection* of the *total space* E onto the space X, and for each $x \in X$ one lets E_x denote the *fiber* $\pi^{-1}(\{x\})$ *over* x.

In §1 we considered arbitrary families $E \xrightarrow{\pi} X$ of fibers, fibre bundles being locally trivial equivalence classes of such families. A coordinate bundle is clearly just a locally trivial family of fibers.

2.2 Proposition: *Let ξ be a fibre bundle, consisting of equivalence classes of families $E \xrightarrow{\pi} X$ of fibers as in Definition* 1.9; *then every family $E \xrightarrow{\pi} X$ representing ξ is a coordinate bundle.*

PROOF: Let $E \xrightarrow{\pi} X$ represent ξ. By definition there is an open covering $\{U_i \mid i \in I\}$ of X such that each $\xi \mid U_i$ is a trivial bundle, so that for $E \mid U_i = \pi^{-1}(U_i)$ one has a homeomorphism Ψ_i such that

$$\begin{array}{ccc} E\mid U_i & \xrightarrow{\Psi_i} & U_i \times F \\ & {\scriptstyle\pi}\searrow \quad \swarrow{\scriptstyle\pi_1} & \\ & U_i & \end{array}$$

commutes and induces G-related homeomorphisms $E_x \xrightarrow{\Psi_i} \{x\} \times F$. If $U_i \cap U_j$ is nonvoid, it follows that the homeomorphisms Ψ_i and Ψ_j induce a composition $(U_i \cap U_j) \times F \xrightarrow{\Psi_j \circ \Psi_i^{-1}} (U_i \cap U_j) \times F$ carrying (x, f) into $(x, \psi_i^j(x)(f))$, where $\psi_i^j(x)(f) \in F$ depends continuously on (x, f), and where each $\psi_i^j(x)$ is a homeomorphism. Since each $E_x \xrightarrow{\Psi_i} \{x\} \times F$ is G-related, each composition $\{x\} \times F \xrightarrow{\Psi_i^{-1}} E_x \xrightarrow{\Psi_j} \{x\} \times F$ belongs to G, by definition of G-relatedness. Hence ψ_i^j is a (continuous) map $U_i \cap U_j \to G$, as required.

Thus if one identifies a fibre bundle over X by choosing a single representative $E \xrightarrow{\pi} X$ of an isomorphism class ξ of families of fibers, that representative is always a coordinate bundle.

Proposition 2.2 also provides the desired relation between Definition 1.9 and the informal description of fibre bundles given earlier.

2.3 Corollary: *Any fibre bundle is an equivalence class of coordinate bundles.*

PROOF: The equivalence relation is the isomorphism

$$\begin{array}{ccc} E & \xrightarrow{\mathbf{f}} & E' \\ & {\scriptstyle\pi}\searrow \quad \swarrow{\scriptstyle\pi'} & \\ & X & \end{array}$$

of families of fibers over the same space X, now restricted only to those families which happen to be coordinate bundles.

We already know from Proposition 1.10 that pullbacks of fibre bundles are fibre bundles, so that Proposition 2.2 implies that pullbacks of coordinate bundles are coordinate bundles. Here is a more explicit version of the same result.

2.4 Proposition: *Let $E' \xrightarrow{\pi'} X'$ be a coordinate bundle with respect to an open covering $\{U_i' \mid i \in I\}$ of a space X', let $E \xrightarrow{\pi} X$ be its pullback along a map $X \xrightarrow{g} X'$, and let $U_i = g^{-1}(U_i')$ for each $i \in I$; then $E \xrightarrow{\pi} X$ is a coordinate* *bundle with respect to the open covering $\{U_i \mid i \in I\}$ of X. Specifically, if $U'_i \cap U'_j \xrightarrow{\Psi'^j_i} G$ are the transition functions of $E' \xrightarrow{\pi'} X'$, then the compositions $\psi'^j_i \circ g$ are the transition functions $U_i \cap U_j \xrightarrow{\psi^j_i} G$ of the pullback $E \xrightarrow{\pi} X$.*

PROOF: Let

$$\begin{array}{ccc} E & \xrightarrow{\quad g \quad} & E' \\ \downarrow{\scriptstyle \pi} & & \downarrow{\scriptstyle \pi'} \\ X & \xrightarrow{\quad g \quad} & X' \end{array}$$

be the pullback diagram. Since pullbacks of products are products, the local trivializations $E' \mid U'_i \xrightarrow{\Psi'_i} U'_i \times F$ of $E' \xrightarrow{\pi'} X'$ pull back to corresponding local trivializations $E \mid U_i \xrightarrow{\Psi_i} U_i \times F$ of $E \xrightarrow{\pi} X$. For any nonvoid intersection $U_i \cap U_j \subset X$ this provides a commutative diagram

$$\begin{array}{ccccc} (U_i \cap U_j) \times F & \xleftarrow{\Psi_i} & E \mid U_i \cap U_j & \xrightarrow{\Psi_j} & (U_i \cap U_j) \times F \\ \downarrow{\scriptstyle g \times \mathrm{id}} & & \downarrow{\scriptstyle g \mid U_i \cap U_j} & & \downarrow{\scriptstyle g \times \mathrm{id}} \\ (U'_i \cap U'_j) \times F & \xleftarrow{\Psi'_i} & E' \mid U'_i \cap U'_j & \xrightarrow{\Psi'_j} & (U'_i \cap U'_j) \times F. \end{array}$$

Each of the compositions $\Psi_j \circ \Psi_i^{-1}$ and $\Psi'_j \circ \Psi'^{-1}_i$ is described via a transition function, ψ^j_i and ψ'^j_i, as in Definition 2.1, and the outer rectangle of the diagram maps any $(x, f) \in (U_i \cap U_j) \times F$ as indicated:

$$\begin{array}{ccc} (x, f) & \longmapsto & (x, \psi^j_i(x) f) \\ \downarrow{\scriptstyle g \times \mathrm{id}} & & \downarrow{\scriptstyle g \times \mathrm{id}} \\ (g(x), f) & \longmapsto & (g(x), \psi'^j_i(g(x)) f). \end{array}$$

The result $\psi^j_i(x) f = \psi'^j_i(g(x)) f$ for any $(x, f) \in (U_i \cap U_j) \times F$ implies $\psi^j_i = \psi'^j_i \circ g$, as claimed.

Suppose that $E \xrightarrow{\pi} X$ is a coordinate bundle with an open covering $\{U_i \mid i \in I\}$ of X and a family $\{\Psi_i \mid i \in I\}$ of local trivializations $E \mid U_i \xrightarrow{\Psi_i} U_i \times F$. Then if $U_i \cap U_j \cap U_k$ is nonvoid, the restrictions of Ψ_i, Ψ_j, and Ψ_k trivially satisfy $(\Psi_k \circ \Psi_j^{-1}) \circ (\Psi_j \circ \Psi_i^{-1}) = \Psi_k \circ \Psi_i^{-1}$ as self-homeomorphisms of $(U_i \cap U_j \cap U_k) \times F$ to itself; consequently the corresponding transition functions satisfy $\psi^k_j(x) \psi^j_i(x) = \psi^k_i(x)$ in G for every $x \in U_i \cap U_j \cap U_k$. Conversely, this condition suffices for the construction of a coordinate bundle.

In the following construction we assume as before that a fiber F and structure group G are given, that G acts effectively on F, and that G has an admissible topology.

2.5 Proposition: *Let $\{U_i | i \in I\}$ be an open covering of a space X, and suppose for every nonvoid intersection $U_i \cap U_j$ that there is a map $U_i \cap U_j \xrightarrow{\psi_i^j} G$. The maps ψ_i^j are the transition functions of a unique coordinate bundle with fiber F and the given action $G \times F \to F$ if and only if for every nonvoid intersection $U_i \cap U_j \cap U_k$ and every $x \in U_i \cap U_j \cap U_k$ the identity*

$$\psi_j^k(x)\psi_i^j(x) = \psi_i^k(x)$$

is satisfied in G.

Proof: We have just learned that the condition is necessary. To prove the converse, observe that if $i = j = k$, then the condition becomes $\psi_i^i(x)\psi_i^i(x) = \psi_i^i(x)$, so that ψ_i^i maps any $x \in U_i$ into the identity $1 \in G$; similarly if one merely assumes $i = k$, one has $\psi_i^j(x)\psi_j^i(x) = \psi_i^i(x) = 1$, so that $\psi_j^i(x) = (\psi_i^j(x))^{-1}$ for every $x \in U_i \cap U_j$. Let $\sim$ be the relation in the disjoint union $\bigcup_i(U_i \times F)$ for which $(x_i, f_i) \sim (x_j, f_j)$ whenever $x_i = x_j \in U_i \cap U_j$ and $f_j = \psi_i^j(x_i)f_i \in F$. The hypothesis of the proposition guarantees that $\sim$ is transitive, and the preceding consequences of the hypothesis guarantee that $\sim$ is reflexive and symmetric. Hence $\sim$ is an equivalence relation, and one sets $E = \bigcup_i(U_i \times F)/\sim$ in the quotient topology. The first projections $U_i \times F \xrightarrow{\pi_1} U_i$ induce a map $E \xrightarrow{\pi} X$, and one easily verifies that $E \xrightarrow{\pi} X$ is the desired coordinate bundle, uniqueness being trivial.

In the next chapter we shall use Proposition 2.5 to construct new fibre bundles out of old ones, sometimes changing both the fiber and the structure group.

2.6 Definition: Let $G \times F \to F$ and $G' \times F' \to F'$ be effective actions of transformation groups G and G' on topological spaces F and F', respectively, the topologies of G and G' being admissible. A *morphism of transformation groups* is a pair (Γ, Φ) of maps $G \xrightarrow{\Gamma} G'$ and $F \xrightarrow{\Phi} F'$ such that Γ is a group homomorphism and the diagram

$$\begin{array}{ccc} G \times F & \longrightarrow & F \\ \Big\downarrow{\scriptstyle \Gamma \times \Phi} & & \Big\downarrow{\scriptstyle \Phi} \\ G' \times F' & \longrightarrow & F' \end{array}$$

commutes.

In the following result a morphism of transformation groups is applied to a fibre bundle ξ over a space X to construct a new fibre bundle ξ' over the same space X. One chooses any coordinate bundle $E \xrightarrow{\pi} X$ representing ξ, for a family $\{\Psi_i | i \in I\}$ of local trivializations with respect to an open covering $\{U_i | i \in I\}$ of X, and one constructs a new coordinate bundle $E' \xrightarrow{\pi'} X$, for a family $\{\Psi'_i | i \in I\}$ of local trivializations with respect to the same open covering. One then verifies that the fibre bundle ξ' represented by $E' \xrightarrow{\pi'} X$ depends only on ξ itself.

2.7 Proposition: *Let $(G \xrightarrow{\Gamma} G',\ F \xrightarrow{\Phi} F')$ be a morphism of transformation groups, and let X be a topological space. Then to any fibre bundle ξ over X with structure group G and fiber F the morphism (Γ, Φ) assigns a unique fibre bundle ξ' over X with structure group G' and fiber F', satisfying the following condition: for any coordinate descriptions of ξ and ξ' (as above), there is a projection-preserving map $E \xrightarrow{\mathbf{f}} E'$ whose restriction $\mathbf{f}|U_i$ to each $E|U_i \subset E$ provides a commutative diagram*

$$\begin{array}{ccc} E|U_i & \xrightarrow[\approx]{\Psi_i} & U_i \times F \\ \Big\downarrow{\scriptstyle \mathbf{f}|U_i} & & \Big\downarrow{\scriptstyle \mathrm{id} \times \Phi} \\ E'|U_i & \xrightarrow[\approx]{\Psi'_i} & U_i \times F'. \end{array}$$

PROOF: Let $U_i \cap U_j \xrightarrow{\psi^i_j} G$ be the transition functions corresponding to $\{\Psi_i | i \in I\}$, and let $U_i \cap U_j \xrightarrow{\psi'^i_j} G'$ be the compositions $U_i \cap U_j \xrightarrow{\psi^i_j} G \xrightarrow{\Gamma} G'$, for nonvoid intersections $U_i \cap U_j$. Since Γ is a group homomorphism, the conditions $\psi^k_j(x)\psi^j_i(x) = \psi^k_i(x)$ imply that $\psi'^k_j(x)\psi'^j_i(x) = \psi'^k_i(x)$ whenever $U_i \cap U_j \cap U_k$ is nonvoid, for any $x \in U_i \cap U_j \cap U_k$. By Proposition 2.5, the maps ψ'^i_j are the transition functions of a unique coordinate bundle $E' \xrightarrow{\pi'} X$ with fiber F' and group action $G' \times F' \to F'$. The total spaces E and E' are quotients of disjoint unions $\bigcup_i (U_i \times F)$ and $\bigcup_i (U_i \times F')$, and the local trivializations $E|U_i \xrightarrow{\Psi_i} U_i \times F$ and $E'|U_i \xrightarrow{\Psi'_i} U_i \times F'$ arise from the projections of $\bigcup_i (U_i \times F)$ and $\bigcup_i (U_i \times F')$ onto each $U_i \times F$ and $U_i \times F'$, respectively. The relation $(x_i, f_i) \sim (x_j, f_j)$ in $\bigcup_i (U_i \times F)$ means that $x_i = x_j \in U_i \cap U_j$ and $f_j = \psi^j_i(x_i) f_i \in F$, and since

$$\begin{array}{ccc} G \times F & \longrightarrow & F \\ \Big\downarrow{\scriptstyle \Gamma \times \Phi} & & \Big\downarrow{\scriptstyle \Phi} \\ G' \times F' & \longrightarrow & F' \end{array}$$

commutes, it follows that both $x_i = x_j \in U_i \cap U_j$ and $\Phi f_j = \Phi(\psi_i^j(x_i) f_i) = (\Gamma\psi_i^j(x))(\Phi f_i) = \psi_i'^j(x)(\Phi f_i)$; that is, the relation $(x_i, \Phi f_i) \sim (x_j, \Phi f_j)$ is satisfied in $\bigcup_i (U_i \times F')$. Consequently $F \xrightarrow{\Phi} F'$ induces a map $\bigcup_i (U_i \times F) \to \bigcup_i (U_i \times F')$, which in turn induces a projection-preserving map

$$E = \bigcup_i (U_i \times F)/\sim \xrightarrow{\mathbf{f}} \bigcup_i (U_i \times F')/\sim = E'.$$

The remainder of the proof consists of direct verifications.

2.8 Definition: Let $(G \xrightarrow{\Gamma} G', F \xrightarrow{\Phi} F')$ be a morphism of transformation groups. Then for any fibre bundle ξ over a space X, with structure group G and fiber F, the *induced bundle* with respect to (Γ, Φ) and ξ is the bundle ξ' of Proposition 2.7.

It is clear that the construction of induced bundles is functorial in the following sense, for the morphism (Γ, Φ) and bundle ξ of Definition 2.8: for any map $\tilde{X} \xrightarrow{g} X$ one has $(g^! \xi)' = g^! \xi'$ over $\tilde{X}$. The proof consists of direct verifications, similar to the verifications omitted from the proof of Proposition 2.7. This property will henceforth be used with no further comment.

The following result will be used in the next chapter to verify that different morphisms of transformation groups sometimes lead to the same induced bundle.

2.9 Proposition: *Let (Γ, Φ) be a morphism of transformation groups consisting of a group automorphism $G \xrightarrow{\Gamma} G$ and an action $F \xrightarrow{\Phi} F$ of G, and let ξ be any fibre bundle with fiber F and structure group G over a space X; then the induced bundle ξ' over X satisfies $\xi' = \xi$.*

PROOF: Proposition 2.7 provides a commutative diagram

$$\begin{array}{ccc} E & \xrightarrow{\mathbf{f}} & E' \\ & {\scriptstyle \pi'} \searrow \quad \swarrow {\scriptstyle \pi} & \\ & X & \end{array}$$

of coordinate bundles representing ξ and ξ', respectively, so that according to Proposition 1.12 one need only verify that $\mathbf{f}$ induces a G-related isomorphism $E_x \to E'_x$ for each $x \in X$. It suffices to show that if $x \in U_i \cap U_j \cap U_k$, and if Ψ_j, Ψ'_k, and $\mathbf{f}_i$ denote restrictions of the maps $E|U_j \xrightarrow{\Psi_j} U_j \times F$, $E'|U_k \xrightarrow{\Psi'_k} U_k \times F$, and $E|U_i \xrightarrow{\mathbf{f}|U_i} E'|U_i$ of Proposition 2.7, then the composition $\{x\} \times F \xrightarrow{\Psi_j^{-1}} E_x \xrightarrow{\mathbf{f}_i} E'_x \xrightarrow{\Psi'_k} \{x\} \times F$ is induced by an element of G. Since $\mathbf{f}|U_i = \Psi_i'^{-1} \circ (\mathrm{id} \times \Phi) \circ \Psi_i$ by Proposition 2.7, the preceding composition is the restriction to $\{x\} \times F$ of $\Psi'_k \circ \Psi_i'^{-1} \circ (\mathrm{id} \times \Phi) \circ \Psi_i \circ \Psi_j^{-1}$, so that one must show for the transition functions $U_i \cap U_k \xrightarrow{\psi_i'^k} G$ and

$U_j \cap U_i \xrightarrow{\psi_j^i} G$ that the composition $F \xrightarrow{\psi_i'^k(x) \circ \Phi \circ \psi_j^i(x)} F$ is a transformation in G; but since $\psi_j^i(x) \in G$ and $\psi_i'^k(x) \in G$, this is an immediate consequence of the hypothesis $\Phi \in G$.

3. Bundles over Contractible Spaces

In this section we prove a result which directly implies that any fibre bundle over any contractible space is trivial.

3.1 Lemma: *Let η be a fibre bundle over the product $X \times [0,1]$ of an arbitrary topological space X and the closed unit interval $[0,1]$, and let t be any point in $[0,1]$. If each of the restrictions $\eta|X \times [0,t]$ and $\eta|X \times [t,1]$ is trivial, then η is itself trivial.*

Proof: Let F and G be the fiber and structure group as usual, and let $E \xrightarrow{\pi} X \times [0,1]$ represent η. By hypothesis there are trivializations $E|X \times [0,t] \xrightarrow{\Psi_0} X \times [0,t] \times F$ and $E|X \times [t,1] \xrightarrow{\Psi_1} X \times [t,1] \times F$. The restrictions of Ψ_0 and Ψ_1 to $E|X \times \{t\}$ induce a homeomorphism $X \times \{t\} \times F \xrightarrow{\Psi_1 \circ \Psi_0^{-1}} X \times \{t\} \times F$ carrying any $(x,t,f) \in X \times \{t\} \in F$ into $(x,t,\psi(x)f) \in X \times \{t\} \times F$ for a unique "transition function" $X \times \{t\} \xrightarrow{\psi} G$, just as in the preceding section. (The quotation marks indicate only that the domain $X \times \{t\}$ of ψ is not open in $X \times [0,1]$; however, there is no change in the existence and uniqueness of ψ.) There is then a map $X \times [t,1] \times F \xrightarrow{\Psi} X \times [t,1] \times F$ given by $\Psi(x,s,f) = (x,s,\psi(x)^{-1}f)$ for which the composition $\Psi \circ \Psi_1$ is a new trivialization $E|X \times [t,1] \xrightarrow{\Psi_1'} X \times [t,1] \times F$. The induced homeomorphism $X \times \{t\} \times F \xrightarrow{\Psi_1' \circ \Psi_0^{-1}} X \times \{t\} \times F$ carries $(x,t,f) \in X \times \{t\} \times F$ into $(x,t,\psi'(x)f) \in X \times \{t\} \times F$ for a new "transition function" $X \times \{t\} \xrightarrow{\psi'} G$ with constant value $1 \in G$, so that Ψ_0 and Ψ_1' are restrictions of a common trivialization $E|X \times [0,1] \to X \times [0,1] \times F$, as required.

3.2 Lemma: *Let $\{W_1, \ldots, W_q\}$ be a finite open covering of $[0,1]$ such that each W_i is an interval, and let η be a fibre bundle over $X \times [0,1]$, for any space X. If each of the restrictions $\eta|X \times W_1, \ldots, \eta|X \times W_q$ is trivial, then η is itself trivial.*

Proof: Renumbering $W_1, \ldots, W_q$ if necessary, there are $q+1$ real numbers $t_0, t_1, \ldots, t_q$ with $0 = t_0 < t_1 < \cdots < t_q = 1$ such that $[t_{i-1}, t_i] \subset W_i$ for $i = 1, \ldots, q$. The inclusions $X \times [t_{i-1}, t_i] \subset X \times W_i$ imply that each restriction $\eta|X \times [t_{i-1}, t_i]$ is trivial, and the result is then an obvious iteration of Lemma 3.1.

3.3 Lemma: *Let X be any space, and let η be any fibre bundle over the product $X \times [0,1]$. Then there is at least one open covering $\{U_i | i \in I\}$ of X such that each restriction $\eta | U_i \times [0,1]$ is trivial.*

PROOF: By Definition 1.9 there is an open covering $\{Y_j | j \in J\}$ of $X \times [0,1]$ such that each restriction $\eta | Y_j$ is trivial. Any point $(x, t) \in X \times [0,1]$ has a neighborhood basis of sets of the form $V \times W$ for an open neighborhood V of x and an interval $W \subset [0,1]$ that is open in $[0,1]$ and contains t in its interior. Hence each $(x, t) \in X \times [0,1]$ lies in at least one neighborhood of the form $V \times W$, where $V \times W \subset Y_j$ for at least one $j \in J$. Since restrictions of trivial bundles are trivial (as in Lemma 1.8) it follows that there is a covering of $X \times [0,1]$ by sets of the form $V \times W$ such that each restriction $\eta | V \times W$ is trivial, where W is an interval that is open in $[0,1]$. For each $x \in X$ let $\mathscr{Y}_x$ be the family of all such products $V \times W$ for which $x \in V$ and $V \times W \subset Y_j$ for some $j \in J$. Since $[0,1]$ is compact, there is a finite subfamily $\{V_1 \times W_1, \ldots, V_q \times W_q\}$ of $\mathscr{Y}_x$ such that $\{W_1, \ldots, W_q\}$ covers $[0,1]$, and for $U_x = V_1 \cap \cdots \cap V_q$ each of the restrictions $\eta | U_x \times W_1, \ldots, \eta | U_x \times W_q$ is trivial. Lemma 3.2 then implies that $\eta | U_x \times [0,1]$ is itself trivial, so that $\{U_x | x \in X\}$ is a covering of the desired form.

3.4 Proposition: *Let X be any space, and let η be a fibre bundle over the product $X \times [0,1]$ such that the restriction $\eta | X \times \{0\}$ is trivial. Then η is itself trivial.*

PROOF: Let $E \to X \times [0,1]$ represent η, and let $E | X \times \{0\} \xrightarrow{\Psi} X \times \{0\} \times F$ be a fixed trivialization of $\eta | X \times \{0\}$. By Lemma 3.3 there is an open covering $\{U_i | i \in I\}$ of X such that each $\eta | U_i \times [0,1]$ is trivial, and one can use the procedure of Lemma 3.1 to guarantee that there are trivializations $E | U_i \times [0,1] \xrightarrow{\Psi_i} U_i \times [0,1] \times F$ such that each restriction $E | U_i \times \{0\} \xrightarrow{\Psi_i} U_i \times \{0\} \times F$ coincides with the corresponding restriction of Ψ. Hence the transition functions $U_i \cap U_j \times [0,1] \xrightarrow{\psi_i^j} G$ have the constant value $\psi_i^j(x, 0) = 1 \in G$ on each nonvoid intersection $U_i \cap U_j \times \{0\}$. For each $i \in I$ let $E | U_i \times [0,1] \xrightarrow{\mathbf{f}_i} U_i \times \{0\} \times F$ be the composition of Ψ_i with the map $U_i \times [0,1] \times F \to U_i \times \{0\} \times F$ that carries (x, t, f) into $(x, 0, f)$. Since $\psi_i^j(x, 0) = 1$ for each nonvoid intersection $U_i \cap U_j$, it follows that the restrictions $E | U_i \cap U_j \times [0,1] \xrightarrow{\mathbf{f}_i} U_i \cap U_j \times \{0\} \times F$ and $E | U_i \cap U_j \times [0,1] \xrightarrow{\mathbf{f}_j} U_i \cap U_j \times \{0\} \times F$ agree, so that there is a well-defined map $\mathbf{f}$ for which the diagram

$$\begin{array}{ccc} E & \xrightarrow{\mathbf{f}} & X \times \{0\} \times F \\ \downarrow & & \downarrow \\ X \times [0,1] & \xrightarrow{f} & X \times \{0\} \end{array}$$

commutes, where $f(x, t) = (x, 0)$ for each $(x, t) \in X \times [0, 1]$. By Corollary 1.13, η is then the pullback along $X \times [0, 1] \xrightarrow{f} X \times \{0\}$ of the trivial bundle $\eta | X \times \{0\}$, so that by Lemma 1.8 η is itself trivial, as asserted.

Here is the main result of the section.

3.5 Proposition: *Let ξ be any fibre bundle over any contractible space X; then ξ is trivial.*

PROOF: By hypothesis X is homotopy equivalent to the singleton space $\{*\}$, so that there is a composition $X \xrightarrow{g} \{*\} \xrightarrow{h} X$ that is homotopy equivalent to the identity map $X \to X$; that is, there is a map $X \times [0, 1] \xrightarrow{f} X$ whose restrictions $X \times \{0\} \xrightarrow{f_0} X$ and $X \times \{1\} \xrightarrow{f_1} X$ are the composition $h \circ g$ and the identity map, respectively. Let η be the pullback $f^!\xi$ over $X \times [0, 1]$, and observe that $\eta | X \times \{0\} = f^!_0\xi = g^!h^!\xi$ and $\eta | X \times \{1\} = f^!_1\xi = \xi$. Since the bundle $h^!\xi$ over $\{*\}$ is necessarily trivial, it follows from Lemma 1.8 that its pullback $\eta | X \times \{0\}$ is also trivial. Hence Proposition 3.4 implies that η is trivial, so that Lemma 1.8 implies that its restriction ξ $(= \eta | X \times \{1\})$ is also trivial, as asserted.

4. Pullbacks along Homotopic Maps

Let $X'' \xrightarrow{f_0} X$ and $X'' \xrightarrow{f_1} X$ be homotopic maps into an arbitrary space X, and let ξ be any fibre bundle over X. We shall show that if X'' is homotopy equivalent to a paracompact space then $f^!_0\xi = f^!_1\xi$ over X''. In particular, if f_0 and f_1 are homotopic maps in the category $\mathscr{B}$ of base spaces, then $f^!_0\xi = f^!_1\xi$ over $X'' \in \mathscr{B}$ for any fiber bundle ξ over $X \in \mathscr{B}$.

4.1 Lemma: *Let $E \to Y$ represent any fiber bundle η over an arbitrary space Y, and suppose that $Y \xrightarrow{g} Y$ is a map that restricts to the identity outside of some open set $V \subset Y$. If the restriction $\eta | V$ is trivial, then there is a morphism*

$$\begin{array}{ccc} E & \xrightarrow{\mathbf{g}} & E \\ \downarrow & & \downarrow \\ Y & \xrightarrow{g} & Y \end{array}$$

in the sense of Definition 1.1 *such that* **g** *restricts to the identity outside of $E | V$.*

PROOF: Let the restriction of $\mathbf{g}$ to $E|Y - V$ be the identity, and for any trivialization $E|V \xrightarrow{\Psi} V \times F$ of $\eta|V$ let the restriction of g to $E|V$ be the composition $\Psi^{-1} \circ (g|V \times \mathrm{id}) \circ \Psi$.

4.2 Lemma: *Let X' be a paracompact space, let $E \to X' \times [0,1]$ represent any fibre bundle η over the product $X' \times [0,1]$, and let $X' \times [0,1] \xrightarrow{g} X' \times \{1\}$ carry any (x', t) into $(x', 1)$. Then there is a morphism*

$$\begin{array}{ccc} E & \xrightarrow{\mathbf{g}} & E|X' \times \{1\} \\ \downarrow & & \downarrow \\ X' \times [0,1] & \xrightarrow{g} & X' \times \{1\} \end{array}$$

in the sense of Definition 1.1.

PROOF: According to Lemma 3.3 there is at least one open covering $\{U_i | i \in I\}$ of X' such that each restriction $\eta|U_i \times [0,1]$ is trivial. Since X' is paracompact, one may as well assume that $\{U_i | i \in I\}$ is locally finite and that there is a partition of unity $\{h_i | i \in I\}$ subordinate to $\{U_i | i \in I\}$. For any well-ordering of the index set I, and for each $i \in I$, let $k_{i-1} = \sum_{j<i} h_j$ and let $Y_{i-1} \subset X' \times [0,1]$ consist of those $(x', t) \in X' \times [0,1]$ such that $t \geqq k_{i-1}(x')$; in particular, for the initial element $0 \in I$ one has $Y_0 = X' \times [0,1]$. Define $Y_{i-1} \xrightarrow{g_i} Y_i \subset Y_{i-1}$ by setting $g_i(x', t) = (x', \max(k_i(x'), t))$, so that g_i is the identity outside of the open set $(U_i \times [0,1]) \cap Y_{i-1}$. Since the restriction $\eta|U_i \times [0,1]$ is trivial, it follows from Lemma 4.1 that there is a morphism

$$\begin{array}{ccc} E|Y_{i-1} & \xrightarrow{\mathbf{g}_i} & E|Y_i \\ \downarrow & & \downarrow \\ Y_{i-1} & \xrightarrow{g_i} & Y_i \end{array}$$

in the sense of Definition 1.1. Since $\{U_i | i \in I\}$ is locally finite, each $x' \in X'$ has a neighborhood $U_{x'}$ such that the restrictions of g_i and $\mathbf{g}_i$ to $U_{x'}$ and $E|U_{x'}$, respectively, are identities except for finitely many indices $i_1, \ldots, i_p \in I$. Furthermore $(U_{x'} \times [0,1]) \cap Y_i = U_{x'} \times \{1\}$ whenever $i_1 < i, \ldots, i_p < i$, so that the composition

$$\begin{array}{ccccccc} E = E|Y_0 & \xrightarrow{\mathbf{g}_i} \cdots \longrightarrow & E|Y_{i-1} & \xrightarrow{\mathbf{g}_1} & E|Y_i & \longrightarrow \cdots \\ \downarrow & & \downarrow & & \downarrow & \\ X' \times [0,1] = Y_0 & \xrightarrow{g_1} \cdots \longrightarrow & Y_{i-1} & \xrightarrow{g_i} & Y_i & \longrightarrow \cdots \end{array}$$

is a well-defined morphism of the desired form

$$\begin{array}{ccc} E & \xrightarrow{\mathbf{g}} & E|X' \times \{1\} \\ \downarrow & & \downarrow \\ X' \times [0,1] & \xrightarrow{g} & X' \times \{1\}. \end{array}$$

Here is a special case of the main theorem of the section.

4.3 Proposition: *Let $X' \xrightarrow{f_0} X$ and $X' \xrightarrow{f_1} X$ be homotopic maps of a paracompact space X' into an arbitrary space X. Then $f^!_0\xi = f^!_1\xi$ over X' for any fibre bundle ξ over X.*

PROOF: By hypothesis there is a map $X' \times [0,1] \xrightarrow{f} X$ with restrictions f_0 and f_1 to $X' \times \{0\}$ and $X' \times \{1\}$, respectively. Hence for $\eta = f^!\xi$ one has $\eta|X' \times \{0\} = f^!_0\xi$ and $\eta|X' \times \{1\} = f^!_1\xi$. Let $E \to X' \times [0,1]$ represent η. The left-hand morphism in the composition

$$\begin{array}{ccccc} E|X' \times \{0\} & \xrightarrow{\mathbf{h}} & E & \xrightarrow{\mathbf{g}} & E|X' \times \{1\} \\ \downarrow & & \downarrow & & \downarrow \\ X' \times \{0\} & \xrightarrow{h} & X' \times [0,1] & \xrightarrow{g} & X' \times \{1\} \end{array}$$

is the pullback diagram of an inclusion, and the right-hand morphism is provided by Lemma 4.2. Since $X' \times \{0\} \xrightarrow{g \circ h} X' \times \{1\}$ is the identification homeomorphism and $E|X' \times \{0\} \xrightarrow{\mathbf{g} \circ \mathbf{h}} E|X' \times \{1\}$ induces G-related isomorphisms in each fiber, it follows from Proposition 1.12 that $\eta|X' \times \{0\} = \eta|X' \times \{1\}$; that is, $f^!_0\xi = f^!_1\xi$, as asserted.

In order to extend Proposition 4.3 to the case in which X' is replaced by any space X'' homotopy equivalent to X' we first refine Lemma 3.3, which was used in the proof of Proposition 4.3.

4.4 Lemma: *Let X'' be any space, let η be any fibre bundle over the product $X'' \times [0,1]$, and let $\{U_i | i \in I\}$ be any open covering of X'' such that each restriction $\eta|U_i \times \{0\}$ is trivial. Then each restriction $\eta|U_i \times [0,1]$ is also trivial.*

PROOF: In Proposition 3.4 replace X and η by U_i and $\eta|U_i \times [0,1]$, respectively, for each $i \in I$.

4.5 Lemma: *Let $X'' \xrightarrow{g} X'$ be a homotopy equivalence of a space X'' with a paracompact space X', with homotopy inverse $X' \xrightarrow{h} X''$. Then $g^! h^! \zeta = \zeta$ for any fibre bundle ζ over X''.*

PROOF: Since X' is paracompact, there is a locally finite open cover $\{V_i | i \in I\}$ of X' such that each restriction $h^!\zeta | V_i$ of the pullback $h^!\zeta$ over X' is trivial, and there is a partition of unity $\{k_i | i \in I\}$ subordinate to $\{V_i | i \in I\}$. Let $U_i = g^{-1}(V_i)$, and let $X'' \xrightarrow{h_i} \mathbb{R}$ be the composition $k_i \circ g$, for each $i \in I$. Then, even though X'' is not itself necessarily paracompact, $\{U_i | i \in I\}$ is a locally finite cover of X'' such that each restriction $g^! h^! \zeta | U_i$ is trivial, and $\{h_i | i \in I\}$ is a partition of unity subordinate to $\{U_i | i \in I\}$. Since g and h are homotopy inverses, there is a map $X'' \times [0, 1] \xrightarrow{f} X''$ whose restrictions $X'' \times \{0\} \xrightarrow{f_0} X''$ and $X'' \times \{1\} \xrightarrow{f_1} X''$ are the composition $h \circ g$ and the identity map, respectively. Let η be the pullback $f^!\zeta$ over $X'' \times [0, 1]$, and observe that $\eta | X'' \times \{0\} = f^!_0 \zeta = g^! h^! \zeta$ and $\eta | X'' \times \{1\} = f^!_1 \zeta = \zeta$. Since each restriction $g^! h^! \zeta | U_i$ is trivial, that is, since each restriction $\eta | U_i \times \{0\}$ is trivial, Lemma 4.4 implies that each restriction $\eta | U_i \times [0, 1]$ is trivial. Hence if $E'' \to X'' \times [0, 1]$ represents η, then one can use the partition of unity $\{h_i | i \in I\}$ to construct the right-hand morphism in the composition

$$\begin{array}{ccccc}
E'' | X'' \times \{0\} & \longrightarrow & E'' & \longrightarrow & E'' | X'' \times \{1\} \\
\downarrow & & \downarrow & & \downarrow \\
X'' \times \{0\} & \longrightarrow & X'' \times [0, 1] & \longrightarrow & X'' \times \{1\}
\end{array}$$

exactly as in the proof of Lemma 4.2; the left-hand morphism of the same diagram is the pullback diagram of an inclusion. It follows as in Proposition 4.3 that $\eta | X'' \times \{0\} = \eta | X'' \times \{1\}$; that is, $g^! h^! \zeta = \zeta$ as asserted.

Here is the main theorem of the section.

4.6 Theorem: *Let X'' be homotopy equivalent to a paracompact space, and let $X'' \xrightarrow{f_0} X$ and $X'' \xrightarrow{f_1} X$ be homotopic maps of X'' into an arbitrary space X. Then $f^!_0 \xi = f^!_1 \xi$ for any fibre bundle ξ over X.*

PROOF: Let $X'' \xrightarrow{g} X'$ be a homotopy equivalence from X'' to a paracompact space X', with homotopy inverse $X' \xrightarrow{h} X''$, so that $g^! h^! f^!_0 \xi = f^!_0 \xi$ and $g^! h^! f^!_1 \xi = f^!_1 \xi$ over X'' by Lemma 4.5. However, the compositions $X' \xrightarrow{h} X'' \xrightarrow{f_0} X$ and $X' \xrightarrow{h} X'' \xrightarrow{f_1} X$ are homotopic maps from the paracompact space X' to X, so that Proposition 4.3 yields $h^! f^!_0 \xi = (f_0 \circ h)^! \xi = (f_1 \circ h)^! \xi = h^! f^!_1 \xi$. Consequently $f^!_0 \xi = g^!(h^! f^!_0 \xi) = g^!(h^! f^!_1 \xi) = f^!_1 \xi$ as claimed.

We shall use only the following special case of Theorem 4.6. Recall from Definition 1.9 that a fib*er* bundle is any fib*re* bundle whose base space lies in the category $\mathscr{B}$ of Definition I.1.2.

4.7 Proposition: *Let $X'' \xrightarrow{f_0} X$ and $X'' \xrightarrow{f_1} X$ be homotopic maps in the category $\mathscr{B}$ of base spaces (Definition* I.1.2), *and let ξ be a fiber bundle over X; then $f^!_0\xi = f^!_1\xi$ over X''.*

Proof: One of the provisions of Definition I.1.2 is that any base space $X'' \in \mathscr{B}$ is homotopy equivalent to a metrizable space X'. Since any metrizable space X' is paracompact, by Lemma I.8.2, Theorem 4.6 applies.

The product of any base space $X \in \mathscr{B}$ by the closed unit interval $[0, 1] \in \mathscr{B}$ is also a base space $X \times [0, 1] \in \mathscr{B}$, by Proposition I.1.4.

4.8 Corollary: *Given a base space $X \in \mathscr{B}$, any fiber bundle η over the product $X \times [0,1]$ is the pullback $\pi^!_1\xi$ along the first projection $X \times [0,1] \xrightarrow{\pi_1} X$, for some fiber bundle ξ over X.*

Proof: Let $X \xrightarrow{i_0} X \times [0,1]$ be the inclusion $x \mapsto (x, 0)$ and set $\xi = i^!_0\eta$. Since the composition $X \times [0,1] \xrightarrow{\pi_1} X \xrightarrow{i_0} X \times [0,1]$ is homotopic to the identity, Proposition 4.7 gives $\pi^!_1\xi = \pi^!_1 i^!_0\eta = (i_0 \circ \pi_1)^!\eta = \eta$ as claimed.

4.9 Corollary: *Given a base space $X \in \mathscr{B}$, any fiber bundle over the product $X \times [0,1]$ can be represented by a coordinate bundle of the form $E \times [0,1] \xrightarrow{\pi \times \mathrm{id}} X \times [0,1]$ for a coordinate bundle $E \xrightarrow{\pi} X$ over X.*

Proof: Immediate consequence of Corollary 4.8.

Corollary 4.9 gives a clearer view of Proposition 4.7. Let $X'' \times [0,1] \xrightarrow{f} X$ be a homotopy of maps f_0 and f_1 from $X'' \in \mathscr{B}$ to $X \in \mathscr{B}$, and let ξ be any fiber bundle over X. Then the pullback $f^!\xi$ can be represented by a coordinate bundle of the form $E'' \times [0,1] \xrightarrow{\pi'' \times \mathrm{id}} X'' \times [0,1]$ for some coordinate bundle $E'' \xrightarrow{\pi''} X''$, so that $f^!_0\xi$ and $f^!_1\xi$ are represented by $E'' \times \{0\} \xrightarrow{\pi'' \times \mathrm{id}} X'' \times \{0\}$ and $E'' \times \{1\} \xrightarrow{\pi'' \times \mathrm{id}} X'' \times \{1\}$, respectively; a fortiori $f^!_0\xi = f^!_1\xi$.

4.10 Proposition: *Let*

$$\begin{array}{ccc} E' & \xrightarrow{g} & E \\ \downarrow{\scriptstyle \pi'} & & \downarrow{\scriptstyle \pi} \\ X' & \xrightarrow{g} & X \end{array}$$

be a morphism of coordinate bundles such that g is a homotopy equivalence in the category $\mathscr{B}$ of base spaces; then **g** *is also a homotopy equivalence.*

PROOF: Let ξ be the fiber bundle represented by $E \xrightarrow{\pi} X$, with pullback $g^!\xi$ represented by $E' \xrightarrow{\pi'} X'$, and let $X \xrightarrow{h} X'$ be a homotopy inverse of g. Then $X \xrightarrow{g \circ h} X$ is homotopic to the identity map, so that $h^!g^!\xi = \xi$ by Proposition 4.7; hence there is a pullback diagram

$$\begin{array}{ccc} E & \xrightarrow{\mathbf{h}} & E' \\ {\scriptstyle \pi}\downarrow & & \downarrow{\scriptstyle \pi'} \\ X & \xrightarrow{h} & X' \end{array}$$

for some **h**, which provides a composed morphism

$$\begin{array}{ccc} E & \xrightarrow{\mathbf{g} \circ \mathbf{h}} & E \\ {\scriptstyle \pi}\downarrow & & \downarrow{\scriptstyle \pi} \\ X & \xrightarrow{g \circ h} & X \end{array}$$

of coordinate bundles. Let $X \times [0,1] \xrightarrow{f} X$ be the homotopy from $g \circ h$ to the identity map $X \to X$. By Corollary 4.9 the pullback $f^!\xi$ is represented by a coordinate bundle of the form $E \times [0,1] \xrightarrow{\pi \times \mathrm{id}} X \times [0,1]$, whose restrictions $E \times \{0\} \to X \times \{0\}$ and $E \times \{1\} \to X \times \{1\}$ represent $h^!g^!\xi$ and ξ, respectively. The pullback diagram along f is then of the form

$$\begin{array}{ccc} E \times [0,1] & \xrightarrow{\mathbf{f}} & E \\ {\scriptstyle \pi \times \mathrm{id}}\downarrow & & \downarrow{\scriptstyle \pi} \\ X \times [0,1] & \xrightarrow{f} & X, \end{array}$$

where **f** provides a homotopy from $E \xrightarrow{\mathbf{g} \circ \mathbf{h}} E$ to the identity map $E \to E$. Similarly there is a homotopy from $E' \xrightarrow{\mathbf{h} \circ \mathbf{g}} E'$ to the identity map $E' \to E'$, which completes the proof.

5. Reduction of Structure Groups

Let ξ be a fibre bundle with structure group G and fiber F over a space X, and let K be a subgroup of G. The action $G \times F \to F$ restricts to an action

$K \times F \to F$. Hence, for the inclusion homomorphism $K \xrightarrow{\Gamma} G$ and the identity map $F \xrightarrow{\Phi} F$, the pair (Γ, Φ) is a morphism of transformation groups in the sense of Definition 2.6.

5.1 Definition: The structure group G of a fibre bundle ξ over X can be *reduced* to a subgroup $K \subset G$ if and only if ξ can be induced, as in Definition 2.8, by applying (Γ, Φ) to some fibre bundle over X with structure group K.

For example, the structure group G of any trivial bundle can obviously be reduced to the trivial subgroup $\{1\} \subset G$.

The *reduction theorem* of this section concerns any fiber bundle ξ with structure group G over any $X \in \mathscr{B}$. It asserts that if $H \subset G$ is a sub*space* homeomorphic to a euclidean space $\mathbb{R}^p$, and if $K \subset G$ is a closed sub*group* such that every element of G is uniquely of the form hk for $h \in H$ and $k \in K$, then the structure group G of ξ can be reduced to K. The homeomorphism $H \to \mathbb{R}^p$ need not be a group isomorphism.

A *Lie group* is any topological group G that is also a smooth manifold, for which the group operation $G \times G \to G$ and the operation $G \xrightarrow{(\ )^{-1}} G$ carrying elements into their inverses are both smooth maps. For example, the general linear groups $GL(m, \mathbb{R})$ and $GL(n, \mathbb{C})$ are trivially Lie groups. At the end of this section we quote the very general Iwasawa–Mal'cev theorem, which asserts that if G is a Lie group with only finitely many connected components, then there is a decomposition $G = HK$ as in the preceding paragraph, in which K is any maximal compact subgroup of G; hence the reduction theorem permits one to replace any such structure group G by a *compact* Lie group. Although we do not prove the full Iwasawa–Mal'cev theorem, which can be found in several references given later, we do devote the next section to the special cases required in this book, in which G is either a complex general linear group $GL(n, \mathbb{C})$, a real general linear group $GL(m, \mathbb{R})$, or the subgroup $GL^+(m, \mathbb{R}) \subset GL(m, \mathbb{R})$ of those elements in $GL(m, \mathbb{R})$ with positive determinants; the corresponding maximal compact subgroups are the unitary groups $U(n) \subset GL(n, \mathbb{C})$, the orthogonal groups $O(m) \subset GL(m, \mathbb{R})$, and the rotation groups $O^+(m) \subset GL^+(m, \mathbb{R})$, respectively.

Before embarking on the proof of the reduction theorem, we rephrase Definition 5.1 directly in terms of coordinate bundles. Let $E \xrightarrow{\pi} X$ be a coordinate bundle representing ξ, and let $\{U_i | i \in I\}$ be an open covering of X for which there are trivializations $E|U_i \xrightarrow{\Psi_i} U_i \times F$; the transition functions $U_i \cap U_j \xrightarrow{\psi_i^j} G$ are defined as usual by the requirement that $\Psi_j \circ \Psi_i^{-1}(x, f) = (x, \psi_i^j(x)f) \in (U_i \cap U_j) \times F$ for any $(x, f) \in (U_i \cap U_j) \times F$. For any family $\{\lambda_i | i \in I\}$ of maps $U_i \xrightarrow{\lambda_i} G$ there are new trivializations $E|U_i \xrightarrow{\Psi_i'} U_i \times F$

consisting of compositions $E|U_i \xrightarrow{\Psi_i} U_i \times F \xrightarrow{\Lambda_i^{-1}} U_i \times F$ in which $\Lambda_i(x, f) = (x, \lambda_i(x)f)$. There are then new transition functions $U_i \cap U_j \xrightarrow{\psi'^j_i} G$ for the same coordinate bundle $E \xrightarrow{\pi} X$ and same covering $\{U_i | i \in I\}$ of X, defined by the requirement that $\Psi'_j \circ \Psi'^{-1}_i(x, f) = \Lambda_j^{-1} \circ \Psi_j \circ \Psi_i^{-1} \circ \Lambda_i(x, f) = (x, (\lambda_j(x)^{-1} \cdot \psi_i^j(x) \cdot \lambda_i(x))f) = (x, \psi'^j_i(x)f)$ for any $(x, f) \in (U_i \cap U_j) \times F$; that is, $\psi'^j_i(x) = \lambda_j(x)^{-1} \cdot \psi_i^j(x) \cdot \lambda_i(x) \in G$ for any $x \in U_i \cap U_j$. Clearly the structure group G can be reduced to a subgroup $K \subset G$ if and only if one can find a family $\{\lambda_i | i \in I\}$ of maps $U_i \xrightarrow{\lambda_i} G$ such that $\lambda_j(x)^{-1} \cdot \psi_i^j(x) \cdot \lambda_i(x) \in K$ for each $x \in U_i \cap U_j$. To prove the reduction theorem it therefore suffices to find such a family $\{\lambda_i | i \in I\}$ of maps $U_i \xrightarrow{\lambda_i} G$.

5.2 Definition: Let $E \xrightarrow{\pi} X$ be any coordinate bundle over a space X; a *section* of $E \xrightarrow{\pi} X$ is any (continuous) map $X \xrightarrow{\sigma} E$ such that $X \xrightarrow{\sigma} E \xrightarrow{\pi} X$ is the identity on X.

In the following lemma the fiber is any euclidean space $\mathbb{R}^p$; however, the unspecified structure group does *not* necessarily act linearly on $\mathbb{R}^p$.

5.3 Lemma: *Let X be a paracompact hausdorff space for which there is a countable locally finite open covering $\{U_n | n \in \mathbb{N}\}$ such that each connected component of each U_n is contained in a contractible open set in X. Then for any coordinate bundle $E \xrightarrow{\pi} X$ with fiber $\mathbb{R}^p$ there is a section $X \xrightarrow{\sigma} E$.*

PROOF: Any paracompact space is normal (as in Dugundji [2, p. 163], or Kelley [1, pp. 158–169], for example), so that there is a countable locally finite refinement $\{V_n | n \in \mathbb{N}\}$ of $\{U_n | n \in \mathbb{N}\}$ by open sets V_n whose closures satisfy $\bar{V}_n \subset U_n$. Since each connected component of each U_n is contained in a contractible open set in X it follows from Proposition 3.5 and Lemma 1.8 that there are local trivializations $E|U_n \xrightarrow{\Psi_n} U_n \times \mathbb{R}^p$, hence local trivializations $E|\bar{V}_n \xrightarrow{\Psi_n} \bar{V}_n \times \mathbb{R}^p$. Let $W_n = \bar{V}_0 \cup \cdots \cup \bar{V}_n$ for each $n \in \mathbb{N}$, and observe that since $W_0 = \bar{V}_0$, a section $W_0 \xrightarrow{\sigma_0} E|W_0$ is given by $\sigma_0(x) = \Psi_0^{-1}(x, 0)$ for any $x \in W_0$. For any section $W_{n-1} \xrightarrow{\sigma_{n-1}} E|W_{n-1}$ the restriction $\sigma_{n-1}|\bar{V}_n \cap W_{n-1}$ can be composed with Ψ_n to yield a map

$$\bar{V}_n \cap W_{n-1} \xrightarrow{\sigma_{n-1}} E|\bar{V}_n \cap W_{n-1} \xrightarrow{\Psi_n} (\bar{V}_n \cap W_{n-1}) \times \mathbb{R}^p,$$

and since $\bar{V}_n \cap W_{n-1}$ is a closed subset of a normal space, the Tietze extension theorem (in Dugundji [2, pp. 149–150], for example) permits one to extend $\bar{V}_n \cap W_{n-1} \xrightarrow{\Psi_n \circ \sigma_{n-1}} (\bar{V}_n \cap W_{n-1}) \times \mathbb{R}^p$ to a map $\bar{V}_n \xrightarrow{\tau_n} \bar{V}_n \times \mathbb{R}^p$. It follows that the composition $\bar{V}_n \xrightarrow{\tau_n} \bar{V}_n \times \mathbb{R}^p \xrightarrow{\Psi_n^{-1}} E|\bar{V}_n$ is the restriction to $\bar{V}_n \subset W_n$ of a section $W_n \xrightarrow{\sigma_n} E|W_n$ whose restriction to $W_{n-1} \subset W_n$ is σ_{n-1}. Since $\{\bar{V}_n | n \in \mathbb{N}\}$ covers X each $x \in X$ lies in some W_n, and since $\{U_n | n \in \mathbb{N}\}$

is locally finite, one has $\sigma_{n+1}(x) = \sigma_n(x)$ for n sufficiently large. Hence there is a well-defined section $\sigma = \lim_{n\to\infty} \sigma_n$, as claimed.

Now let K be any closed subgroup of a topological group G, and let $\sim$ be the equivalence relation in G with $g' \sim g$ if and only if $g' = gk$ for some $k \in K$. The quotient $G/\sim$, in the quotient topology, is the *homogeneous space* G/K. Since K is not necessarily a normal subgroup of G, the homogeneous space G/K is not necessarily a group. However, there is a natural action $G \times G/K \to G/K$ of G on G/K given by left-multiplication. The kernel of the natural action is the largest subgroup $K_0 \subset K$ that is normal in G, so that for $\tilde{G} = G/K_0$ there is an effective transformation group $\tilde{G} \times G/K \to G/K$.

Let $E \xrightarrow{\pi} X$ be any coordinate bundle with fiber F and structure group G, with respect to some open covering $\{U_i | i \in I\}$ of X. Any local trivializations $E|U_i \xrightarrow{\Psi_i} U_i \times F$ provide transition functions $U_i \cap U_j \xrightarrow{\psi_i^j} G$ as always: $\Psi_j \circ \Psi_i^{-1}(x, f) = (x, \psi_i^j(x)f)$ for $(x, f) \in (U_i \cap U_j) \times F$. For any closed subgroup $K \subset G$ let $G \xrightarrow{\Gamma} G/K_0 = \tilde{G}$ and $G \xrightarrow{\Phi} G/K$ denote the canonical surjections, where Γ is in fact a group epimorphism. If one sets $\tilde{\psi}_i^j(x) = \Gamma(\psi_i^j(x)) \in \tilde{G}$ for any $x \in U_i \cap U_j$, then the conditions $\psi_j^k(x) \cdot \psi_i^j(x) = \psi_i^k(x)$ imply $\tilde{\psi}_j^k(x) \cdot \tilde{\psi}_i^j(x) = \tilde{\psi}_i^k(x)$ for any $x \in U_i \cap U_j \cap U_k$; hence according to Proposition 2.5 there is a unique coordinate bundle $\tilde{E} \xrightarrow{\tilde{\pi}} X$ with fiber G/K and structure group $\tilde{G}$ defined with respect to the covering $\{U_i | i \in I\}$ by means of the transition functions $U_i \cap U_j \xrightarrow{\tilde{\psi}_i^j} \tilde{G}$. We temporarily call $\tilde{E} \xrightarrow{\tilde{\pi}} X$ the *associated bundle* with respect to the given coordinate bundle $E \xrightarrow{\pi} X$ and given closed subgroup K of the structure group G.

5.4 Lemma: *Let X satisfy the conditions of Lemma 5.3, let $E \xrightarrow{\pi} X$ be any coordinate bundle with structure group G, and suppose that $H \subset G$ is a sub***space** *homeomorphic to a euclidean space $\mathbb{R}^p$ and that $K \subset G$ is a closed sub* **group** *such that every element of G is uniquely of the form hk for $h \in H$ and $k \in K$. Then there is a section $X \xrightarrow{\tilde{\sigma}} \tilde{E}$ of the associated bundle $\tilde{E} \xrightarrow{\tilde{\pi}} X$.*

PROOF: The hypotheses guarantee that the homogeneous space G/K is homeomorphic to H, hence also to $\mathbb{R}^p$. Hence by Lemma 5.3 there is a section $X \xrightarrow{\sigma} \tilde{E}$, as claimed.

5.5 Lemma: *Let G be a topological group with a sub***space** *$H \subset G$ and closed sub***group** *$K \subset G$ as in Lemma 5.4, and let $G \xrightarrow{\Phi} G/K$ be the canonical surjection. Then there is a map $G/K \xrightarrow{\tau} G$ such that the composition $G/K \xrightarrow{\tau} G \xrightarrow{\Phi} G/K$ is the identity.*

PROOF: It has already been noted that G/K is homeomorphic to H, and one can identify any $h \in H$ as an element of G; let τ denote the composition $G/K \to H \subset G$.

If the coordinate bundle $E \xrightarrow{\pi} X$ is described in terms of a given covering $\{U_i \mid i \in I\}$ of the space X, then the associated bundle $\tilde{E} \xrightarrow{\tilde{\pi}} X$ can clearly be described in terms of the same covering; that is, if there are local trivializations $E|U_i \xrightarrow{\Psi_i} U_i \times F$ of $E \xrightarrow{\pi} X$, then there are also local trivializations $\tilde{E}|U_i \xrightarrow{\tilde{\Psi}_i} U_i \times G/K$ of $\tilde{E} \xrightarrow{\tilde{\pi}} X$. Although Ψ_i and $\tilde{\Psi}_i$ are not directly related, the compositions $\Psi_j \circ \Psi_i^{-1}$ and $\tilde{\Psi}_j \circ \tilde{\Psi}_i^{-1}$ provide transition functions $U_i \cap U_j \xrightarrow{\psi_i^j} G$ and $U_i \cap U_j \xrightarrow{\tilde{\psi}_i^j} \tilde{G}$ that are related by the very definition of $\tilde{E} \xrightarrow{\tilde{\pi}} X$; specifically, $\tilde{\psi}_i^j$ is the composition $U_i \cap U_j \xrightarrow{\psi_i^j} G \xrightarrow{\Gamma} \tilde{G}$. For each $i \in I$ let $\tilde{E}|U_i \xrightarrow{\tilde{\psi}_i} G/K$ be the composition $\tilde{E}|U_i \xrightarrow{\tilde{\Psi}_i} U_i \times G/K \xrightarrow{\pi_2} G/K$, where π_2 is the second projection.

5.6 Lemma: *For any $x \in U_i \cap U_j$ and any $\tilde{e} \in \tilde{E}_x$ one has $\tilde{\psi}_i^j(x)\tilde{\psi}_i(\tilde{e}) = \tilde{\psi}_j(\tilde{e})$ in G/K.*

Proof: By definition of $\tilde{\psi}_i$ and $\tilde{\psi}_j$ one has $\tilde{\Psi}_i(\tilde{e}) = (x, \tilde{\psi}_i(\tilde{e}))$ and $\tilde{\Psi}_j(\tilde{e}) = (x, \tilde{\psi}_j(\tilde{e}))$, and by definition of the associated bundle $\tilde{E} \xrightarrow{\pi} X$ the trivializations $\tilde{\Psi}_i$ and $\tilde{\Psi}_j$ are related only by the requirement that $(U_i \cap U_j) \times G/K \xrightarrow{\tilde{\Psi}_j \circ \tilde{\Psi}_i^{-1}} (U_i \cap U_j) \times G/K$ carry any $(x, \tilde{f})$ into $(x, \tilde{\psi}_i^j(x)\tilde{f})$, for $\tilde{f} \in G/K$. Hence

$$(x, \tilde{\psi}_i^j(x)\tilde{\psi}_i(\tilde{e})) = \tilde{\Psi}_j\tilde{\Psi}_i^{-1}(x, \tilde{\psi}_i(\tilde{e})) = \tilde{\Psi}_j(\tilde{e}) = (x, \tilde{\psi}_j(\tilde{e})),$$

to which one applies π_2 to complete the proof.

5.7 Definition: Let the coordinate bundle $E \xrightarrow{\pi} X$ satisfy the conditions of Lemma 5.4, and let $\tilde{E} \xrightarrow{\tilde{\pi}} X$ be the associated bundle with fiber G/K and structure group $\tilde{G}$. If $E \xrightarrow{\pi} X$ and $\tilde{E} \xrightarrow{\tilde{\pi}} X$ are defined with respect to an open covering $\{U_i \mid i \in I\}$ of X, and if the local trivializations $\tilde{E}|U_i \xrightarrow{\tilde{\Psi}_i} U_i \times G/K$ are given by $\tilde{\Psi}_i(\tilde{e}) = (\tilde{\pi}(\tilde{e}), \tilde{\psi}_i(\tilde{e}))$, then the *reducing maps* $U_i \xrightarrow{\lambda_i} G$ are the compositions

$$U_i \xrightarrow{\tilde{\sigma}|U_i} E|U_i \xrightarrow{\tilde{\psi}_i} G/K \xrightarrow{\tau} G,$$

where $\tilde{\sigma}|U_i$ is the restriction to U_i of the section $X \xrightarrow{\tilde{\sigma}} \tilde{E}$ of Lemma 5.4, and where $G/K \xrightarrow{\tau} G$ is defined in Lemma 5.5.

5.8 Lemma: *Let $U_i \cap U_j \xrightarrow{\psi_i^j} G$ be the transition functions and $U_i \xrightarrow{\lambda_i} G$ the reducing maps of the coordinate bundle $E \xrightarrow{\pi} X$ of Definition 5.7; then*

$$\Phi(\psi_i^j(x) \cdot \lambda_i(x)) = \Phi(\lambda_j(x)) \in G/K \qquad \textit{for any} \quad x \in U_i \cap U_j,$$

where $G \xrightarrow{\Phi} G/K$ is the canonical surjection. Hence $\lambda_j(x)^{-1} \cdot \psi_i^j(x) \cdot \lambda_i(x)$ lies in the kernel of Φ.

PROOF: Since $G/K \xrightarrow{\tau} G \xrightarrow{\Phi} G/K$ is the identity by Lemma 5.5, it follows from Lemma 5.6 that

$$\begin{aligned}\Phi(\psi_i^j(x)\cdot\lambda_i(x)) &= \tilde{\psi}_i^j(x)\Phi(\lambda_i(x))\\ &= \tilde{\psi}_i^j(x)\tilde{\psi}_i(\tilde{\sigma}(x)) = \tilde{\psi}_j(\tilde{\sigma}(x)) = \Phi(\lambda_j(x)).\end{aligned}$$

5.9 Lemma: *Let the coordinate bundle $E \xrightarrow{\pi} X$ satisfy the conditions of Lemma 5.4; then its structure group G can be reduced to the given closed subgroup $K \subset G$.*

PROOF: If $U_i \cap U_j \xrightarrow{\psi_i^j} G$ are transition functions for $E \xrightarrow{\pi} X$ with respect to the covering $\{U_n | n \in \mathbb{N}\}$ of X, then for any maps $U_i \xrightarrow{\lambda_i} G$ whatsoever one obtains new transition functions $U_i \cap U_j \xrightarrow{\psi_i'^j} G$ for the same coordinate bundle by setting $\psi_i'^j(x) = \lambda_j(x)^{-1}\cdot\psi_i^j(x)\cdot\lambda_i(x) \in G$ for any $x \in U_i \cap U_j$. In particular, if one chooses $U_i \xrightarrow{\lambda_i} G$ to be the reducing maps of Definitions 5.7, then it follows from Lemma 5.8 that the canonical surjection $G \xrightarrow{\Phi} G/K$ carries each $\psi_i'^j(x)$ into the point $\Phi(K) \in G/K$, hence that $\psi_i'^j(x) \in K$, as required.

5.10 Theorem (Reduction Theorem): *Let ξ be any fiber bundle with structure group G over any base space $X \in \mathscr{B}$, and suppose that there is a subspace $H \subset G$ homeomorphic to a euclidean space $\mathbb{R}^p$, and a closed subgroup $K \subset G$, such that every element of G is uniquely of the form hk for $h \in H$ and $k \in K$. Then the structure group G of ξ can be reduced to the subgroup $K \subset G$.*

PROOF: By Proposition I.8.3 there is a homotopy equivalence $X \xrightarrow{g} X'$ of any $X \in \mathscr{B}$ with a space $X' \in \mathscr{B}$ for which there is a countable locally finite covering $\{U_n | n \in \mathbb{N}\}$ such that each connected component of each U_n is contained in a contractible open set in X'; that is, X' satisfies the conditions of Lemma 5.3. If $X' \xrightarrow{h} X$ is a homotopy inverse of g, then the pullback $h^!\xi$ of ξ over X' is represented by a coordinate bundle $E' \xrightarrow{\pi'} X'$ that satisfies the conditions of Lemma 5.4. Hence by Lemma 5.9 the structure group G of the fiber bundle $h^!\xi$ can be reduced to the subgroup $K \subset G$. Since $\xi = g^!h^!\xi$, by Lemma 4.5, it follows that the structure group G of ξ can also be reduced to the subgroup $K \subset G$, as claimed.

Suppose that ξ is a fiber bundle whose structure group is an arbitrary Lie group G with only finitely many connected components. We shall quote one of the major triumphs of the theory of Lie groups, which permits one to apply the preceding reduction theorem to obtain a very satisfying result. As promised earlier, however, explicit proofs will be given in the next section for the groups $GL(n, \mathbb{C})$, $GL(m, \mathbb{R})$, and $GL^+(m, \mathbb{R})$; in these cases Theorem 5.10 will permit one to regard any fiber bundle with one of these groups for its structure group as a fiber bundle whose structure group is one of the *compact* Lie groups $U(n) \subset GL(n, \mathbb{C})$, $O(m) \subset GL(m, \mathbb{R})$, or $O^+(m) \subset GL^+(m, \mathbb{R})$, respectively.

Any Lie group G with only finitely many connected components possesses a *maximal* compact subgroup $K \subset G$, and one can show that any other maximal compact subgroup of G is of the form $g^{-1}Kg$ for some $g \in G$. This property of Lie groups is closely associated with the following theorem.

5.11 Theorem (Iwasawa–Mal'cev Decomposition Theorem): *Let G be any Lie group with only finitely many connected components, and let $K \in G$ be any maximal compact subgroup of G. Then there are one-parameter subgroups $H_1, \ldots, H_p$ of G, each isomorphic to the additive group $\mathbb{R}^1$, such that every element of G is uniquely of the form $(h_1 \cdots h_p)k$ for $h_1 \in H_1, \ldots, h_p \in H_p$, and $k \in K$.*

Proofs of Theorem 5.11 can be found in Iwasawa [1], Cartier [1, pp. 22-15–22-16], Mostow [1, pp. 47–48], and in Hochschild [1, pp. 180–186], for example. Theorem 5.11 is usually called the "Iwasawa decomposition theorem"; however, a rationale for including Mal'cev's name is given in Remark 8.22.

5.12 Proposition: *Let ξ be a fiber bundle over any base space $X \in \mathscr{B}$, whose structure group G is any Lie group with only finitely many connected components; then the structure group G can be reduced to a compact subgroup $K \subset G$.*

PROOF: This is an immediate consequence of Theorems 5.11 and 5.10.

One can obtain a slightly weaker version of Proposition 5.12 without appealing to the Iwasawa–Mal'cev decomposition theorem: see Corollary 6.14.

6. Polar Decompositions

In this section we show that if G is any of the linear groups $GL(n, \mathbb{C})$, $GL(m, \mathbb{R})$, or $GL^+(m, \mathbb{R})$, then there is a subspace $H \subset G$ diffeomorphic to a euclidean space $\mathbb{R}^p$, and a (maximal) compact subgroup $K \subset G$, such that every element of G is uniquely of the form hk, for $h \in H$ and $k \in K$. The compact subgroups K are the unitary groups $U(n) \subset GL(n, \mathbb{C})$, the orthogonal groups $O(m) \subset GL(m, \mathbb{R})$, and the rotation groups $O^+(m) \subset GL^+(m, \mathbb{R})$, respectively, which are easily shown to be compact. Hence the reduction theorem (Theorem 5.10), and the results of this section, imply for any fiber bundle whose structure group G is one of the groups $GL(n, \mathbb{C})$, $GL(m, \mathbb{R})$, or $GL^+(m, \mathbb{R})$ that G can be reduced to one of the compact subgroups $U(n)$, $O(m)$, or $O^+(m)$, respectively.

Let $GL(n, \mathbb{C})$ be the general linear group of invertible $n \times n$ matrices of complex numbers, acting as usual on the left of the complex vector space $\mathbb{C}^n$ of column vectors. One assigns $GL(n, \mathbb{C})$ the relative topology as an open subset of $\mathbb{C}^{n^2}$, so that it is a complex manifold of complex dimension n^2, hence real dimension $2n^2$, covered by a single coordinate neighborhood. The group product $GL(n, \mathbb{C}) \times GL(n, \mathbb{C}) \to GL(n, \mathbb{C})$ and group inverse $GL(n, \mathbb{C}) \xrightarrow{(\ )^{-1}} GL(n, \mathbb{C})$ are trivially smooth, so that $GL(n, \mathbb{C})$ is a complex Lie group.

The usual *hermitian inner product* $\mathbb{C}^n \times \mathbb{C}^n \xrightarrow{\langle\,,\,\rangle} \mathbb{C}$ is given by setting $\langle x, y \rangle = x_1\bar{y}_1 + \cdots + x_n\bar{y}_1 + \cdots + x_n\bar{y}_n$ for column vectors x and y with entries $x_1, \ldots, x_n$ and $\bar{y}_1, \ldots, \bar{y}_n$, where $\bar{y}_1, \ldots, \bar{y}_n$ are complex conjugates of $y_1, \ldots, y_n$. The *adjoint* $A^* \in \operatorname{End} \mathbb{C}^n$ of any $A \in \operatorname{End} \mathbb{C}^n$ is defined by requiring that $\langle Ax, y \rangle = \langle x, A^*y \rangle$ for all $(x, y) \in \mathbb{C}^n \times \mathbb{C}^n$, and A is *self-adjoint* with respect to $\langle\,,\,\rangle$ if $A^* = A$. A self-adjoint element $A \in \operatorname{End} \mathbb{C}^n$ is *positive* if $\langle Ax, x \rangle \geqq 0$ for all $x \in \mathbb{C}^n$, with equality holding only for $x = 0$. It is clear that positive elements $A, B \in \operatorname{End} \mathbb{C}^n$ belong in the general linear group $GL\,(n, \mathbb{C}) \subset \operatorname{End}\, \mathbb{C}^n$ and that the sum of two positive elements $A,B \in GL(n, \mathbb{C})$ is a positive element $A + B \in GL(n, \mathbb{C})$.

For any endomorphism A of $\mathbb{C}^n$, a classical existence and uniqueness theorem for ordinary differential equations provides a unique map $\mathbb{R} \xrightarrow{Y} \operatorname{End} \mathbb{C}^n$ such that $(d/dt)Y = AY$, with prescribed initial value $Y(0) \in \operatorname{End} \mathbb{C}^n$; the uniqueness theorem also implies that $Y(s)Y(t) = Y(s + t)$ for any $s, t \in \mathbb{R}$. The solution Y of the initial value problem $(d/dt)Y = AY$ for $Y(0) = I \in GL(n, \mathbb{C})$ is the *exponental* of A, written $Y(t) = e^{At}$. Thus $e^{As}e^{At} = e^{A(s+t)}$ for all $s, t \in \mathbb{R}$, and since $e^{Ao} = I$ it follows that $e^{-At}e^{At} = I$, hence that $e^{At} \in GL(n, \mathbb{C})$ for all $t \in \mathbb{R}$.

If $B \in \operatorname{End} \mathbb{C}^n$ is self-adjoint, then the differential equations $(d/dt)Y = BY$ and $(d/dt)Y = B^*Y$ are one and the same, so that $e^{Bt} = e^{B^*t} = (e^{Bt})^*$ for all $t \in \mathbb{R}$. Since $e^{Bt} = e^{Bt/2}e^{Bt/2} = (e^{Bt/2})^*e^{Bt/2}$, it follows for any $x \in \mathbb{C}^n$ that $\langle e^{Bt}x, x \rangle = \langle e^{Bt/2}x, e^{Bt/2}x \rangle \geqq 0$, with equality if and only if $x = 0$. Hence $e^{Bt} \in GL(n, \mathbb{C})$ is positive for any self-adjoint $B \in \operatorname{End} \mathbb{C}^n$ and any $t \in \mathbb{C}$.

For any positive $A \in GL(n, \mathbb{C})$ and any $s \in [0, 1]$ the sum $I(1 - s) + As$ is positive, hence an element of $GL(n, \mathbb{C})$, and the *logarithm* of A is defined by setting

$$\ln A = (A - I) \int_{s=0}^{1} (I(1 - s) + As)^{-1}\, ds.$$

6.1 Lemma: *For any positive $A \in GL(n, \mathbb{C})$ one has $e^{\ln A} = A$.*

PROOF: Set $B = A - I$, so that $I + Bst$ is positive for any $s \in [0, 1]$ and $t \in [0, 1]$. Then $\ln(I + Bt) = Bt \int_{s=0}^{1} (I + Bst)^{-1}\, ds$, the value at $t = 1$ being $\ln A$. Since

$$t \frac{d}{dt} Bt(I + Bst)^{-1} + \frac{d}{ds}(I + Bst)^{-1} = 0,$$

it follows that

$$t \frac{d}{dt} \ln(I + Bt) = -\int_{s=0}^{1} \frac{d}{ds}(I + Bst)^{-1}\, ds = Bt(I + Bt)^{-1}$$

hence that

$$\frac{d}{dt} \ln(I + Bt) = B(I + Bt)^{-1};$$

consequently

$$\frac{d}{dt} e^{\ln(I + Bt)} = B(I + Bt)^{-1} e^{\ln(I + Bt)}$$

for all $t \in [0, 1]$, the value of $e^{\ln(I + Bt)}$ at $t = 0$ being I. Since $I + Bt$ is another solution Y of the initial value problem $(d/dt)Y = B(I + Bt)^{-1} Y$ for $Y(0) = I$, the uniqueness theorem for ordinary differential equations gives $e^{\ln(I + Bt)} = I + Bt$ for all $t \in [0, 1]$; in particular, for $t = 1$ one has $e^{\ln A} = A$, as claimed.

6.2 Lemma: *For any self-adjoint* $B \in \operatorname{End} \mathbb{C}^n$ *one has* $\ln e^B = B$.

PROOF: Compute

$$\begin{aligned} \frac{d}{dt} \ln e^{Bt} &= \frac{d}{dt} \int_{s=0}^{1} (e^{Bt} - I)(I(1 - s) + e^{Bt}s)^{-1}\, ds \\ &= Be^{Bt} \int_{s=0}^{1} (I(1 - s) + e^{Bt}s)^{-2}\, ds \\ &= Be^{Bt} \int_{s=0}^{1} \frac{d}{ds} s(I(1 - s) + e^{Bt}s)^{-1}\, ds \\ &= Be^{Bt}e^{-Bt} = B \qquad \text{and} \qquad \ln e^{Bo} = 0; \end{aligned}$$

another solution Y of the initial value problem $(d/dt)Y = B$ for $Y(0) = 0$ is given by setting $Y(t) = Bt$, so that the uniqueness theorem for ordinary differential equations gives $\ln e^{Bt} = Bt$ for all $t \in R$, including the case $t = 1$.

Lemmas 6.1 and 6.2 together imply that the exponential map is a diffeomorphism from the self-adjoint elements $B \in \operatorname{End} \mathbb{C}^n$ onto the positive elements $e^B \in GL(n, \mathbb{C})$; the logarithm is the inverse map.

The *square root* of any positive element of $GL(n, \mathbb{C})$ is given by setting $\sqrt{e^B} = e^{B/2}$, where $B \in \operatorname{End} \mathbb{C}^n$ is self-adjoint; it follows from Lemmas 6.1 and 6.2 that $\sqrt{e^B}\sqrt{e^B} = e^B$. For any $A \in GL(n, \mathbb{C})$ one has $\langle AA^*x, x\rangle = \langle A^*x, A^*x\rangle \geqq 0$ for any $x \in \mathbb{C}^n$, with equality only for $x = 0$, so that AA^* is positive. The square root $|A| = \sqrt{AA^*} \in GL(n, \mathbb{C})$ is the *modulus* of A. Clearly $|A|$ is a positive element $e^B \in GL(n, \mathbb{C})$, where $B = \ln AA^*/2$, by Lemmas 6.1 and 6.2.

An element $B \in GL(n, \mathbb{C})$ is *unitary* whenever $BB^* = I$; that is, $B^{-1} = B^*$. If B and C are both unitary, then $(BC)(BC)^* = BCC^*B^* = BB^* = I$, and also $B^{-1}(B^{-1})^* = B^{-1}B = I$, so that the unitary elements of $GL(n, \mathbb{C})$ form a subgroup $U(n) \subset GL(n, \mathbb{C})$, the *unitary subgroup*.

6.3 Proposition (Polar Decomposition of $GL(n, \mathbb{C})$): *Any element $A \in GL(n, \mathbb{C})$ is uniquely of the form e^BC, where e^B is positive and $C \in U(n)$.*

PROOF: The modulus $|A|$ is uniquely of the form e^B for the self-adjoint element $B = \ln AA^*/2$, so that $A = e^BC$ for $C = |A|^{-1}A$. Since $|A|^* = |A|$, one has $CC^* = |A|^{-1}AA^*|A|^{-1} = |A|^{-1}|A|^2|A|^{-1} = I$, so that C is unitary. For any other such decomposition e^FG *of* A the equality $e^BC = A = e^FG$ gives $e^{-F}e^B = GC^{-1}$, which is unitary, so that $e^{-F}e^{2B}e^{-F} = (e^{-F}e^B)(e^{-F}e^B)^* = I$. Hence $e^{2B} = e^{2F}$, so that $B = \frac{1}{2}\ln e^{2B} = \frac{1}{2}\ln e^{2F} = F$, which in turn implies $C = G$.

6.4 Corollary: *For any $n > 0$ there is a subspace $H \subset GL(n, \mathbb{C})$ diffeomorphic to $\mathbb{R}^{n^2}$ such that every element of $GL(n, \mathbb{C})$ is uniquely of the form hk for $h \in H$ and an element k in the unitary subgroup $U(n) \subset GL(n, \mathbb{C})$.*

PROOF: We have already observed that Lemmas 6.1 and 6.2 imply that the exponential map is a diffeomorphism from the self-adjoint elements $B \in \operatorname{End} \mathbb{C}^n$ to the positive elements $e^B \in GL(n, \mathbb{C})$, and the self-adjoint elements $B \in \operatorname{End} \mathbb{C}^n$ form a real vector space $\mathbb{R}^{n^2}$.

Since $GL(n, \mathbb{C})$ is of real dimension $2n^2$, it follows from Corollary 6.4 that the unitary group $U(n)$ is of real dimension n^2; it is not itself a complex manifold.

6.5 Corollary: *For any $n > 0$ the inclusion $U(n) \to GL(n, \mathbb{C})$ is a homotopy equivalence.*

PROOF: Let $GL(n, \mathbb{C}) \to U(n)$ project $e^BC \in GL(n, \mathbb{C})$ onto $C \in U(n)$. Then $U(n) \to GL(n, \mathbb{C}) \to U(n)$ is the identity, and the map $GL(n, \mathbb{C}) \times [0, 1] \to GL(n, \mathbb{C})$ taking (e^BC, t) into $e^{Bt}C$ is a homotopy from the identity $GL(n, \mathbb{C}) \to GL(n, \mathbb{C})$ to the composition $GL(n, \mathbb{C}) \to U(n) \to GL(n, \mathbb{C})$.

We now consider real analogs of the preceding results, replacing the hermitian inner product $\mathbb{C}^n \times \mathbb{C}^n \xrightarrow{\langle , \rangle} \mathbb{C}$ by the usual euclidean inner product $\mathbb{R}^m \times \mathbb{R}^m \xrightarrow{\langle , \rangle} \mathbb{R}$. The *adjoint* $A^* \in \operatorname{End} \mathbb{R}^m$ of any $A \in \operatorname{End} \mathbb{R}^m$ satisfies $\langle Ax, y\rangle = \langle x, A^*y\rangle$ for all $(x, y) \in \mathbb{R}^m \times \mathbb{R}^m$, and *self-adjoint* elements $B \in \operatorname{End} \mathbb{R}^m$ and *positive* elements $A \in GL(m, \mathbb{R})$ are defined exactly as in the complex case, as are the *exponential* $e^B \in GL(m, \mathbb{R})$ of any $B \in \operatorname{End} \mathbb{R}^m$ and the *logarithm* $\ln A \in \operatorname{End} \mathbb{R}^m$ of any positive $A \in GL(m, \mathbb{R})$. Real analogs of Lemmas 6.1 and 6.2 then imply that the exponential map is a diffeomorphism from the self-adjoint elements of $\operatorname{End} \mathbb{R}^m$ to the positive elements of $GL(m, \mathbb{R})$. The *orthogonal subgroup* $O(m) \subset GL(m, \mathbb{R})$ consists of those $A \in GL(m, \mathbb{R})$ such that $AA^* = I$; that is, $A^{-1} = A^*$.

6.6 Proposition (Polar Decomposition of $GL(m, \mathbb{R})$): *Any element $A \in GL(m, \mathbb{R})$ is uniquely of the form e^BC, where e^B is positive and $C \in O(m)$.*

PROOF: This is a real analog of Proposition 6.3.

6.7 Corollary: *For any $m > 0$ there is a subspace $H \subset GL(m, \mathbb{R})$ diffeomorphic to $\mathbb{R}^{m(m+1)/2}$ such that every element of $GL(m, \mathbb{R})$ is uniquely of the form hk for $h \in H$ and an element k in the orthogonal subgroup $O(m) \subset GL(m, \mathbb{R})$.*

PROOF: The real analogs of Lemmas 6.1 and 6.2 imply that the exponential map is a diffeomorphism from the self-adjoint elements $B \in \operatorname{End} \mathbb{R}^m$ to the positive elements $e^B \in GL(m, \mathbb{R})$, and the self-adjoint elements $B \in \operatorname{End} \mathbb{R}^m$ form a real vector space $\mathbb{R}^{m(m+1)/2}$.

6.8 Corollary: *For any $m > 0$ the inclusion $O(m) \to GL(m, \mathbb{R})$ is a homotopy equivalence.*

PROOF: This is an obvious real analog of Corollary 6.5.

Recall that $GL^+(m, \mathbb{R})$ is the subgroup of those elements in $GL(m, \mathbb{R})$ with positive determinants; the *rotation group* is the subgroup $O^+(m) = O(m) \cap GL^+(m, \mathbb{R}) \subset GL^+(m, \mathbb{R})$.

6.9 Proposition (Polar Decomposition of $GL^+(m, \mathbb{R})$): *Any element $A \in GL^+(m, \mathbb{R})$ is uniquely of the form e^BC, where e^B is positive and $C \in O^+(m)$.*

PROOF: This is an immediate corollary of Proposition 6.6.

6.10 Corollary: *For any $m > 0$ there is a subspace $H \subset GL^+(m, \mathbb{R})$ diffeomorphic to $\mathbb{R}^{m(m+1)/2}$ such that every element of $GL^+(m, \mathbb{R})$ is uniquely of the form hk for $h \in H$ and an element k in the rotation subgroup $O^+(m) \subset GL^+(m, \mathbb{R})$.*

PROOF: See Corollary 6.7.

6.11 Corollary: *For any $m > 0$ the inclusion $O^+(m) \to GL^+(m, \mathbb{R})$ is a homotopy equivalence.*

PROOF: See Corollaries 6.5 and 6.8.

In order to present the main result of this section, it is necessary to know that the subgroups $U(n) \subset GL(n, \mathbb{C})$, $O(m) \subset GL(m, \mathbb{R})$, and $O^+(m) \subset GL^+(m, \mathbb{R})$ are closed; in fact, they are even compact.

6.12 Proposition: *For any $n > 0$ and $m > 0$ the unitary group $U(n)$, the orthogonal group $O(m)$, and the rotation group $O^+(m)$ are compact.*

PROOF: For any matrix $(a_p^r) \in U(n)$ one has $a_p^1\bar{a}_q^1 + \cdots + a_p^n\bar{a}_q^n = \delta_{pq}$ for the Kronecker delta δ_{pq}, by definition of $U(n)$. For $p = q$ it follows that $|a_p^r|^2 \leqq 1$, so that $(a_p^r) \in \mathbb{C}^{n^2}$ lines in the compact subset $(D^2)^{n^2} \subset \mathbb{C}^{n^2}$, where $D^2 \subset \mathbb{C}$ is the closed unit disk. Since limits of points $(a_p^r) \in (D^2)^{n^2}$ satisfying the algebraic relations $a_p^1\bar{a}_q^1 + \cdots + a_p^n\bar{a}_q^n = \delta_{pq}$ themselves satisfy the same relations, $U(n)$ is a closed subset of the compact set $(D^2)^{n^2}$, hence compact. Analogous proofs apply to $O(m)$ and $O^+(m)$.

6.13 Theorem (Linear Reduction Theorem): *Let ξ be any fiber bundle over any base space $X \in \mathscr{B}$ whose structure group G is one of the linear groups* $GL(n, \mathbb{C})$, $GL(m, \mathbb{R})$, or $GL^+(m, \mathbb{R})$; *then G can be reduced to one of the compact subgroups $U(n) \subset GL(n, \mathbb{C})$, $O(m) \subset GL(m, \mathbb{R})$, or $O^+(m) \subset GL^+(m, \mathbb{R})$, respectively.*

PROOF: By Proposition 6.12 the given subgroups are compact, a fortiori closed, and it remains to apply the general reduction theorem (Theorem 5.10) to Corollaries 6.4, 6.7, or 6.10, respectively.

6.14 Corollary: *Let ξ be any fiber bundle over any base space $X \in \mathscr{B}$ whose structure group G is a linear Lie group with only finitely many connected components; specifically, G is a subgroup of $GL(m, \mathbb{R})$ for some $m > 0$. Then ξ can be regarded as a fiber bundle with a compact structure group.*

PROOF: One simply regards ξ as a fiber bundle with structure group $GL(m, \mathbb{R})$, to which Theorem 6.13 applies.

Corollary 6.14 is essentially a weak version of Proposition 5.12. However, its proof does not require the Iwasawa–Mal'cev decomposition theorem, and in any event most interesting Lie groups with at most finitely many connected components *are* linear. (The first example of a connected nonlinear Lie group appeared in Birkhoff [1]; the example appears as an exercise in Hochschild [1, p. 225]. A method for constructing other nonlinear Lie groups is given in J. F. Price [1, pp. 119–121, 156–157].)

Fiber bundles with structure groups $GL(n, \mathbb{C})$, $GL(m, \mathbb{R})$, $GL^+(m, \mathbb{R})$, or equivalently $U(n)$, $O(m)$, $O^+(m)$, will occupy the rest of the book. It is therefore reasonable to begin looking at the topologies of these groups. For the moment we merely count their connected components.

6.15 Lemma: *Let H be a connected closed subgroup of a topological group G such that the homogeneous space G/H is connected; then G is connected.*

PROOF: If G is covered by two nonvoid open sets U and V, then their images U' and V' in G/H are two nonvoid open sets covering G/H. Since G/H is connected, there is an $x \in G$ whose image $x' \in G/H$ lies in the intersection $U' \cap V'$. The inverse image W of the set $\{x'\} \subset G/H$ is trivially homeomorphic to H, hence connected, so that the two nonvoid sets $U \cap W$ and $V \cap W$, which cover W, have a nonvoid intersection; a fortiori $U \cap V$ is nonvoid.

6.16 Lemma: *The rotation group $O^+(m) \subset O(m)$ is connected for every $m > 0$.*

PROOF: We proceed by induction of m, observing that $O^+(1)$ consists of a single point. Regard $O^+(m)$ as a transformation group, acting via rotations of the $(m-1)$-sphere $S^{m-1} \subset \mathbb{R}^m$, and embed $O^+(m-1)$ in $O^+(m)$ as the subgroup leaving some fixed unit vector $e \in S^{m-1}$ invariant. For any other $f \in S^{m-1}$ there is at least one $g \in O^+(m)$ with $ge = f \in S^{m-1}$, and an easy verification shows for any $h \in O^+(m)$ that $he = f \in S^{m-1}$ if and only if h and g have the same image in $O^+(m)/O^+(m-1)$. Since the image S^{m-1} of the induced one-to-one map $O^+(m)/O^+(m-1) \to S^{m-1}$ is compact, the map is a homeomorphism. Since S^{m-1} is connected, so is $O^+(m)/O^+(m-1)$, and since $O^+(m-1)$ is connected by the inductive hypothesis, $O^+(m)$ is connected by Lemma 6.15, as required.

6.17 Lemma: *The orthogonal group $O(m)$ has two connected components for every $m > 0$.*

PROOF: Multiplication of the connected component $O^+(m) \subset O(m)$ by an element in $O(m)$ with determinant -1 provides a diffeomorphism from $O^+(m)$ to the complement $O(m) - O^+(m)$.

6.18 Lemma: *The unitary group $U(n)$ is connected for every $n > 0$.*

PROOF: The construction of Lemma 6.16 is easily modified, beginning with the circle $U(1) = S^1$, to yield a homeomorphism of the homogeneous space $U(n)/U(n-1)$ with the $(2n-1)$-sphere $S^{2n-1} \subset \mathbb{C}^n$; since $U(n-1)$ is connected by the inductive hypothesis, $U(n)$ is connected by Lemma 6.15.

6.19 Proposition: *For any $n > 0$ and any $m > 0$ the group $GL(n, \mathbb{C})$ has one connected component, and $GL^+(m, \mathbb{R})$ is one of the two connected components of $GL(m, \mathbb{R})$.*

PROOF: By Lemmas, 6.16–6.18, the statements are valid for the subgroups $U(n) \subset GL(n, \mathbb{C})$, $O(m) \subset GL(m, \mathbb{R})$, and $O^+(m) \subset GL^+(m, \mathbb{R})$, and by Corollaries 6.5, 6.8, and 6.11 the inclusions are homotopy equivalences.

7. The Leray–Hirsch Theorem

We henceforth consider only fib*er* bundles rather than fib*re* bundles, both of which appeared in Definition 1.9; that is, all base spaces, now and forever more, belong to the category $\mathscr{B}$ of Chapter I.

Suppose that $E \xrightarrow{\pi} X$ is a coordinate bundle for some $X \in \mathscr{B}$, representing a fiber bundle ξ over X. For any commutative ring Λ with unit let $H^*(X)$ and $H^*(E)$ be the singular cohomology rings $H^*(X; \Lambda)$ and $H^*(E; \Lambda)$, respectively, with coefficients in Λ. Since $H^*(-)$ is a contravariant functor, there is an induced ring homomorphism $H^*(X) \xrightarrow{\pi^*} H^*(E)$, and one can use π^* to regard $H^*(E)$ as a left $H^*(X)$-module, the product $\beta \cdot \alpha \in H^*(E)$ of $\alpha \in H^*(E)$ by a scalar $\beta \in H^*(X)$ being the cup product $\pi^*\beta \cup \alpha \in H^*(E)$.

Up to canonical isomorphisms the $H^*(X)$-module $H^*(E)$ depends only on the fiber bundle ξ, and not upon the particular coordinate bundle $E \xrightarrow{\pi} X$ chosen to represent ξ. For if $E' \xrightarrow{\pi'} X$ is any other coordinate bundle representing ξ there is a homeomorphism f such that

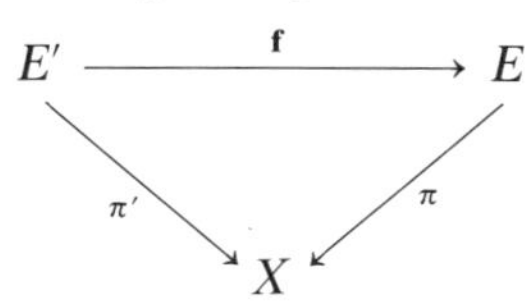

commutes, so that $H^*(E) \xrightarrow{f^*} H^*(E')$ is an isomorphism of $H^*(X)$-modules, as claimed.

There is even more latitude in the construction of the $H^*(X)$-module $H^*(E)$, up to canonical isomorphisms. For any map $X' \xrightarrow{g} X$ in the category $\mathscr{B}$ of base spaces, and for any coordinate bundle $E \xrightarrow{\pi} X$, there is a pullback diagram

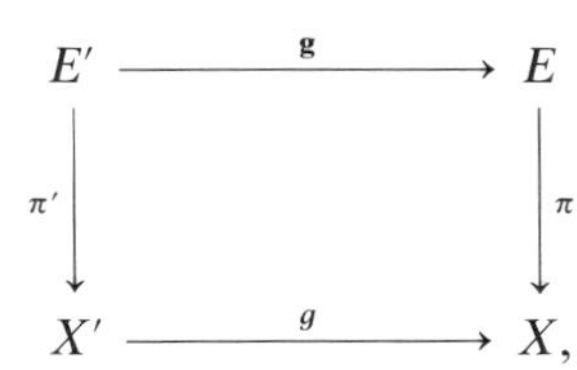

hence a commutative diagram

$$\begin{array}{ccc} H^*(X) & \xrightarrow{g^*} & H^*(X') \\ \Big\downarrow{\scriptstyle \pi^*} & & \Big\downarrow{\scriptstyle \pi'^*} \\ H^*(E) & \xrightarrow{\mathbf{g}^*} & H^*(E') \end{array}$$

of ring homomorphisms. One can regard $\mathbf{g}^*$ as a module homomorphism over the ring homomorphism g^*; that is, $\mathbf{g}^*(\beta \cdot \alpha) = \mathbf{g}^*(\pi^*\beta \cup \alpha) = \mathbf{g}^*\pi^*\beta \cup \mathbf{g}^*\alpha = \pi'^*g^*\beta \cup \mathbf{g}^*\alpha = (g^*\beta) \cdot (\mathbf{g}^*\alpha)$ for any $\alpha \in H^*(E)$ and $\beta \in H^*(X)$. If g is a homotopy equivalence, then $H^*(X) \xrightarrow{g^*} H^*(X')$ is a ring isomorphism: this is an immediate consequence of the homotopy axiom for singular cohomology.

7.1 Lemma: *If $X' \xrightarrow{g} X$ is a homotopy equivalence in $\mathscr{B}$, then for any coordinate bundle $E \xrightarrow{\pi} X$ the homomorphism $H^*(E) \xrightarrow{\mathbf{g}^*} H^*(E')$ is an isomorphism over the ring isomorphism $H^*(X) \xrightarrow{g^*} H^*(X')$.*

Proof: In the pullback diagram

$$\begin{array}{ccc} E' & \xrightarrow{\mathbf{g}} & E \\ \Big\downarrow{\scriptstyle \pi'} & & \Big\downarrow{\scriptstyle \pi} \\ X' & \xrightarrow{g} & X \end{array}$$

the map $\mathbf{g}$ is also a homotopy equivalence, by Proposition 4.10.

In the following result $H^*(\ )$ continues to denote singular cohomology with coefficients in a fixed commutative ring Λ with unit. For any coordinate bundle $E \xrightarrow{\pi} X$ over any $X \in \mathscr{B}$, and for any $x \in X$, let $E_x \xrightarrow{j_x} E$ denote the inclusion of the fiber E_x over x; then $H^*(E) \xrightarrow{j_x^*} H^*(E_x)$ is a homomorphism of modules over the ground ring Λ. An element $\alpha \in H^*(E)$ is *homogeneous* whenever $\alpha \in H^q(E)$ for some fixed index $q \in \mathbb{Z}$, in which case $j_x^*\alpha \in H^q(E_x)$ for each $x \in X$.

7.2 Theorem (Absolute Leray–Hirsch Theorem): *Let $E \xrightarrow{\pi} X$ be a coordinate bundle over any base space $X \in \mathscr{B}$, let $H^*(-)$ be singular cohomology $H^*(-; \Lambda)$ with coefficients in a commutative ground ring Λ with unit, and suppose that there are finitely many homogeneous elements $\alpha_1, \ldots, \alpha_r \in H^*(E)$ such that for each $x \in X$ the Λ-module $H^*(E_x)$ is free on the basis $\{j_x^*\alpha_1, \ldots, j_x^*\alpha_r\}$; then the $H^*(X)$-module $H^*(E)$ is free on the basis $\{\alpha_1, \ldots, \alpha_r\}$.*

PROOF: First suppose that $X' \xrightarrow{g} X$ is any homotopy equivalence, with pullback diagram

$$\begin{array}{ccc} E' & \xrightarrow{\mathbf{g}} & E \\ {\scriptstyle \pi'}\downarrow & & \downarrow{\scriptstyle \pi} \\ X' & \xrightarrow{g} & X \end{array}$$

as usual. For any $x' \in X'$ the map $\mathbf{g}$ induces a G-related isomorphism $E'_{x'} \xrightarrow{\mathbf{g}_{x'}} E_{g(x')}$ of fibers, hence a Λ-module isomorphism $H^*(E_{g(x')}) \xrightarrow{\mathbf{g}^*_{x'}} H^*(E'_{x'})$. Since $j^*_{x'}(\mathbf{g}^*\alpha) = \mathbf{g}^*_{x'} j^*_{g(x')}\alpha$ for any $\alpha \in H^*(E)$, it follows that each Λ-module $H^*(E'_{x'})$ is free on the basis $\{j^*_{x'}(\mathbf{g}^*\alpha_1), \ldots, j_{x'}(\mathbf{g}^*\alpha_r)\}$. Hence by Lemma 7.1 one can substitute $E' \xrightarrow{\pi'} X'$ for the given coordinate bundle $E \xrightarrow{\pi} X$ whenever $X' \xrightarrow{g} X$ is a homotopy equivalence. In particular, since $X \in \mathscr{B}$, there is a homotopy equivalence $X' \xrightarrow{g} X$ such that X' is metrizable and of finite type, by Definition I.1.2, so that one may as well assume throughout the remainder of the proof that X is itself of finite type, as in Definition I.1.1.

Suppose that $\alpha_1 \in H^{q_1}(E), \ldots, \alpha_r \in H^{q_r}(E)$ for integers $q_1, \ldots, q_r$, let q be any integer, and for each open set $U \subset X$ let $h^q(U)$ be the direct sum $H^{q-q_1}(U) \oplus \cdots \oplus H^{q-q_r}(U)$ of Λ-modules, where $H^{q-q_i}(U) = 0$ for $q - q_i < 0$. If R is the cohomology ring $H^*(X)$ of the base space X itself, then one can use the ring homomorphism $R = H^*(X) \to H^*(U)$ induced by the inclusion $U \subset X$ to regard the direct sum $\coprod_{q \in Z} h^q(U)$ as a graded R-module, via cup product in $H^*(U)$. It follows from the classical Mayer–Vietoris cohomology exact sequence that one thereby obtains a Mayer–Vietoris functor $\coprod_{q \in Z} h^q$ on the category $\mathcal{O}(X)$ of open sets $U \subset X$ to the category $\mathfrak{M}_R^{\oplus}$ of $\mathbb{Z}$-graded R-modules, as in Corollary I.9.4.

We now construct another Mayer–Vietoris functor $\coprod_{q \in Z} k^q$ from $\mathcal{O}(X)$ to $\mathfrak{M}_R^{\oplus}$, and we later construct a natural transformation $\coprod_{q \in \mathbb{Z}} h^q \xrightarrow{\theta} \coprod_{q \in \mathbb{Z}} k^q$ to which to apply Corollary I.9.4. For any open set $U \subset X$ let $E|U = \pi^{-1}(U)$ as usual, and for any integer q let $k^q(U)$ be the Λ-module $H^q(E|U)$. If $E|U \xrightarrow{\pi_U} U$ is the restriction to $E|U$ of the original coordinate bundle $E \xrightarrow{\pi} X$ (over the space X of finite type), one can then combine the homomorphism $R = H^*(X) \to H^*(U) \xrightarrow{\pi_U^*} H^*(E|U)$ with cup product in $H^*(E|U)$ to regard the direct sum $\coprod_{q \in \mathbb{Z}} k^q(U)$ $(= H^*(E|U))$ as a graded R-module. It follows from the classical Mayer–Vietoris cohomology sequence that one thereby obtains another Mayer–Vietoris functor, $\coprod_{q \in \mathbb{Z}} k^q$, from $\mathcal{O}(X)$ to $\mathfrak{M}_R^{\oplus}$, as promised.

Now recall that $h^q(U) = H^{q-q_1}(U) \oplus \cdots \oplus H^{q-q_r}(U)$ for each $q \in \mathbb{Z}$ and $U \in \mathcal{O}(X)$, and that the hypothesis of the theorem provides elements $\alpha_1 \in H^{q_1}(E), \ldots, \alpha_r \in H^{q_r}(E)$. Let $E|U \xrightarrow{j_U} E$ be the inclusion map, and recall that $E|U \xrightarrow{\pi_U} U$ denotes the restriction of $E \xrightarrow{\pi} X$. Then for each $\beta_i \in H^{q-q_i}(U)$ one has $\pi_U^*\beta_i \in H^{q-q_i}(E|U)$ and $j_U^*\alpha_i \in H^{q_i}(E|U)$, for $i = 1, \ldots, r$, so that there is a well-defined map $h^q(U) \to k^q(U)$ carrying each $\beta_1 \oplus \cdots \oplus \beta_r \in h^q(U)$ into the sum $\sum_{i=1}^r \pi_U^*\beta_i \cup j_U^*\alpha_i \in k^q(U)$. One easily verifies that the direct sum of such maps, over all $q \in \mathbb{Z}$, is an R-module homomorphism $\coprod_{q \in \mathbb{Z}} h^q(U) \xrightarrow{\theta_U} \coprod_{q \in \mathbb{Z}} k^q(U)$ and that the family $\{\theta_U\}_U$ of such homomorphisms is a natural transformation $\coprod_{q \in \mathbb{Z}} h^q \xrightarrow{\theta} \coprod_{q \in \mathbb{Z}} k^q$ of Mayer–Vietoris functors on $\mathcal{O}(X)$ to $\mathfrak{M}_R^{\oplus}$.

The final step in the proof of Theorem 7.2 is to show that θ is a natural equivalence, a fortiori that θ_X is an $H^*(X)$-module isomorphism. If $U \in \mathcal{O}(X)$ is contractible, then the restriction $E|U \xrightarrow{\pi_U} U$ represents a trivial bundle over U, by Proposition 3.5, so that there is a homeomorphism $\mathbf{f}$ such that

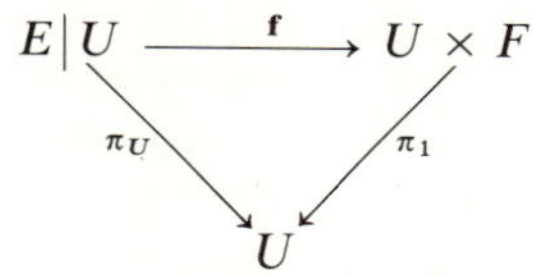

commutes, where F is the fiber and π_1 is the first projection. If U contracts to $x \in U$, then $\coprod_{q \in \mathbb{Z}} h^q(U) \xrightarrow{\theta_U} \coprod_{q \in \mathbb{Z}} k^q(U)$ is canonically equivalent to the homomorphism $H^*(\{x\}) \oplus \cdots \oplus H^*(\{x\}) \to H^*(E_x)$, which carries $\beta_1 \oplus \cdots \oplus \beta_r$ into $\sum_{i=1}^n \pi_1^* \beta_i \cup j_x^*\alpha_i \in H^*(E_x)$, where $H^*(\{x\}) = H^0(\{x\}) = \Lambda$. Since $H^*(E_x)$ is free on the basis $\{j_x^*\alpha_1, \ldots, j_x^*\alpha_r\}$, by hypothesis, it follows that θ_U is an isomorphism for contractible $U \in \mathcal{O}(X)$. Since X is of finite type, the Mayer–Vietoris technique applies in the form of Corollary I.9.4, with the consequence that the $H^*(X)$-module homomorphism $H^*(X) \oplus \cdots \oplus H^*(X) \xrightarrow{\theta_X} H^*(E)$ carrying any $\beta_1 \oplus \cdots \oplus \beta_r \in H^{q-q_1}(X) \oplus \cdots \oplus H^{q-q_r}(X)$ into $\pi^*\beta_1 \cup \alpha_1 + \cdots + \pi^*\beta_r \cup \alpha_r \in H^q(E)$ is an isomorphism; that is, $H^*(E)$ is a free $H^*(X)$-module with basis $\{\alpha_1, \ldots, \alpha_r\}$, as asserted.

In a certain sense Theorem 7.2 is a generalization of a Künneth theorem, as follows.

7.3 Corollary: *Let $X \in \mathscr{B}$ be any base space, and let F be any space whose cohomology $H^*(F)$ $(=H^*(F; \Lambda))$ is a free Λ-module on finitely many homogeneous generators; then for the projections π_1 and π_2 of $X \times F$ onto X and F, respectively, the map $H^*(X) \otimes_\Lambda H^*(F) \to H^*(X \times F)$ carrying $\beta \otimes \alpha$ into $\pi_1^*\beta \cup \pi_2^*\alpha$ is an $H^*(X)$-module isomorphism.*

PROOF: Apply Theorem 7.2 to the trivial coordinate bundle $X \times F \xrightarrow{\pi_1} X$.

The label attached to Theorem 7.2 suggests that there is a corresponding *relative Leray–Hirsch theorem.* There is such a theorem, its proof being virtually identical to that of Theorem 7.2 itself. We require only a very specialized relative Leray–Hirsch theorem, however, which will be formulated in a later chapter.

8. Remarks and Exercises

8.1 Remark: The first organized account of the material of this chapter appeared in Steenrod [4, Part I], in 1951. Steenrod profoundly influenced the development of fiber bundles, and Part I of his book, at least, remains surprisingly modern.

More recent introductory accounts of fiber bundles (in general) can be found in Borel and Hirzebruch [1, Chapter II], Auslander and MacKenzie [2, Chapter 9], Holmann [1], Husemoller [1, Part I], Liulevicius [2, Chapter I] and Liulevicius [3, Chapter I], in the beginning pages of Lees [1], Eells [1], Porter [2, Chapter 2], and Rohlin and Fuks [1, Chapter IV], for example.

Expository accounts of more general *fiber spaces* (or *fibrations*), considered later in these Remarks, can be found in Cartan [3, Exposés 6, 7, 8], Schwartz [1, Part I], Hu [2, Chapter III], May [1, pp. 1–30], Switzer [1, Chapter 4], and G. W. Whitehead [1, pp. 29–75], for example.

8.2 Remark: The first explicit definition of a coordinate bundle is due to Whitney [2], and it was further amplified in Whitney [4, 5, 6]. The importance of Whitney's brief original paper and its successors was quickly recognized. Related notions of *fiber spaces* (without structure groups) were soon independently introduced by Hurewicz and Steenrod [1] and by Eckmann [1], followed by Fox [3]; fiber spaces will be discussed further in Remarks 8.5–8.8. Simultaneously Ehresmann and Feldbau [1], and later Ehresmann [3, 4] considered alternatives to Whitney's construction, *with* structure groups. The general definition of coordinate bundles with arbitrary fibers F and appropriate structure groups G had become mathematical folklore before its first appearance as an incidental feature of Steenrod [2], and the equivalence relation providing fib*re* bundles (and fib*er*) bundles in the sense of the present chapter was equally well understood before its first publication in Steenrod [4].

8.3 Remark: The distinction between fib*re* bundles and fib*er* bundles is a convenient technical device whose introduction is justified by the appearance of the category $\mathscr{B}$ in several major results of this chapter: Proposition 4.7, the general reduction theorem (Theorem 5.10) and its application (Proposition 5.12), the linear reduction theorem (Theorem 6.13), and the Leray–Hirsch theorem (Theorem 7.2). These are some of the reasons for introducing the category $\mathscr{B}$ in the first place. The author apologizes to those readers who might prefer an interchange of the spellings "fib*re*" and "fib*er*."

8.4 Remark: One can achieve some of the results of the preceding remark by restricting the bundles themselves, rather than their base spaces. According to Dold [5] a fib*re* bundle ξ over a base space X (not necessarily in $\mathscr{B}$) is *numerable* if there is a covering $\{U_i | i \in I\}$ of X and a corresponding locally finite partition of unity $\{h_i | i \in I\}$ such that (i) the covering $\{h_i^{-1}(0,1] | i \in I\}$ refines $\{U_i | i \in I\}$ and (ii) each restriction $\xi | U_i$ is trivial. Clearly if X is metric (a fortiori paracompact) and of finite type (as in Definition I.1.1), then Proposition 3.5 implies that any fib*re* bundle over X is numerable. By Definition I.1.2 any space in $\mathscr{B}$ is homotopy equivalent to such a space X, so that one can use Lemma 4.5 to conclude that any fib*er* bundle whatsoever is automatically numerable. An equivalent definition of numerability occurs in Derwent [3].

8.5 Remark: A map $E \xrightarrow{\pi} X$ has the *covering homotopy property* with respect to a space Y if any commutative diagram

$$\begin{array}{ccc} Y \times \{0\} & \longrightarrow & E \\ \downarrow & & \downarrow \pi \\ Y \times [0,1] & \longrightarrow & X \end{array}$$

(with left-hand inclusion map) admits a *homotopy lifting* f for which

$$\begin{array}{ccc} Y \times \{0\} & \longrightarrow & E \\ \downarrow & \overset{f}{\nearrow} & \downarrow \pi \\ Y \times [0,1] & \longrightarrow & X \end{array}$$

is also commutative. Borsuk [1] introduced homotopy liftings and the covering homotopy property. One of the principal features of the fiber spaces $E \xrightarrow{\pi} X$ of Hurewicz and Steenrod [1] is the *covering homotopy theorem*, which asserts that such fiber spaces have the covering homotopy property with respect to any space Y whatsoever; the proof requires restrictions on the base space X'. Fox [4] showed conversely that any map $E \xrightarrow{\pi} X$

satisfying the conclusion of the covering homotopy theorem is a fiber space in the sense of Hurewicz and Steenrod; Fox's result requires further restrictions on X.

The covering homotopy theorem is also valid for the more general fiber spaces of Hu [1]. Accordingly, Hurewicz [1] (and later, Curtis [1]), defined even more general fiber spaces as "locally trivial" maps $E \xrightarrow{\pi} X$ satisfying the conclusion of the covering homotopy theorem; the base space X is required to be paracompact.

8.6 Remark: A *Hurewicz fibration* is any projection $E \xrightarrow{\pi} X$ whatsoever that satisfies the covering homotopy property with respect to any space Y. This definition ignores the "local triviality" condition of Hurewicz [1]; one can easily construct Hurewicz fibrations which are not "locally trivial," as in Husch [1], for example. However, there are easily defined conditions which *do* guarantee "local triviality" of Hurewicz fibrations, some of which can be found in Raymond [1]. Other recent information about Hurewicz fibrations is in Arnold [1] and in Jaber and Alkutibi [1].

8.7 Remark: Serre [1] considers projections $E \xrightarrow{\pi} X$ that satisfy the covering homotopy property just with respect to polyhedra Y; in fact, it suffices to restrict the polyhedra Y to be cubes $[0, 1]^n$ of arbitrary dimension n, as in Spanier [4, pp. 374–376], for example. R. Brown [1] gives an example of such a *Serre fibration* that is not a Hurewicz fibration.

8.8 Remark: One of the implications of Hu [1] is that every fib*re* bundle over a suitable base space satisfies the covering homotopy theorem; an independent proof of the same result appears in Steenrod [4, pp. 50–54]. Since then Huebsch [1, 2] and Derwent [1] provided simplified proofs of the covering homotopy theorem for fib*re* bundles over arbitrary paracompact base spaces; a special case (of interest for Chapter III) is in Szigeti [1].

8.9 Remark: We have deliberately avoided some of the most natural and beautiful fiber bundles since they will not be used explicitly in the sequel; however, they will appear in some later exercises. Any topological group G has an obvious effective action $G \times G \to G$ upon itself, for which the topology of G is trivially admissible: left-multiplication. A *principal G-bundle* ξ_P is any fiber bundle in the sense of Definition 1.9, in which a topological group G serves both as the structure group and as the fiber, the action $G \times G \to G$ being left-multiplication.

If $E \xrightarrow{\pi} X$ is any coordinate bundle with structure group G acting on a fiber F, and with transition functions $U_i \cap U_j \xrightarrow{\psi_i^j} G$ defined with respect to an open covering $\{U_i | i \in I\}$ of the base space X, then there is a *principal coordinate bundle* $E_P \xrightarrow{\pi_P} X$ with fiber G, obtained by letting the transition functions ψ_i^j act on G rather than F.

Principal G-bundles were first suggested by Ehresmann [3, 4], in terms of the fibrations of Ehresmann and Feldbau [1].

8.10 Exercise: Show that the preceding construction can be used to replace any fiber bundle ξ with structure group G by an *associated principal G-bundle* ξ_P, which is independent of the specific coordinate representations $E \xrightarrow{\pi} X$ and $E_P \xrightarrow{\pi_P} X$.

8.11 Exercise: Let $E_P \xrightarrow{\pi_P} X$ represent a principal G-bundle ξ_P, and observe that there is a well-defined action $E_P \times G \to E_P$ in which G acts on the *right* of E_P. Suppose that there is also an effective action $G \times F \to F$ of G on the *left* of a topological space F, for which the topology of G is admissible. Then the product $E_P \times F$ has an equivalence relation $\sim$ with $(e', f') \sim (e, f)$ whenever $(e', f') = (eg^{-1}, gf)$ for some $g \in G$, and there is a quotient $E = E_P \times F/\sim$ in the quotient topology; furthermore, the projection $E_P \xrightarrow{\pi_P} X$ induces an obvious projection $E \xrightarrow{\pi} X$. Show that $E \xrightarrow{\pi} X$ is a coordinate bundle with structure group G and fiber F.

8.12 Exercise: Let $E_P \xrightarrow{\pi_P} X$ represent a principal G-bundle ξ_P, and let ξ be the fiber bundle represented by the coordinate bundle $E \xrightarrow{\pi} X$ at the end of the preceding exercise. Show that ξ depends only on the given principal G-bundle ξ_P, independently of the coordinate bundle $E_P \xrightarrow{\pi_P} X$ representing ξ_P.

8.13 Exercise: Show that the constructions of Exercises 8.10 and 8.12 are inverse, in the obvious sense. That is, show that the composition of the construction of Exercises 8.10 and 8.12 yields the given bundle ξ of Exercise 8.10, and that the composition of the constructions of Exercises 8.12 and 8.10 yields the given principal G-bundle ξ_P of Exercise 8.12. (*Caution*: It is always assumed that the action of G on the fiber F is effective.)

8.14 Remark: To some extent Exercise 8.13 permits one to concentrate exclusively on principal G-bundles, and many books on differential geometry (and other subjects) use this consequence to advantage. Principal G-bundles are defined and studied in Bishop and Crittenden [1, pp. 41–45], Kobayashi and Nomizu [1, pp. 50–54], and Koszul [1, Chapter II], for example, in each case *before* the general definition of a fiber bundle is introduced.

However, the choice of G itself as a fiber is not always a virtue, for the same reason that the study of an effective transformation group $G \times F \to F$ is frequently easier than the study of the left-multiplication $G \times G \to G$. For example, the usual action $GL(n, \mathbb{R}) \times \mathbb{R}^n \to \mathbb{R}^n$ of the general linear group is much clearer, in terms of the geometry of $\mathbb{R}^n$, than is the left-multiplication $GL(n, \mathbb{R}) \times GL(n, \mathbb{R}) \to GL(n, \mathbb{R})$.

8.15 Exercise: Let G be a topological group and let $E \xrightarrow{\pi} X$ represent a principal G-bundle ξ with a section $X \xrightarrow{\sigma} E$. Show that ξ is the trivial principal G-bundle, represented by the first projection $X \times G \xrightarrow{pr_1} X$. (See Steenrod [4, p. 36].)

8.16 Exercise: Let G be a topological group and let $E \xrightarrow{\pi} X$ represent *any* principal G-bundle ξ. Show that the pullback $\pi^!\xi$ is the trivial principal G-bundle over E.

8.17 Exercise: According to the preceding exercise, $\pi^!\xi$ can be represented by a coordinate bundle $E \times G \xrightarrow{pr_1} E$, and the composition $E \times G \xrightarrow{pr_1} E \xrightarrow{\pi} X$ represents a principal $(G \times G)$-bundle η over X, not necessarily trivial. Observe that the pullback diagram

$$\begin{array}{ccc} E \times G & \xrightarrow{\pi \times \mathrm{id}} & X \times G \\ {\scriptstyle pr_1}\downarrow & & \downarrow{\scriptstyle pr_1} \\ E & \xrightarrow{\pi} & X \end{array}$$

for the trivial principal G-bundle over X also provides a principal $(G \times G)$-bundle η over X with the *same* representation $E \times G \xrightarrow{pr_1} E \xrightarrow{\pi} X$. Show that in general η and ξ are nevertheless *different* principal $(G \times G)$-bundles. (See Akin [2].)

8.18 Remark: One of the most useful features of principal fiber bundles is that they can be *classified* in the following sense. For any topological group G there exists a *universal principal G-bundle* γ_G over a *classifying space* $BG \in \mathscr{B}$ such that any principal G-bundle over any base space $X \in \mathscr{B}$ is a pullback $f^!\gamma_G$, with respect to a map $X \xrightarrow{f} BG$ that is uniquely defined up to homotopy. We shall establish an analogous *homotopy classification theorem* in the next chapter, which will suffice for all the remaining chapters.

Chern and Sun [1] provided the first homotopy classification theorem, for any classical linear Lie group G, applying to principal G-bundles over certain base spaces. Heller [1] described an algebraic setting which established the existence (but not the construction) of universal principal G-bundles for an arbitrary topological group G, base spaces being suitably restricted. The first general construction of a universal principal G-bundle for an arbitrary topological group G was given by Milnor [2], thereby providing a homotopy classification theorem for principal G-bundles over moderately restricted base spaces; the restrictions were somewhat relaxed by Lusztig [1], and further relaxed by Gel'fand and Fuks [1] and Segal [1]. An entirely different construction of universal principal G-bundles appears in Cartan [4], in a simplicial setting.

It was observed in tom Dieck [1] that Milnor's construction applies to *numerable* principal bundles over arbitrary base spaces. Since any fibre bundle over a base space $X \in \mathscr{B}$ is automatically numerable, by Remark 8.4, it follows that Milnor's construction classifies any principal bundle over any $X \in \mathscr{B}$.

8.19 Remark: If $EG \to BG$ represents a universal principal G-bundle ξ_G, then the total space EG is contractible. Conversely, if the total space EG of a principal G-bundle ξ, is contractible then ξ is universal. An early version of this characterization of universality appears in Steenrod [4, pp. 102–103], and the general case is in Dold [5]. Also see tom Dieck [1] and Liulevicius [2, p. 53] for portions of Dold's result.

8.20 Remark: One can pass from the category of topological groups to the larger category of associative H-spaces with unit in carrying out Milnor's construction. This was done by Dold and Lashof [1] (who had further applications in mind), and improvements to their work were added by Fuchs [1] and Gottlieb [1]. The Dold–Lashof construction led to an entirely different construction of universal bundles, by Milgram [1], and it was observed in Steenrod [8] (and by an earlier reviewer) that Milgram's construction could be reformulated to provide a satisfying identity for classifying spaces: $BG \times BH = B(G \times H)$. Steenrod also observed that in case G is an honest topological group, then the total space EG of Milgram's universal principal G-bundle is itself a group containing G as a subgroup, for which the homogeneous space EG/G is precisely the classifying space BG. Milgram and Steenrod *define* the universality of $EG \to BG$ via contractibility of EG, as in Remark 8.19, for any associative H-space G.

An excellent alternative version of the Milgram–Steenrod construction is given in May [1, pp. 31–54], and a sketch of the original version is given in Porter [2, pp. 132–165].

One can even pass beyond the category of associative H-spaces with unit to the category of *all* topological spaces, for which there exist the "classifying spaces" of Segal [1]. A very minor alteration of Segal's construction, applied to an honest topological group G, simply reproduces the original construction of Milnor [2]; this was observed by Accascina [1] and others.

Incidentally, if G is a topological group, and if BG and $B'G$ are classifying spaces in the category $\mathscr{B}$ of base spaces, each satisfying the homotopy classification theorem with respect to principal G-bundles over base spaces in $\mathscr{B}$, then BG and $B'G$ are trivially homotopy equivalent. However, if BG and $B'G$ apply to different categories of base spaces, then their homotopy equivalence is less clear. A very general homotopy equivalence theorem is developed in tom Dieck [2, 3], to assist anyone with a desire to stray outside the category $\mathscr{B}$.

8.21 Remark: In Thom [4] two coordinate bundles $E \xrightarrow{\pi} X$ and $E' \xrightarrow{\pi'} X$ over the same base space X are *fiber homotopy equivalent* (or of the same *fiber homotopy type*) whenever there is a homotopy equivalence of E and E' via maps that preserve projections onto X. The definition applies equally well to more general fibrations, and fiber homotopy equivalence has been studied extensively by Dold [1, 5], Fadell [1, 2, 3], and Stasheff [1, 2, 3], for example. Specifically, Fadell [3] shows that any Hurewicz fibration $E \xrightarrow{\pi} X$ over any $X \in \mathscr{W}_0$ (Definition I.3.11) is fiber homotopy equivalent to a coordinate bundle, for some fiber F and structure group G. According to Allaud [1] the same result is true for Hurewicz fibrations over any $X \in \mathscr{W}$ (Definition I.3.8), and it is probable that the same result is also true even more generally for Hurewicz fibrations over any base space $X \in \mathscr{B}$ whatsoever (Definition I.1.2).

The classification of fiber homotopy types of fiber bundles began in Dold [1] and continued in Curtis and Lashof [1], Dold and Lashof [1], Fuchs [1], Stasheff [1, 2, 3], and Siegel [1]. Although we shall not attempt to describe any classifications here, the reader is hereby warned that many results of later chapters concern properties of fiber bundles that are in fact merely properties of their fiber homotopy types. This phenomenon will be identified as it arises.

8.22 Remark: The Iwasawa–Mal'cev decomposition theorem (Theorem 5.11) is the culmination of many efforts by Chevalley and others to generalize the familiar polar decompositions of Propositions 6.3, 6.6, and 6.9. Theorem 5.11 was first published in Mal'cev [1], in 1945; however, Chevalley, who reviewed Mal'cev's paper, identified several gaps and obscurities. Iwasawa, who knew Chevalley's review, but who could not obtain a copy of Mal'cev's paper even as late as 1949, published a new proof of Theorem 5.11, which is formulated in Iwasawa [1, p. 530]. Iwasawa's proof can be found in Cartier [1, pp. 22–15, 22–16], in Mostow [1, pp. 47–48], and in Hochschild [1, pp. 180–186].

Since Mal'cev directly inspired Iwasawa's work, and since Mal'cev's efforts probably contain the bulk of a correct proof, Mal'cev certainly deserves at least hyphenated credit for the result. Hence Theorem 5.11 is the *Iwasawa–Mal'cev theorem.*

8.23 Remark: The Leray–Hirsch theorem (Theorem 7.2) was established by Leray [1, 2, 3, 4, 5] and G. Hirsch [1, 2, 3, 4, 5, 6], with mild restrictions on the base space. Leray's work coincides with the development of spectral sequences; Hirsch's constructions are somewhat more geometric. The clarification of spectral sequences by Serre [1, 2] paved the way for simplifications of Leray's technique, with results which can be found in Kudo [1, 2], Blanchard [1], E. H. Brown [1], Dold [4], Vastersavendts [1], and Dold [6], for example.

A particularly important special case of the Leray–Hirsch theorem (for sphere bundles) is implicit in much earlier work of Gysin [1]; explicit geometric proofs of this special case were given by Thom [1] and by Chern and Spanier [1]; the latter proof was extended by Spanier [1]. Another historically important special case (for bundles *over* spheres) appears in Wang [1].

Since the Leray–Hirsch theorem is independent of the structure group, it is clear that it applies equally well to more general fibrations; for example, a corresponding result for Serre fibrations is given in G. Hirsch [8].

The first application of a Mayer–Vietoris method to prove even a special case of the Leray–Hirsch theorem appears in Milnor [3, pp. 136–142], material which is repeated in Milnor and Stasheff [1, pp. 105–114]; the base space is severely restricted. The same technique is applied in Spanier [4, pp. 258–259] and in Husemoller [1, 1st ed., pp. 229–230] to prove the Leray–Hirsch theorem in general, but with the same restrictions on the base space. The Mayer–Vietoris proof of the Leray–Hirsch theorem over any base space $X \in \mathscr{B}$ first appeared in Osborn [6]; it is based upon the method of Connell [1, pp. 499–501], which Connell attributes to E. H. Brown.

8.24 Remark: According to Proposition 3.5, any fibre bundle over a contractible base is trivial. Hence, if $E \xrightarrow{\pi} X$ represents a fibre bundle, then the Ljusternik–Schnirelmann category of the total space E is clearly related to the Ljusternik–Schnirelmann category of the base space X. This observation leads to several results, some of which are in Švarc [1, 2, 3], Ginsburg [1], Varadarajan [1], Hardie [1], Palais [1], Ono [2], and Moran [1, 2].

CHAPTER III

Vector Bundles

0. Introduction

A *vector bundle* is any fiber bundle whose fiber and structure group are a vector space and its general linear group, respectively, the general linear group acting in the usual way on the vector space. Vector bundles arise naturally in differential topology. For example, if X is a smooth m-dimensional manifold, then there is a well-defined *tangent bundle* over X with fiber $\mathbb{R}^m$ and structure group $GL(m, \mathbb{R})$.

For notational convenience we first consider real vector bundles, described and illustrated in the first ten sections of the chapter. The main result is that there is a "universal" real vector bundle γ^m over the base space $G^m(\mathbb{R}^\infty) \in \mathscr{B}$, such that any vector bundle with fiber $\mathbb{R}^m$ and structure group $GL(m, \mathbb{R})$ over any base space $X \in \mathscr{B}$ is a pullback $f^!\gamma^m$ of γ^m along some map $X \xrightarrow{f} G^m(\mathbb{R}^\infty)$; the map f is uniquely defined up to homotopy.

Analogous descriptions and illustrations of complex vector bundles are provided in §11 merely by substituting the complex field $\mathbb{C}$ for the real field $\mathbb{R}$ throughout, with occasional complex conjugations; in particular, there is a "universal" complex vector bundle over the base space $G^m(\mathbb{C}^\infty) \in \mathscr{B}$. Some relations between real and complex vector bundles are explored in §12.

1. Real Vector Bundles

This section consists of definitions and the rationale for the name "vector bundle."

1.1 Definition: A *real vector bundle of rank m*, or simply a *real m-plane bundle*, is any fiber bundle whose fiber is the real m-dimensional vector space $\mathbb{R}^m$ and whose structure group is the general linear group $GL(m, \mathbb{R})$ of invertible $m \times m$ matrices, acting in the usual way on $\mathbb{R}^m$. A *real line bundle* is a real vector bundle of rank 1.

Throughout the bulk of this chapter we frequently omit the adjective "real"; complex vector bundles are first introduced in §11. Since the fiber $\mathbb{R}^m$ of an m-plane bundle ξ has the structure of a vector space, which is preserved under the action of the structure group $GL(m, \mathbb{R})$, one expects for any coordinate bundle $E \xrightarrow{\pi} X$ representing ξ that the fibers E_x are also m-dimensional vector spaces.

1.2 Proposition: *If a coordinate bundle $E \xrightarrow{\pi} X$ represents a real m-plane bundle ξ over a base space $X \in \mathscr{B}$, then the fibers E_x are real m-dimensional vector spaces in a natural way; furthermore, if $E' \xrightarrow{\pi'} X$ also represents ξ, then there is a natural vector space isomorphism $E_x \to E'_x$ for each $x \in X$.*

PROOF: As in §II.1, $E \xrightarrow{\pi} X$ is a family of fibers $F = \mathbb{R}^m$ with structure group $G = GL(m, \mathbb{R})$, and for each $x \in X$ there is a $GL(m, \mathbb{R})$-related family S_x of homeomorphisms $E_x \xrightarrow{h} \mathbb{R}^m$; that is, if $h \in S_x$, $h' \in S_x$, and $g \in GL(m, \mathbb{R})$, then the compositions $E_x \xrightarrow{h} \mathbb{R}^m \xrightarrow{g} \mathbb{R}^m$ and $\mathbb{R}^m \xrightarrow{h^{-1}} E_x \xrightarrow{h'} \mathbb{R}^m$ belong to S_x and $GL(m, \mathbb{R})$, respectively. For any $r_1, r_2 \in \mathbb{R}$ and $e_1, e_2 \in E_x$ one can use any $h \in S_x$ and the linear structure of $\mathbb{R}^m$ to obtain $h^{-1}(r_1 h(e_1) + r_2 h(e_2)) \in E_x$. For any other $h' \in S_x$ the composition $h'h^{-1} \in GL(m, \mathbb{R})$ preserves the linear structure of $\mathbb{R}^m$ so that $h'h^{-1}(r_1 h(e_1) + r_2 h(e_2)) = r_1 h'(e_1) + r_2 h'(e_2) \in \mathbb{R}^m$; that is, $h^{-1}(r_1 h(e_1) + r_2 h(e_2)) = h'^{-1}(r_1 h'(e_1) + r_2 h'(e_2)) \in E_x$. Thus $h^{-1}(r_1 h(e_1) + r_2 h(e_2))$ is an element $r_1 e_1 + r_2 e_2 \in E_x$ that is independent of the choice of $h \in S_x$, and one thereby obtains a linear structure in E_x; trivially $E_x \xrightarrow{h} \mathbb{R}^m$ is itself an isomorphism with respect to the linear structures of E_x and $\mathbb{R}^m$, which completes the proof of the first assertion. If $E' \xrightarrow{\pi'} X$ also represents ξ, then there is a commutative diagram

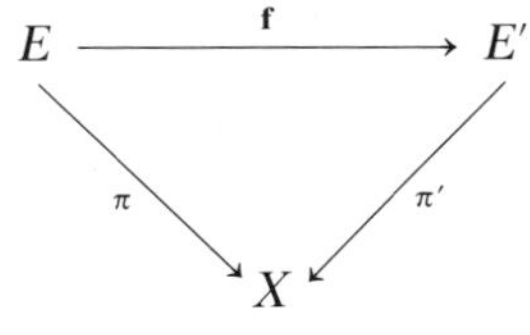

for a homeomorphism $\mathbf{f}$ that induces a $GL(m, \mathbb{R})$-related isomorphism $E_x \xrightarrow{\mathbf{f}_x} E'_x$ for each $x \in X$, as in §II.1; that is, if $h \in S_x$ and $h' \in S'_x$, then the composition $\mathbb{R}^m \xrightarrow{h^{-1}} E_x \xrightarrow{\mathbf{f}_x} E'_x \overset{h'}{\to} \mathbb{R}^m$ belongs to $GL(m, \mathbb{R})$. It follows from the preceding definition of the linear structures of E_x and E'_x that $E_x \xrightarrow{\mathbf{f}_x} E'_x$ is a vector space isomorphism, as required.

1.3 Proposition: *Let $E \overset{\pi}{\to} X$ and $\tilde{E} \overset{\tilde{\pi}}{\to} Y$ be coordinate bundles representing a real m-plane bundle ξ over $X \in \mathscr{B}$ and a real n-plane bundle η over $Y \in \mathscr{B}$, and suppose for some $p \geqq 0$ that there is a commutative diagram*

$$\begin{array}{ccc} E & \xrightarrow{\mathbf{g}} & \tilde{E} \\ {\scriptstyle \pi}\downarrow & & \downarrow{\scriptstyle \tilde{\pi}} \\ X & \xrightarrow{g} & Y \end{array}$$

such that each induced map $E_x \xrightarrow{\mathbf{g}_x} \tilde{E}_{g(x)}$ is linear of rank p; then if $E' \overset{\pi'}{\to} X$ and $\tilde{E}' \overset{\tilde{\pi}'}{\to} Y$ also represent ξ and η, there is a commutative diagram

$$\begin{array}{ccc} E' & \xrightarrow{\mathbf{g}'} & \tilde{E}' \\ {\scriptstyle \pi'}\downarrow & & \downarrow{\scriptstyle \tilde{\pi}'} \\ X & \xrightarrow{g} & Y \end{array}$$

such that each induced map $E'_x \xrightarrow{\mathbf{g}'_x} \tilde{E}'_{g(x)}$ is also linear of rank p.

PROOF: As in Proposition 1.2 there are commutative diagrams

$$\begin{array}{ccc} E & \xrightarrow{\mathbf{f}} & E' \\ & {\scriptstyle \pi}\searrow \quad \swarrow{\scriptstyle \pi'} & \\ & X & \end{array} \qquad \text{and} \qquad \begin{array}{ccc} \tilde{E} & \xrightarrow{\tilde{\mathbf{f}}} & \tilde{E}' \\ & {\scriptstyle \tilde{\pi}}\searrow \quad \swarrow{\scriptstyle \tilde{\pi}'} & \\ & Y & \end{array}$$

for homeomorphisms $\mathbf{f}$ and $\tilde{\mathbf{f}}$ such that each $E_x \xrightarrow{\mathbf{f}_x} E'_x$ and $\tilde{E}_y \xrightarrow{\tilde{\mathbf{f}}_y} \tilde{E}'_y$ is a vector space isomorphism. It suffices to set $\mathbf{g}' = \tilde{\mathbf{f}} \circ \mathbf{g} \circ \mathbf{f}^{-1}$ and to observe that each $\tilde{\mathbf{f}}_{g(x)} \circ \mathbf{g}_x \circ \mathbf{f}_x^{-1}$ is of rank p.

Proposition 1.3 leads unambiguously to morphisms in a category whose objects are real vector bundles.

1.4 Definition: Let $X \overset{g}{\to} Y$ be a map of base spaces and let ξ and η be vector bundles over X and Y, represented by coordinate bundles $E \overset{\pi}{\to} X$ and

$\tilde{E} \xrightarrow{\tilde{\pi}} Y$, respectively. A *bundle homomorphism* $\xi \to \eta$ *of rank* $p \geqq 0$ is represented by any commutative diagram

$$\begin{array}{ccc} E & \xrightarrow{\quad \mathbf{g} \quad} & \tilde{E} \\ {\scriptstyle \pi}\downarrow & & \downarrow{\scriptstyle \tilde{\pi}} \\ X & \xrightarrow{\quad g \quad} & Y \end{array}$$

such that each induced map $E_x \xrightarrow{\mathbf{g}_x} \tilde{E}_{g(x)}$ is linear of rank p.

2. Whitney Sums and Products

For any real vector spaces $\mathbb{R}^m$ and $\mathbb{R}^n$ the direct sum $\mathbb{R}^m \oplus \mathbb{R}^n$ and tensor product $\mathbb{R}^m \otimes \mathbb{R}^n$ are also vector spaces, $\mathbb{R}^{m+n}$ and $\mathbb{R}^{mn}$, and one expects corresponding constructions for vector bundles; these constructions and their properties are the goal of this section. Specifically, let ξ be an m-plane bundle over a base space $X \in \mathscr{B}$, let ξ' be an n-plane bundle over a base space $X' \in \mathscr{B}$, and recall from Proposition I.1.4 that $\mathscr{B}$ is closed with respect to products; we shall construct an $(n + m)$-plane bundle $\xi + \xi'$ and an nm-plane bundle $\xi \times \xi'$ over the base space $X \times X' \in \mathscr{B}$. In case $X = X'$ these constructions will lead to an $(m + n)$-plane bundle $\xi \oplus \xi'$ and an mn-plane bundle $\xi \otimes \xi'$ over X itself, and for any vector bundles ξ, ξ', ξ'' over the same $X \in \mathscr{B}$ the identities $(\xi \oplus \xi') \oplus \xi'' = \xi \oplus (\xi' \oplus \xi'')$, $\xi \oplus \xi' = \xi' \oplus \xi$, $(\xi \otimes \xi') \otimes \xi'' = \xi \otimes (\xi' \otimes \xi'')$, $\xi \otimes \xi' = \xi' \otimes \xi$, and $(\xi \oplus \xi') \otimes \xi'' = (\xi \otimes \xi'') \oplus (\xi' \otimes \xi'')$ are easily verified.

We start with addition of vector bundles. Given coordinate bundles $E \xrightarrow{\pi} X$ and $E' \xrightarrow{\pi'} X'$ that represent an m-plane bundle ξ over $X \in \mathscr{B}$ and an n-plane bundle ξ' over $X' \in \mathscr{B}$, respectively, the cartesian product $E \times E' \xrightarrow{\pi \times \pi'} X \times X'$ is a coordinate bundle with fiber $\mathbb{R}^m \times \mathbb{R}^n$ and structure group $GL(m, \mathbb{R}) \times GL(n, \mathbb{R})$. Since $X \times X' \in \mathscr{B}$ by Proposition I.1.4, $E \times E' \xrightarrow{\pi \times \pi'} X \times X'$ represents a fiber bundle η over $X \times X'$, in the sense of Definition II.1.9, which is trivially independent of the choices of coordinate bundles representing ξ and ξ'. The fiber $\mathbb{R}^m \times \mathbb{R}^n$ of η is the underlying topological space of the direct sum $\mathbb{R}^m \oplus \mathbb{R}^n$, so that there is a homeomorphism $\mathbb{R}^m \times \mathbb{R}^n \xrightarrow{\Phi} \mathbb{R}^m \oplus \mathbb{R}^n$ of topological spaces. There is also a group homomorphism $GL(m, \mathbb{R}) \times GL(n, \mathbb{R}) \xrightarrow{\Gamma} GL(m + n, \mathbb{R})$ with

$$\Gamma(A, B) = \begin{pmatrix} A & 0 \\ 0 & B \end{pmatrix}$$

for any invertible $m \times m$ matrix $A \in GL(m, \mathbb{R})$ and invertible $n \times n$ matrix $B \in GL(n, \mathbb{R})$. Clearly Γ and Φ form a morphism (Γ, Φ) of transformation groups in the sense of Definition II.2.6, which one can apply to η to induce a new bundle over $X \times X'$, as in Proposition II.2.7.

2.1 Definition: Let ξ be an m-plane bundle over $X \in \mathscr{B}$, let ξ' be an n-plane bundle over $X' \in \mathscr{B}$, and let η be the fiber bundle with fiber $\mathbb{R}^m \times \mathbb{R}^n$ and structure group $GL(m, \mathbb{R}) \times GL(n, \mathbb{R})$ over $X \times X' \in \mathscr{B}$, as in the preceding discussion. The *sum* $\xi + \xi'$ is the $(m+n)$-plane bundle over $X \times X'$ induced from η by the preceding morphism (Γ, Φ) of transformation groups, as in Definition II.2.8, in which $GL(m, \mathbb{R}) \times GL(n, \mathbb{R}) \xrightarrow{\Gamma} GL(m+m, \mathbb{R})$ and $\mathbb{R}^m \times \mathbb{R}^n \xrightarrow{\Phi} \mathbb{R}^m \oplus \mathbb{R}^n$.

One can obviously reduce the verbiage of Definition 2.1. As it stands, however, there is an entirely analogous construction of a product of the m-plane bundle ξ and the n-plane bundle ξ' beginning with the same fiber bundle η over $X \times X'$, but using a different morphism of transformation groups. Let $\mathbb{R}^m \times \mathbb{R}^n \xrightarrow{\Phi} \mathbb{R}^m \otimes \mathbb{R}^n$ $(= \mathbb{R}^{mn})$ be the map

$$\Phi((x^1, \ldots, x^m), (y^1, \ldots, y^n)) = (x^1y^1, \ldots, x^1y^n; \ldots; x^my^1, \ldots, x^my^n)$$

and let $GL(m, \mathbb{R}) \times GL(n, \mathbb{R}) \xrightarrow{\Gamma} GL(mn, \mathbb{R})$ be the group homomorphism with

$$\Gamma(A, B) = \begin{pmatrix} a_1^1B & \cdots & a_m^1B \\ \vdots & & \vdots \\ a_1^mB & \cdots & a_m^mB \end{pmatrix}$$

for any invertible $m \times m$ matrix

$$\begin{pmatrix} a_1^1 & \cdots & a_m^1 \\ \vdots & & \vdots \\ a_1^m & \cdots & a_m^m \end{pmatrix} = A \in GL(m, \mathbb{R})$$

and any invertible $n \times n$ matrix $B \in GL(n, \mathbb{R})$. Clearly the new pair, Γ and Φ, also forms a morphism (Γ, Φ) of transformation groups in the sense of Definition II.2.6.

2.2 Definition: Let ξ be an m-plane bundle over $X \in \mathscr{B}$, let ξ' be an n-plane bundle over $X' \in \mathscr{B}$, and let η be the fiber bundle with fiber $\mathbb{R}^m \times \mathbb{R}^n$ and structure group $GL(m, \mathbb{R}) \times GL(n, \mathbb{R})$ over $X \times X' \in \mathscr{B}$ as in the earlier discussion. The *product* $\xi \times \xi'$ is the mn-plane bundle over $X \times X'$ induced from η by the preceding morphism (Γ, Φ) of transformation groups, as in Definition II.2.8, where

$$GL(m, \mathbb{R}) \times GL(n, \mathbb{R}) \xrightarrow{\Gamma} GL(mn, \mathbb{R}) \qquad \text{and} \qquad \mathbb{R}^m \times \mathbb{R}^n \xrightarrow{\Phi} \mathbb{R}^m \otimes \mathbb{R}^n.$$

There is a noticeable asymmetry in Definition 2.2. What happens if one replaces the given morphism (Γ, Φ) of transformation groups by another obvious choice (Γ', Φ')? Specifically, let $\mathbb{R}^m \times \mathbb{R}^n \xrightarrow{\Phi'} \mathbb{R}^m \otimes \mathbb{R}^n \, (= \mathbb{R}^{mn})$ be the map

$$\Phi'((x^1, \ldots, x^m), (y^1, \ldots, y^n)) = (x^1y^1, \ldots, x^my^1; \ldots; x^1y^n, \ldots, x^my^n),$$

and let $GL(m, \mathbb{R}) \times GL(n, \mathbb{R}) \xrightarrow{\Gamma'} GL(mn, \mathbb{R})$ be the group homomorphism with

$$\Gamma'(A, B) = \begin{pmatrix} Ab_1^1 & \cdots & Ab_n^1 \\ \vdots & & \vdots \\ Ab_1^n & \cdots & Ab_n^n \end{pmatrix}$$

for any invertible $m \times m$ matrix $A \in GL(m, \mathbb{R})$ and any invertible $n \times n$ matrix

$$\begin{pmatrix} b_1^1 & \cdots & b_n^1 \\ \vdots & & \vdots \\ b_1^n & \cdots & b_n^n \end{pmatrix} = B \in GL(n, \mathbb{R}).$$

Clearly the pair (Γ', Φ') forms a morphism of transformation groups in the sense of Definition II.2.6.

2.3 Proposition *If one substitutes the preceding morphism* (Γ', Φ') *of transformation groups for the morphism* (Γ, Φ) *of transformation groups used in Definition* 2.2, *one obtains the same mn-plane bundle* $\xi \times \xi'$ *over* $X \times X'$.

PROOF: Observe that there are factorizations $\Gamma' = \Gamma'' \circ \Gamma$ and $\Phi' = \Phi'' \circ \Phi$ for a morphism (Γ'', Φ'') induced by a change of basis in $\mathbb{R}^{mn}$, where Γ'' is an automorphism of $GL(mn, \mathbb{R})$ and Φ'' itself belongs to $GL(mn, \mathbb{R})$. It follows from Proposition II.2.9 that (Γ'', Φ'') leaves $\xi \times \xi'$ unchanged, as required.

For any topological space X the *diagonal map* $X \xrightarrow{\Delta} X \times X$ carries each $x \in X$ into $(x, x) \in X \times X$. In the following definition we implicitly use the following special case of Proposition I.1.4: if $X \in \mathscr{B}$, then $X \times X \in \mathscr{B}$.

2.4 Definition: Let ξ and ξ' be vector bundles over the same base space $X \in \mathscr{B}$, and let $\xi + \xi'$ and $\xi \times \xi'$ be their sum and product over $X \times X$, as in Definitions 2.1 and 2.2, respectively. The *Whitney sum* $\xi \oplus \xi'$ over X and the *Whitney product* $\xi \otimes \xi'$ over X are the pullbacks $\Delta^!(\xi + \xi')$ and $\Delta^!(\xi \times \xi')$ along the diagonal map $X \xrightarrow{\Delta} X \times X$.

There are many equivalent ways of defining the Whitney sum $\xi \oplus \xi'$ and Whitney product $\xi \otimes \xi'$ of an m-plane bundle ξ and an n-plane bundle ξ' over the same $X \in \mathscr{B}$. For example, if $E \xrightarrow{\pi} X$ and $E' \xrightarrow{\pi'} X$ represent ξ and ξ', respectively, so that $E \times E' \xrightarrow{\pi \times \pi'} X \times X$ represents a fiber bundle η over

$X \times X \in \mathscr{B}$ with fiber $R^m \times R^n$ and structure group $GL(m, \mathbb{R}) \times GL(n, \mathbb{R})$, then there is a pullback $\zeta = \Delta^! \eta$ along the diagonal map $X \xrightarrow{\Delta} X \times X$, which can itself be used to define $\xi \oplus \xi'$ and $\xi \otimes \xi'$. Let $(\Gamma^{\oplus}, \Phi^{\oplus})$ denote the morphism of transformation groups used in Definition 2.1, with $GL(m, \mathbb{R}) \times GL(n, \mathbb{R}) \xrightarrow{\Gamma^{\oplus}} GL((m+n, \mathbb{R})$ and $\mathbb{R}^m \times \mathbb{R}^n \xrightarrow{\Phi^{\oplus}} \mathbb{R}^m \oplus \mathbb{R}^n$, and let $(\Gamma^{\otimes}, \Phi^{\otimes})$ denote the morphism of transformation groups used in Definition 2.2, with $(GL(m, \mathbb{R}) \times GL(n, \mathbb{R}) \xrightarrow{\Gamma^{\oplus}} GL(mn, \mathbb{R})$ and $\mathbb{R}^m \times \mathbb{R}^n \xrightarrow{\Phi^{\otimes}} \mathbb{R}^m \otimes \mathbb{R}^n$.

2.5 Proposition: *Given an m-plane bundle ξ and an n-plane bundle ξ' over $X \in \mathscr{B}$, let ζ be the preceding bundle $\Delta^! \eta$ over X, with fiber $\mathbb{R}^m \times \mathbb{R}^n$ and structure group $GL(m, \mathbb{R}) \times GL(n, \mathbb{R})$. Then $\xi \oplus \xi'$ is the bundle induced from ζ by the morphism $(\Gamma^{\oplus}, \Phi^{\oplus})$ of transformation groups, and $\xi \otimes \xi'$ is the bundle induced from ζ by the morphism $(\Gamma^{\otimes}, \Phi^{\otimes})$ of transformation groups.*

PROOF: According to the discussion following Definition II.2.8, the construction of fiber bundles induced by a given morphism of transformation groups commutes with the construction of pullbacks.

2.6 Proposition: *For any real vector bundles ξ, ξ', ξ'' over the same base space $X \in \mathscr{B}$ one has $(\xi \oplus \xi') \oplus \xi'' = \xi \oplus (\xi' \oplus \xi'')$, $\xi \oplus \xi' = \xi' \oplus \xi$, $(\xi \otimes \xi') \otimes \xi'' = \xi \otimes (\xi' \otimes \xi'')$, $\xi \otimes \xi' = \xi' \otimes \xi$,* and *$(\xi \oplus \xi') \otimes \xi'' = (\xi \otimes \xi'') \oplus (\xi' \otimes \xi'')$ over X.*

PROOF: These identities are all immediate consequences of Proposition II.2.9. We prove the commutative law $\xi \oplus \xi' = \xi' \oplus \xi$, for example. If ξ is an m-plane bundle and ξ' is an n-plane bundle, then $\xi \oplus \xi'$ is induced from the bundle ζ of Proposition 2.5, with fiber $\mathbb{R}^m \times \mathbb{R}^n$ and structure group $GL(m, \mathbb{R}) \times GL(n, \mathbb{R})$, by applying the morphism $(\Gamma^{\oplus}, \Phi^{\oplus})$ of transformation groups. Let $\mathbb{R}^{m+n} \xrightarrow{\Phi} \mathbb{R}^{m+n}$ be the change of basis $\Phi(x_1, \ldots, x_m, y_1, \ldots, y_n) = (y_1, \ldots, y_n, x_1, \ldots, x_m)$, inducing an inner automorphism $GL(m+n, \mathbb{R}) \xrightarrow{\Gamma} GL(m+n, \mathbb{R})$, and observe that Φ itself belongs to $GL(m+n, \mathbb{R})$. Then $\xi' \oplus \xi$ is induced from the same bundle ζ of Proposition 2.5 by applying the composition $(\Gamma \circ \Gamma^{\oplus}, \Phi \circ \Phi^{\oplus}) = (\Gamma, \Phi) \circ (\Gamma^{\oplus}, \Phi^{\oplus})$, and since (Γ, Φ) is a morphism of transformation groups satisfying the hypotheses of Proposition II.2.9, one has $\xi \oplus \xi' = \xi' \oplus \xi$ as claimed. (Incidentally, the commutative law $\xi \otimes \xi' = \xi' \otimes \xi$ for products was essentially proved by a similar method in Proposition 2.3, except for pulling back the resulting identity along the diagonal map $X \xrightarrow{\Delta} X \times X$.)

Proposition 2.6 asserts that the set of real vector bundles over a fixed $X \in \mathscr{B}$ satisfies most of the axioms required of a commutative ring, with respect to the operations $\oplus$ and $\otimes$. In fact, there are even neutral elements with respect to sums and products.

2.7 Definition: For any base space $X \in \mathscr{B}$ and any natural number $m \geqq 0$, the *trivial real m-plane bundle* ε^m over X is the vector bundle represented by the first projection $X \times \mathbb{R}^m \xrightarrow{\pi_1} X$.

Clearly the trivial 0-plane bundle ε^0 and the trivial line bundle ε^1 satisfy $\xi \oplus \varepsilon^0 = \xi$ and $\xi \otimes \varepsilon^1 = \xi$ for any real vector bundle ξ whatsoever over the same base space. Consequently, *except for additive inverses* the real vector bundles over a given base space $X \in \mathscr{B}$ form a commutative ring with unit, with respect to $\oplus$ and $\otimes$. There is an analogous situation for complex vector bundles, which will be further exploited in Volumes 2 and 3.

3. Riemannian Metrics

An inner product on a real vector space $\mathbb{R}^m$ is any symmetric bilinear map $\mathbb{R}^m \times \mathbb{R}^m \xrightarrow{\langle\,,\,\rangle} \mathbb{R}$ such that $\langle e, e \rangle > 0$ for all nonzero vectors $e \in \mathbb{R}^m$. We shall establish the existence and demonstrate the usefulness of corresponding "inner products" for any vector bundle over any base space $X \in \mathscr{B}$.

For any real m-plane bundle ξ over $X \in \mathscr{B}$ let ζ be the bundle $\Delta^! \eta$ over X with fiber $\mathbb{R}^m \times \mathbb{R}^m$ and structure group $GL(m, \mathbb{R}) \times GL(m, \mathbb{R})$, as in Proposition 2.5, for $\xi' = \xi$. If $\{U_i | i \in I\}$ is an open covering of X, with local trivializations $E|U_i \xrightarrow{\Psi_i} U_i \times \mathbb{R}^m$ of a coordinate bundle $E \xrightarrow{\pi} X$ representing ξ, then $\{U_i \times U_j | (i, j) \in I \times I\}$ is a corresponding covering of $X \times X$ for the bundle η. Since the subfamily $\{U_i \times U_i | i \in I\}$ covers the image of the diagonal $X \xrightarrow{\Delta} X \times X$, it follows that there are corresponding local trivializations $E''|U_i \xrightarrow{\Psi_i''} U_i \times (\mathbb{R}^m \times \mathbb{R}^m)$ of a corresponding representation $E'' \xrightarrow{\pi''} X$ of the bundle ζ. Specifically, transition functions $U_i \cap U_j \xrightarrow{\psi_i^j} GL(m, \mathbb{R})$ describing the coordinate bundle $E \xrightarrow{\pi} X$ provide transition functions $U_i \cap U_j \xrightarrow{(\psi_i^j, \psi_i^j)} GL(m, \mathbb{R}) \times GL(m, \mathbb{R})$ describing the coordinate bundle $E'' \xrightarrow{\pi''} X$.

If E_x is the fiber over $x \in X$ of the coordinate bundle $E \xrightarrow{\pi} X$, then $E_x \times E_x$ is the corresponding fiber E_x'' of the coordinate bundle $E'' \xrightarrow{\pi''} X$. Suppose that $E'' \xrightarrow{\langle\,,\,\rangle} \mathbb{R}$ is any map which restricts to an inner product $E_x \times E_x = E_x'' \xrightarrow{\langle\,,\,\rangle_x} \mathbb{R}$ over each $x \in X$. If $E' \xrightarrow{\pi'} X$ is any other representation of ξ, and if $E''' \xrightarrow{\pi'''} X$ is the corresponding representation of ζ, then the fiber $E_x''' = E_x' \times E_x'$ over each $x \in X$ is related to the fiber $E_x'' = E_x \times E_x$ by an element in the image of the diagonal $GL(m, \mathbb{R}) \to GL(m, \mathbb{R}) \times GL(m, \mathbb{R})$. Hence if $E''' \xrightarrow{\mathbf{f}} E''$ is an isomorphism of coordinate bundles representing ζ, and if $E''' \xrightarrow{\langle\,,\,\rangle'} \mathbb{R}$ is the composition $E''' \xrightarrow{\mathbf{f}} E'' \xrightarrow{\langle\,,\,\rangle} \mathbb{R}$, it follows that $\langle\,,\,\rangle'$ also restricts to an inner product $E'_x \times E'_x = E'''_x \xrightarrow{\langle\,,\,\rangle'_x} \mathbb{R}$ over each $x \in X$. Consequently the following definition is independent of the coordinate bundle $E \xrightarrow{\pi} X$ chosen to represent ξ.

3.1 Definition: Let ξ be a real m-plane bundle over $X \in \mathscr{B}$, represented by a coordinate bundle $E \xrightarrow{\pi} X$, and let $E'' \xrightarrow{\pi''} X$ represent the preceding bundle ζ, with fiber $\mathbb{R}^m \times \mathbb{R}^m$ and structure group $GL(m, \mathbb{R}) \times GL(m, \mathbb{R})$. A *riemannian metric* on ξ is any map $E'' \xrightarrow{\langle\,,\,\rangle} \mathbb{R}$ that restricts to an inner product $E_x \times E_x \to \mathbb{R}$ for each $x \in X$.

3.2 Lemma: *If $X \in \mathscr{B}$ is paracompact, then there is a riemannian metric on any real vector bundle ξ over X.*

Proof: Let $E \xrightarrow{\pi} X$ and $E'' \xrightarrow{\pi''} X$ represent ξ and ζ as before, and suppose that $\{U_i | i \in I\}$ is an open covering of X with local trivializations $E|U_i \xrightarrow{\Psi_i} U_i \times \mathbb{R}^m$. Since X is paracompact, one may as well suppose that $\{U_i | i \in I\}$ is locally finite, and that there is a partition of unity $\{h_i | i \in I\}$ subordinate to $\{U_i | i \in I\}$. For the corresponding local trivializations $E''|U_i \xrightarrow{\Psi''_i} U_i \times (\mathbb{R}^m \times \mathbb{R}^m)$, and for any fixed inner product $\mathbb{R}^m \times \mathbb{R}^m \xrightarrow{\langle\,,\,\rangle_i} \mathbb{R}$, there is a map $U_i \times (\mathbb{R}^m \times \mathbb{R}^m) \xrightarrow{h_i\langle\,,\,\rangle_i} \mathbb{R}$ carrying each $(x, (e_1, e_2)) \in U_i \times (\mathbb{R}^m \times \mathbb{R}^m)$ into the real number $h_i(x)\langle e_1, e_2\rangle_i$. Since h_i vanishes outside U_i, each composition $h_i\langle\,,\,\rangle_i \circ \Psi''_i$ extends to a map $E'' \xrightarrow{h_i\langle\,,\,\rangle_i \circ \Psi''_i} \mathbb{R}$ that vanishes outside U_i, and one trivially verifies that the well-defined sum $\sum_{i \in I} h_i\langle\,,\,\rangle_i \circ \Psi''_i$ is a riemannian metric.

3.3 Lemma: *Let $\langle\,,\,\rangle$ be a riemannian metric on an m-plane bundle ξ over any $X \in \mathscr{B}$, and let $X' \xrightarrow{f} X$ be any map in $\mathscr{B}$; then there is a riemannian metric $\langle\,,\,\rangle'$ on the pullback $f^!\xi$ over $X' \in \mathscr{B}$.*

Proof: If $E \xrightarrow{\pi} X$ and $E'' \xrightarrow{\pi''} X$ represent the bundles ξ and ζ, as before, and if $E' \xrightarrow{\pi'} X'$ and $E''' \xrightarrow{\pi'''} X'$ represent the pullbacks $f^!\xi$ and $f^!\zeta$, then there is a commutative diagram

$$\begin{array}{ccc} E''' & \xrightarrow{\quad \mathbf{f} \quad} & E'' \\ \Big\downarrow{\scriptstyle \pi'''} & & \Big\downarrow{\scriptstyle \pi''} \\ X' & \xrightarrow{\quad f \quad} & X, \end{array}$$

as in Lemma II.1.3, for which the composition $E''' \xrightarrow{\mathbf{f}} E'' \xrightarrow{\langle\,,\,\rangle} \mathbb{R}$ is the desired riemannian metric on $f^!\xi$.

3.4 Proposition: *Given any real vector bundle ξ over any base space $X \in \mathscr{B}$, there is a riemannian metric on ξ.*

Proof: By Definition I.1.2 the space X is homotopy equivalent to a metrizable space X' of finite type; by Lemma I.8.2 (Stone [1]) the metrizable space X' is paracompact. Thus there is a homotopy equivalence $X \xrightarrow{g} X'$ of X with a paracompact space X'. Let $X' \xrightarrow{h} X$ be a homotopy inverse of g. Since X' is paracompact, the pullback $h^!\xi$ over X' has a riemannian metric by Lemma 3.2, so that the pullback $g^!h^!\xi$ has a riemannian metric by Lemma 3.3; but $g^!h^!\xi = \xi$ by Lemma II.4.5.

We now set the scene for a useful application of Proposition 3.4. Recall from Definition 1.4 that if ξ and η are vector bundles over base spaces X and Y, represented by coordinate bundles $E \xrightarrow{\pi} X$ and $\tilde{E} \xrightarrow{\tilde{\pi}} Y$, respectively, then a vector bundle morphism $\xi \to \eta$ of rank $p \geqq 0$ is represented by any commutative diagram

$$\begin{array}{ccc} E & \xrightarrow{\mathrm{g}} & \tilde{E} \\ {\scriptstyle \pi}\downarrow & & \downarrow{\scriptstyle \tilde{\pi}} \\ X & \xrightarrow{g} & Y \end{array}$$

such that each restriction $E_x \xrightarrow{\mathrm{g}_x} E_{g(x)}$ is linear of rank p. In particular, if $E^1 \xrightarrow{\pi^1} X$ and $E \xrightarrow{\pi} X$ represent vector bundles ξ^1 and ξ over the same $X \in \mathscr{B}$, then a vector bundle morphism $\xi^1 \to \xi$ of rank p is represented by any commutative diagram

$$\begin{array}{ccc} E^1 & \xrightarrow{\mathrm{g}} & E \\ & {\scriptstyle \pi^1}\searrow \quad \swarrow{\scriptstyle \pi} & \\ & X & \end{array}$$

such that each restriction $E_x^1 \xrightarrow{\mathrm{g}_x} E_x$ is linear of rank p.

3.5 Definition: Let $\xi^1 \to \xi$ be a vector bundle morphism

$$\begin{array}{ccc} E^1 & \xrightarrow{\mathrm{g}} & E \\ & {\scriptstyle \pi^1}\searrow \quad \swarrow{\scriptstyle \pi} & \\ & X & \end{array}$$

such that each restriction $E_x^1 \xrightarrow{\mathrm{g}_x} E_x$ is a monomorphism; then ξ^1 is a *subbundle* of ξ.

For example, let ξ^1 and ξ^2 be real vector bundles of ranks p and q, respectively, over $X \in \mathscr{B}$, with Whitney sum $\xi^1 \oplus \xi^2$ of rank $p + q$. If $E^1 \xrightarrow{\pi^1} X$ and $E^2 \xrightarrow{\pi^2} X$ represent ξ^1 and ξ^2, then $E^1 \times E^2 \xrightarrow{\pi^1 \times \pi^2} X \times X$ represents a fiber bundle η with pullback $\zeta = \Delta^! \eta$ along the diagonal map $X \xrightarrow{\Delta} X \times X$ as before; the fiber and structure group of ζ are $\mathbb{R}^p \times \mathbb{R}^q$ and $GL(p, \mathbb{R}) \times GL(q, \mathbb{R})$. One obtains a coordinate bundle $E \xrightarrow{\pi} X$ representing $\xi^1 \oplus \xi^2$ by applying $(\Gamma^\oplus, \Phi^\oplus)$ to ζ, as in Proposition 2.5. The fibers E_x are direct sums $E_x^1 \oplus E_x^2$ for each $x \in X$, and there is a well-defined map $E^1 \xrightarrow{\mathbf{g}} E$ that identifies E_x^1 as a direct summand of E_x for each $x \in X$; continuity of $\mathbf{g}$ is an easy exercise. Thus ξ^1 is a subbundle of $\xi^1 \oplus \xi^2$, as expected.

3.6 Proposition: *Let ξ^1 be a subbundle of a vector bundle ξ over a base space $X \in \mathscr{B}$; then there is another subbundle ξ^2 of ξ such that $\xi = \xi^1 \oplus \xi^2$.*

PROOF: Suppose that ξ^1 is of rank p, represented by a coordinate bundle $E^1 \xrightarrow{\pi^1} X$, and that ξ is of rank m, represented by a coordinate bundle $E \xrightarrow{\pi} X$. By definition there is a commutative diagram

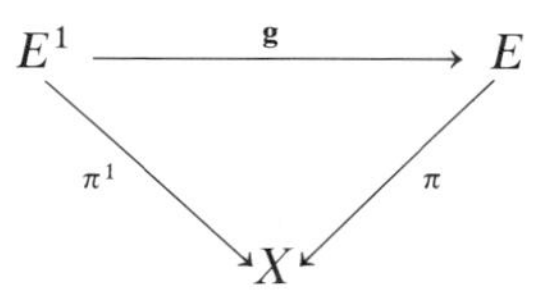

such that each restriction $E_x^1 \xrightarrow{\mathbf{g}_x} E_x$ is a monomorphism. Let $E'' \xrightarrow{\langle\,,\,\rangle} \mathbb{R}$ represent a riemannian metric on ξ, inducing an inner product $E_x \times E_x = E_x'' \xrightarrow{\langle\,,\,\rangle_x} \mathbb{R}$ over each $x \in X$. One identifies E_x^1 as a linear subspace of E_x via the monomorphism $\mathbf{g}_x$, and there is an orthogonal complement E_x^2 of E_x^1 with respect to $\langle\,,\,\rangle_x$, of rank $q = m - p$. The projection $E \xrightarrow{\pi} X$ restricts to a projection $E^2 \xrightarrow{\pi^2} X$ of the union $E^2 = \bigcup_{x \in X} E_x^2 \subset E$. We shall show that $E^2 \xrightarrow{\pi^2} X$ is a coordinate bundle representing a vector bundle ξ^2; trivially one then has $\xi^1 \oplus \xi^2 = \xi$.

Let $\{U_i \mid i \in I\}$ be any open covering of X for which there are local trivializations $E \mid U_i \xrightarrow{\Psi_i} U_i \times \mathbb{R}^m$, so that there are corresponding local trivializations $E'' \mid U_i \xrightarrow{\Psi_i''} U_i \times (\mathbb{R}^m \times \mathbb{R}^m)$. The riemannian metric $E'' \xrightarrow{\langle\,,\,\rangle} \mathbb{R}$ induces a riemannian metric $U_i \times (\mathbb{R}^m \times \mathbb{R}^m) \xrightarrow{\langle\,,\,\rangle_i} \mathbb{R}$ for each $i \in I$, which can be regarded as a nonsingular symmetric matrix whose entries are real-valued continuous functions on U_i. By the classical Gram–Schmidt process one can choose an orthonormal basis of $\mathbb{R}^m$, for each $i \in I$, such that the first p basis elements span the image under Ψ_i of $E^1 \mid U_i \subset E \mid U_i$, scalars being real-valued continuous functions on U_i; the remaining q basis elements, for each $i \in I$, span the image under Ψ_i of $E^2 \mid U_i \subset E \mid U_i$.

Suppose that $U_i \cap U_j$ is nonempty and that the preceding bases are chosen for each of the indices $i \in I$ and $j \in I$. The transition functions $U_i \cap U_j \xrightarrow{\psi_i^j} GL(m, \mathbb{R})$ for the coordinate bundle $E \xrightarrow{\pi} X$ can then be regarded as a family of linear transformations $\mathbb{R}^p \oplus \mathbb{R}^q \xrightarrow{\psi_i^j(x)} \mathbb{R}^p \oplus \mathbb{R}^q$ that depend continuously on $x \in U_i \cap U_j$ and preserve the summands $\mathbb{R}^p$ and $\mathbb{R}^q$. The restrictions $\mathbb{R}^q \xrightarrow{\psi_i^j(x)} \mathbb{R}^q$ then also depend continuously on $x \in U_i \cap U_j$, and they provide the transition functions $U_i \cap U_j \xrightarrow{\psi_i^{j,2}} GL(q, \mathbb{R})$ of a coordinate bundle representing a q-plane bundle ξ^2. Since the transition functions $\psi_i^{j,2}$ arise from local trivializations $E^2 | U_i \xrightarrow{\Psi_i^2} U_i \times \mathbb{R}^q$ of the projection $E^2 \xrightarrow{\pi^2} X$, it follows that $E^2 \xrightarrow{\pi^2} X$ is itself a coordinate bundle representing a q-plane bundle ξ^2 as required.

Proposition 3.6 is one of several classical applications of riemannian metrics. One can also use riemannian metrics to show, for example, that the structure group $GL(m, \mathbb{R})$ of any real m-plane bundle over any $X \in \mathscr{B}$ can be reduced to the orthogonal subgroup $O(m) \subset GL(m, \mathbb{R})$; however, since this is a special case of the linear reduction theorem (Theorem II.6.13), the details will be left as an exercise. (See Remark 13.17.)

4. Sections of Vector Bundles

For any base space $X \in \mathscr{B}$ let $C^0(X)$ be the ring of real-valued continuous functions $X \to \mathbb{R}$. In this section we develop a one-to-one correspondence between real vector bundles over X and (isomorphism classes of) certain $C^0(X)$-modules.

We begin with a special case of Definition II.5.2.

4.1 Definition: Let $E \xrightarrow{\pi} X$ represent a vector bundle over $X \in \mathscr{B}$; a *section* of $E \xrightarrow{\pi} X$ is any (continuous) map $X \xrightarrow{\sigma} E$ such that $X \xrightarrow{\sigma} E \xrightarrow{\pi} X$ is the identity on X.

According to Proposition 1.2, if $E \xrightarrow{\pi} X$ represents a vector bundle, then the fiber E_x over each $x \in X$ is a vector space in a natural way. We use this structure to turn the set of sections $X \xrightarrow{\sigma} E$ of such a bundle into a $C^0(X)$-module.

4.2 Lemma: *If $E \xrightarrow{\pi} X$ represents a vector bundle, then for any two sections $X \xrightarrow{\sigma} E$ and $X \xrightarrow{\sigma'} E$, and for any $f \in C^0(X)$, there are unique sections $X \xrightarrow{\sigma + \sigma'} E$ and $X \xrightarrow{f\sigma} E$ such that $(\sigma + \sigma')(x) = \sigma(x) + \sigma'(x)$ and $(f\sigma)(x) = f(x)\sigma(x)$ in E_x for each $x \in X$.*

Proof: Although uniqueness is automatic, one must verify that $\sigma + \sigma'$ and $f\sigma$ are continuous. If $E \xrightarrow{\pi} X$ represents a real m-plane bundle, for example, there is an open covering $\{U_i | i \in I\}$ of X with local trivializations $E|U_i \xrightarrow{\Psi_i} U_i \times \mathbb{R}^m$, and it suffices to verify for the restrictions $\sigma|U_i$, $\sigma'|U_i$, and $f|U_i$ that

$$U_i \xrightarrow{\Psi_i \circ (\sigma|U_i + \sigma'|U_i)} U_i \times \mathbb{R}^m$$

and

$$U_i \xrightarrow{\Psi_i \circ (f|U_i)(\sigma|U_i)} U_i \times \mathbb{R}^m$$

are continuous for each $i \in I$. However, $\Psi_i \circ \sigma|U_i$ carries $x \in U_i$ into $(x, (\sigma_1(x), \ldots, \sigma_m(x))) \in U_i \times \mathbb{R}^m$ for $\sigma_1, \ldots, \sigma_m \in C^0(U_i)$, and $\Psi_i \circ \sigma'|U_i$ carries $x \in U_i$ into $(x, (\sigma'_1(x), \ldots, \sigma'_m(x)) \in U_i \times \mathbb{R}^m$ for $\sigma'_1, \ldots, \sigma'_m \in C^0(U_i)$, and sums $\sigma_1 + \sigma'_1, \ldots, \sigma_m + \sigma'_m$ and products $f\sigma_1, \ldots, f\sigma_m$ of continuous functions $U_i \to \mathbb{R}$ are continuous.

4.3 Proposition: *If $E \xrightarrow{\pi} X$ represents a vector bundle, then the set $\mathscr{F}$ of sections $X \xrightarrow{\sigma} E$ forms a $C^0(X)$-module with respect to the addition and scalar multiplication of Lemma* 4.2.

Proof: Let the *zero-section* $X \xrightarrow{0} E$ carry each $x \in X$ into $0 \in E_x$, and for any section $X \xrightarrow{\sigma} E$ let $X \xrightarrow{-\sigma} E$ carry each $x \in X$ into $-\sigma(x) \in E_x$. One verifies as in Lemma 4.2 that 0 and $-\sigma$ are continuous, and that $\sigma + (-\sigma) = 0$. The remaining module axioms are trivially satisfied.

4.4 Definition: If $E \xrightarrow{\pi} X$ represents a vector bundle over X, then the $C^0(X)$-module $\mathscr{F}$ of Proposition 4.3 is the corresponding *module of sections.*

Clearly if $E \xrightarrow{\pi} X$ and $E' \xrightarrow{\pi'} X$ represent the same vector bundle ξ over $X \in \mathscr{B}$, then the isomorphism

$$\begin{array}{ccc} E & \longrightarrow & E' \\ & {\scriptstyle\pi}\searrow \quad \swarrow{\scriptstyle\pi} & \\ & X & \end{array}$$

of coordinate bundles induces a $C^0(X)$-module isomorphism $\mathscr{F} \to \mathscr{F}'$ of the corresponding modules of sections. Hence a vector bundle ξ over $X \in \mathscr{B}$ induces an isomorphism class of $C^0(X)$-modules. By abuse of language one frequently picks a particular coordinate bundle $E \xrightarrow{\pi} X$ representing ξ and calls the corresponding $C^0(X)$-module $\mathscr{F}$ *the* module of sections of ξ.

One feature of Lemma 4.2 and Proposition 4.3 deserves further comment. If $C^0(X; U_i)$ denotes the ring of restrictions to $U_i \subset X$ of continuous functions $X \to R$, then the local trivialization $E|U_i \xrightarrow{\Psi_i} U_i \times \mathbb{R}^m$ turns the $C^0(X)$-module $\mathscr{F}$ into the free $C^0(X; U_i)$-module $(C^0(X; U_i))^m$ of rank m. Furthermore, there is another way of describing both $C^0(X; U_i)$ and $(C^0(X; U_i))^m$. Let $I(U_i) \subset C^0(X)$ be the ideal of all continuous functions $X \xrightarrow{f} \mathbb{R}$ with $f(x) = 0$ for $x \in U_i$; then $C^0(X; U_i)$ is isomorphic to the quotient $C^0(X)/I(U_i)$, and $(C^0(X; U_i))^m$ is isomorphic to the quotient of the $C^0(X)$-module $(C^0(X))^m$ by the submodule $I(U_i)(C^0(X))^m$. More generally, a $C^0(X)$-module $\mathscr{F}$ is *locally free of rank m* if there is an open covering $\{U_i | i \in I\}$ of X such that each $C^0(X)/I(U_i)$-module $\mathscr{F}/I(U_i)\mathscr{F}$ is free of rank m. Thus if $E \xrightarrow{\pi} X$ represents an m-plane bundle, then the corresponding $C^0(X)$-module of sections is locally free of rank m.

4.5 Proposition: *For any $X \in \mathscr{B}$ let $\mathscr{F}$ be a locally free $C^0(X)$-module of rank m. Then $\mathscr{F}$ is the module of sections of a coordinate bundle $E \xrightarrow{\pi} X$ representing a real m-plane bundle ξ over X; furthermore, ξ is unique.*

PROOF: Let $\{U_i | i \in I\}$ be an open covering of X such that each $\mathscr{F}/I(U_i)\mathscr{F}$ is free of rank m, and fix a basis for each $\mathscr{F}/I(U_i)\mathscr{F}$; the choices of bases induce a $C^0(X; U_i)$-module isomorphism $\mathscr{F}/I(U_i)\mathscr{F} \to (C^0(X; U_i))^m$ for each $i \in I$. If $U_i \cap U_j$ is nonempty, then $\mathscr{F}/I(U_i \cap U_j)\mathscr{F}$ is also free of rank m, and it inherits a basis from $\mathscr{F}/I(U_i)\mathscr{F}$ and a basis from $\mathscr{F}/I(U_j)\mathscr{F}$; hence there is an invertible $m \times m$ matrix of elements in $C^0(X; U_i \cap U_j)$ relating the two bases. Such a matrix is precisely a continuous map $U_i \cap U_j \xrightarrow{\psi_i^j} GL(m, \mathbb{R})$, and since the basis of each $\mathscr{F}/I(U_i)\mathscr{F}$ is fixed, the conditions $\psi_j^k(x)\psi_i^j(x) = \psi_i^k(x)$ of Proposition II.2.5 are trivially satisfied; that is, the maps ψ_i^j are the transition functions of a coordinate bundle $E \xrightarrow{\pi} X$ representing a real m-plane bundle ξ over X. Uniqueness of ξ is clear.

4.6 Theorem: *For any $X \in \mathscr{B}$ and any $m \geqq 0$ there is a one-to-one correspondence between the real m-plane bundles ξ over X and the isomorphism classes of locally free $C^0(X)$-modules $\mathscr{F}$ of rank m.*

PROOF: Since any vector bundle is itself defined as an isomorphism class of coordinate bundles, this is an immediate consequence of Propositions 4.3 and 4.5.

If $\mathscr{F}$ is a module of sections of a real m-plane bundle over $X \in \mathscr{B}$, then the dual $C^0(X)$-module $\mathscr{F}^* = \mathrm{Hom}_{C^0(X)}(\mathscr{F}, C^0(X))$ is itself locally free of rank m. In light of Proposition 4.5 it is reasonable to identify the m-plane bundle that has $\mathscr{F}^*$ as its module of sections.

4.7 Proposition: *For any $X \in \mathscr{B}$ let $\mathscr{F}$ be a locally free $C^0(X)$-module of rank m, and let $\mathscr{F}^*$ be its dual $\mathrm{Hom}_{C^0(X)}(\mathscr{F}, C^0(X))$. Then $\mathscr{F}$ and $\mathscr{F}^*$ are modules of sections of coordinate bundles representing the same real m-plane bundle over X.*

PROOF: By Proposition 4.5, $\mathscr{F}$ is the $C^0(X)$-module of a coordinate bundle $E \xrightarrow{\pi} X$ representing a real m-plane bundle ξ over X, and by Proposition 3.4 there is a riemannian metric on ξ. As in Definition 3.1 the riemannian metric is defined in terms of inner products $E_x \times E_x \xrightarrow{\langle\,,\,\rangle_x} \mathbb{R}$ that depend continuously on $x \in X$, so that any two sections $\sigma \in \mathscr{F}$ and $\tau \in \mathscr{F}$ determine an element $\langle \sigma, \tau \rangle \in C^0(X)$ with value $\langle \sigma(x), \tau(x) \rangle_x$ on $x \in X$. Consequently there is a nondegenerate bilinear map $\mathscr{F} \times \mathscr{F} \xrightarrow{\langle\,,\,\rangle} C^0(X)$ over $C^0(X)$, which induces an isomorphism $\mathscr{F} \to \mathscr{F}^*$ carrying any $\tau \in \mathscr{F}$ into the element $\mathscr{F} \xrightarrow{\langle\,,\tau\rangle} C^0(X)$ of $\mathrm{Hom}_{C^0(X)}(\mathscr{F}, C^0(X))$. Thus $\mathscr{F}$ and $\mathscr{F}^*$ are isomorphic $C^0(X)$-modules, and the result follows from Theorem 4.6.

A *nowhere-vanishing* section of a vector bundle ξ over $X \in \mathscr{B}$ is any section $X \xrightarrow{\sigma} E$ of a coordinate bundle $E \xrightarrow{\pi} X$ representing ξ such that $\sigma(x) \neq 0 \in E_x$ for each $x \in X$.

4.8 Proposition: *If an m-plane bundle ξ has a nowhere-vanishing section, then $\xi = \varepsilon^1 \oplus \eta$ for the trivial line bundle ε^1 and some $(m-1)$-plane bundle η.*

PROOF: Let $E \xrightarrow{\pi} X$ represent ξ, let $X \xrightarrow{\sigma} E$ be the nowhere-vanishing section, let $E' \subset E$ be the set of all points of the form $r\sigma(x)$ for any real number $r \in \mathbb{R}$, and let $E' \xrightarrow{\pi'} X$ be the restriction to E' of $E \xrightarrow{\pi} X$. Since every $\sigma(x)$ is nonzero, it follows that every element of E' is *uniquely* of the form $r\sigma(x)$ for some real number $r \in \mathbb{R}$, so that there is a coordinate bundle isomorphism

$$\begin{array}{ccc} E' & \longrightarrow & X \times \mathbb{R}^1 \\ & {\scriptstyle \pi} \searrow \quad \swarrow {\scriptstyle \pi_1} & \\ & X. & \end{array}$$

Thus the inclusion $E' \to E$ identifies ε^1 as a subbundle of ξ, and Proposition 3.6 implies the desired result.

5. Smooth Vector Bundles

For any smooth manifold X the algebra $C^\infty(X)$ of *smooth* functions $X \to \mathbb{R}$ can be substituted in the preceding discussion for the algebra $C^0(X)$ of continuous functions $X \to \mathbb{R}$, locally free $C^\infty(X)$-modules being substituted for locally free $C^0(X)$-modules. A smooth analog of Theorem 4.6 then relates isomorphism classes of locally free $C^\infty(X)$-modules to corresponding *smooth* vector bundles. Some smooth vector bundles of primary interest in differential topology are defined via this relation in the next section.

We begin with a discussion which leads directly to the definition of smooth vector bundles.

The underlying topological space of the general linear group $GL(m, \mathbb{R})$ of invertible $m \times m$ matrices is an open subset of $\mathbb{R}^{m^2}$, in the relative topology, so that $GL(m, \mathbb{R})$ is itself a smooth manifold of dimension m^2: it can be covered by an atlas consisting of a single coordinate neighborhood. Furthermore, the group product $GL(m, \mathbb{R}) \times GL(m, \mathbb{R}) \to GL(m, \mathbb{R})$ and group inverse $GL(m, \mathbb{R}) \xrightarrow{(\;)^{-1}} GL(m, \mathbb{R})$ are trivially smooth maps, so that $GL(m, \mathbb{R})$ is a *Lie group*. Clearly the usual action $GL(m, \mathbb{R}) \times \mathbb{R}^m \to \mathbb{R}^m$ of $GL(m, \mathbb{R})$ on the vector space $\mathbb{R}^m$ is also smooth.

Recall that coordinate bundles $E \xrightarrow{\pi} X$ were described in general via local trivializations in Definition II.2.1. The local trivializations were homeomorphisms, transition functions were required to be continuous maps of open subsets of X into the given structure group G, and G was required to act continuously on the given fiber. According to the preceding paragraph, if X is a smooth manifold one can reasonably specialize Definition II.2.1 to coordinate bundles $E \xrightarrow{\pi} X$ with fiber $\mathbb{R}^m$ and structure group $GL(m, \mathbb{R})$, but with the additional requirements that local trivializations be diffeomorphisms and that transition functions be smooth maps of open subsets of X into $GL(m, \mathbb{R})$.

Specifically, suppose that $E \xrightarrow{\pi} X$ is a smooth map of smooth manifolds, and that there is an open covering $\{U_i | i \in I\}$ of X with local trivializations Ψ_i that are *diffeomorphisms* such that each

$$\begin{array}{ccc} E|U_i & \xrightarrow{\Psi_i} & U_i \times \mathbb{R}^m \\ & {\scriptstyle\pi}\searrow \quad \swarrow{\scriptstyle\pi_1} & \\ & U_i & \end{array}$$

commutes, where $E|U_i = \pi^{-1}(U_i)$, as always. If the intersection $U_i \cap U_j$ of two sets in the covering is nonvoid, then there is a composed diffeomorphism $\Psi_j \circ \Psi_i^{-1}$ such that

$$\begin{array}{ccc} (U_i \cap U_j) \times \mathbb{R}^m & \xrightarrow{\Psi_j \circ \Psi_i^{-1}} & (U_i \cap U_j) \times \mathbb{R}^m \\ & {\scriptstyle\pi_1}\searrow \quad \swarrow{\scriptstyle\pi_1} & \\ & U_i \cap U_j & \end{array}$$

commutes. One necessarily has $\Psi_j \circ \Psi_i^{-1}(x, f) = (x, \psi_i^j(x)f)$ for each $(x, f) \in (U_i \cap U_j) \times \mathbb{R}^m$, where $\mathbb{R}^m \xrightarrow{\psi_i^j(x)} \mathbb{R}^m$ a diffeomorphism for each $x \in U_i \cap U_j$.

5.1 Definition: If each of the preceding ψ_i^j's is a smooth map from $U_i \cap U_j$ to the group $GL(m, \mathbb{R})$, acting in the usual way on $\mathbb{R}^m$, then $E \xrightarrow{\pi} X$ is a *smooth coordinate bundle* over the smooth manifold X, with fiber $\mathbb{R}^m$ and structure group $GL(m, \mathbb{R})$.

The terminology of coordinate bundles in the sense of Definition II.2.1 applies equally well to the specialized smooth coordinate bundles of Definition 5.1. Specifically, the diffeomorphisms $E|U_i \xrightarrow{\Psi_i} U_i \times \mathbb{R}^m$ are *local trivializations* and the smooth maps $U_i \cap U_j \xrightarrow{\psi_i^j} GL(m, \mathbb{R})$ are *transition functions*, which automatically satisfy the conditions $\psi_j^k(x)\psi_i^j(x) = \psi_i^k(x)$ of Proposition II.2.5.

Two smooth coordinate bundles $E \xrightarrow{\pi} X$ and $E' \xrightarrow{\pi'} X$ with fiber $\mathbb{R}^m$ and structure group $GL(m, \mathbb{R})$ over the same smooth manifold X are *isomorphic* if there is a diffeomorphism **f** such that the diagram

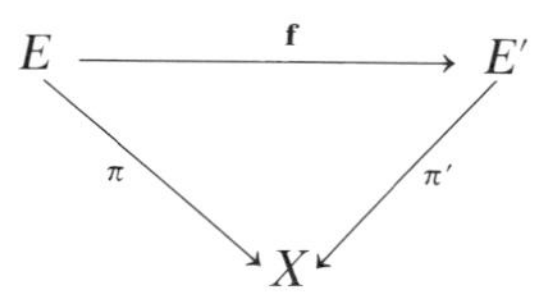

commutes and induces a vector space isomorphism $E_x \xrightarrow{\mathbf{f}_x} E'_x$ for each $x \in X$. Isomorphism of such smooth coordinate bundles is trivially an equivalence relation.

5.2 Definition: A *smooth real vector bundle of rank m* over a smooth manifold X is any isomorphism class of smooth coordinate bundles $E \xrightarrow{\pi} X$ with fiber $\mathbb{R}^m$ and structure group $GL(m, \mathbb{R})$, for the usual action $GL(m, \mathbb{R}) \times \mathbb{R}^m \to \mathbb{R}^m$.

Since smooth maps are also continuous, it follows that smooth vector bundles can also be considered as vector bundles in the sense of Definition 1.1. In fact, we shall learn in §9 that *any* vector bundle whatsoever over a smooth manifold can be represented by a smooth coordinate bundle, a result which suggests that Definition 5.2 is superfluous. However, if one is given a smooth vector bundle, one can study its smooth properties as such.

5.3 Proposition: *Let $E' \xrightarrow{\pi'} X'$ be a smooth coordinate bundle representing a smooth m-plane bundle ξ' over X'. Then the pullback $E \xrightarrow{\pi} X$ of $E' \xrightarrow{\pi'} X'$ along any smooth map $X \xrightarrow{g} X'$ is also a smooth coordinate bundle; a fortiori, the pullback $g^!\xi'$ is a smooth m-plane bundle over X.*

PROOF: Let $E' \overset{\pi'}{\to} X'$ be defined as a coordinate bundle with respect to an open covering $\{U_i' | i \in I\}$ of X'; by Definition 5.1 it is smooth if and only if the transition functions $U_i' \cap U_j' \xrightarrow{\psi_i'^j} GL(m, \mathbb{R})$ are smooth. If one sets $U_i = g^{-1}(U_i')$ for each $i \in I$, then according to Proposition II.2.4 $E \overset{\pi}{\to} X$ is a coordinate bundle with respect to the open covering $\{U_i | i \in I\}$ of X, its transition functions $U_i \cap U_j \xrightarrow{\psi_i^j} GL(m, \mathbb{R})$ being the compositions $\psi_i'^j \circ g | U_i \cap U_j$. Since g is smooth, the latter compositions are smooth, so that $E \overset{\pi}{\to} X$ is smooth by Definition 5.1.

5.4 Definition: Let $E \overset{\pi}{\to} X$ be a smooth coordinate bundle, representing a smooth vector bundle over a smooth manifold X. A *smooth section* of $E \overset{\pi}{\to} X$ is any smooth map $X \overset{\sigma}{\to} E$ such that $X \overset{\sigma}{\to} E \overset{\pi}{\to} X$ is the identity on X.

The set $\mathscr{F}$ of smooth sections $X \overset{\sigma}{\to} E$ of any smooth coordinate bundle $E \overset{\pi}{\to} X$ representing a vector bundle forms a $C^\infty(X)$-module, as in Proposition 4.3. Specifically, for any $\sigma \in \mathscr{F}$, $\sigma' \in \mathscr{F}$, and $f \in C^\infty(X)$ one defines $\sigma + \sigma'$ and $f\sigma$ by setting $(\sigma + \sigma')(x) = \sigma(x) + \sigma'(x) \in E_x$ and $(f\sigma)(x) = f(x)\sigma(x) \in E_x$ for each $x \in X$; the verifications that $\sigma + \sigma' \in \mathscr{F}$ and $f\sigma \in \mathscr{F}$ are analogous to corresponding verifications in Lemma 4.2, with $C^\infty(X)$ instead of $C^0(X)$.

5.5 Definition: If $E \overset{\pi}{\to} X$ represents a smooth vector bundle over a smooth manifold X, then the preceding $C^\infty(X)$-module $\mathscr{F}$ is the corresponding *module of smooth sections of* $E \overset{\pi}{\to} X$.

For any smooth vector bundle ξ over a smooth manifold X one frequently chooses a particular smooth coordinate bundle $E \overset{\pi}{\to} X$ to represent ξ, calling the corresponding $C^\infty(X)$-module $\mathscr{F}$ *the* module of smooth sections of ξ, by abuse of language.

A $C^\infty(X)$-module $\mathscr{F}$ is *locally free of rank* $m \geqq 0$ if there is an open covering $\{U_i | i \in I\}$ of X such that each $C^\infty(X)/I(U_i)$-module $\mathscr{F}/I(U_i)\mathscr{F}$ is free of rank m, where $I(U_i) \subset C^\infty(X)$ is the ideal of those functions that vanish on U_i.

5.6 Proposition: *For any smooth manifold X and any $m \geqq 0$ there is a one-to-one correspondence between the smooth real vector bundles ξ of rank m over X and the isomorphism classes of locally free $C^\infty(X)$-modules $\mathscr{F}$ of rank m.*

PROOF: This is the smooth version of Theorem 4.6, which one proves merely by substituting $C^\infty(X)$ for $C^0(X)$ throughout the proofs of Propositions 4.3 and 4.5.

We shall later illustrate Proposition 5.6 with specific examples.

An alternative characterization of riemannian metrics was used, without identifying it as such, in the proof of Proposition 4.7. We shall soon prove a smooth analog of Proposition 4.7, and for variety we begin with the smooth analog of the alternative characterization of riemannian metrics.

5.7 Definition: For any smooth manifold X, and for any locally free $C^\infty(X)$-module $\mathscr{F}$ of fixed rank $m \geqq 0$, a *smooth riemannian metric* is a symmetric bilinear map $\mathscr{F} \times \mathscr{F} \xrightarrow{\langle\,,\,\rangle} C^\infty(X)$ such that for each $\sigma \in \mathscr{F}$ and each $x \in X$ one has $\langle \sigma, \sigma \rangle(x) > 0$ whenever $\sigma(x) \neq 0$.

The proof of the following existence theorem is essentially a smooth version of the proof of Lemma 3.2.

5.8 Proposition: *For any smooth manifold X, and for any locally free $C^\infty(X)$-module $\mathscr{F}$ of fixed rank $m \geqq 0$, there is at least one smooth riemannian metric $\mathscr{F} \times \mathscr{F} \xrightarrow{\langle\,,\,\rangle} C^\infty(X)$.*

PROOF: Let $\{U_i \mid i \in I\}$ be an open covering of X such that each $C^\infty(X)/I(U_i)$-module $\mathscr{F}/I(U_i)\mathscr{F}$ is free, where $I(U_i) \subset C^\infty(X)$ is the ideal of smooth functions vanishing on U_i. Since X is paracompact, by Lemma I.6.1, one may as well suppose that $\{U_i \mid i \in I\}$ is locally finite; by Lemma I.6.2 there is a smooth partition of unity $\{h_i \mid i \in I\}$ subordinate to $\{U_i \mid i \in I\}$. For each $i \in I$ the $C^\infty(X)/I(U_i)$-module $\mathscr{F}/I(U_i)\mathscr{F}$ is isomorphic to $(C^\infty(X)/I(U_i))^m$, and one constructs a symmetric bilinear map

$$\mathscr{F}/I(U_i)\mathscr{F} \times \mathscr{F}/I(U_i)\mathscr{F} \xrightarrow{\langle\,,\,\rangle_i} C^\infty(X)/I(U_i)$$

via the usual $m \times m$ matrix with 1's down the diagonal and 0's elsewhere. Since h_i vanishes outside U_i, each $\langle\,,\,\rangle_i$ induces a symmetric bilinear map $\mathscr{F} \times \mathscr{F} \xrightarrow{h_i\langle\,,\,\rangle_i} C^\infty(X)$, for which $h_i(x)\langle \sigma, \sigma \rangle(x) > 0$ whenever both $h_i(x) > 0$ and $\sigma(x) \neq 0$. Since $\{h_i \mid i \in I\}$ is a partition of unity, the sum $\sum_{i \in I} h_i \langle\,,\,\rangle_i$ is then a well-defined map $\mathscr{F} \times \mathscr{F} \xrightarrow{\langle\,,\,\rangle} C^\infty(X)$ of the desired type.

5.9 Proposition: *For any smooth manifold X let $\mathscr{F}$ be a locally free $C^\infty(X)$-module of fixed rank $m \geqq 0$, and let $\mathscr{F}^*$ be the dual,* $\mathrm{Hom}_{C^\infty(X)}(\mathscr{F}, C^\infty(X))$. *Then $\mathscr{F}$ and $\mathscr{F}^*$ are modules of smooth sections of smooth coordinate bundles representing the same smooth m-plane bundle over X.*

PROOF: This is the smooth version of Proposition 4.7, and its proof is identical to that of Proposition 4.7, except that one uses the smooth riemannian metric $\mathscr{F} \times \mathscr{F} \xrightarrow{\langle\,,\,\rangle} C^\infty(X)$ of Proposition 5.8 in place of a merely continuous riemannian metric.

Let $E \xrightarrow{\pi} X$ be any smooth coordinate bundle that represents a vector bundle over a smooth manifold X, and let $X \xrightarrow{\sigma} E$ be a *continuous* section of $E \xrightarrow{\pi} X$. Can one approximate σ in some sense by a *smooth* section $X \xrightarrow{\tau} E$ of $E \xrightarrow{\pi} X$?

In order to pose a more precise question let $\mathscr{F} \times \mathscr{F} \xrightarrow{\langle\,,\rangle} C^\infty(X)$ $(\subset C^0(X))$ be a riemannian metric for the $C^\infty(X)$-module $\mathscr{F}$ of sections of $E \xrightarrow{\pi} X$, as in Proposition 5.8, and observe that $\mathscr{F}$ can also be regarded in the obvious way as a $C^0(X)$-module. The riemannian metric induces a *euclidean norm* $\mathscr{F} \xrightarrow{\|\,\|} C^0(X)$, which carries each section $\sigma \in \mathscr{F}$ into the positive square root $\sqrt{\langle\sigma,\sigma\rangle} \in C^0(X)$, assigning the classical euclidean norm $\|\sigma(x)\|_x = \langle\sigma(x), \sigma(x)\rangle_x$ to the vector $\sigma(x) \in E_x$ for each $x \in X$. Clearly $\|\sigma\| = 0 \in C^0(X)$ if and only if $\sigma \in \mathscr{F}$ is the zero-section; furthermore the Schwarz inequality $|\langle\sigma,\tau\rangle| \leqq \|\sigma\|\,\|\tau\|$ and resulting triangle inequality $\|\sigma + \tau\| \leqq \|\sigma\| + \|\tau\|$ are satisfied in $C^0(X)$ for any sections $\sigma, \tau \in \mathscr{F}$ because they are satisfied in each E_x. We shall show for any *continuous* section $X \xrightarrow{\sigma} E$ and any strictly positive function $\varepsilon \in C^0(X)$ that there is a *smooth* section $X \xrightarrow{\tau} E$ such that $\|\tau - \sigma\| < \varepsilon$; that is, we shall show that $\|\tau(x) - \sigma(x)\|_x < \varepsilon(x)$ for each $x \in X$.

5.10 Lemma: *Let $\bar{U} \times \mathbb{R}^m \xrightarrow{\pi_1} \bar{U}$ be the product m-plane bundle over a smooth compact manifold $\bar{U}$, let $\mathscr{F}$ be the $C^0(\bar{U})$-module of continuous sections $\bar{U} \to \bar{U} \times \mathbb{R}^m$, and let $\mathscr{F} \xrightarrow{\|\,\|} C^0(\bar{U})$ be any euclidean norm. For any strictly positive function $\varepsilon \in C^0(\bar{U})$ and any continuous section $\bar{U} \xrightarrow{\sigma} \bar{U} \times \mathbb{R}^m$ there is then a smooth section $\bar{U} \xrightarrow{\tau} \bar{U} \times \mathbb{R}^m$ such that $\|\tau - \sigma\| < \varepsilon$.*

PROOF: The given section σ can be regarded as an m-tuple $(\sigma^1, \ldots, \sigma^m)$ of elements $\sigma^i \in C^0(\bar{U})$, and the euclidean norm arises from a riemannian metric which can be regarded as a corresponding $m \times m$ symmetric matrix (g_{ij}) of functions $g_{ij} \in C^0(\bar{U})$ such that each $(g_{ij}(x))$ is positive-definite. In particular $g_{ii} > 0$ in $C^0(\bar{U})$, and one can set $M_i = \max_{x \in \bar{U}} g_{ii}(x)$; the maximum value exists because $\bar{U}$ is compact. Similarly one sets $\varepsilon_0 = \min_{x \in \bar{U}} \varepsilon(x) > 0$. The Stone–Weierstrass theorem (Theorem I.6.12) applies to the algebra $A(\bar{U}) = C^\infty(\bar{U})$, so that for each $\sigma^i \in C^0(\bar{U})$ there is a $\tau^i \in C^\infty(\bar{U})$ such that $|\tau^i(x) - \sigma^i(x)| < \varepsilon_0/mM_i$ for all $x \in \bar{U}$. Let $\bar{U} \xrightarrow{\sigma_i} \bar{U} \times R^m$ be the section corresponding to the m-tuple $(\tau^1, \ldots, \tau^{i-1}, \sigma^i, \ldots, \sigma^m)$, for $i = 1, \ldots, m$; in particular $\sigma = \sigma_0$ and σ_m is a smooth section τ, so that it remains only to show that $\|\tau - \sigma\| < \varepsilon$. However,

$$\langle\sigma_i - \sigma_{i-1}, \sigma_i - \sigma_{i-1}\rangle = g_{ii}(\tau^i - \sigma^i)^2,$$

so that

$$\|\sigma_i - \sigma_{i-1}\| = \sqrt{g_{ii}}|\tau^i - \sigma^i| < \varepsilon_0/m;$$

the triangle inequality then gives

$$\|\tau - \sigma\| = \|\sigma_m - \sigma_0\| \leqq \|\sigma_1 - \sigma_0\| + \cdots + \|\sigma_m - \sigma_{m-1}\| < \varepsilon_0 \leqq \varepsilon,$$

as required.

5.11 Proposition: *Let $E \xrightarrow{\pi} X$ be any smooth coordinate bundle representing an m-plane bundle over a smooth manifold X, let $\mathscr{F}$ be the $C^0(X)$-module of (continuous) sections $X \xrightarrow{\sigma} E$, and let $\mathscr{F} \xrightarrow{\|\,\|} C^0(X)$ be any euclidean norm. For any strictly positive function $\varepsilon \in C^0(X)$ and any continuous section $X \xrightarrow{\sigma} E$ there is then a smooth section $X \xrightarrow{\tau} E$ such that $\|\tau - \sigma\| < \varepsilon$.*

PROOF: For any covering of X by open sets U such that each restriction $E|U$ is trivial, Lemma I.6.1 provides a countable locally finite refinement $\{U'_n | n \in \mathbb{N}\}$ by open sets U'_n with compact closures $\bar{U}'_n$, and it follows that there is a smooth local trivialization $E|U'_n \xrightarrow{\Psi_n} U'_n \times \mathbb{R}^m$ for each $n \in \mathbb{N}$. By Proposition I.6.10 one can shrink $\{U'_n | n \in \mathbb{N}\}$ to a new open covering $\{U_n | n \in \mathbb{N}\}$ with compact closures $\bar{U}_n \subset U'_n$, and by Corollary I.6.11 there is a smooth partition of unity $\{h_n | n \in \mathbb{N}\}$ subordinate to $\{U_n | n \in \mathbb{N}\}$. For each $n \in \mathbb{N}$ the local trivialization Ψ_n and the given euclidean norm $\|\,\|$ induce a euclidean norm $\mathscr{F}|\bar{U}_n \xrightarrow{\|\,\|_n} C^0(\bar{U}_n)$ on the $C^0(\bar{U}_n)$-module $\mathscr{F}|\bar{U}_n$ of continuous sections $\bar{U}_n \xrightarrow{\Psi_n \circ \sigma|\bar{U}_n} \bar{U}_n \times \mathbb{R}^m$, with $\|\Psi_n \circ \sigma|\bar{U}_n\|_n(x) = \|\sigma|\bar{U}_n\|(x)$ for each restriction $\bar{U}_n \xrightarrow{\sigma|\bar{U}_n} E|\bar{U}_n$ and each $x \in \bar{U}_n$. Since $\bar{U}_n$ is compact, Lemma 5.10 provides a smooth section $\bar{U}_n \xrightarrow{\tau_n} E|\bar{U}_n$ such that $\|\tau_n - \sigma|\bar{U}_n\|(x) < \varepsilon(x)$, for each $n \in \mathbb{N}$ and each $x \in \bar{U}_n$. The partition of unity $\{h_n | n \in \mathbb{N}\}$ consists of *smooth* functions $X \xrightarrow{h_n} [0, 1]$ such that h_n vanishes outside of U_n for each $n \in \mathbb{N}$, so that each smooth section τ_n can be extended to a smooth section $X \xrightarrow{h_n\tau_n} E$ that vanishes outside U_n. Since $\{U_n | n \in \mathbb{N}\}$ is locally finite, the sum $\sum_{n \in \mathbb{N}} h_n\tau_n$ is a well-defined smooth section $X \xrightarrow{\tau} E$ of $E \xrightarrow{\pi} X$. Finally, for any $x \in X$ let N_x be the finite subset of those indices $n \in \mathbb{N}$ with $x \in U_n$, and observe that since $\sum_{n \in \mathbb{N}_x} h_n(x) = 1$, one has

$$\begin{aligned} \|\tau - \sigma\|(x) &= \left\| \sum_{n \in \mathbb{N}} h_n\tau_n - \sum_{n \in \mathbb{N}} h_n\sigma \right\|(x) \\ &= \sum_{n \in \mathbb{N}_x} h_n(x)\|\tau_n - \sigma|\bar{U}_n\|(x) < \sum_{n \in \mathbb{N}_x} h_n(x)\varepsilon(x) = \varepsilon(x); \end{aligned}$$

that is, $\|\tau - \sigma\| < \varepsilon$ for a smooth section $X \xrightarrow{\tau} E$, as claimed.

As in Proposition 4.8, a section $X \xrightarrow{\sigma} E$ of a coordinate bundle $E \xrightarrow{\pi} X$ representing a vector bundle over a base space $X \in \mathscr{B}$ is *nowhere-vanishing* whenever $\sigma(x) \neq 0 \in E_x$ for each $x \in X$.

5.12 Proposition: *Let $E \xrightarrow{\pi} X$ be any smooth coordinate bundle representing a vector bundle over a smooth manifold X. Then if there is a continuous nowhere-vanishing section $X \xrightarrow{\sigma} E$ there is also a smooth nowhere-vanishing section $X \xrightarrow{\tau} E$.*

Proof: Let $\mathscr{F} \xrightarrow{\|\,\|} C^0(X)$ be a euclidean norm as before. Since $X \xrightarrow{\sigma} E$ is nowhere-vanishing, one has $\|\sigma\|(x) > 0$ for every $x \in X$, so that there is a strictly positive function $\varepsilon = \frac{1}{2}\|\sigma\| \in C^0(X)$. By Proposition 5.11 there is then a smooth section $X \xrightarrow{\tau} E$ with $\|\tau - \sigma\| < \varepsilon = \frac{1}{2}\|\sigma\|$, for which the triangle inequality gives $\|\tau\| \geqq \|\sigma\| - \|\tau - \sigma\| > \frac{1}{2}\|\sigma\| > 0$, as required.

6. Vector Fields and Tangent Bundles

There is an obvious contravariant functor from the category of smooth manifolds to the category of real-valued function algebras, which we have already used: to any smooth manifold X one assigns the algebra $C^\infty(X)$ of smooth real-valued functions $X \xrightarrow{f} \mathbb{R}$, and to any smooth map $Y \to X$ one assigns the algebra homomorphism $C^\infty(X) \to C^\infty(Y)$ carrying $X \xrightarrow{f} \mathbb{R}$ into the composition $Y \to X \xrightarrow{f} \mathbb{R}$. There is also a contravariant functor that assigns a specific $C^\infty(X)$-module $\mathscr{E}(X)$ to each smooth manifold X, and that assigns a module homomorphism $\mathscr{E}(X) \to \mathscr{E}(Y)$ over the ring homomorphism $C^\infty(X) \to C^\infty(Y)$ to each smooth map $Y \to X$. We shall show that if X is m-dimensional, then $\mathscr{E}(X)$ is a locally free $C^\infty(X)$-module of rank m, so that Proposition 5.6 provides a canonical smooth m-plane bundle $\tau(X)$ over X, assigned to X itself. The properties of $\tau(X)$ characterize much of the differential topology of X itself.

The next few paragraphs are devoted to the construction of the dual $\mathscr{E}^*(X)$ of $\mathscr{E}(X)$, which will lead to $\mathscr{E}(X)$ itself, and to some elementary properties of the resulting bundle $\tau(X)$ over X.

6.1 Definition: For any smooth manifold X, a *smooth vector field on X* is any real linear map $C^\infty(X) \xrightarrow{L} C^\infty(X)$ such that $L(fg) = (Lf)g + f(Lg)$ for any $f \in C^\infty(X)$ and $g \in C^\infty(X)$.

For example, if X is the smooth manifold $\mathbb{R}^m$, and if $x^1, \ldots, x^m \in C^\infty(\mathbb{R}^m)$ are the projections $\mathbb{R}^m \to \mathbb{R}$ that provide the usual coordinate functions on

$\mathbb{R}^m$, then the partial derivations $\partial/\partial x^1, \ldots, \partial/\partial x^m$ are vector fields on $\mathbb{R}^m$. Similarly, if X is the smooth submanifold $(\mathbb{R}^m)^+ \subset \mathbb{R}^m$ with nonnegative mth coordinate, then $\partial/\partial x^1, \ldots, \partial/\partial x^m$ are vector fields on $(\mathbb{R}^m)^+$.

If L and M are vector fields on a given smooth manifold X, and if $g \in C^\infty(X)$, then there are vector fields $L + M$ and gL on X that carry any $f \in C^\infty(X)$ into $(Lf) + (Mf) \in C^\infty(X)$ and $g(Lf) \in C^\infty(X)$, respectively. It is clear that the set of all vector fields on X forms a $C^\infty(X)$-module with respect respect to these definitions. For example, if $g^1, \ldots, g^m \in C^\infty(\mathbb{R}^m)$, then $g^1\, \partial/\partial x^1 + \cdots + g^m\, \partial/\partial x^m$ is a vector field on $\mathbb{R}^m$.

6.2 Definition: For any smooth manifold X, the *$C^\infty(X)$-module $\mathscr{E}^*(X)$ of smooth vector fields on X* consists of smooth vector fields L, with respect to the preceding addition and scalar multiplication.

The goal of the next few lemmas is to show that if X is an m-dimensional smooth manifold, then $\mathscr{E}^*(X)$ is a locally free $C^\infty(X)$-module of rank m.

6.3 Lemma: *Vector fields annihilate constant functions.*

PROOF: Let $1 \in C^\infty(X)$ be the constant function on X with value $1 \in \mathbb{R}$. For any vector field $C^\infty(X) \xrightarrow{L} C^\infty(X)$ on X one then has $L(1) = L(1^2) = 2L(1)$; hence $L(1) = 0$, so that $L(c1) = cL(1) = 0$ for any $c \in \mathbb{R}$.

6.4 Lemma: *For any smooth manifold X and any open subset $U \subset X$, let $f \in C^\infty(X)$ and $g \in C^\infty(X)$ have common restrictions $f|U = g|U \in C^\infty(U)$. Then for any vector field $C^\infty(X) \xrightarrow{L} C^\infty(X)$ on X the images $Lf \in C^\infty(X)$ and $Lg \in C^\infty(X)$ have common restrictions $Lf|U = Lg|U \in C^\infty(U)$.*

PROOF: For any $x \in U$ Lemma I.6.2 provides an $h \in C^\infty(X)$ that vanishes outside U and satisfies $h(x) = 1$. Then $(f - g)h = 0$, so that $(Lf - Lg)h + (f - g)Lh = 0$. Evaluation at x yields $(Lf)(x) - (Lg)(x) = 0$, and since x is an arbitrary point of U, one has $Lf|U = Lg|U$.

Lemma 6.4 asserts that any vector field L is local in the sense that $Lf|U$ depends only on $f|U$; we shall rephrase the result with this in mind.

If U is any open set of a smooth manifold X, and if $I(U) \subset C^\infty(X)$ is the ideal of smooth functions $X \to \mathbb{R}$ that vanish on U, then the inclusion $U \subset X$ induces a homomorphism $C^\infty(X) \to C^\infty(U)$ whose image is the quotient algebra $C^\infty(X)/I(U)$. One defines *smooth vector fields* $C^\infty(X)/I(U) \xrightarrow{M} C^\infty(X)/I(U)$ by the obvious requirement that $M((f|U)(g|U)) = M(f|U)(g|U) + (f|U)M(g|U)$ for any $f|U$ and $g|U$ in $C^\infty(X)/I(U)$.

6.5 Lemma: *Let U be any open set of a smooth manifold X, and let $C^\infty(X) \to C^\infty(X)/I(U)$ be the restriction map carrying any $f \in C^\infty(X)$ into $f|U \in$*

$C^\infty(X)/I(U)$. Then any vector field $C^\infty(X) \xrightarrow{L} C^\infty(X)$ on X induces a unique vector field $L|U$ such that

$$\begin{array}{ccc} C^\infty(X) & \xrightarrow{\quad L \quad} & C^\infty(X) \\ \downarrow & & \downarrow \\ C^\infty(X)/I(U) & \xrightarrow{\quad L|U \quad} & C^\infty(X)/I(U) \end{array}$$

commutes.

PROOF: This is just a rephrasing of Lemma 6.4, as promised.

6.6 Lemma: *Let $x^1, \ldots, x^m \in C^\infty((\mathbb{R}^m)^+)$ be the usual coordinate functions on the submanifold $(\mathbb{R}^m)^+ \subset \mathbb{R}^m$ with nonnegative mth coordinate. Then for any vector field $C^\infty((\mathbb{R}^m)^+) \xrightarrow{L} C^\infty((\mathbb{R}^m)^+)$ one has $L = (Lx^1)\partial/\partial x^1 + \cdots + (Lx^m)\partial/\partial x^m$ for the functions $Lx^1, \ldots, Lx^m \in C^\infty((\mathbb{R}^m)^+)$.*

PROOF: The line segment joining any two points $x_0 \in (\mathbb{R}^m)^+$ and $x_1 \in (\mathbb{R}^m)^+$ lies entirely in $(\mathbb{R}^m)^+$, so that for any $f \in C^\infty((\mathbb{R}^m)^+)$ and each index $i = 1, \ldots, m$ one can define a function $f_i \in C^\infty((\mathbb{R}^m)^+ \times (\mathbb{R}^m)^+)$ by setting

$$f_i(x_0, x_1) = \int_{t=0}^{1} \frac{\partial f}{\partial x^i}(x_0 + t(x_1 - x_0))\,dt.$$

Then

$$f_i(x_0, x_0) = \frac{\partial f}{\partial x^i}(x_0)$$

and

$$\begin{aligned} f(x_1) - f(x_0) &= \int_{t=0}^{1} \frac{\partial f}{\partial t}(x_0 + t(x_1 - x_0))\,dt \\ &= \sum_{i=1}^{m} (x_1^i - x_0^i) f_i(x_0, x_1), \end{aligned}$$

for $x_0 = (x_0^1, \ldots, x_0^m)$ and $x_1 = (x_1^1, \ldots, x_1^m)$. For fixed $x_0 \in (\mathbb{R}^m)^+$ the latter identity becomes

$$f - f(x_0) = \sum_{i=1}^{m} (x^i - x_0^i) f_i(x_0, \cdot)$$

in $C^\infty(\mathbb{R}^m)^+)$, to which one applies L and Lemma 6.3 to conclude that

$$(Lf)(x_1) = \sum_{i=1}^{m} (Lx^i)(x_1) f_i(x_0, x_1) + \sum_{i=1}^{m} (x_1^i - x_0^i) L f_i(x_0, x_1)$$

for any $x_1 \in (\mathbb{R}^m)^+$. In particular, if x_0 is any point of $(\mathbb{R}^m)^+$ and $x_1 = x_0$, then

$$(Lf)(x_0) = \sum_{i=1}^{m} (Lx^i)(x_0) f_i(x_0, x_0) = \sum_{i=1}^{m} (Lx^i)(x_0) \frac{\partial f}{\partial x^i}(x_0),$$

so that $Lf = \sum_{i=1}^{m} (Lx^i)(\partial f/\partial x^i)$ for any $f \in C^\infty((\mathbb{R}^m)^+)$: that is, $L = \sum_{i=1}^{m} (Lx^i)(\partial/\partial x^i)$ as claimed.

If U and U' are open sets of a smooth manifold X such that the closure $\bar{U}$ of U satisfies $\bar{U} \subset U'$, then every restriction $f|U \in C^\infty(U)$ of an element $f \in C^\infty(X)$ is trivially the restriction $g|U \in C^\infty(U)$ of the element $g = f|U' \in C^\infty(U')$. The following lemma provides a useful converse assertion.

6.7 Lemma: *Let U and U' be open sets of a smooth manifold X such that $\bar{U} \subset U'$; then the inclusion $U' \subset X$ induces an isomorphism $C^\infty(X)/I(U) \to C^\infty(U')/I(U)$ of the restrictions to U of the algebras of smooth functions on X and U'.*

PROOF: The induced homomorphism $C^\infty(X)/I(U) \to C^\infty(U')/I(U)$ is trivially injective, and it remains to show that the restriction $g|U \in C^\infty(U)$ of any $g \in C^\infty(U')$ is also the restriction $f|U \in C^\infty(U)$ of some $f \in C^\infty(X)$. Since X is paracompact, by Lemma I.6.1, it is also normal. (See page 163 of Dugundji [2], for example.) Hence there is an open set $U'' \subset X$ such that $\bar{U} \subset U''$ and $\bar{U}'' \subset U'$. By Lemma I.6.2 there is a smooth partition of unity subordinate to the covering of X by the two open sets U'' and $X \backslash \bar{U}$; in particular, there is an $h \in C^\infty(X)$ such that $h|U = 1$ and $h(X \backslash \bar{U}'') = 0$. For any $g \in C^\infty(U')$ the product $(h|U')g \in C^\infty(U')$ vanishes on the open set $U' \backslash \bar{U}''$, and since $\bar{U}'' \subset U'$, the function $(h|U')g$ is the restriction $f|U'$ of a function $f \in C^\infty(X)$ that vanishes on $X \backslash \bar{U}''$. Since $h|U = 1$, it follows that $f|U = g|U$ as required.

6.8 Proposition: *For any smooth m-dimensional manifold X the $C^\infty(X)$-module $\mathscr{E}^*(X)$ of smooth vector fields on X is locally free of rank m.*

PROOF: By Proposition I.6.10 there is a countable open covering $\{U'_n | n \in \mathbb{N}\}$ of X and an atlas $\{\Phi_n | n \in \mathbb{N}\}$, which consists of homeomorphisms $U'_n \xrightarrow{\Phi_n} V'_n$ onto open sets $\Phi_n(U'_n) = V'_n \subset (\mathbb{R}^m)^+$; the homeomorphisms Φ_n are in fact diffeomorphisms, by definition of the smooth structure of X. By Proposition I.6.10 one can also shrink $\{U'_n | n \in \mathbb{N}\}$ to a new open covering $\{U_n | n \in \mathbb{N}\}$ with $\bar{U}_n \subset U'_n$ for each $n \in \mathbb{N}$, and the diffeomorphisms Φ_n provide new open subsets $\Phi_n(U_n) = V_n \subset (\mathbb{R}^m)^+$ with $\bar{V}_n \subset V'_n$ for each $n \in \mathbb{N}$.

We shall show for each $n \in \mathbb{N}$ that the $C^\infty(X)/I(U_n)$-module $\mathscr{E}^*(X)/I(U_n)\mathscr{E}^*(X)$ is free of rank m. Since $C^\infty(X)/I(U_n) \to C^\infty(U'_n)/I(U_n)$ is an isomorphism, by Lemma 6.7, and since vector fields are local, by Lemma 6.5, it suffices to show for each $n \in \mathbb{N}$ that the $C^\infty(U'_n)/I(U_n)$-module $\mathscr{E}^*(U'_n)/I(U_n)\mathscr{E}^*(U'_n)$ is free of rank m. Since the diffeomorphisms $U'_n \xrightarrow{\Phi_n} V'_n \subset (\mathbb{R}^m)^+$ restrict to diffeomorphisms $U_n \xrightarrow{\Phi_n} V_n \subset (\mathbb{R}^m)^+$, it therefore suffices to show for any open sets $V \subset (\mathbb{R}^m)^+$ and $V' \subset (\mathbb{R}^m)^+$ with $\bar{V} \subset V'$ that the $C^\infty(V')/I(V)$-module $\mathscr{E}^*(V')/I(V)\mathscr{E}^*(V')$ is free of rank m.

Since $C^\infty((\mathbb{R}^m)^+)/I(V) \to C^\infty(V')/I(V)$ is an isomorphism, by Lemma 6.7, and since vector fields are local, by Lemma 6.5, we are left with the task of showing for any open set $V \subset (\mathbb{R}^m)^+$ that the $C^\infty((\mathbb{R}^m)^+)/I(V)$-module $\mathscr{E}^*((\mathbb{R}^m)^+)/I(V)\mathscr{E}((\mathbb{R}^m)^+)$ is free of rank m. However, this is an immediate consequence of Lemma 6.6, which asserts that the $C^\infty((\mathbb{R}^m)^+)$-module $\mathscr{E}^*((\mathbb{R}^m)^+)$ is free of rank m, one basis being $\partial/\partial x^1, \ldots, \partial/\partial x^m$.

According to Proposition 5.6, if X is a smooth manifold, then any locally free $C^\infty(X)$-module $\mathscr{F}$ of rank m determines a unique real m-plane bundle ξ over X; specifically, the elements of $\mathscr{F}$ are the smooth sections of some coordinate bundle $E \xrightarrow{\pi} X$ representing ξ. According to Proposition 6.8, if X is of dimension m, then a particular locally free $C^\infty(X)$-module $\mathscr{E}^*(X)$ of rank m is available for such an application.

6.9 Definition: For any smooth manifold X the *tangent bundle* $\tau(X)$ is the smooth real vector bundle over X represented by the smooth coordinate bundle whose sections are vector fields $L \in \mathscr{E}^*(X)$.

We shall give a more direct construction of a particular smooth coordinate bundle $E \xrightarrow{\pi} X$ whose sections are vector fields, which will give a better understanding of the tangent bundle $\tau(X)$ it represents.

Let $\{U_i \mid i \in I\}$ be an open covering of the smooth m-dimensional manifold X by coordinate neighborhoods U_i, chosen as in Proposition 6.8, and let $\{\Phi_i \mid i \in I\}$ be a corresponding atlas of diffeomorphisms $U_i \xrightarrow{\Phi_i} \Phi_i(U_i)$, where each $\Phi_i(U_i)$ is open in $(\mathbb{R}^m)^+$. For any nonvoid intersection $U_i \cap U_j$ the composition $\Phi_i(U_i \cap U_j) \xrightarrow{\Phi_j \circ \Phi_i^{-1}} \Phi_j(U_i \cap U_j)$ is a diffeomorphism of open sets in $(\mathbb{R}^m)^+$ that can be described directly in terms of coordinate functions: for any $(x^1, \ldots, x^m) \in \Phi_i(U_i \cap U_j)$ one has

$$\Phi_j \circ \Phi_i^{-1}(x^1, \ldots, x^m) = (\bar{y}^1(x^1, \ldots, x^m), \ldots, \bar{y}^m(x^1, \ldots, x^m)) \in \Phi_j(U_i \cap U_j)$$

for uniquely defined smooth real-valued functions $\bar{y}^1, \ldots, \bar{y}^m$ on $\Phi_i(U_i \cap U_j) \subset (\mathbb{R}^m)^+$. The inverse diffeomorphism $\Phi_j(U_i \cap U_j) \xrightarrow{\Phi_i \circ \Phi_j^{-1}} \Phi_i(U_i \cap U_j)$ has a similar description: for any $(y^1, \ldots, y^m) \in \Phi_j(U_i \cap U_j)$ one has

$$\Phi_i \circ \Phi_j^{-1}(y^1, \ldots, y^m) = (\bar{x}^1(y^1, \ldots, y^m), \ldots, \bar{x}^m(y^1, \ldots, y^m)) \in \Phi_i(U_i \cap U_j)$$

for uniquely defined smooth real-valued functions $\bar{x}^1, \ldots, \bar{x}^m$ on $\Phi_j(U_i \cap U_j) \subset (\mathbb{R}^m)^+$. The partial derivatives $\partial\bar{y}^q/\partial x^p$ and $\partial\bar{x}^p/\partial y^q$ provide $m \times m$ *jacobian matrices*

$$\left(\frac{\partial \bar{y}^q}{\partial x^p}\circ \Phi_i\right) = \begin{pmatrix} \dfrac{\partial \bar{y}^1}{\partial x^1}\circ\Phi_i & \cdots & \dfrac{\partial \bar{y}^1}{\partial x^m}\circ\Phi_i \\ \vdots & & \vdots \\ \dfrac{\partial \bar{y}^m}{\partial x^1}\circ\Phi_i & \cdots & \dfrac{\partial \bar{y}^m}{\partial x^m}\circ\Phi_i \end{pmatrix}$$

and

$$\left(\frac{\partial \bar{x}^p}{\partial y^q}\circ \Phi_j\right) = \begin{pmatrix} \dfrac{\partial \bar{x}^1}{\partial y^1}\circ\Phi_j & \cdots & \dfrac{\partial \bar{x}^1}{\partial y^m}\circ\Phi_j \\ \vdots & & \vdots \\ \dfrac{\partial \bar{x}^m}{\partial y^1}\circ\Phi_j & \cdots & \dfrac{\partial \bar{x}^m}{\partial y^m}\circ\Phi_j \end{pmatrix}$$

of smooth functions on $U_i \cap U_j \subset X$, each being the inverse of the other. In particular, since $(\partial \bar{y}^q/\partial x^p \circ \Phi_i)$ is invertible, it can be regarded as a smooth map $U_i \cap U_j \xrightarrow{\psi_i^j} GL(m,\mathbb{R})$ carrying any $u \in U_i \cap U_j$ into $((\partial \bar{y}^q/\partial x^p)(\Phi_i(u))) \in GL(m,\mathbb{R})$. If $U_i \cap U_j \cap U_k$ is nonvoid, the chain rule implies for any $u \in U_i \cap U_j \cap U_k$ that $\psi_j^k(u)\cdot\psi_i^j(u) = \psi_i^k(u) \in GL(m,\mathbb{R})$, so that by Proposition II.2.5 the smooth maps $U_i \cap U_j \xrightarrow{\psi_i^j} GL(m,\mathbb{R})$ are the transition functions of a unique coordinate bundle with structure group $GL(m,\mathbb{R})$ and fiber $\mathbb{R}^m$. Since the structure group $GL(m,\mathbb{R})$ acts on the left, as always, we regard elements of $\mathbb{R}^m$ as column vectors.

6.10 Proposition: *Let X be a smooth m-dimensional manifold with an open coordinate covering $\{U_i \mid i \in I\}$ and atlas $\{\Phi_i \mid i \in I\}$. For each nonvoid $U_i \cap U_j \subset X$ let*

$$(x^1,\ldots,x^m) \mapsto (\bar{y}^1(x^1,\ldots,x^m),\ldots,\bar{y}^m(x^1,\ldots,x^m))$$

be the coordinate description of the diffeomorphism

$$\Phi_i(U_i \cap U_j) \xrightarrow{\Phi_j \circ \Phi_i^{-1}} \Phi_j(U_i \cap U_j),$$

and let $U_i \cap U_j \xrightarrow{\psi_i^j} GL(m,\mathbb{R})$ be the transition functions given by the jacobian matrices, $\psi_i^j(u) = ((\partial \bar{y}^q/\partial x^p)(\Phi_i(u)))$ for $u \in U_i \cap U_j$, as above. The smooth sections of the resulting smooth coordinate bundle $E \xrightarrow{\pi} X$ are the elements of the $C^\infty(X)$-module $\mathscr{E}^*(X)$ of vector fields on X, as in Definition 6.2.

PROOF: It remains only to identify the sections $X \to E$ of the resulting bundle with the vector fields on X. The usual coordinate functions on $(\mathbb{R}^m)^+$ provide a distinguished basis of vector fields on $(\mathbb{R}^m)^+$, the partial derivations with respect to the coordinates, which restrict to corresponding bases

$(\partial/\partial x^1, \ldots, \partial/\partial x^m)$ and $(\partial/\partial y^1, \ldots, \partial/\partial y^m)$ of vector fields on $\Phi_i(U_i \cap U_j) \subset (\mathbb{R}^m)^+$ and $\Phi_j(U_i \cap U_j) \subset (\mathbb{R}^m)^+$, respectively. For any $f \in C^\infty(U_i \cap U_j)$ let

$$L_p f = \frac{\partial(f \circ \Phi_i^{-1})}{\partial x^p} \circ \Phi_i$$

and let

$$M_q f = \frac{\partial(f \circ \Phi_j^{-1})}{\partial y^q} \circ \Phi_j,$$

so that $(L_1, \ldots, L_m)$ and $(M_1, \ldots, M_m)$ each form a basis of the vector fields on $U_i \cap U_j$, which are restrictions to $U_i \cap U_j$ of corresponding bases of the vector fields on U_i and U_j, respectively. The restrictions satisfy relations

$$M_q f = \sum_{p=1}^{m} \left(\frac{\partial \bar{x}^p}{\partial y^q} \circ \Phi_j \right) L_p f$$

for any $f \in C^\infty(U_i \cap U_j)$; that is,

$$(M_1, \ldots, M_m) = (L_1, \ldots, L_m) \begin{pmatrix} \dfrac{\partial \bar{x}^1}{\partial y^1} \circ \Phi_j & \cdots & \dfrac{\partial \bar{x}^1}{\partial y^m} \circ \Phi_j \\ \vdots & & \vdots \\ \dfrac{\partial \bar{x}^m}{\partial y^1} \circ \Phi_j & \cdots & \dfrac{\partial \bar{x}^m}{\partial y^m} \circ \Phi_j \end{pmatrix},$$

where the matrix entries merely serve as coefficients for the vector fields $L_1, \ldots, L_m$. Suppose that a section $X \to E$ is given, inducing restrictions $U_i \xrightarrow{\sigma} E|U_i$ and $U_j \xrightarrow{\tau} E|U_j$. For fixed local trivializations $E|U_i \xrightarrow{\Psi_i} U_i \times \mathbb{R}^m$ and $E|U_j \xrightarrow{\Psi_j} U_j \times \mathbb{R}^m$ each of the restrictions σ and τ can be regarded as a column vector

$$\begin{pmatrix} \sigma^1 \\ \vdots \\ \sigma^m \end{pmatrix} \quad \text{and} \quad \begin{pmatrix} \tau^1 \\ \vdots \\ \tau^m \end{pmatrix}$$

of smooth real-valued functions on U_i and U_j, which can be identified with

$$\sigma^1 L_1 + \cdots + \sigma^m L_m = (L_1, \ldots, L_m) \begin{pmatrix} \sigma^1 \\ \vdots \\ \sigma^m \end{pmatrix}$$

and with

$$\tau^1 M_1 + \cdots + \tau^m M_m = (M_1, \ldots, M_m) \begin{pmatrix} \tau^1 \\ \vdots \\ \tau^m \end{pmatrix}$$

as vector fields on U_i and U_j, respectively. One must verify that these vector fields have the same restrictions to $U_i \cap U_j$. Since

$$\begin{pmatrix} \tau^1 \\ \vdots \\ \tau^m \end{pmatrix} = \begin{pmatrix} \dfrac{\partial \bar{y}^1}{\partial x^1} \circ \Phi_i & \cdots & \dfrac{\partial \bar{y}^1}{\partial x^m} \circ \Phi_i \\ \vdots & & \vdots \\ \dfrac{\partial \bar{y}^m}{\partial x^1} \circ \Phi_i & \cdots & \dfrac{\partial \bar{y}^m}{\partial x^m} \circ \Phi_i \end{pmatrix} \begin{pmatrix} \sigma^1 \\ \vdots \\ \sigma^m \end{pmatrix},$$

by definition of the transition functions $\psi_i^j = (\partial \bar{y}^q / \partial x^p \circ \Phi_i)$, one has

$$\begin{aligned} (M_1, \ldots, M_m) \begin{pmatrix} \tau^1 \\ \vdots \\ \tau^m \end{pmatrix} &= (L_1, \ldots, L_m) \left(\frac{\partial \bar{x}^p}{\partial y^q} \circ \Phi_j \right) \left(\frac{\partial \bar{y}^q}{\partial x^p} \circ \Phi_i \right) \begin{pmatrix} \sigma^1 \\ \vdots \\ \sigma^m \end{pmatrix} \\ &= (L_1, \ldots, L_m) \begin{pmatrix} \sigma^1 \\ \vdots \\ \sigma^m \end{pmatrix} \end{aligned}$$

over $U_i \cap U_j$, as desired, the two jacobian matrices being inverses. Thus any section of the coordinate bundle $E \xrightarrow{\pi} X$ constructed from the given transition functions determines a unique vector field on X, as required; the converse assertion is trivial.

One can regard Proposition 6.10 as an alternative description of the tangent bundle $\tau(X)$ of the smooth manifold X. However, both 6.9 and Proposition 6.10 describe a specific coordinate bundle whose isomorphism class is the tangent bundle $\tau(X)$; one can easily represent $\tau(X)$ by other coordinate bundles. For example, Proposition 5.9 asserts that one can replace the $C^\infty(X)$-module $\mathscr{E}^*(X)$ of vector fields $C^\infty(X) \xrightarrow{L} C^\infty(X)$ by its dual $\mathscr{E}^{**}(X) = \mathrm{Hom}_{C^\infty(X)}(\mathscr{E}^*(X), C^\infty(X))$; we shall do so in the next result.

For any module $\mathscr{F}$ whatsover over a commutative ring, $C^\infty(X)$, for example, there is always a canonical homomorphism from $\mathscr{F}$ into the second conjugate $\mathscr{F}^{**} = \mathrm{Hom}_{C^\infty(X)}(\mathrm{Hom}_{C^\infty(X)}(\mathscr{F}, C^\infty(X)), C^\infty(X))$, carrying any $\theta \in \mathscr{F}$ into that element $\theta^{**} \in \mathscr{F}^{**}$ that maps any $L \in \mathscr{E}^*$ into the value of L on θ. If $\mathscr{F}$ is a free $C^\infty(X)$-module of finite rank, then the canonical homomorphism $\mathscr{F} \to \mathscr{F}^{**}$ is trivially an isomorphism; similarly, if $\mathscr{F}$ is merely locally free of finite rank, then one can use smooth partitions of unity in an obvious way to conclude that $\mathscr{F} \to \mathscr{F}^{**}$ is an isomorphism. For this reason we shall use the notation $\mathscr{E}(X)$ rather than $\mathscr{E}^{**}(X)$ to describe the dual of $\mathscr{E}^*(X)$; in Remark 13.20 and Exercise 13.21 the notation $\mathscr{E}(X)$ is justified by a direct construction.

6.11 Definition: For any smooth manifold X the *$C^\infty(X)$-module $\mathscr{E}(X)$ of differentials on X* is the dual $\operatorname{Hom}_{C^\infty(X)}(\mathscr{E}^*(X), C^\infty(X))$ of the $C^\infty(X)$-module $\mathscr{E}^*(X)$ of smooth vector fields on X.

There are some classical differentials. For any $g \in C^\infty(X)$ there is a $C^\infty(X)$-linear map $\mathscr{E}^*(X) \xrightarrow{dg} C^\infty(X)$ carrying any vector field $L \in \mathscr{E}^*(X)$ into the function $Lg \in C^\infty(X)$; consequently $dg \in \mathscr{E}(X)$. Similarly, for any natural number $p \geqq 0$ and any $2p$ elements $f_1, \ldots, f_p, g_1, \ldots, g_p \in C^\infty(X)$ there is a differential $f_1\, dg_1 + \cdots + f_p\, dg_p \in \mathscr{E}(X)$ carrying any $L \in \mathscr{E}^*(X)$ into $f_1(Lg_1) + \cdots + f_p(Lg_p) \in C^\infty(X)$.

In case $X = \mathbb{R}^m$ the m coordinate functions $x^1, \ldots, x^m$ lead to m differentials $dx^1, \ldots, dx^m$. Since $\partial x^i/\partial x^j = 1$ or $\partial x^i/\partial x^j = 0$ according as $i = j$ or $i \neq j$, it follows that $(dx^1, \ldots, dx^m)$ is the basis of $\mathscr{E}(\mathbb{R}^m)$ dual to the basis $(\partial/\partial x^1, \ldots, \partial/\partial x^m)$ of $\mathscr{E}^*(\mathbb{R}^m)$. A similar remark applies to $(\mathbb{R}^m)^+$ and to any open subset of $(\mathbb{R}^m)^+$.

6.12 Proposition: *For any smooth manifold X the tangent bundle $\tau(X)$ is the smooth real vector bundle over X represented by the smooth coordinate bundle whose sections form the $C^\infty(X)$-module $\mathscr{E}(X)$ of differentials on X.*

Proof: Since $\mathscr{E}(X)$ is defined as the dual $\mathscr{E}^{**}(X) = \operatorname{Hom}_{C^\infty(X)}(\mathscr{E}^*(X), C^\infty(X))$ of the $C^\infty(X)$-module $\mathscr{E}^*(X)$ of vector fields on X, this is an immediate consequence of Definition 6.9 and Proposition 5.9.

In view of Proposition 6.12 it is of interest to have an analog of Proposition 6.10, using $\mathscr{E}(X)$ in place of $\mathscr{E}^*(X)$. We use the notation of the discussion preceding Proposition 6.10.

In place of the jacobian matrices $(\partial \bar{y}^q/\partial x^p \circ \Phi_i)$ and $(\partial \bar{x}^p/\partial y^q \circ \Phi_j)$ displayed earlier, we shall use the transposed jacobian matrices

$$ {}^t\!\left(\frac{\partial \bar{x}^p}{\partial y^q} \circ \Phi_j\right) = \begin{pmatrix} \dfrac{\partial \bar{x}^1}{\partial y^1} \circ \Phi_j & \cdots & \dfrac{\partial \bar{x}^m}{\partial y^1} \circ \Phi_j \\ \vdots & & \vdots \\ \dfrac{\partial \bar{x}^1}{\partial y^m} \circ \Phi_j & \cdots & \dfrac{\partial \bar{x}^m}{\partial y^m} \circ \Phi_j \end{pmatrix} $$

and

$$ {}^t\!\left(\frac{\partial \bar{y}^q}{\partial x^p} \circ \Phi_i\right) = \begin{pmatrix} \dfrac{\partial \bar{y}^1}{\partial x^1} \circ \Phi_i & \cdots & \dfrac{\partial \bar{y}^m}{\partial x^1} \circ \Phi_i \\ \vdots & & \vdots \\ \dfrac{\partial \bar{y}^1}{\partial x^m} \circ \Phi_i & \cdots & \dfrac{\partial \bar{y}^m}{\partial x^m} \circ \Phi_i \end{pmatrix}, $$

each of which is the inverse of the other. In particular, since ${}^t(\partial\bar{x}^p/\partial y^q \circ \Phi_j)$ is invertible, it can be regarded as a smooth map $U_i \cap U_j \xrightarrow{\psi_i^j} GL(m, \mathbb{R})$ carrying any $u \in U_i \cap U_j$ into ${}^t((\partial\bar{x}^p/\partial y^q)(\Phi_j(u))) \in GL(m, \mathbb{R})$. If the intersection $U_i \cap U_j \cap U_k$ is nonvoid, the chain rule implies for any $u \in U_i \cap U_j \cap U_k$ that $\psi_j^k(u).\ \psi_i^j(u) = \psi_i^k(u) \in GL(m, \mathbb{R})$, so that by Proposition II.2.5 the smooth maps $U_i \cap U_j \xrightarrow{\psi_i^j} GL(m, \mathbb{R})$ are the transition functions of a unique coordinate bundle with structure group $GL(m, \mathbb{R})$ and fiber $\mathbb{R}^m$. As before, since the structure group $GL(m, \mathbb{R})$ acts on the left, the elements of R^m will be regarded as column vectors.

6.13 Proposition: *Let X be a smooth manifold with an open coordinate covering $\{U_i | i \in I\}$ and atlas $\{\Phi_i | i \in I\}$. For each nonvoid $U_i \cap U_j$ let $(y^1, \ldots, y^m) \mapsto (\bar{x}^1(y^1, \ldots, y^m), \ldots, \bar{x}^m(y^1, \ldots, y^m))$ be the coordinate description of the diffeomorphism $\Phi_j(U_i \cap U_j) \xrightarrow{\Phi_i \circ \Phi_j^{-1}} \Phi_i(U_i \cap U_j)$, and let $U_i \cap U_j \xrightarrow{\psi_i^j} GL(m, R)$ be the transition functions given by the transposed jacobian matrices, $\psi_i^j(u) = {}^t((\partial\bar{x}^p/\partial y^q)(\Phi_j(u)))$ for $u \in U_i \cap U_j$, as above. The smooth sections of the resulting smooth coordinate bundle $E \xrightarrow{\pi} X$ are the elements of the $C^\infty(X)$-module $\mathscr{E}(X)$ of differentials on X, as in Definition* 6.11.

PROOF: The coordinate functions on $(\mathbb{R}^m)^+$ restrict to bases $(dx^1, \ldots, dx^m)$ and $(dy^1, \ldots, dy^m)$ of the differentials on $\Phi_i(U_i) \subset (\mathbb{R}^m)^+$ and $\Phi_j(U_j) \subset (\mathbb{R}^m)^+$, and we use their further restrictions to $\Phi_i(U_i \cap U_j)$ and $\Phi_j(U_i \cap U_j)$. In particular, $(d(x^1 \circ \Phi_i), \ldots, d(x^m \circ \Phi_i))$ and $(d(y^1 \circ \Phi_j), \ldots, d(y^m \circ \Phi_j))$ are bases of the differentials on $U_i \cap U_j$, and by the chain rule they satisfy the relations

$$d(y^q \circ \Phi_j) = \sum_{p=1}^{m} \left(\frac{\partial\bar{y}^q}{\partial x^p} \circ \Phi_i\right) d(x^p \circ \Phi_i) \qquad \text{for} \quad q = 1, \ldots, m;$$

that is,

$$(d(y^1 \circ \Phi_j), \ldots, d(y^m \circ \Phi_j))$$
$$= (d(x^1 \circ \Phi_i), \ldots, d(x^m \circ \Phi_i)) \begin{pmatrix} \dfrac{\partial\bar{y}^1}{\partial x^1} \circ \Phi_i & \cdots & \dfrac{\partial\bar{y}^m}{\partial x^1} \circ \Phi_i \\ \vdots & & \vdots \\ \dfrac{\partial\bar{y}^1}{\partial x^m} \circ \Phi_i & \cdots & \dfrac{\partial\bar{y}^m}{\partial x^m} \circ \Phi_i \end{pmatrix}.$$

Suppose that a section $X \to E$ is given, inducing restrictions $U_i \xrightarrow{\sigma} E | U_i$ and $U_j \xrightarrow{\tau} E | U_j$. For the fixed local trivializations $E | U_i \xrightarrow{\Psi_i} U_i \times \mathbb{R}^m$ and $E | U_j \xrightarrow{\Psi_j} U_j \times \mathbb{R}^m$ each of the restrictions σ and τ can be regarded as a column vector,

$$\begin{pmatrix}\sigma^1\\ \vdots\\ \sigma^m\end{pmatrix} \quad\text{and}\quad \begin{pmatrix}\tau^1\\ \vdots\\ \tau^m\end{pmatrix},$$

of smooth real-valued functions on U_i and U_j, which can be identified with

$$\sigma^1\, d(x^1 \circ \Phi_i) + \cdots + \sigma^m\, d(x^m \circ \Phi_i) = (d(x^1 \circ \Phi_i), \ldots, d(x^m \circ \Phi_i))\begin{pmatrix}\sigma^1\\ \vdots\\ \sigma^m\end{pmatrix}$$

and

$$\tau^1\, d(y^1 \circ \Phi_j) + \cdots + \tau^m\, d(y^m \circ \Phi_j) = (d(y^1 \circ \Phi_j), \ldots, d(y^m \circ \Phi_j))\begin{pmatrix}\tau^1\\ \vdots\\ \tau^m\end{pmatrix}$$

as differentials on U_i and U_j, respectively. One must verify that these differentials have the same restrictions to $U_i \cap U_j$. Since

$$\begin{pmatrix}\tau^1\\ \vdots\\ \tau^m\end{pmatrix} = \begin{pmatrix}\dfrac{\partial \bar{x}^1}{\partial y^1} \circ \Phi_j & \cdots & \dfrac{\partial \bar{x}^m}{\partial y^1} \circ \Phi_j\\ \vdots & & \vdots\\ \dfrac{\partial \bar{x}^1}{\partial y^m} \circ \Phi_j & \cdots & \dfrac{\partial \bar{x}^m}{\partial y^m} \circ \Phi_j\end{pmatrix}\begin{pmatrix}\sigma^1\\ \vdots\\ \sigma^m\end{pmatrix}$$

by definition of the transition functions $\psi_i^j = {}^t(\partial \bar{x}^p/\partial y^q \circ \Phi_j)$, one has

$$\begin{aligned}
&(d(y^1 \circ \Phi_j), \ldots, d(y^m \circ \Phi_j))\begin{pmatrix}\tau^1\\ \vdots\\ \tau^m\end{pmatrix}\\
&\quad = (d(x^1 \circ \Phi_i), \ldots, d(x^m \circ \Phi_i))\, {}^t\!\left(\frac{\partial \bar{y}^q}{\partial x^p} \circ \Phi_i\right) {}^t\!\left(\frac{\partial \bar{x}^p}{\partial y^q} \circ \Phi_j\right)\begin{pmatrix}\sigma^1\\ \vdots\\ \sigma^m\end{pmatrix}\\
&\quad = (d(x^1 \circ \Phi_i), \ldots, d(x^m \circ \Phi_i))\begin{pmatrix}\sigma^1\\ \vdots\\ \sigma^m\end{pmatrix}
\end{aligned}$$

over $U_i \cap U_j$, as desired, the two transposed jacobian matrices being inverses. Thus any section of the coordinate bundle $E \xrightarrow{\pi} X$ constructed from the given transition functions determines a unique differential on X, as required; the converse assertion is trivial.

It is worth noting that the differing transition functions $U_i \cap U_j \xrightarrow{\Psi_i^j} GL(m, \mathbb{R})$ appearing in Propositions 6.10 and 6.13 are transposed inverses of each other, where the resulting smooth coordinate bundles both represent the *same* tangent bundle $\tau(X)$: in Proposition 6.10 one has $\psi_i^j = (\partial \bar{y}^q/\partial x^p \circ \Phi_i)$ and in Proposition 6.13 one has $\psi_i^j = {}^t(\partial \bar{x}^p/\partial y^q \circ \Phi_j)$. Although the map $GL(m, \mathbb{R}) \xrightarrow{\Gamma} GL(m, \mathbb{R})$ carrying each invertible $m \times m$ matrix into its transposed inverse is indeed a group automorphism, Γ is not part of a morphism (Γ, Φ) of transformation groups to which one can apply Proposition II.2.9 to conclude directly that the resulting coordinate bundles represent the same vector bundle. On the other hand, the preceding situation is part of a more general phenomenon: if coordinate bundles $E \xrightarrow{\pi} X$ and $E' \xrightarrow{\pi'} X$ are described by transition functions $U_i \cap U_j \xrightarrow{\psi_i^j} GL(m, \mathbb{R})$ and $U_i \cap U_j \xrightarrow{\psi_i'^j} GL(m, R)$ that are transposed inverses of each other, then $E \xrightarrow{\pi} X$ and $E' \xrightarrow{\pi'} X$ represent the same real m-plane bundle over X, for any $X \in \mathscr{B}$. (See Exercise 13.24.)

Throughout this section we have considered the $C^\infty(X)$-modules $\mathscr{E}^*(X)$ and $\mathscr{E}(X)$ associated with *smooth* representations of the tangent bundle $\tau(X)$ of a smooth manifold X. One can replace $\mathscr{E}^*(X)$ and $\mathscr{E}(X)$ by $C^0(X) \otimes_{C^\infty(X)} \mathscr{E}^*(X)$ and $C^0(X) \otimes_{C^\infty(X)} \mathscr{E}(X)$, however, which represent the same tangent bundle $\tau(X)$; that is, one can replace smooth sections of coordinate bundles representing $\tau(X)$ by continuous sections of the same coordinate bundles. Thanks to Propositions 5.11 and 5.12, the use of continuous sections presents no serious handicap, however, as the following result suggests.

Suppose that $E \xrightarrow{\pi} X$ is the smooth coordinate bundle whose sections $X \xrightarrow{\tau} E$ are vector fields $C^\infty(X) \xrightarrow{M} C^\infty(X)$ on a given smooth manifold X. The sections $X \xrightarrow{\tau} E$ are smooth maps, by construction; however, one can also consider sections $X \xrightarrow{\sigma} E$ that are merely continuous, as just noted. Accordingly, we distinguish between *smooth* vector fields and *continuous* vector fields on the same smooth manifold X.

6.14 Proposition: *Suppose that X is a smooth manifold that admits a nowhere-vanishing continuous vector field L; then X also admits a nowhere-vanishing smooth vector field M.*

PROOF: Continuous and smooth vector fields are continuous and smooth sections $X \xrightarrow{\sigma} E$ and $X \xrightarrow{\tau} E$, respectively, of that smooth coordinate bundle $E \xrightarrow{\pi} X$ whose smooth sections are smooth vector fields, elements of $\mathscr{E}^*(X)$. Hence the result is a special case of Proposition 5.12.

In the opening paragraph of this section we indicated that there is a contravariant functor on the category of smooth manifolds that assigns to each smooth manifold X and to each smooth map $Y \to X$ a $C^\infty(X)$-module $\mathscr{E}(X)$

and a module homomorphism $\mathscr{E}(X) \to \mathscr{E}(Y)$ over the algebra homomorphism $C^\infty(X) \to C^\infty(Y)$, respectively. We have constructed the $C^\infty(X)$-module $\mathscr{E}(X)$, somewhat indirectly; however, we have not yet constructed the homomorphism $\mathscr{E}(X) \to \mathscr{E}(Y)$. Since both constructions will be carried out more directly in Remark 13.20 and Exercises 13.21 and 13.22, we only sketch the latter homomorphism here; it will not be required in the sequel.

Suppose that $Y \xrightarrow{\Phi} X$ is a smooth map of smooth manifolds. The induced algebra homomorphism $C^\infty(X) \xrightarrow{\Phi^*} C^\infty(Y)$ carries any smooth function $x \xrightarrow{f} \mathbb{R}$ into the composition $Y \xrightarrow{f \circ \Phi} \mathbb{R}$; that is, $\Phi^* f = f \circ \Phi$. Following Definition 6.11 it was noted that for any natural number $p \geqq 0$ and any $2p$ elements $f_1, \ldots, f_p, g_1, \ldots, g_p \in C^\infty(X)$ there is a differential $f_1\, dg_1 + \cdots + f_p dg_p \in \mathscr{E}(X)$ carrying any $L \in \mathscr{E}^*(X)$ into $f_1(Lg_1) + \cdots + f_p(Lg_p) \in C^\infty(X)$. There is a corresponding differential $(f_1 \circ \Phi)\, d(g_1 \circ \Phi) + \cdots + (f_p \circ \Phi)\, d(g_p \circ \Phi) \in \mathscr{E}(Y)$, and one easily verifies via local coordinate systems and smooth partitions of unity that there is then a unique module homomorphism $\mathscr{E}(X) \xrightarrow{\Phi^*} \mathscr{E}(Y)$ over $C^\infty(X) \xrightarrow{\Phi^*} C^\infty(Y)$ such that $\Phi^*(f_1 dg_1 + \cdots + f_p dg_p) = (f_1 \circ \Phi)\, d(g_1 \circ \Phi) + \cdots + (f_p \circ \Phi)\, d(g_p \circ \Phi)$ for any $f_1, \ldots, f_p, g_1, \ldots, g_p \in C^\infty(X)$. A more satisfying construction of $\mathscr{E}(X) \xrightarrow{\Phi^*} \mathscr{E}(Y)$ will appear in Remark 13.20 and Exercises 13.21 and 13.22, as promised earlier.

We now compute a concrete example of a tangent bundle. Let $\mathbb{R}^* \subset \mathbb{R}$ denote the nonzero real numbers and let $\mathbb{R}^{(n+1)*}$ denote the nonzero vectors in $\mathbb{R}^{n+1}$, in the relative topologies. The real projective space RP^n is the quotient $\mathbb{R}^{(n+1)*}/\sim$ of $\mathbb{R}^{(n+1)*}$ by the equivalence relation $\sim$ with $x' \sim x$ whenever $x' = ax \in \mathbb{R}^{(n+1)*}$ for some $a \in \mathbb{R}^*$, as in §I.5; alternatively, RP^n is the Grassmann manifold $G^1(\mathbb{R}^{n+1})$, which is smooth by Proposition I.7.3. Let $E \subset \mathbb{R}^{(n+1)*} \times \mathbb{R}^{n+1}$ consist of those pairs $(x, y) \in \mathbb{R}^{(n+1)*} \times \mathbb{R}^{n+1}$ such that y is orthogonal to x in the usual inner product of $\mathbb{R}^{n+1}$. There is then a quotient $E/\approx$ of E by the equivalence relation $\approx$ with $(x', y') \approx (x, y)$ whenever $x' = ax \in \mathbb{R}^{(n+1)*}$ and $y' = ay \in \mathbb{R}^{n+1}$ for the same $a \in \mathbb{R}^*$. If $(x'\ y') \approx (x, y)$, then $x' \sim x$, so that the first projection $\mathbb{R}^{(n+1)*} \times \mathbb{R}^{n+1} \xrightarrow{\pi_1} \mathbb{R}^{(n+1)*}$ induces a projection $E/\approx \xrightarrow{\pi} RP^n$ of quotient spaces, which is continuous in the quotient topologies.

6.15 Proposition: *The preceding projection $E/\approx \xrightarrow{\pi} RP^n$ is a smooth coordinate bundle that represents the tangent bundle $\tau(RP^n)$ of the real projective space RP^n.*

Proof: According to Definition 6.9, $\tau(RP^n)$ is represented by the smooth coordinate bundle whose sections are vector fields $C^\infty(RP^n) \xrightarrow{L} C^\infty(RP^n)$; we shall identify $E/\approx \xrightarrow{\pi} RP^n$ with that coordinate bundle. First observe that

$C^\infty(RP^n)$ can be regarded as the subalgebra of those functions $f \in C^\infty(\mathbb{R}^{(n+1)*})$ that are homogeneous of degree 0. Thus if $\mathbb{R}^{(n+1)*} \xrightarrow{m_a} \mathbb{R}^{(n+1)*}$ is multiplication by $a \in \mathbb{R}^*$, one has $f \in \mathbb{C}^\infty(\mathbb{R}P^n)$ ($\subset \mathbb{C}^\infty(\mathbb{R}^{(n+1)*})$) if and only if $f \circ m_a = f$ for each $a \in \mathbb{R}^*$; equivalently $f \in \mathbb{C}^\infty(RP^n)(\subset \mathbb{C}^\infty(\mathbb{R}^{(n+1)*}))$ if and only if f is constant on each line through $0 \in \mathbb{R}^{n+1}$. For any

$$y = \begin{pmatrix} y^0 \\ \vdots \\ y^n \end{pmatrix} \in \mathbb{R}^{n+1},$$

one then has

$$\begin{aligned} a(y^0\,\partial/\partial x^0 + \cdots + y^n\,\partial/\partial x^n) f \circ m_a &= (y^0\,\partial/\partial x^0 + \cdots + y^n\,\partial/\partial x^n)(f \circ m_a) \\ &= (y^0\,\partial/\partial x^0 + \cdots + y^n\,\partial/\partial x^n) f \end{aligned}$$

for any $f \in \mathbb{C}^\infty(RP^n)(\subset \mathbb{C}^\infty(\mathbb{R}^{(n+1)*}))$, by the chain rule. Hence, if $x' = ax \in \mathbb{R}^{(n+1)*}$, one obtains a vector field $\mathbb{C}^\infty(RP^n) \xrightarrow{L_y} \mathbb{C}^\infty(RP^n)$ only by imposing the second condition $y' = ay \in \mathbb{R}^{n+1}$ in the equivalence relation $(x', y') \approx (x, y)$ described earlier, for the same $a \in R^*$. Since any $f \in C^\infty(RP^n)$ is constant along each line through $0 \in \mathbb{R}^{n+1}$, it follows that $L_x f(x) = 0$ for each $x \in \mathbb{R}^{(n+1)*}$, so that L_x vanishes at the equivalence class in RP^n of $x \in \mathbb{R}^{(n+1)*}$. However, the values of the vectors L_y for any $y \in \mathbb{R}^{n+1}$ orthogonal to x span an n-dimensional vector space at the preceding point of RP^n, and since RP^n is an n-dimensional smooth manifold this completes the proof: if $E \subset \mathbb{R}^{(n+1)*} \times \mathbb{R}^{n+1}$ consists of those pairs $(x, y) \in \mathbb{R}^{(n+1)*} \times \mathbb{R}^{n+1}$ such that y is orthogonal to x, then $E/\approx \xrightarrow{\pi} RP^n$ represents the tangent bundle $\tau(RP^n)$, as claimed.

There is a useful corollary of Proposition 6.15. This time we replace $E \subset \mathbb{R}^{(n+1)*} \times \mathbb{R}^{n+1}$ by the *entire space* $\mathbb{R}^{(n+1)*} \times \mathbb{R}^{n+1}$ and let E'' be the quotient $\mathbb{R}^{(n+1)*} \times \mathbb{R}^{n+1}/\approx$ by the equivalence relation $\approx$ with $(x, y') \approx (x, y)$ whenever $x' = ax \in \mathbb{R}^{(n+1)*}$ and $y' = ay \in \mathbb{R}^{n+1}$ for the same $a \in \mathbb{R}^*$, as before. The first projection $\mathbb{R}^{(n+1)*} \times \mathbb{R}^{n+1} \xrightarrow{\pi_1} \mathbb{R}^{(n+1)*}$ still induces a projection $E'' \xrightarrow{\pi''} RP^n$ of quotient spaces, which is continuous in the quotient topologies.

6.16 Corollary: *The preceding projection $E'' \xrightarrow{\pi''} RP^n$ is a smooth coordinate bundle that represents the Whitney sum $\tau(RP^n) \oplus \varepsilon^1$ of the tangent bundle $\tau(RP^n)$ and the trivial line bundle ε^1 over the real projective space RP^n.*

PROOF: The product representation $RP^n \times \mathbb{R}^1 \xrightarrow{\pi_1} RP^n$ of the trivial bundle can be regarded as a projection of quotients in an obvious fashion: there is an equivalence relation $\approx$ in $\mathbb{R}^{(n+1)*} \times \mathbb{R}^1$ with $(x', s') \approx (x, s)$ whenever $x' = ax \in \mathbb{R}^{(n+1)*}$, for some $a \in \mathbb{R}^*$, and $s' = s$, independently of $a \in \mathbb{R}^*$, and the first projection $R^{(n+1)*} \times \mathbb{R}^1 \xrightarrow{\pi_1} \mathbb{R}^{(n+1)*}$ induces $RP^n \times \mathbb{R}^1 \xrightarrow{\pi_1} RP^n$. However, in place of $\mathbb{R}^{(n+1)*} \times \mathbb{R}^1$ one can substitute the space $E' \subset \mathbb{R}^{(n+1)*} \times \mathbb{R}^{n+1}$ of points of the form (x, z) for any $x \in \mathbb{R}^{(n+1)*}$ and any scalar multiple $z = sx$ of x. The corresponding equivalence relation $\approx$ in E' then sets $(x', z') \approx (x, z)$ whenever $x' = ax$ and $z' = az$ for the same $a \in \mathbb{R}^*$. The first projection $\mathbb{R}^{(n+1)*} \times \mathbb{R}^{n+1} \xrightarrow{\pi_1} \mathbb{R}^{(n+1)*}$ induces a projection $E'/\!\approx \xrightarrow{\pi'} RP^n$ that can be identified with $RP^n \times \mathbb{R}^1 \xrightarrow{\pi_1} RP^n$ by observing that each $(x, z) \in E'$ is of the form (x, sx) for a unique $s \in \mathbb{R}$. The coordinate bundle $E/\!\approx \xrightarrow{\pi} RP^n$ representing $\tau(RP^n)$ in Proposition 6.15 and the coordinate bundle $E'/\!\approx \xrightarrow{\pi'} RP^n$ just constructed to represent ε^1 are of the same nature; the only difference is that $E \subset \mathbb{R}^{(n+1)*} \times \mathbb{R}^{n+1}$ consists of those pairs (x, y) such that y is orthogonal to x, while $E' \subset \mathbb{R}^{(n+1)*} \times \mathbb{R}^{n+1}$ consists of those pairs (x, z) such that z is a scalar multiple of x. The equivalence relation $\approx$ is defined in the same way in each case, and since any element of $\mathbb{R}^{n+1}$ is uniquely of the form $y + z$ for y orthogonal to $x \in \mathbb{R}^{(n+1)*}$ and z a scalar multiple of the same $x \in \mathbb{R}^{(n+1)*}$, it follows that the Whitney sum $\tau(RP^n) \oplus \varepsilon^1$ is represented by the quotient map $\mathbb{R}^{(n+1)*} \times \mathbb{R}^{n+1}/\!\approx \xrightarrow{\pi''} \mathbb{R}^{(n+1)*}/\!\sim$ induced by the first projection $\mathbb{R}^{(n+1)*} \times \mathbb{R}^{n+1} \xrightarrow{\pi_1} \mathbb{R}^{(n+1)*}$, as claimed.

7. Canonical Vector Bundles

For any natural numbers $m > 0$ and $n > 0$ the real Grassmann manifold $G^m(\mathbb{R}^{m+n})$ is a smooth closed mn-dimensional manifold, as in Proposition I.7.3; in particular, the real projective space $RP^n = G^1(\mathbb{R}^{n+1})$ is a smooth closed n-dimensional manifold. In this section we construct a smooth coordinate bundle representing a particularly useful nontrivial m-plane bundle γ_n^m over each $G^m(\mathbb{R}^{m+n})$; applications will occur later. We also show that $(n + 1)\gamma_n^1 = \tau(RP^n) \oplus \varepsilon^1$, where $(n + 1)\gamma_n^1$ is the Whitney sum of $n + 1$ copies of the canonical line bundle γ_n^1 over RP^n, where $\tau(RP^n)$ is the tangent bundle of RP^n, and where ε^1 is the trivial line bundle over RP^n.

We briefly recall the construction of the Grassmann manifolds $G^m(\mathbb{R}^{m+n})$, which first appear (for $V = \mathbb{R}^{m+n}$) in Definition I.7.1. Let $(\mathbb{R}^{m+n})^{m*}$ denote the set of ordered m-tuples $(x_1, \ldots, x_m)$ of linearly independent vectors $x_1, \ldots, x_m \in \mathbb{R}^{m+n}$, in the relative topology of the m-fold product $\mathbb{R}^{m+n} \times \cdots \times \mathbb{R}^{m+n}$, and let $\sim$ be the equivalence relation in $(\mathbb{R}^{m+n})^{m*}$ with $(x_1, \ldots, x_m) \sim (y_1, \ldots, y_m)$ whenever $(x_1, \ldots, x_m)$ and $(y_1, \ldots, y_m)$ span the same m-plane in $\mathbb{R}^{m+n}$. Then $G^m(\mathbb{R}^{m+n})$ is the quotient $(\mathbb{R}^{m+n})^{m*}/\sim$ in the quotient topology; that is, $G^m(\mathbb{R}^{m+n})$ is the set of m-planes in $\mathbb{R}^{m+n}$ in an appropriate topology.

If $(x_1, \ldots, x_m)$ and $(y_1, \ldots, y_m)$ span the same m-plane in $\mathbb{R}^{m+n}$, then there is an $m \times m$ invertible matrix $(a_p^q) \in GL(m, \mathbb{R})$ such that $x_p = \sum_{q=1}^m a_p^q y_q$ for each $p = 1, \ldots, m$; that is,

$$(x_1, \ldots, x_m) = (y_1, \ldots, y_m)\begin{pmatrix} a_1^1 & \cdots & a_m^1 \\ \vdots & & \vdots \\ a_1^m & \cdots & a_m^m \end{pmatrix},$$

where each of $x_1, \ldots, x_m, y_1, \ldots, y_m$ is a column vector with $m + n$ entries. We shall write this relation in either of the equivalent forms $(x_1, \ldots, x_m) = (y_1, \ldots, y_m)(a_p^q)$ or $(y_1, \ldots, y_m) = (x_1, \ldots, x_m)(a_p^q)^{-1}$ for $(a_p^q) \in GL(m, \mathbb{R})$.

Elements of the product space $(\mathbb{R}^{m+n})^{m*} \times \mathbb{R}^m$ are of the form

$$\left((x_1, \ldots, x_m), \begin{pmatrix} s^1 \\ \vdots \\ s^m \end{pmatrix}\right)$$

for

$$(x_1, \ldots, x_m) \in (\mathbb{R}^{m+n})^{m*} \qquad \text{and} \qquad \begin{pmatrix} s^1 \\ \vdots \\ s^m \end{pmatrix} \in \mathbb{R}^m,$$

and we write

$$\left((x_1, \ldots, x_m), \begin{pmatrix} s^1 \\ \vdots \\ s^m \end{pmatrix}\right) \approx \left((y_1, \ldots, y_m), \begin{pmatrix} t^1 \\ \vdots \\ t^m \end{pmatrix}\right)$$

whenever both $(y_1, \ldots, y_m) = (x_1, \ldots, x_m)(a_p^q)^{-1}$ and

$$\begin{pmatrix} t^1 \\ \vdots \\ t^m \end{pmatrix} = (a_p^q)\begin{pmatrix} s^1 \\ \vdots \\ s^m \end{pmatrix}$$

for the same $(a_p^q) \in GL(m, \mathbb{R})$. Trivially $\approx$ is an equivalence relation in $(\mathbb{R}^{m+n})^{m*} \times \mathbb{R}^m$, and we temporarily let E denote the quotient space $(\mathbb{R}^{m+n})^{m*} \times \mathbb{R}^m/\approx$, in the quotient topology. Clearly if

$$\left((x_1, \ldots, x_m), \begin{pmatrix} s^1 \\ \vdots \\ s^m \end{pmatrix}\right) \approx \left((y_1, \ldots, y_m), \begin{pmatrix} t^1 \\ \vdots \\ t^m \end{pmatrix}\right)$$

in $(\mathbb{R}^{m+n})^{m*} \times \mathbb{R}^m$, then $(x_1, \ldots, x_m) \sim (y_1, \ldots, y_m)$ in $(\mathbb{R}^{m+n})^{m*}$, so that the first projection $(\mathbb{R}^{m+n})^{m*} \times \mathbb{R}^m \xrightarrow{\pi_1} (\mathbb{R}^{m+n})^{m*}$ induces a projection $E \xrightarrow{\pi}$

$G^m(\mathbb{R}^{m+n})$ of quotient spaces. Furthermore, if

$$\left((x_1, \ldots, x_m), \begin{pmatrix} s^1 \\ \vdots \\ s^m \end{pmatrix}\right) \approx \left((y_1, \ldots, y_m), \begin{pmatrix} t^1 \\ \vdots \\ t^m \end{pmatrix}\right)$$

then for some $(a_p^q) \in GL(m, \mathbb{R})$ one has

$$\begin{aligned} t^1 y_1 + \cdots + t^m y_m &= (y_1, \ldots, y_m)\begin{pmatrix} t^1 \\ \vdots \\ t^m \end{pmatrix} \\ &= (x_1, \ldots, x_m)(a_p^q)^{-1}(a_p^q)\begin{pmatrix} s^1 \\ \vdots \\ s^m \end{pmatrix} \\ &= (x_1, \ldots, x_m)\begin{pmatrix} s^1 \\ \vdots \\ s^m \end{pmatrix} \\ &= s^1 x_1 + \cdots + s^m x_m, \end{aligned}$$

so that the equivalence class $e \in E$ of

$$\left((x_1, \ldots, x_m), \begin{pmatrix} s^1 \\ \vdots \\ s^m \end{pmatrix}\right) \in (\mathbb{R}^{m+n})^{m*} \times \mathbb{R}^m$$

defines a specific point $s^1 x_1 + \cdots + s^m x_m$ in the m-plane $\pi(e) \subset \mathbb{R}^{m+n}$ spanned by $(x_1, \ldots, x_m) \in (\mathbb{R}^{m+n})^{m*}$. Thus E consists of pairs $(\pi(e), e)$ in which $\pi(e) \in G^m(\mathbb{R}^{m+n})$ is an m-plane in $\mathbb{R}^{m+n}$ and e is a point in the m-plane $\pi(e)$.

7.1 Lemma: *The projection $E \xrightarrow{\pi} G^m(\mathbb{R}^{m+n})$ is a smooth coordinate bundle with structure group $GL(m, \mathbb{R})$ and fiber $\mathbb{R}^m$, representing a smooth real m-plane bundle over the Grassmann manifold $G^m(\mathbb{R}^{m+n})$.*

PROOF: Recall from Proposition I.7.3 that the smooth structure of $G^m(\mathbb{R}^{m+n})$ was obtained via the canonical embedding $G^m(\mathbb{R}^{m+n}) \xrightarrow{F} G^1(\bigwedge^m \mathbb{R}^{m+n})$ and a fixed basis $\{x_1, \ldots, x_{m+n}\}$ of $\mathbb{R}^{m+n}$. Each open set $U_{i_1, \ldots, i_m} \subset \operatorname{Im} F$ consisted of projective equivalence classes of points

$$\textstyle\sum_{j_1 < \cdots < j_m} a(j_1, \ldots, j_m) x_{j_1} \wedge \cdots \wedge x_{j_m} \in \bigwedge^m \mathbb{R}^{m+n}$$

such that $a(i_1, \ldots, i_m) \neq 0$, and the Plücker relations implied that all ratios $a(j_1, \ldots, j_m)/a(i_1, \ldots, i_m)$ are polynomial functions of the mn ratios

$a(i_1, \ldots, \hat{i}_k, \ldots, i_m, j)/a(i_1, \ldots, i_m)$. Here is another description of the same set $U_{i_1, \ldots, i_m}$ in terms of the inner product $\mathbb{R}^{m+n} \times \mathbb{R}^{m+n} \xrightarrow{\langle\,,\,\rangle} \mathbb{R}$ with $\langle x_i, x_j \rangle = \delta_{ij}$: $U_{i_1, \ldots, i_m}$ consists of the projective equivalence classes of all points of the form $(x_{i_1} + y_{i_1}) \wedge \cdots \wedge (x_{i_m} + y_{i_m})$, with automatic Plücker relations, where the nonzero elements among $y_{i_1}, \ldots, y_{i_m}$ are orthogonal to the subspace spanned by $x_{i_1}, \ldots, x_{i_m}$. In this description one has $a(i_1, \ldots, i_m) = 1$, each $a(j_1, \ldots, j_m)$ is the coefficient of $x_{j_1} \wedge \cdots \wedge x_{j_m}$ in the expansion of $(x_{i_1} + y_{i_1}) \wedge \cdots \wedge (x_{i_m} + y_{i_m})$, and each $a(j_1, \ldots, j_m)$ is a polynomial function of the mn coordinates $a(i_1, \ldots, \hat{i}_k, \ldots, i_m, j)$.

As above, each point of $E \,|\, U_{i_1, \ldots, i_m}$ is a pair $(\pi(e), e)$ in which $\pi(e)$ is the m-plane spanned by $x_{i_1} + y_{i_1}, \ldots, x_{i_m} + y_{i_m}$ and e is a point

$$(x_{i_1} + y_{i_1}, \ldots, x_{i_m} + y_{i_m}) \begin{pmatrix} s^1 \\ \vdots \\ s^m \end{pmatrix} \in \pi(e);$$

there is a local trivialization $E \,|\, U_{i_1, \ldots, i_m} \xrightarrow{\Psi_i} U_{i_1, \ldots, i_m} \times R^m$ carrying $(\pi(e), e)$ into

$$\left(\pi(e), \begin{pmatrix} s^1 \\ \vdots \\ s^m \end{pmatrix}\right).$$

If $U_{j_1, \ldots, j_m} \subset \operatorname{Im} F$ is any other set of the preceding form, then $U_{i_1, \ldots, i_m} \cap U_{j_1, \ldots, j_m}$ is automatically nonvoid, and one must show that the transition function $U_{i_1, \ldots, i_m} \cap U_{j_1, \ldots, j_m} \xrightarrow{\psi_i^j} GL(m, \mathbb{R})$ induced by $\Psi_j \circ \Psi_i^{-1}$ is smooth.

Suppose that the intersection $\{i_1, \ldots, i_m\} \cap \{j_1, \ldots, j_m\}$ contains $m - p$ indices, which we temporarily assume to be $i_{p+1} = j_{p+1}, \ldots, i_m = j_m$ for notational convenience; the adjustment required for the general case will be indicated later. In the notation relative to $U_{i_1, \ldots, i_m}$ one has

$$(x_{i_1} + y_{i_1}) \wedge \cdots \wedge (x_{i_m} + y_{i_m}) \in U_{i_1, \ldots, j_m} \cap U_{j_1, \ldots, j_m}$$

if and only if the coefficient $a(j_1, \ldots, j_m)$ of $x_{j_1} \wedge \cdots \wedge x_{j_m}$ in the expansion of $(x_{i_1} + y_{i_1}) \wedge \cdots \wedge (x_{i_m} + y_{i_m})$ is nonzero. This happens if and only if there is a nonsingular matrix

$$B = \left(\begin{array}{ccc|c} b_1^1 & \cdots & b_p^1 & \\ \vdots & & \vdots & 0 \\ b_1^p & \cdots & b_p^p & \\ \hline b_1^{p+1} & \cdots & b_p^{p+1} & \\ \vdots & & \vdots & I \\ b_1^m & \cdots & b_p^m & \end{array}\right)$$

for the $(m-p) \times (m-p)$ identity matrix I, such that $(x_{i_1} + y_{i_1}, \ldots, x_{i_m} + y_{i_m}) = (x_{j_1}, \ldots, x_{j_m})B$ modulo the subspace $\mathbb{R}^{n-p} \subset \mathbb{R}^{m+n}$ orthogonal to the subspace spanned by $x_{i_1}, \ldots, x_{i_m}, x_{j_1}, \ldots, x_{j_p}$. Up to a $\pm$ sign each b_k^j is the coefficient $a(i_1, \ldots, \hat{i}_k, \ldots, i_m, j)$ of $x_{i_1} \wedge \cdots \wedge \hat{x}_k \wedge \cdots \wedge x_{i_m} \wedge x_j$ in the expansion of $(x_{i_1} + y_{i_1}) \wedge \cdots \wedge (x_{i_m} + y_{i_m})$. Since $a(j_1, \ldots, j_m)$ is a homogeneous polynomial of degree p in the mn coordinates $a(i_1, \ldots, \hat{i}_k, \ldots, i_m, j)$, it follows that $a(j_1, \ldots, j_m)B^{-1}$ has homogeneous rational entries of degree zero in the same coordinates. Furthermore, if $A = (a(j_1, \ldots, j_m)B^{-1})^{-1}$, and if $(x_{i_1} + y_{i_1}, \ldots, x_{i_m} + y_{i_m}) \in (\mathbb{R}^{m+n})^{m*}$ is a $U_{i_1, \ldots, i_m}$-description representing a point in the intersection $U_{i_1, \ldots, i_m} \cap U_{j_1, \ldots, j_m}$, then $(x_{i_1} + y_{i_1}, \ldots, x_{i_m} + y_{i_m})A^{-1}$ is the corresponding $U_{j_1, \ldots, j_m}$-description representing the same point.

Now recall that the total space E of γ_n^m is a quotient $(\mathbb{R}^{m+n})^{m*} \times \mathbb{R}^m/\approx$, where

$$\left((u_1, \ldots, u_m), \begin{pmatrix} s^1 \\ \vdots \\ s^m \end{pmatrix}\right) \approx \left((v_1, \ldots, v_m), \begin{pmatrix} t^1 \\ \vdots \\ t^m \end{pmatrix}\right)$$

whenever $(v_1, \ldots, v_m) = (u_1, \ldots, u_m)A^{-1}$ and

$$\begin{pmatrix} t^1 \\ \vdots \\ t^m \end{pmatrix} = A \begin{pmatrix} s^1 \\ \vdots \\ s^m \end{pmatrix}$$

for the same nonsingular matrix A. Consequently the transition function $U_{i_1, \ldots, i_m} \cap U_{j_1, \ldots, j_m} \xrightarrow{\psi_i^j} GL(m, \mathbb{R})$ carries any point with $U_{i_1, \ldots, i_m}$-coordinates $a(i_1, \ldots, \hat{i}_k, \ldots, i_m, j)$ into the preceding nonsingular matrix A with homogeneous rational entries of degree zero in the same coordinates; in particular ψ_i^j is smooth.

To complete the proof of Lemma 7.1 it remains only to discard the specialized hypothesis $i_{p+1} = j_{p+1}, \ldots, i_m = j_m$. Whenever $\{i_1, \ldots, i_m\}$ and $\{j_1, \ldots, j_m\}$ have $m - p$ elements in common, one can reorder the entire set $\{1, \ldots, m + n\}$ to produce the specialized hypothesis, and the only effect of undoing the reordering is a permutation of the rows of A and a possibly distinct permutation of the columns of A. This does not affect the assertion of the lemma.

7.2 Definition: The smooth coordinate bundle $E \xrightarrow{\pi} G^m(\mathbb{R}^{m+n})$ of Lemma 7.1 represents the *canonical real m-plane bundle* γ_n^m over the Grassmann manifold $G^m(\mathbb{R}^{m+n})$. In the case $m = 1$, for which $G^1(\mathbb{R}^{n+1}) = RP^n$, the smooth coordinate bundle $E \xrightarrow{\pi} G^1(\mathbb{R}^{n+1})$ of Lemma 7.1 represents the *canonical real line bundle* γ_n^1 over the real projective space RP^n.

We next establish the identity $(n+1)\gamma_n^1 = \tau(RP^n) \oplus \varepsilon^1$ mentioned at the beginning of the section. The method will be to find a coordinate bundle $E' \xrightarrow{\pi'} RP^n$ representing $(n+1)\gamma_n^1$ that can be compared to the coordinate bundle $E'' \xrightarrow{\pi''} RP^n$ of Corollary 6.16, representing $\tau(RP^n) \oplus \varepsilon^1$. Although the construction of $E' \xrightarrow{\pi'} RP^n$ will appear to differ from that of $E'' \xrightarrow{\pi''} RP^n$, we shall show that they are nevertheless identical coordinate bundles.

Let $\mathbb{R}^{(n+1)*}$ be the space of nonzero vectors in $\mathbb{R}^{n+1}$, in the relative topology, let $\mathbb{R}^*$ be the nonzero real numbers, let $\approx'$ be the equivalence relation in $\mathbb{R}^{(n+1)*} \times \mathbb{R}^{n+1}$ with $(x', y') \approx' (x, y)$ whenever both $x' = ax \in \mathbb{R}^{(n+1)*}$ and $y' = a^{-1}y \in \mathbb{R}^{n+1}$ for the same $a \in \mathbb{R}^*$, and let E' be the quotient space $\mathbb{R}^{(n+1)*} \times \mathbb{R}^{n+1}/\approx'$, in the quotient topology. If $(x', y') \approx' (x, y)$, then x' and x satisfy the equivalence relation $x' \sim x$ used to define the real projective space $\mathbb{R}^{(n+1)*}/\sim\, = RP^n$, so that the first projection $\mathbb{R}^{(n+1)*} \times \mathbb{R}^{n+1} \xrightarrow{\pi_1} \mathbb{R}^{(n+1)*}$ induces a projection $E' \xrightarrow{\pi'} RP^n$.

7.3 Lemma: *The preceding projection $E' \xrightarrow{\pi'} RP^n$ is a smooth coordinate bundle that represents the Whitney sum $(n+1)\gamma_n^1$ of $n+1$ copies of the canonical line bundle γ_n^1 over RP^n.*

PROOF: We briefly recall the general construction of the total space of the coordinate bundle $E \xrightarrow{\pi} G^m(\mathbb{R}^{m+n})$ representing γ_n^m. One introduces an equivalence relation $\approx$ in $\mathbb{R}^{(m+n)m*} \times \mathbb{R}^m$ with

$$\left((x_1, \ldots, x_m), \begin{pmatrix} s^1 \\ \vdots \\ s^m \end{pmatrix}\right) \approx \left((y_1, \ldots, y_m), \begin{pmatrix} t^1 \\ \vdots \\ t^m \end{pmatrix}\right) \in \mathbb{R}^{(m+n)m*} \times \mathbb{R}^m$$

whenever both $(y_1, \ldots, y_m) = (x_1, \ldots, x_m)(a_p^q)^{-1}$ and

$$\begin{pmatrix} t^1 \\ \vdots \\ t^m \end{pmatrix} = (a_p^q) \begin{pmatrix} s^1 \\ \vdots \\ s^m \end{pmatrix}$$

for the same $(a_p^q) \in GL(m, \mathbb{R})$, and one sets $E = \mathbb{R}^{(m+n)m*} \times \mathbb{R}^m/\approx$. In case $m = 1$ one can regard multiplication by $a^{-1} \in \mathbb{R}^*$ as the action of $(a_p^q) \in GL(m, \mathbb{R})$, so that the total space of the coordinate bundle $E \xrightarrow{\pi} RP^n$ representing γ_n^1 can be described as follows: one introduces an equivalence relation $\approx$ in $\mathbb{R}^{(n+1)*} \times \mathbb{R}^1$ with $(x, s) \approx (y, t)$ whenever $(y, t) = (ax, a^{-1}s)$ for some $a \in \mathbb{R}^*$, and one sets $E = \mathbb{R}^{(n+1)*} \times \mathbb{R}^1/\approx$. A point of E is then a pair $(\pi(e), e)$, consisting of a 1-dimensional subspace $\pi(e) \subset \mathbb{R}^{n+1}$ spanned by some $x \in \mathbb{R}^{(n+1)*}$ and a point $sx = e \in \pi(e)$; if $(y, t) = (ax, a^{-1}s)$, then one has the same subspace $\pi(e) \in \mathbb{R}^{n+1}$ and the same point $ty = (a^{-1}s)(ax) = sx = e \in \pi(e)$. In order to conform to the notation of Corollary 6.16 we write x' in place of y, so that γ_n^1 has a total space $E = \mathbb{R}^{(n+1)*} \times \mathbb{R}^1/\approx$ with $(x', t) \approx (x, s)$ whenever $(x', t) = (ax, a^{-1}s)$ for some $a \in \mathbb{R}^*$. The corresponding total space E' of the Whitney sum $(n+1)\gamma_n^1$ of $n+1$ copies of γ_n^1 is then the quotient of $\mathbb{R}^{(n+1)*} \times \mathbb{R}^{n+1}$ by the equivalence relation $\approx'$ with

$$\left(x', \begin{pmatrix} t^0 \\ \vdots \\ t^n \end{pmatrix}\right) \approx' \left(x, \begin{pmatrix} s^0 \\ \vdots \\ s^n \end{pmatrix}\right)$$

whenever both $x' = ax \in \mathbb{R}^{(n+1)*}$ and

$$\begin{pmatrix} t^0 \\ \vdots \\ t^n \end{pmatrix} = a^{-1} \begin{pmatrix} s^0 \\ \vdots \\ s^n \end{pmatrix} \in \mathbb{R}^{n+1}$$

for the same $a \in \mathbb{R}^*$. The quotient $E' = \mathbb{R}^{(n+1)*} \times \mathbb{R}^{n+1}/\approx'$ and the obvious projection $E' \xrightarrow{\pi'} RP^n$ are precisely the definitions used in the statement of the lemma, in the notation

$$y = \begin{pmatrix} s^0 \\ \vdots \\ s^n \end{pmatrix} \qquad \text{and} \qquad y' = \begin{pmatrix} t^0 \\ \vdots \\ t^n \end{pmatrix}.$$

7.4 Proposition: *Let $\tau(RP^n) \oplus \varepsilon^1$ be the Whitney sum of the tangent bundle $\tau(RP^n)$ and the trivial line bundle ε^1 over the real projective space RP^n, and let $(n+1)\gamma_n^1$ be the Whitney sum of $n+1$ copies of the canonical line bundle γ_n^1 over RP^n; then $\tau(RP^n) \oplus \varepsilon^1 = (n+1)\gamma_n^1$.*

PROOF: According to Corollary 6.16 $\tau(RP^n) \oplus \varepsilon^1$ is represented by a coordinate bundle $E'' \xrightarrow{\pi''} RP^n$ with $E'' = \mathbb{R}^{(n+1)*} \times \mathbb{R}^{n+1}/\approx$, where $(x', y') \approx (x, y) \in \mathbb{R}^{(n+1)*} \times \mathbb{R}^{n+1}$ whenever $(x', y') = (ax, ay)$ for some $a \in \mathbb{R}^*$. According to Lemma 7.3, $(n+1)\gamma_n^1$ is similarly represented by a coordinate bundle $E' \xrightarrow{\pi'} RP^n$ with $E' = \mathbb{R}^{(n+1)*} \times \mathbb{R}^{n+1}/\approx'$, where $(x', y') \approx' (x, y) \in \mathbb{R}^{(n+1)*} \times \mathbb{R}^{n+1}$ whenever $(x', y') = (ax, a^{-1}y)$ for some $a \in \mathbb{R}^*$. Each of the projections $E'' \xrightarrow{\pi''} RP^n$ and $E' \xrightarrow{\pi'} RP^n$ is induced by the first projection $\mathbb{R}^{(n+1)*} \times \mathbb{R}^{n+1} \xrightarrow{\pi_1} \mathbb{R}^{(n+1)*}$; however, one must reconcile the evident difference in the equivalence relations $\approx$ and $\approx'$. Observe that if $S^n \subset \mathbb{R}^{(n+1)*}$ is the usual unit sphere, then one can alternatively describe RP^n as the quotient $S^n/\sim$, where $x' \sim x \in S^n$ whenever $x' = ax$ for some $a \in \mathbb{R}^*$ as before, and since x and x' are both unit vectors, one either has $a = 1$ or $a = -1$; that is, in this description of RP^n one can replace the group $GL(1, \mathbb{R}) = \mathbb{R}^*$ by the subgroup

$O(1) = \{+1, -1\} \subset \mathbb{R}^*$. Thus $E'' = S^n \times \mathbb{R}^{n+1}/\approx$, where $(x', y') \approx (x, y) \in S^n \times \mathbb{R}^{n+1}$ whenever $(x', y') = (ax, ay)$ for some $a \in O(1)$, and similarly $E' = S^n \times \mathbb{R}^{n+1}/\approx'$, where $(x', y') \approx' (x, y) \in S^n \times \mathbb{R}^{n+1}$ whenever $(x', y') = (ax, a^{-1}y)$ for some $a \in O(1)$, the projections $E'' \xrightarrow{\pi''} RP^n$ and $E' \xrightarrow{\pi'} RP^n$ both being induced by the first projection $S^n \times \mathbb{R}^{n+1} \xrightarrow{\pi_1} S^n$. Since $a^{-1} = a$ for both elements $a \in O(1)$, it follows that $E'' \xrightarrow{\pi''} RP^n$ and $E' \xrightarrow{\pi'} RP^n$ are the same coordinate bundle, hence that $\tau(RP^n) \oplus \varepsilon^1 = (n+1)\gamma_n^1$, as claimed.

8. The Homotopy Classification Theorem

One can replace the canonical real m-plane bundle γ_n^m over the Grassmann manifold $G^m(\mathbb{R}^{m+n}) \in \mathscr{B}$ by a corresponding real m-plane bundle γ^m over the Grassmann manifold $G^m(\mathbb{R}^\infty) \in \mathscr{B}$. We shall show that if ξ is any real m-plane bundle whatsoever over an arbitrary base space $X \in \mathscr{B}$, then ξ is the pullback $f^!\gamma^m$ of γ^m along a map $X \xrightarrow{f} G^m(\mathbb{R}^\infty)$ that is uniquely determined up to homotopy. For any real m-plane bundle ξ over a compact space $X \in \mathscr{B}$ (or over a finite-dimensional simplicial space $X = |K| \in \mathscr{B}$, or over a smooth manifold $X \in \mathscr{M} \subset \mathscr{B}$), one can substitute an appropriate finite-dimensional Grassmann manifold $G^m(\mathbb{R}^{m+n})$ for $G^m(\mathbb{R}^\infty)$; the homotopy uniqueness of $X \xrightarrow{f} G^m(\mathbb{R}^\infty)$ is then replaced by an "ersatz homotopy uniqueness" of the corresponding map $X \to G^m(\mathbb{R}^{m+n})$. In the latter cases it follows for some $n > 0$ that there is an n-plane bundle η over X for which $\xi \oplus \eta = \varepsilon^{m+n}$.

The real linear space $\mathbb{R}^\infty$ was defined in §1.5 as the direct limit $\lim_{n\to\infty} \mathbb{R}^n$, in the weak topology; elements of $\mathbb{R}^\infty$ are sequences of real numbers that contain only finitely many nonzero entries. It will be useful, although not essential, to regard elements $x \in \mathbb{R}^\infty$ as infinite column vectors

$$\begin{pmatrix} x^0 \\ x^1 \\ x^2 \\ \vdots \end{pmatrix};$$

for evident typographical reasons we shall occasionally abandon this convention. The Grassmann manifolds $G^m(\mathbb{R}^{m+n})$ and $G^m(\mathbb{R}^\infty)$ were both defined in Definition I.7.1; by Proposition I.7.5 they both belong to the category $\mathscr{B}$ of base spaces. Briefly, if $(\mathbb{R}^\infty)^{m*}$ is the set of ordered m-tuples $(x_1, \ldots, x_m)$ of linearly independent vectors $x_1, \ldots, x_m$ in $\mathbb{R}^\infty$, then there is an equivalence relation $\sim$ in $(\mathbb{R}^\infty)^{m*}$ with $(x_1, \ldots, x_m) \sim (y_1, \ldots, y_m)$ whenever $(x_1, \ldots, x_m)$ and $(y_1, \ldots, y_m)$ span the same m-plane in $\mathbb{R}^\infty$, and one sets $G^m(\mathbb{R}^\infty) = (\mathbb{R}^\infty)^{m*}/\sim$ in the quotient topology. As in the previous section, the relation

$(x_1, \ldots, x_m) \sim (y_1, \ldots, y_m)$ is the same as requiring that $(x_1, \ldots, x_m) = (y_1, \ldots, y_m)(a_p^q)$ for a unique invertible $m \times m$ matrix $(a_p^q) \in GL(m, \mathbb{R})$; equivalently, $(x_1, \ldots, x_m) \sim (y_1, \ldots, y_m) \in (\mathbb{R}^\infty)^{m*}$ whenever $(y_1, \ldots, y_m) = (x_1, \ldots, x_m)(a_p^q)^{-1}$ for a unique $(a_p^q) \in GL(m, \mathbb{R})$.

We mimic the construction of the canonical m-plane bundle γ_n^m over $G^m(\mathbb{R}^{m+n})$, which will lead to the corresponding m-plane bundle γ^m over $G^m(\mathbb{R}^\infty)$. Elements of the product space $(\mathbb{R}^\infty)^{m*} \times \mathbb{R}^m$ are of the form

$$\left((x_1, \ldots, x_m), \begin{pmatrix} s^1 \\ \vdots \\ s^m \end{pmatrix}\right)$$

for $(x_1, \ldots, x_m) \in (\mathbb{R}^\infty)^{m*}$ and

$$\begin{pmatrix} s^1 \\ \vdots \\ s^m \end{pmatrix} \in \mathbb{R}^m.$$

We write

$$\left((x_1, \ldots, x_m), \begin{pmatrix} s^1 \\ \vdots \\ s^m \end{pmatrix}\right) \approx \left((y_1, \ldots, y_m), \begin{pmatrix} t^1 \\ \vdots \\ t^m \end{pmatrix}\right)$$

whenever both $(y_1, \ldots, y_m) = (x_1, \ldots, x_m)(a_p^q)^{-1}$ and

$$\begin{pmatrix} t^1 \\ \vdots \\ t^q \end{pmatrix} = (a_p^q) \begin{pmatrix} s^1 \\ \vdots \\ s^m \end{pmatrix}$$

for the same $(a_p^q) \in GL(m, \mathbb{R})$, and we let E^∞ denote the quotient space $(\mathbb{R}^\infty)^{m*} \times \mathbb{R}^m/\approx$ in the quotient topology. As in the finite-dimensional case, if

$$\left((x_1, \ldots, x_m), \begin{pmatrix} s^1 \\ \vdots \\ s^m \end{pmatrix}\right) \approx \left((y_1, \ldots, y_m), \begin{pmatrix} t^1 \\ \vdots \\ t^m \end{pmatrix}\right)$$

in $(\mathbb{R}^\infty)^{m*} \times \mathbb{R}^m$, then $(x_1, \ldots, x_m) \sim (y_1, \ldots, y_m)$ in $(\mathbb{R}^\infty)^{m*}$, so that the first projection $(\mathbb{R}^\infty)^{m*} \times \mathbb{R}^m \xrightarrow{\pi_1} (\mathbb{R}^\infty)^{m*}$ induces a projection $E^\infty \xrightarrow{\pi^\infty} G^m(\mathbb{R}^\infty)$; as before, the preceding hypotheses imply $t^1 y_1 + \cdots + t^m y_m = s^1 x_1 + \cdots + s^m x_m \in \mathbb{R}^\infty$, so that elements of E^∞ may be regarded as pairs $(\pi^\infty(e), e)$ in which e is a point $s^1 x_1 + \cdots + x^m x_m$ in the m-plane $\pi^\infty(e) \subset \mathbb{R}^\infty$ spanned by $(x_1, \ldots, x_m)$.

8.1 Lemma: *The projection $E^\infty \xrightarrow{\pi^\infty} G^m(\mathbb{R}^\infty)$ is a coordinate bundle with structure group $GL(m, \mathbb{R})$ and fiber $\mathbb{R}^m$, representing a real m-plane bundle over the Grassmann manifold $G^m(\mathbb{R}^\infty)$.*

PROOF: One constructs local trivializations $E|U_{i_1 \cdots i_m} \xrightarrow{\Psi_i} U_{i_1 \cdots i_m} \times \mathbb{R}^m$ with resulting transition functions $U_{i_1 \cdots i_m} \cap U_{j_1 \cdots j_m} \xrightarrow{\psi_i^j} GL(m, \mathbb{R})$ that consist of invertible $m \times m$ matrices exactly as in the proof of Lemma 7.1; although smoothness of the latter entries is not defined, one easily verifies that they are continuous, as required.

8.2 Definition: The coordinate bundle $E^\infty \xrightarrow{\pi^\infty} G^m(\mathbb{R}^\infty)$ of Lemma 8.1 represents the *universal real m-plane bundle* γ^m over the Grassmann manifold $G^m(\mathbb{R}^\infty)$. In the case $m = 1$, for which $G^1(\mathbb{R}^\infty) = RP^\infty$, the coordinate bundle $E^\infty \xrightarrow{\pi^\infty} G^1(\mathbb{R}^\infty)$ of Lemma 8.1 represents the *universal real line bundle* over the real projective space RP^∞.

We now embark upon a sequence of lemmas that will lead to the main theorem of the section. They will be formulated in the following terminology.

8.3 Definition: Let $E \xrightarrow{\pi} X$ be a coordinate bundle with structure group $GL(m, \mathbb{R})$ and fiber $\mathbb{R}^m$ over a base space $X \in \mathscr{B}$, representing a real m-plane bundle over X. A *Gauss map* is any map $E \xrightarrow{\mathbf{g}} \mathbb{R}^\infty$ that restricts to a linear monomorphism $E_x \xrightarrow{\mathbf{g}_x} \mathbb{R}^\infty$ for the fiber $E_x \subset E$ over each $x \in X$.

Morphisms of arbitrary families of fibers were characterized in Definition II.1.1, and coordinate bundles are merely locally trivial families of fibers. In particular, for the structure group $GL(m, \mathbb{R})$ and fiber $\mathbb{R}^m$, a morphism from the coordinate bundle $E \xrightarrow{\pi} X$ of Definition 8.3 to the coordinate bundle $E^\infty \xrightarrow{\pi^\infty} G^m(\mathbb{R}^\infty)$ of Definition 8.2 is a commutative diagram

$$\begin{array}{ccc} E & \xrightarrow{\mathbf{f}} & E^\infty \\ \Big\downarrow{\scriptstyle \pi} & & \Big\downarrow{\scriptstyle \pi^\infty} \\ X & \xrightarrow{f} & G^m(\mathbb{R}^\infty) \end{array}$$

such that for each $x \in X$ the restriction $E_x \xrightarrow{\mathbf{f}_x} E^\infty_{f(x)}$ is a linear isomorphism of real n-dimensional vector spaces.

8.4 Lemma: *Given a coordinate bundle $E \xrightarrow{\pi} X$ as in Definition 8.3, there is a one-to-one correspondence between Gauss maps $E \xrightarrow{\mathbf{g}} \mathbb{R}^\infty$ and morphisms from $E \xrightarrow{\pi} X$ to $E^\infty \xrightarrow{\pi^\infty} G^m(\mathbb{R}^\infty)$.*

PROOF: Let $E \xrightarrow{\mathbf{g}} \mathbb{R}^\infty$ be a Gauss map for the given coordinate bundle. Then for any $e \in E$ the fiber $E_{\pi(e)}$ over $\pi(e) \in X$ is mapped isomorphically onto an m-dimensional subspace $\mathbf{g}(E_{\pi(e)}) \subset \mathbb{R}^\infty$, and $e \in E$ is itself mapped into a point $\mathbf{g}(e) \in \mathbf{g}(E_{\pi(e)})$. Since elements of $G^m(\mathbb{R}^\infty)$ are m-planes in $\mathbb{R}^\infty$, one can define $X \xrightarrow{f} G^m(\mathbb{R}^\infty)$ by setting $f(x) = \mathbf{g}(E_x)$ for each $x \in X$, and since elements of the fibers $E_{f(x)}$ over $f(x) \in G^m(\mathbb{R}^\infty)$ are points of the m-plane $\mathbf{g}(E_x) \subset \mathbb{R}^\infty$, one can define $E \xrightarrow{\mathbf{f}} E^\infty$ by setting $\mathbf{f}(e) = (\mathbf{g}(E_{\pi(e)}), \mathbf{g}(e))$ for each $e \in E$. The resulting commutative diagram

$$\begin{array}{ccc} E & \xrightarrow{\mathbf{f}} & E^\infty \\ \Big\downarrow{\scriptstyle \pi} & & \Big\downarrow{\scriptstyle \pi^\infty} \\ X & \xrightarrow{f} & G^m(\mathbb{R}^\infty) \end{array}$$

is the corresponding morphism of coordinate bundles with structure group $GL(m, \mathbb{R})$ and fiber $\mathbb{R}^m$. Conversely, suppose that one is given such a morphism of coordinate bundles. Then the composition $E \xrightarrow{\pi} X \xrightarrow{f} G^m(\mathbb{R}^\infty)$ carries any $e \in E$ into an m-dimensional subspace $f(\pi(e)) \subset \mathbb{R}^\infty$, and by commutativity of the diagram defining the morphism one has $\mathbf{f}(e) \in f(\pi(e))$; the composition $E \xrightarrow{\mathbf{f}} f(\pi(e)) \subset \mathbb{R}^\infty$ is itself the corresponding Gauss map $E \xrightarrow{\mathbf{g}} \mathbb{R}^\infty$.

In the next two lemmas we shall show for any Gauss maps $E \xrightarrow{\mathbf{g}_0} \mathbb{R}^\infty$ and $E \xrightarrow{\mathbf{g}_1} \mathbb{R}^\infty$ that there is a Gauss map $E \times [0, 1] \xrightarrow{\mathbf{g}} \mathbb{R}^\infty$ for the coordinate bundle $E \times [0, 1] \xrightarrow{\pi \times \mathrm{id}} X \times [0, 1]$, such that $\mathbf{g}|E \times \{0\} = \mathbf{g}_0$ and such that $\mathbf{g}|E \times \{1\} = \mathbf{g}_1$. We temporarily write elements of $\mathbb{R}^\infty$ as row vectors rather than column vectors, for typographical reasons.

8.5 Lemma: *Let $E \xrightarrow{\pi} X$ represent a real m-plane bundle over a base space $X \in \mathscr{B}$, let $E \xrightarrow{\mathbf{g}_0} \mathbb{R}^\infty$ be a Gauss map carrying any $e \in E$ into $(g^0(e), g^1(e), g^2(e), \ldots) \in \mathbb{R}^\infty$, and let $E \xrightarrow{\mathbf{g}_1} \mathbb{R}^\infty$ be the Gauss map carrying any $e \in E$ into $(0, g^0(e), 0, g^1(e), 0, g^2(e), \ldots)$; then there is a Gauss map $E \times [0, 1] \xrightarrow{\mathbf{g}} \mathbb{R}^\infty$ for the coordinate bundle $E \times [0, 1] \xrightarrow{\pi \times \mathrm{id}} X \times [0, 1]$, carrying any point $(e, t) \in E \times [0, 1]$ into $(1 - t)\,\mathbf{g}_0(e) + t\mathbf{g}_1(e) \in \mathbb{R}^\infty$.*

PROOF: The map $\mathbf{g}$ is trivially linear on each fiber $(E \times [0, 1])_{(x,t)}$, and the restrictions $\mathbf{g}|E \times \{0\}$ and $\mathbf{g}|E \times \{1\}$ are monomorphisms since $\mathbf{g}_0$ and $\mathbf{g}_1$ are Gauss maps; it remains to show for any $t \in (0, 1)$ that the restriction $\mathbf{g}|E \times \{t\}$ is a monomorphism. Suppose that $\mathbf{g}(e, t) = 0 \in \mathbb{R}^\infty$ for some $e \in E$ and some $t \in [0, 1]$ such that $t \neq 0$ and $1 - t \neq 0$. The vanishing of the 0th entry of $\mathbf{g}(e, t)$ would imply that $g^0(e) = 0$; hence the vanishing of the first entry of $\mathbf{g}(e, t)$ would imply that $g^1(e) = 0$; hence the vanishing of the second entry of $\mathbf{g}(e, t)$ would imply that $g^2(e) = 0$; and so on, which would imply the contradiction $\mathbf{g}(e) = 0 \in \mathbb{R}^\infty$.

8.6 Lemma: *Let $E \xrightarrow{\pi} X$ represent a real m-plane bundle over $X \in \mathscr{B}$, and let $E \xrightarrow{\mathbf{g}_0} \mathbb{R}^\infty$ and $E \xrightarrow{\mathbf{g}_1} \mathbb{R}^\infty$ be any Gauss maps. Then there is a Gauss map $E \times [0,1] \xrightarrow{\mathbf{g}} \mathbb{R}^\infty$ for the coordinate bundle $E \times [0,1] \xrightarrow{\pi \times \mathrm{id}} X \times [0,1]$, such that $\mathbf{g}|E \times \{0\} = \mathbf{g}_0$ and $\mathbf{g}|E \times \{1\} = \mathbf{g}_1$.*

PROOF: For any $e \in E$ let $\mathbf{g}_0(e) = (g_0^0(e), g_0^1(e), g_0^2(e), \ldots)$ and let $\mathbf{g}_1(e) = (g_1^0(e), g_1^1(e), g_1^2(e), \ldots)$. By Lemma 8.5 one can replace $\mathbf{g}_0$ by a Gauss map $E \xrightarrow{\tilde{\mathbf{g}}_0} \mathbb{R}^\infty$ with $\tilde{\mathbf{g}}_0(e) = (0, g_0^0(e), 0, g_0^1(e), 0, \ldots)$, and one can similarly replace $\mathbf{g}_1$ by a Gauss map $E \xrightarrow{\tilde{\mathbf{g}}_1} \mathbb{R}^\infty$ with $\mathbf{g}_1(e) = (g_1^0(e), 0, g_1^1(e), 0, g_1^2(e), \ldots)$. There is then a Gauss map $E \times [0,1] \xrightarrow{\tilde{\mathbf{g}}} \mathbb{R}^\infty$ given by setting $\tilde{\mathbf{g}}(e, t) = (1 - t)\tilde{\mathbf{g}}_0(e) + t\tilde{\mathbf{g}}_1(e)$ for any $(e, t) \in X \times [0,1]$, for which $\tilde{\mathbf{g}}|E \times \{0\} = \tilde{\mathbf{g}}_0$ and $\tilde{\mathbf{g}}|E \times \{1\} = \tilde{\mathbf{g}}_1$, as required.

8.7 Lemma: *Let $X \xrightarrow{f_0} G^m(\mathbb{R}^\infty)$ and $X \xrightarrow{f_1} G^m(\mathbb{R}^\infty)$ be maps of an arbitrary base space $X \in \mathscr{B}$ into the Grassmann manifold $G^m(\mathbb{R}^\infty)$, and let $f_0^!\gamma^m$ and $f_1^!\gamma^m$ be the corresponding pullbacks over X of the universal m-plane bundle γ^m over $G^m(\mathbb{R}^\infty)$; then $f_0^!\gamma^m = f_1^!\gamma^m$ if and only if f_0 is homotopic to f_1.*

PROOF: If f_0 is homotopic to f_1, then $f_0^!\gamma^m = f_1^!\gamma^m$ by Proposition II.4.7. Conversely, if $f_0^!\gamma^m = f_1^!\gamma^m$, then there is a coordinate bundle $E \xrightarrow{\pi} X$ that represents both $f_0^!\gamma^m$ and $f_1^!\gamma^m$, and by Lemma II.1.3 there are morphisms

$$\begin{array}{ccc} E & \xrightarrow{\mathbf{f}_0} & E^\infty \\ {\scriptstyle\pi}\downarrow & & \downarrow{\scriptstyle\pi^\infty} \\ X & \xrightarrow{f_0} & G^m(\mathbb{R}^\infty) \end{array} \qquad \text{and} \qquad \begin{array}{ccc} E & \xrightarrow{\mathbf{f}_1} & E^\infty \\ {\scriptstyle\pi}\downarrow & & \downarrow{\scriptstyle\pi^\infty} \\ X & \xrightarrow{f_1} & G^m(\mathbb{R}^\infty) \end{array}$$

from $E \xrightarrow{\pi} X$ to $E^\infty \xrightarrow{\pi^\infty} G^m(\mathbb{R}^\infty)$. Let $E \xrightarrow{\mathbf{g}_0} \mathbb{R}^\infty$ and $E \xrightarrow{\mathbf{g}_1} \mathbb{R}^\infty$ be the corresponding Gauss maps, as in Lemma 8.4. By Lemma 8.6 there is then a Gauss map $E \times [0,1] \xrightarrow{\mathbf{g}} \mathbb{R}^\infty$ for the coordinate bundle $E \times [0,1] \xrightarrow{\mathrm{id}} X \times [0,1]$ such that $\mathbf{g}|E \times \{0\} = \mathbf{g}_0$ and $\mathbf{g}|E \times \{1\} = \mathbf{g}_1$, and Lemma 8.4 provides a corresponding morphism

$$\begin{array}{ccc} E \times [0,1] & \xrightarrow{\mathbf{f}} & E^\infty \\ {\scriptstyle\pi \times \mathrm{id}}\downarrow & & \downarrow{\scriptstyle\pi^\infty} \\ X \times [0,1] & \xrightarrow{f} & G^m(\mathbb{R}^\infty). \end{array}$$

Since $\mathbf{g}|E \times \{0\} = \mathbf{g}_0$ and $\mathbf{g}|E \times \{1\} = \mathbf{g}_1$, the restrictions

$$\begin{array}{ccc} E \times \{0\} & \xrightarrow{\mathbf{f}_0} & E^\infty \\ {\scriptstyle\pi}\downarrow & & \downarrow{\scriptstyle\pi^\infty} \\ X \times \{0\} & \xrightarrow{f_0} & G^m(\mathbb{R}^\infty) \end{array} \quad \text{and} \quad \begin{array}{ccc} E \times \{1\} & \xrightarrow{\mathbf{f}_1} & E^\infty \\ {\scriptstyle\pi}\downarrow & & \downarrow{\scriptstyle\pi^\infty} \\ X \times \{1\} & \xrightarrow{f_1} & G^m(\mathbb{R}^\infty) \end{array}$$

are the given morphisms, so that $X \times [0,1] \xrightarrow{f} G^m(\mathbb{R}^\infty)$ is a homotopy from f_0 to f_1.

8.8 Lemma: *Let $E \xrightarrow{\pi} X$ represent any real m-plane bundle ξ over a base space $X \in \mathscr{B}$; then there is at least one Gauss map $E \xrightarrow{\mathbf{g}} \mathbb{R}^\infty$.*

PROOF: According to Proposition I.8.3 any base space $X \in \mathscr{B}$ is homotopy equivalent to a paracompact hausdorff space $X' \in \mathscr{B}$ with a countable open covering $\{U'_n | n \in \mathbb{N}\}$ such that each connected component of each U'_n is contained in a contractible open set in X'. Let $X \xrightarrow{f} X'$ be the homotopy equivalence, with homotopy inverse $X' \xrightarrow{g} X$, and let ξ' be the pullback $g^!\xi$. Since X' is paracompact, one has $f^!\xi' = f^!g^!\xi = \xi$ over X by Lemma II.4.5, so that if $E' \xrightarrow{\pi'} X'$ represents ξ', then there is a morphism

$$\begin{array}{ccc} E & \xrightarrow{\mathbf{f}} & E' \\ {\scriptstyle\pi}\downarrow & & \downarrow{\scriptstyle\pi'} \\ X & \xrightarrow{f} & X' \end{array}$$

of coordinate bundles, as in Lemma II.1.3. Since each restriction $E_x \xrightarrow{\mathbf{f}_x} E'_{f(x)}$ is an isomorphism of real m-dimensional vector spaces, the composition $E \xrightarrow{\mathbf{f}} E' \xrightarrow{\mathbf{g}'} \mathbb{R}^\infty$ of $\mathbf{f}$ with any Gauss map $E' \xrightarrow{\mathbf{g}'} \mathbb{R}^\infty$ for $E' \xrightarrow{\pi'} X'$ will be a Gauss map $E \xrightarrow{\mathbf{g}'} \mathbb{R}^\infty$ for $E \xrightarrow{\pi} X$, as desired.

Thus it remains to construct a Gauss map $E' \xrightarrow{\mathbf{g}'} \mathbb{R}^\infty$; that is, one may as well suppose at the outset that $E \xrightarrow{\pi} X$ itself represents a real m-plane bundle bundle ξ over a paracompact hausdorff space $X \in \mathscr{B}$ with a countable locally finite open covering $\{U_n | n \in \mathbb{N}\}$ such that each connected component of each U_n is contained in a contractible open set in X. Since any fibre bundle over any contractible space is trivial, by Proposition II.3.5, each restriction $\xi | U_n$ is trivial, and one chooses a trivialization $E|U_n \xrightarrow{\Psi_n} U_n \times \mathbb{R}^m$ for each $n \in \mathbb{N}$. Since X is paracompact, there is a partition of unity $\{h_n | n \in \mathbb{N}\}$

subordinate to the countable locally finite open covering $\{U_n | n \in \mathbb{N}\}$, and there is then a well-defined map $E \xrightarrow{\sum_{n \in \mathbb{N}} h_n \Psi_n} (\mathbb{R}^m)^\infty = \mathbb{R}^\infty$ carrying any $e \in E$ into the sequence

$$(h_0(\pi(e))\Psi_0(e), h_1(\pi(e))\Psi_1(e), h_2(\pi(e))\Psi_2, \ldots) \in (\mathbb{R}^m)^\infty;$$

trivially the map $\sum_{n \in \mathbb{N}} h_n \Psi_n$ is itself a Gauss map $\mathbf{g}$.

8.9 Theorem (Homotopy Classification Theorem): *Any real m-plane bundle ξ over a base space $X \in \mathscr{B}$ is a pullback $f^! \gamma^m$ of the universal real m-plane bundle γ^m over $G^m(\mathbb{R}^\infty) \in \mathscr{B}$, along a map $X \xrightarrow{f} G^m(\mathbb{R}^\infty)$ that is unique up to homotopy.*

PROOF: Let $E \xrightarrow{\pi} X$ be any coordinate bundle representing ξ. By Lemma 8.8 there is a Gauss map $E \xrightarrow{\mathbf{g}} \mathbb{R}^\infty$, so that Lemma 8.4 provides a morphism

$$\begin{array}{ccc} E & \xrightarrow{\mathbf{f}} & E^\infty \\ \downarrow{\scriptstyle \pi} & & \downarrow{\scriptstyle \pi^\infty} \\ X & \xrightarrow{f} & G^m(\mathbb{R}^\infty) \end{array}$$

to the coordinate bundle $E^\infty \xrightarrow{\pi^\infty} G^m(\mathbb{R}^\infty)$ representing γ^m; hence $\xi = f^! \gamma^m$ for the map $X \xrightarrow{f} G^m(\mathbb{R}^\infty)$, by Corollary II.1.13. The uniqueness of f up to homotopy was proved in Lemma 8.7.

The homotopy classification theorem justifies the extravagant language of Definition 8.2: the m-plane bundle γ^m is indeed universal. The base space $G^m(\mathbb{R}^\infty)$ of γ^m is itself dignified by a suitable name: it is the *classifying space* for real m-plane bundles. Specifically, according to the homotopy classification theorem there is a one-to-one correspondence between the real m-plane bundles over any base space $X \in \mathscr{B}$ and the set $[X, G^m(\mathbb{R}^\infty)]$ of homotopy classes of maps $X \to G^m(\mathbb{R}^\infty)$.

In the proof of the next result we use an alternative description of the coordinate bundle $E^\infty \xrightarrow{\pi^\infty} G^m(\mathbb{R}^\infty)$ of Lemma 8.1, which represents the universal real m-plane bundle γ^m. The total space E^∞ is the quotient $(\mathbb{R}^\infty)^{m*} \times \mathbb{R}^m/\approx$, as before; however, there is another way of expressing the equivalence relation $\approx$. One has

$$\left((x_1, \ldots, x_m), \begin{pmatrix} s^1 \\ \vdots \\ s^m \end{pmatrix} \right) \approx \left((y_1, \ldots, y_m), \begin{pmatrix} t^1 \\ \vdots \\ t^m \end{pmatrix} \right)$$

in $(\mathbb{R}^\infty)^{m*} \times \mathbb{R}^m$ if and only if $x_1 \wedge \cdots \wedge x_m = ay_1 \wedge \cdots \wedge y_m \in \bigwedge^m \mathbb{R}^\infty$, for some $a \in \mathbb{R}^*$, and $s^1x_1 + \cdots + s^mx_m = t^1y_1 + \cdots + t^my_m \in \mathbb{R}^\infty$. The subspace $E^{\infty *}$ of nonzero fibers in E^∞ is then the quotient $(\mathbb{R}^\infty)^{m*} \times \mathbb{R}^{m*}/\approx$, for the same equivalence relation $\approx$. It follows that one can use a change of basis to represent any equivalence class

$$\left[\left((x_1, \ldots, x_m), \begin{pmatrix} s^1 \\ \vdots \\ s^m \end{pmatrix}\right)\right] \in E^{\infty *}$$

in the fashion

$$\left[\left((y_1, \ldots, y_m), \begin{pmatrix} 0 \\ \vdots \\ 0 \\ 1 \end{pmatrix}\right)\right] \in E^{\infty *}$$

for the (nonzero) vector $y_m = s^1x_1 + \cdots + s^mx_m \in \mathbb{R}^{\infty *}$. Thus $E^{\infty *}$ consists of pairs (y_m, P_{m-1}) for $y_m \in \mathbb{R}^{\infty *}$, where P_{m-1} is the $(m-1)$-plane spanned by $y_1, \ldots, y_{m-1}$; that is, P_{m-1} is any $(m-1)$-plane not containing y_m.

8.10 Proposition: *For any $m > 1$ let $E^\infty \xrightarrow{\pi^\infty} G^m(\mathbb{R}^\infty)$ represent the universal real m-plane bundle γ^m, and let $E^{\infty *}$ be the space of nonzero fibers in E^∞. Then there is a homotopy equivalence $G^{m-1}(\mathbb{R}^\infty) \xrightarrow{h} E^{\infty *}$ such that the composition $G^{m-1}(\mathbb{R}^\infty) \xrightarrow{h} E^{\infty *} \xrightarrow{\pi^\infty} G^m(\mathbb{R}^\infty)$ classifies the Whitney sum $\gamma^{m-1} \oplus \varepsilon^1$ over $G^{m-1}(\mathbb{R}^\infty)$.*

PROOF: Let $\mathbb{R}_0^\infty$ be the subspace of all vectors in $\mathbb{R}^\infty$ with 0th entry $0 \in \mathbb{R}$; then the inclusion $\mathbb{R}_0^\infty \subset \mathbb{R}^\infty$ is a homeomorphism whose inverse is the obvious translation of coefficients. Since the inclusion is linear, there is an induced inclusion $G^{m-1}(\mathbb{R}_0^\infty) \xrightarrow{i} G^{m-1}(\mathbb{R}^\infty)$ that is also a homeomorphism. If $y_0 \in \mathbb{R}^{\infty *}$ is the fixed vector with 0th entry $1 \in \mathbb{R}$ and all other entries $0 \in \mathbb{R}$, there is a map $G^{m-1}(\mathbb{R}_0^\infty) \xrightarrow{h_0} E^{\infty *}$ that carries any $(m-1)$-plane $P_{m-1} \in G^{m-1}(\mathbb{R}_0^\infty)$ into the pair (y_0, P_{m-1}).

Let $G^{m-1}(\mathbb{R}^\infty) \xrightarrow{h} E^{\infty *}$ be the composition $h_0 \circ i^{-1}$. The second projection, mapping $(y_m, P_{m-1}) \in E^{m*}$ to $P_{m-1} \in G^{m-1}(\mathbb{R}^\infty)$, is a map $E^{m*} \xrightarrow{k} G^{m-1}(\mathbb{R}^\infty)$ for which the composition $k \circ h$ is trivially the identity map on $G^{m-1}(\mathbb{R}^\infty)$. To show that $h \circ k$ is homotopic to the identity map on E^{m*} one lets $E^{m*} \times [0, 1] \xrightarrow{F} E^{m*}$ carry $((y_m, P_{m-1}), t)$ into $((1-t)y_m + ty_0, P_{m-1})$. Then F_0 is the identity map on E^{m*} and F_1 is the composition $h \circ k$; thus h is a homotopy equivalence.

Since $\pi^\infty \circ h$ is induced by the inclusion $\mathbb{R}^{(m-1)*} \to \mathbb{R}^{m*}$ carrying $(y_1, \ldots, y_{m-1})$ into $(y_0, y_1, \ldots, y_{m-1})$, which induces a (linear) isomorphism

$$(t^0, t^1 y_1 + \cdots + t^{m-1} y_{m-1}) \to (t^0 y_0 + t^1 y_1 + \cdots + t^{m-1} y_{m-1})$$

over each

$$[(y_1, \ldots, y_{m-1})] \mapsto [(y_0, y_1, \ldots, y_{m-1})],$$

it follows that there is a pullback diagram

$$\begin{array}{ccc} E & \xrightarrow{\pi^\infty \circ h} & E^\infty \\ \downarrow{\scriptstyle \pi} & & \downarrow{\scriptstyle \pi^\infty} \\ G^{m-1}(\mathbb{R}^\infty) & \xrightarrow{\pi^\infty \circ h} & G^m(\mathbb{R}^\infty) \end{array}$$

in which the left-hand vertical arrow represents $\varepsilon^1 \oplus \gamma^{m-1}$ $(=\gamma^{m-1} \oplus \varepsilon^1)$. Hence $(\pi^\infty \circ h)^! \gamma^m = \gamma^{m-1} \oplus \varepsilon^1$; that is, $\pi^\infty \circ h$ classifies $\gamma^{m-1} \oplus \varepsilon^1$, which completes the proof.

Proposition 8.10 omits the case $m = 1$; here it is.

8.11 Proposition: *If $E^\infty \xrightarrow{\pi^\infty} RP^\infty$ represents the universal real line bundle γ^1, then the space $E^{\infty *}$ of nonzero fibers in E^∞ is homeomorphic to $\mathbb{R}^{\infty *}$; furthermore, $\mathbb{R}^{\infty *}$ is contractible.*

PROOF: This time $E^{\infty *}$ is a quotient $\mathbb{R}^{\infty *} \times \mathbb{R}^*/\approx$ with $(x, s) \approx (y, t)$ if and only if $sx = ty \in \mathbb{R}^{\infty *}$; that is, $E^{\infty *} = \mathbb{R}^{\infty *}$. Now let $\mathbb{R}^{\infty *} \xrightarrow{j} \mathbb{R}^{\infty *}$ be the obvious shift such that for every $x \in \mathbb{R}^{\infty *}$ the initial coefficient of $j(x) \in \mathbb{R}^{\infty *}$ is $0 \in \mathbb{R}$. One easily verifies that $(1 - 2t)x + 2tj(x) \in \mathbb{R}^{\infty *}$ for every $(x, t) \in \mathbb{R}^{\infty *} \times [0, \frac{1}{2}]$, and, if $x_0 \in \mathbb{R}^{\infty *}$ has initial coefficient $1 \in \mathbb{R}$, then $(2t - 1)x_0 + (2 - 2t)j(x) \in \mathbb{R}^{\infty *}$ for every $(x, t) \in \mathbb{R}^{\infty *} \times [\frac{1}{2}, 1]$. Hence there is a contraction $\mathbb{R}^{\infty *} \times [0, 1] \to \mathbb{R}^{\infty *}$ to $x_0 \in \mathbb{R}^{\infty *}$.

There are other homotopy classification theorems that apply to restricted categories of base spaces, or to vector bundles other than the real m-plane bundles of Definition 1.1. Here is one example of such a theorem; there will be other examples later.

8.12 Proposition: *For any compact base space $X \in \mathscr{B}$ and any $m > 0$ there is an $n > 0$ with the following property: any real m-plane bundle ξ over X is a* *pullback $f^!\gamma_n^m$ of the canonical m-plane bundle γ_n^m over the Grassmann manifold $G^m(\mathbb{R}^{m+n}) \in \mathscr{B}$, along some map $X \xrightarrow{f} G^m(\mathbb{R}^{m+n})$.*

PROOF: Let $E \xrightarrow{\pi} X$ represent ξ. Since X is compact, one can replace the *countable* open covering $\{U_n | n \in \mathbb{N}\}$ used in Lemma 8.8 by a *finite* open covering $\{U_1, \ldots, U_p\}$ of X, with the same properties. In this way one obtains a Gauss map $E \xrightarrow{\mathbf{g}} \mathbb{R}^{mp} \subset \mathbb{R}^\infty$, and for $n = m(p-1)$ the evident analog of Lemma 8.4 then provides a corresponding morphism

$$\begin{array}{ccc} E & \xrightarrow{\mathbf{f}} & E' \\ \downarrow{\scriptstyle\pi} & & \downarrow{\scriptstyle\pi'} \\ X & \xrightarrow{f} & G^m(\mathbb{R}^{m+n}), \end{array}$$

where $E' \xrightarrow{\pi'} G^m(\mathbb{R}^{m+n})$ represents γ_n^m.

8.13 Corollary: *Let ξ be any real m-plane bundle over a compact base space $X \in \mathscr{B}$; then for some $n > 0$ there is an n-plane bundle η over X such that the Whitney sum $\xi \oplus \eta$ is the trivial bundle ε^{m+n} over X.*

PROOF: Recall that $G^m(\mathbb{R}^{m+n})$ consists of m-planes $V \subset \mathbb{R}^{m+n}$, so that if $V^\perp \subset \mathbb{R}^{m+n}$ is the orthogonal complement of V with respect to a fixed inner product on $\mathbb{R}^{m+n}$, then there is a diffeomorphism $G^n(\mathbb{R}^{m+n}) \xrightarrow{g} G^m(\mathbb{R}^{m+n})$ carrying $V^\perp$ into V, as one easily verifies. Hence the total space of the bundle $g^!\gamma_m^n$ over $G^m(\mathbb{R}^{m+n})$ consists of pairs $(\pi(g(e')), e')$ in which $\pi(g(e'))$ is an m-plane $V \subset \mathbb{R}^{m+n}$ and e' is a point of the orthogonal n-plane $V^\perp \subset \mathbb{R}^{m+n}$. Since the total space of γ_n^m consists of pairs $(\pi(e), e)$ in which $\pi(e)$ is an m-plane $V \subset \mathbb{R}^{m+n}$ and e is a point of V, it follows that the total space of $\gamma_n^m \oplus g^!\gamma_m^n$ consists of pairs $(\pi(e), e + e')$ in which $\pi(e)$ is an m-plane $V \subset \mathbb{R}^{m+n}$ and $e + e'$ is *any* point of $\mathbb{R}^{m+n}$; this is precisely the description of the trivial bundle ε^{m+n} over $G^m(\mathbb{R}^{m+n})$, so that $\gamma_n^m \oplus g^!\gamma_m^n = \varepsilon^{m+n}$. Consequently if $\xi = f^!\gamma_n^m$, as in Proposition 8.12, then one can set $\eta = f^!g^!\gamma_m^n$ to conclude that

$$\xi \oplus \eta = f^!(\gamma_n^m \oplus g^!\gamma_m^n) = f^!\varepsilon^{m+n} = \varepsilon^{m+n},$$

as desired.

Proposition 8.12 lacks one feature of the homotopy classification theorem (Theorem 8.9). The *classifying map* $X \xrightarrow{f} G^m(\mathbb{R}^\infty)$ of Theorem 8.9 is unique up to homotopy; but the *finite classifying map* $X \xrightarrow{f} G^m(\mathbb{R}^{m+n})$ of Proposition 8.12 is in general *not* unique up to homotopy. Fortunately there is a simple remedy in the latter case.

Let $E \xrightarrow{\pi} G^m(R^{m+n})$ represent the canonical m-plane bundle γ_n^m over $G^m(\mathbb{R}^{m+n})$, and for any $n'' \geqq n$ let $E'' \xrightarrow{\pi''} G^m(\mathbb{R}^{m+n''})$ represent the canonical m-plane bundle $\gamma_{n''}^m$ over $G^m(\mathbb{R}^{m+n''})$. The usual linear inclusion $\mathbb{R}^{m+n} \to \mathbb{R}^{m+n''}$ then provides a commutative diagram

$$\begin{array}{ccc} E & \xrightarrow{g_{n,n''}} & E \\ \downarrow \pi & & \downarrow \pi'' \\ G^m(\mathbb{R}^{m+n}) & \xrightarrow{g_{n,n''}} & G^m(\mathbb{R}^{m+n''}) \end{array}$$

that is a (linear) isomorphism in each fiber, so that $g_{n,n''}$ is a finite classifying map for γ_n^m itself; that is $\gamma_n^m = g_{n,n''}^! \gamma_{n''}^m$. Clearly if $n \leqq n' \leqq n''$, then one has $g_{n',n''} \circ g_{n,n'} = g_{n,n''}$ and $\gamma_n^m = g_{n,n'}^! g_{n',n''}^! \gamma_{n''}^m$. The maps $g_{n,n'}$, $g_{n,n''}$, and $g_{n',n''}$ are *finite classifying extensions.*

8.14 Proposition (Ersatz Homotopy Uniqueness Theorem): *Let ξ be a real m-plane bundle over $X \in \mathscr{B}$ such that $\xi = f_0^! \gamma_n^m$ and $\xi = f_1^! \gamma_{n'}^m$ for maps $X \xrightarrow{f_0} G^m(\mathbb{R}^{m+n})$ and $X \xrightarrow{f_1} G^m(\mathbb{R}^{m+n'})$. Then there is an $n'' \geqq \max(n, n')$, with finite classifying extensions $g_{n,n''}$ and $g_{n',n''}$, respectively, such that the compositions*

$$X \xrightarrow{f_0} G^m(\mathbb{R}^{m+n}) \xrightarrow{g_{n,n''}} G^m(\mathbb{R}^{m+n''})$$

and

$$X \xrightarrow{f_1} G^m(\mathbb{R}^{m+n'}) \xrightarrow{g_{n',n''}} G^m(\mathbb{R}^{m+n''})$$

are homotopic maps from X to $G^m(\mathbb{R}^{m+n''})$, for which

$$(g_{n,n''} \circ f_0)^! \gamma_{n''}^m = \xi = (g_{n',n''} \circ f_1)^! \gamma_{n''}^m.$$

PROOF: One may as well assume that $n' \geqq n$, in which case one sets $n'' = m + 2n'$; then $m + n'' = 2(m + n') \geqq 2(m + n)$. If $E \xrightarrow{\pi} X$ represents ξ, then the maps f_0 and f_1 are equivalent to finite-dimensional Gauss maps $E \to \mathbb{R}^{m+n} \subset \mathbb{R}^{m+n'}$ and $E \to \mathbb{R}^{m+n'}$, respectively, so that the compositions $g_{n,n''} \circ f_0$ and $g_{n',n''} \circ f_1$ both correspond to finite-dimensional Gauss maps $E \to \mathbb{R}^{2(m+n')}$ in which the last $m + n'$ entries are identically zero. The latter condition provides just the right amount of elbow room for obvious finite-dimensional analogs of Lemmas 8.5 and 8.6; the homotopy from the composition $g_{n,n''} \circ f_0$ to the composition $g_{n',n''} \circ f_1$ is then constructed as in Lemma 8.7.

Proposition 8.14 was discovered in 1949, among the unpublished papers of the late Friedrich Adolph Karl Ersatz.

9. More Smooth Vector Bundles

We prove that any vector bundle whatsoever over a smooth manifold can be represented by a smooth coordinate bundle, as announced in §5.

The first step is a simplicial analog of Proposition 8.12, which does not require compactness.

9.1 Proposition: *Given natural numbers $m > 0$ and $q > 0$, set $n = mq$. Then any real m-plane bundle ξ over any q-dimensional metric simplicial space $|K|$ is a pullback $f^!\gamma_n^m$ of the canonical m-plane bundle γ_n^m over the Grassmann manifold $G^m(\mathbb{R}^{m+n})$, along some map $|K| \xrightarrow{f} G^m(\mathbb{R}^{m+n})$.*

PROOF: Since there is a canonical homeomorphism $|K'| \xrightarrow{\Phi} |K|$, for the first barycentric subdivision K' of the simplicial complex K, one may as well regard ξ as a bundle over $|K'|$. According to Proposition I.2.1, $|K'|$ is of first type. Specifically, since K is q-dimensional, there is a covering $\{U_{0,\alpha}\}_\alpha, \ldots, \{U_{q,\gamma}\}_\gamma$ of $|K'|$ by $q+1$ families of contractible open sets $U_{p,\beta} \subset |K'|$, the sets in each family $\{U_{p,\beta}\}_\beta$ being mutually disjoint. Let $U_p = \bigcup_\beta U_{p,\beta} \subset |K'|$ for each $p = 0, \ldots, q$, so that $\{U_0, \ldots, U_q\}$ is a finite open covering of $|K'|$. Since any fibre bundle over a contractible space is trivial, by Proposition II.3.5, it follows that each restriction $\xi|U_{p,\beta}$ is trivial, and since each family $\{U_{p,\beta}\}_\beta$ is mutually disjoint, it follows that each restriction $\xi|U_p$ is trivial. Thus if $E \xrightarrow{\pi} |K'|$ represents ξ, then one can choose $q+1$ specific trivializations $E|U_0 \xrightarrow{\Psi_0} U_0 \times \mathbb{R}^m, \ldots, E|U_q \xrightarrow{\Psi_q} U_q \times \mathbb{R}^m$ and a partition of unity $\{h_0, \ldots, h_q\}$ subordinate to $\{U_0, \ldots, U_q\}$ to obtain a Gauss map $E \xrightarrow{h_0\Psi_0 + \cdots + h_q\Psi_q} R^{m(q+1)}$ as in Lemma 8.8. The analog of Lemma 8.4 then provides a corresponding morphism

$$\begin{array}{ccc} E & \xrightarrow{\mathbf{f}} & E' \\ \Big\downarrow{\scriptstyle\pi} & & \Big\downarrow{\scriptstyle\pi'} \\ |K'| & \xrightarrow{f} & G^m(\mathbb{R}^{m+n}), \end{array}$$

where $E' \xrightarrow{\pi'} G^m(\mathbb{R}^{m+n})$ represents γ_n^m.

Since Proposition I.4.6 (Dowker [1]) provides a homotopy equivalence $|K|_m \to |K|_w$ from any metric simplicial space $|K|_m$ to its weak counterpart $|K|_w$, Proposition 9.1 applies equally well to q-dimensional *weak* simplicial spaces. Hence, one may as well omit the metric condition from the statement of Proposition 9.1.

9.2 Corollary: *Let ξ be any real m-plane bundle over any finite-dimensional simplicial space $|K|$; then for some $n > 0$ there is an n-plane bundle η over $|K|$ such that the Whitney sum $\zeta \oplus \eta$ is the trivial bundle ε^{m+n} over $|K|$.*

PROOF: Substitute Proposition 9.1 for Proposition 8.12 in the proof of Corollary 8.13.

If the simplicial space $|K|$ in Corollary 9.2 is q-dimensional, then Proposition 9.1 permits one to set $n = mq$. However, there is also a direct proof, using a general position argument, in which one can set $n = q$.

The second step toward the main theorem of this section combines Proposition 9.1 with Theorem I.6.7 (the Cairns–Whitehead triangulation theorem) and Theorem I.6.19 (that any map of smooth manifolds is homotopic to a smooth map); one also uses Proposition I.7.3 (that the Grassmann manifold $G^m(\mathbb{R}^{m+n})$ is a smooth closed mn-dimensional manifold).

9.3 Proposition: *Given natural numbers $m > 0$ and $q > 0$, set $n = mq$. Then any real m-plane bundle ξ over any smooth q-dimensional manifold X is a pullback $f^!_1\gamma^m_n$ of the canonical m-plane bundle γ^m_n over $G^m(\mathbb{R}^{m+n})$ along a smooth map $X \xrightarrow{f_1} G^m(\mathbb{R}^{m+n})$.*

PROOF: By the Cairns–Whitehead theorem X is homeomorphic to a q-dimensional metric simplicial space $|K|$, so that by Proposition 9.1 ξ is a pullback $f^!_0\gamma^m_n$ along some map $X \xrightarrow{f_0} G^m(\mathbb{R}^{m+n})$. By Proposition II.4.7 one can replace f_0 by any map $X \xrightarrow{f_1} G^m(\mathbb{R}^{m+n})$ homotopic to f_0, and by Theorem I.6.19 one can choose a smooth such map f_1.

9.4 Corollary: *Let ξ be any real m-plane bundle over any smooth manifold X; then for some $n > 0$ there is an n-plane bundle η over X such that the Whitney sum $\xi \oplus \eta$ is the trivial bundle ε^{m+n} over X.*

PROOF: Substitute Proposition 9.3 for Proposition 8.12 in the proof of Corollary 8.13.

If the smooth manifold X in Corollary 9.4 is q-dimensional, then Proposition 9.3 permits one to set $n = mq$. However, there is also a direct proof, using the transversality theorem, in which one can set $n = q$. (See pages 99–101 of M. W. Hirsch [4], for example.)

The main result of this section is another corollary of Proposition 9.3.

9.5 Theorem: *Given a smooth manifold X, any real vector bundle ξ over X can be represented by a smooth coordinate bundle $E \overset{\pi}{\to} X$; that is, any real vector bundle over X is itself smooth, as in Definition 5.2.*

PROOF: If ξ is an m-plane bundle, then $\xi = f^! \gamma_n^m$ for some $n > 0$ and some smooth map $X \xrightarrow{f} G^m(\mathbb{R}^{m+n})$, by Proposition 9.3. Since γ_n^m is smooth by Lemma 7.1, it follows from Proposition 5.3 that its pullback $f^! \gamma_n^m$ is smooth.

10. Orientable Vector Bundles

The familiar parlor-trick "one-sidedness" of the Möbius band E reflects the failure of the tangent bundle $\tau(E)$ to be orientable, in the sense described in this section. To define orientability of any real m-plane bundle ξ over any $X \in \mathscr{B}$ one first constructs a real line bundle $\bigwedge^m \xi$ over X, from which one obtains a fiber bundle $o(\xi)$ over X with structure group $O(1)$ $(=\mathbb{Z}/2)$ and fiber S^0 $(=\mathbb{Z}/2)$; the bundle ξ is *orientable* if and only if $o(\xi)$ is trivial. Following the construction of $o(\xi)$ we shall show that ξ is orientable if and only if it is of the form $f^! \gamma^m$ for a map $X \xrightarrow{f} G^m(\mathbb{R}^\infty)$ that factors through the total space of the bundle $o(\gamma^m)$ over $G^m(\mathbb{R}^\infty)$. Finally we show that the canonical line bundle γ_1^1 over RP^1 $(=S^1)$ is not orientable, and that the tangent bundle $\tau(E)$ of the Möbius band E is not orientable.

Let $E \xrightarrow{\pi} X$ represent a real m-plane bundle ξ over $X \in \mathscr{B}$, and for any natural number $p > 0$ let $E \times \cdots \times E \xrightarrow{\pi \times \cdots \times \pi} X \times \cdots \times X$ be the product of p copies of $E \xrightarrow{\pi} X$. The pullback of the product along the diagonal map $X \xrightarrow{\Delta} X \times \cdots \times X$ is a coordinate bundle $E' \xrightarrow{\pi'} X$ with fiber $\mathbb{R}^m \times \cdots \times \mathbb{R}^m$ and structure group $GL(m, \mathbb{R}) \times \cdots \times GL(m, \mathbb{R})$, and we recall from Proposition 2.5 that one can construct the Whitney sum $\xi \oplus \cdots \oplus \xi$ and product $\xi \otimes \cdots \otimes \xi$ by applying the morphisms $(\Gamma^\oplus, \Phi^\oplus)$ and $(\Gamma^\otimes, \Phi^\otimes)$ to $E' \xrightarrow{\pi'} X$; these bundles have structure groups $GL(mp, \mathbb{R})$, $GL(m^p, \mathbb{R})$ and fibers $\mathbb{R}^m \oplus \cdots \oplus \mathbb{R}^m$, $\mathbb{R}^m \otimes \cdots \otimes \mathbb{R}^m$, respectively. We now modify $E' \xrightarrow{\pi'} X$ for another purpose.

Let $GL(m, \mathbb{R})$ itself act on the left of $\mathbb{R}^m \times \cdots \times \mathbb{R}^m$, with $g(x_1, \ldots, x_p) = (gx_1, \ldots, gx_p)$ for every $g \in GL(m, \mathbb{R})$ and $(x_1, \ldots, x_p) \in \mathbb{R}^m \times \cdots \times \mathbb{R}^m$. If the original coordinate bundle $E \xrightarrow{\pi} X$ is defined with respect to a covering $\{U_i \mid i \in I\}$ and transition functions $U_i \cap U_j \xrightarrow{\psi_i^j} GL(m, \mathbb{R})$, one can then construct a new coordinate bundle $E'' \xrightarrow{\pi''} X$ with respect to $\{U_i \mid i \in I\}$, with fiber $\mathbb{R}^m \times \cdots \times \mathbb{R}^m$ and structure group $GL(m, \mathbb{R})$: one uses the same transition functions ψ_i^j and the preceding action of $GL(m, \mathbb{R})$ on $\mathbb{R}^m \times \cdots \times \mathbb{R}^m$.

Now let $\mathbb{R}^m \times \cdots \times \mathbb{R}^m \xrightarrow{\Phi} \bigwedge^p \mathbb{R}^m$ be the map carrying $(x_1, \ldots, x_p) \in \mathbb{R}^m \times \cdots \times \mathbb{R}^m$ into the exterior product $x_1 \wedge \cdots \wedge x_p \in \bigwedge^p \mathbb{R}^m$, and let $GL(m, \mathbb{R}) \xrightarrow{\Gamma} GL(\binom{m}{p}, \mathbb{R})$ be the group homomorphism carrying $g \in GL(m, \mathbb{R})$

into that element of $GL(\binom{m}{p}, \mathbb{R})$ with value $gx_1 \wedge \cdots \wedge gx_p$ on any $x_1 \wedge \cdots \wedge x_p \in \bigwedge^p \mathbb{R}^m$. The pair (Γ, Φ) is a morphism of transformation groups in the sense of Definition II.2.6, so that by Proposition II.2.7 one can apply (Γ, Φ) to the coordinate bundle $E'' \xrightarrow{\pi''} X$ to obtain an $\binom{m}{p}$-plane bundle over X, which is independent of the coordinate bundle $E \xrightarrow{\pi} X$ chosen to represent the m-plane bundle ξ.

10.1 Definition: Given an m-plane bundle ξ over $X \in \mathscr{B}$, the pth *exterior power* $\bigwedge^p \xi$ is the preceding $\binom{m}{p}$-plane bundle over X.

If $m = p$, then $\bigwedge^m \xi$ is a line bundle over X, and since $gx_1 \wedge \cdots \wedge gx_m = (\det g)x_1 \wedge \cdots \wedge x_m$ for any $(x_1, \ldots, x_m) \in \mathbb{R}^m \times \cdots \times \mathbb{R}^m$ the group homomorphism $GL(m, \mathbb{R}) \xrightarrow{\Gamma} GL(1, \mathbb{R})$ merely carries $g \in GL(m, \mathbb{R})$ into its determinant $\det g$, acting via scalar multiplication on $\bigwedge^m \mathbb{R}^m$.

Let $(\bigwedge^m \mathbb{R}^m)^* \subset \bigwedge^m \mathbb{R}^m$ consist of the nonzero elements $y \in \bigwedge^m \mathbb{R}^m$, observe that $(\bigwedge^m \mathbb{R}^m)^*$ is preserved under the action of $GL(1, \mathbb{R})$, and let $\sim$ be the equivalence relation in $(\bigwedge^m \mathbb{R}^m)^*$ with $y' \sim y$ if and only if $y' = ay$ for some $a > 0$. The canonical surjection $(\bigwedge^m \mathbb{R}^m)^* \xrightarrow{\Phi'} (\bigwedge^m \mathbb{R}^m)^*/\sim$ maps $(\bigwedge^m \mathbb{R}^m)^*$ onto a space $(\bigwedge^m \mathbb{R}^m)^*/\sim$ with just two elements, and if $GL^+(1, \mathbb{R}) \subset GL(1, \mathbb{R})$ is the subgroup consisting of multiplications by positive real numbers the canonical epimorphism $GL(1, \mathbb{R}) \xrightarrow{\Gamma'} GL(1, \mathbb{R})/GL^+(1, \mathbb{R})$ has image $\mathbb{Z}/2$, which acts on $(\bigwedge^m \mathbb{R}^m)^*/\sim$ by interchanging the two elements. The pair (Γ', Φ') is another morphism of transformation groups in the sense of Definition II.2.6, carrying the transformation group $GL(1, \mathbb{R}) \times (\bigwedge^m \mathbb{R}^m)^* \to (\bigwedge^m \mathbb{R}^m)^*$ into the transformation group $\mathbb{Z}/2 \times (\bigwedge^m \mathbb{R}^m)^*/\sim \to (\bigwedge^m \mathbb{R}^m)^*/\sim$.

10.2 Definition: Let ξ be any real m-plane bundle over any $X \in \mathscr{B}$, and let $(\bigwedge^m \xi)^*$ be the fiber bundle with fiber $(\bigwedge^m \mathbb{R}^m)^*$ and structure group $GL(1, \mathbb{R})$, obtained from the mth exterior power $\bigwedge^m \xi$ by removing the zero-section. The *orientation bundle* $o(\xi)$ over X is induced by applying the preceding morphism (Γ', Φ') of transformation groups to $(\bigwedge^m \xi)^*$, as in Definition II.2.8.

There is another way of describing $o(\xi)$, which we sketch. According to the linear reduction theorem (Theorem II.6.13) one can reduce the structure group $GL(m, \mathbb{R})$ of ξ to the orthogonal subgroup $O(m) \subset GL(m, \mathbb{R})$, and Proposition 3.4 provides a riemannian metric $\langle\ ,\ \rangle$ for any coordinate bundle representing ξ. One easily alters $\langle\ ,\ \rangle$, if necessary, in such a way that the action of $O(m)$ on each fiber E_x preserves the inner product $E_x \times E_x \xrightarrow{\langle\ ,\ \rangle x} \mathbb{R}$; in fact, this is automatically the case if one uses $\langle\ ,\ \rangle$ itself to reduce $GL(m, \mathbb{R})$ to $O(m) \subset GL(m, \mathbb{R})$, as suggested at the end of §3. Hence one can replace $GL(m, \mathbb{R})$ by $O(m)$, and E by the subspace of fibers of unit length, to obtain the *sphere bundle associated to* ξ, with structure group $O(m)$ and fiber S^{m-1}.

As part of Definition 10.1, we introduced a transformation group $GL(m, \mathbb{R}) \times (R^m \times \cdots \times \mathbb{R}^m) \to \mathbb{R}^m \times \cdots \times \mathbb{R}^m$, and we applied a morphism (Γ, Φ) carrying any $g \in GL(m, \mathbb{R})$ into $\det g \in GL(1, \mathbb{R})$ and any $(x_1, \ldots, x_m) \in \mathbb{R}^m \times \cdots \times \mathbb{R}^m$ into the exterior power $x_1 \wedge \cdots \wedge x_m \in \bigwedge^m \mathbb{R}^m = \mathbb{R}^1$; in effect, however, we replaced $\mathbb{R}^m \times \cdots \times \mathbb{R}^m$ by the subspace $(\mathbb{R}^m)^{m*} \subset \mathbb{R}^m \times \cdots \times \mathbb{R}^m$ of ordered bases of $\mathbb{R}^m$. One can apply the same morphism (Γ, Φ) to the transformation group $O(m) \times (S^{m-1})^{m*} \to (S^{m-1})^{m*}$, where $(S^{m-1})^{m*} \subset (\mathbb{R}^m)^{m*}$ consists of orthonormal bases of $\mathbb{R}^m$; in this case (Γ, Φ) carries any $g \in O(m)$ into $\det g \in O(1)$ and any $(x_1, \ldots, x_m) \in (S^{m-1})^{m*}$ into a point $x_1 \wedge \cdots \wedge x_m$ in the 0-sphere $S^0 \in \mathbb{R}^1$. The orientation bundle $o(\xi)$ of Definition 10.2 is clearly the result of applying the preceding morphism (Γ, Φ) of transformation groups to the sphere bundle associated to the given m-plane bundle.

The preceding description of the orientation bundle $o(\xi)$ suggests that one should regard its structure group as the orthogonal group $O(1)$ $(=\mathbb{Z}/2)$ and its fiber as the 0-sphere S^0 $(=\mathbb{Z}/2)$. In any event, the coordinate bundle representing $o(\xi)$ will be denoted $O(\xi) \xrightarrow{\pi(\xi)} X$; it is the *double covering* of X associated to ξ.

10.3 Proposition: *Let* $X' \xrightarrow{f} X$ *be any map of base spaces, and let* ξ *be any real m-plane bundle over* X. *Then* $o(f^!\xi) = f^! o(\xi)$ *over* X'; *that is, there is a map* $O(f)$ *preserving the action of* $O(1)$ *such that the diagram*

$$\begin{array}{ccc} O(f^!\xi) & \xrightarrow{O(f)} & O(\xi) \\ \Big\downarrow{\scriptstyle \pi(f^!\xi)} & & \Big\downarrow{\scriptstyle \pi(\xi)} \\ X' & \xrightarrow{f} & X \end{array}$$

commutes.

Proof: Any pullback diagram

$$\begin{array}{ccc} E' & \xrightarrow{\mathbf{f}} & E \\ \Big\downarrow{\scriptstyle \pi'} & & \Big\downarrow{\scriptstyle \pi} \\ X' & \xrightarrow{f} & X \end{array}$$

for ξ and $f^!\xi$ induces a corresponding pullback diagram for $(\bigwedge^m \xi)^*$ and $f^!(\bigwedge^m \xi)^*$, and one obtains the result by applying (Γ', Φ').

Observe that since the fiber of $o(\xi)$ consists of two points, $o(\xi)$ is trivial if and only if there is a section $X \xrightarrow{\sigma} O(\xi)$ of the coordinate bundle representing $o(\xi)$.

10.4 Definition: A real m-plane ξ is *orientable* if and only if the orientation bundle $o(\xi)$ is trivial; that is, ξ is orientable if and only if the mth exterior power $\bigwedge^m \xi$ is the trivial line bundle ε^1. An *orientation* of ξ is a specific section $X \xrightarrow{\sigma} O(\xi)$, which *orients* ξ.

We shall later describe orientability of ξ more concretely in terms of the transition functions $U_i \cap U_j \xrightarrow{\Psi_i^j} GL(m, \mathbb{R})$ for any coordinate bundle representing ξ.

10.5 Proposition: *Any orientation σ of a real m-plane bundle ξ over $X \in \mathscr{B}$ induces an orientation σ' of the pullback $f^! \xi$ of ξ along a map $X' \xrightarrow{f} X$ in $\mathscr{B}$; thus, if ξ is oriented, then so is any pullback $f^! \xi$.*

PROOF: We use the diagram of Proposition 10.3. For any $x' \in X'$ the point $\sigma(f(x')) \in O(\xi)$ is the image under $O(f)$ of precisely one of those two points of $O(f^!\xi)$ that project onto x', and one lets $\sigma'(x') \in O(f^! \xi)$ be that point. The composition $X' \xrightarrow{\sigma'} O(f^! \xi) \xrightarrow{\pi(f^!\xi)} X'$ is the identity, by construction, and continuity of σ' is trivial.

Since the total space $O(\xi)$ of the orientation bundle $o(\xi)$ of a real m-plane bundle ξ over $X \in \mathscr{B}$ is a double covering of X, it is trivial to verify that $O(\xi)$ itself belongs to $\mathscr{B}$. Hence it makes sense to pull back bundles along the projection map $O(\xi) \xrightarrow{\pi(\xi)} X$; one can even pull ξ itself back along $\pi(\xi)$ to obtain an m-plane bundle $\pi(\xi)^!\xi$ over $O(\xi) \in \mathscr{B}$.

10.6 Proposition: *If $O(\xi) \xrightarrow{\pi(\xi)} X$ represents the orientation bundle $o(\xi)$ of a real m-plane bundle ξ over X, then the pullback $\pi(\xi)^!\xi$ is an orientable m-plane bundle over $O(\xi)$, oriented in a canonical way.*

PROOF: The commutative diagram of Proposition 10.3 becomes

$$\begin{array}{ccc} O(\pi(\xi)^!\xi) & \xrightarrow{O(\pi(\xi))} & O(\xi) \\ \Big\downarrow{\scriptstyle \pi(\pi(\xi)^!\xi)} & & \Big\downarrow{\scriptstyle \pi(\xi)} \\ O(\xi) & \xrightarrow{\pi(\xi)} & X \end{array}$$

in this case. Let y be any point in the lower left-hand copy of $O(\xi)$. The same point y in the upper right-hand copy of $O(\xi)$ is then the image of precisely one of the two points of $O(\pi(\xi)^!\xi)$ whose projection under $\pi(\pi(\xi)^!\xi)$ is y, and one lets $\sigma(y) \in O(\pi(\xi)^!\xi)$ be that point. The composition $O(\xi) \xrightarrow{\sigma} O(\pi(\xi)^!\xi) \xrightarrow{\pi(\pi(\xi)^!\xi)} O(\xi)$ is the identity, by construction, and continuity of σ is trivial.

We have observed that if ξ is a real m-plane bundle over $X \in \mathscr{B}$, then $O(\xi) \in \mathscr{B}$. In particular, if γ^m is the universal m-plane bundle over the Grassmann manifold $G^m(\mathbb{R}^\infty)$, as in Definition 8.2, then $O(\gamma^m) \in \mathscr{B}$.

10.7 Definition: For any $m > 0$ let $\tilde{G}^m(\mathbb{R}^\infty)$ be the total space $O(\gamma^m)$ of the orientation bundle $o(\gamma^m)$ of the universal real m-plane bundle γ^m over $G^m(\mathbb{R}^\infty)$; that is, $\tilde{G}^m(\mathbb{R}^\infty) \xrightarrow{\pi(\gamma^m)} G^m(\mathbb{R}^\infty)$ represents $o(\gamma^m)$. The pullback $\pi(\gamma^m)^!\gamma^m$ over $\tilde{G}^m(\mathbb{R}^\infty)$ is the *universal oriented m-plane bundle* $\tilde{\gamma}^m$.

According to Proposition 10.6, $\tilde{\gamma}^m$ has a canonical orientation.

10.8 Proposition (Homotopy Classification Theorem): *Let ξ be a real m-plane bundle over $X \in \mathscr{B}$. Then ξ is orientable if and only if $\xi = \tilde{f}^!\tilde{\gamma}^m$ for some map $X \xrightarrow{\tilde{f}} \tilde{G}^m(\mathbb{R}^\infty)$.*

PROOF: If $\xi = \tilde{f}^!\tilde{\gamma}^m$, then ξ is orientable by Proposition 10.5. Conversely, suppose that ξ is any orientable bundle over $X \in \mathscr{B}$, and let $X \xrightarrow{f} G^m(\mathbb{R}^\infty)$ classify ξ as a real m-plane bundle, so that $\xi = f^!\gamma^m$; such an f exists by Theorem 8.9. The pullback diagram of Proposition 10.3 is then of the form

$$\begin{array}{ccc} O(\xi) & \xrightarrow{O(f)} & \tilde{G}^m(\mathbb{R}^\infty) \\ {\scriptstyle \pi(\xi)}\downarrow & & \downarrow{\scriptstyle \pi(\gamma^m)} \\ X & \xrightarrow{f} & G^m(\mathbb{R}^\infty). \end{array}$$

Since any orientation of ξ is a section $X \xrightarrow{\sigma} O(\xi)$ of the bundle $o(\xi)$, by definition, the composition $X \xrightarrow{\sigma} O(\xi) \xrightarrow{\pi(\xi)} X$ is the identity, so that $f = \pi(\gamma^m) \circ O(f) \circ \sigma$. If $X \xrightarrow{\tilde{f}} \tilde{G}^m(\mathbb{R}^\infty)$ is the composition $O(f) \circ \sigma$, then $\xi = (\pi(\gamma^m) \circ \tilde{f})^!\gamma^m = \tilde{f}^!\pi(\gamma^m)^!\gamma^m = \tilde{f}^!\tilde{\gamma}^m$ as desired.

Briefly, Proposition 10.8 asserts that ξ is orientable if and only if it can be classified by a map of the form $X \xrightarrow{\tilde{f}} \tilde{G}^m(\mathbb{R}^\infty) \xrightarrow{\pi(\gamma^m)} G^m(\mathbb{R}^\infty)$, where $\tilde{G}^m(\mathbb{R}^\infty)$ is the double covering of $G^m(\mathbb{R}^\infty)$ with respect to γ^m. The map $\tilde{f}$ is the *oriented classifying map*, determining a specific orientation of ξ.

One can give direct constructions of $\tilde{G}^m(\mathbb{R}^\infty)$ and the universal oriented m-plane bundle $\tilde{\gamma}^m$ over $\tilde{G}^m(\mathbb{R}^\infty)$ that superficially resemble the corresponding unoriented constructions more closely than does Definition 10.7. According to Definition I.7.1 the unoriented Grassmann manifold $G^m(\mathbb{R}^\infty)$ can be regarded as the quotient $(\mathbb{R}^\infty)^{m*}/\sim$ of $(\mathbb{R}^\infty)^{m*}$ by the equivalence relation $\sim$ with $(x_1, \ldots, x_m) \sim (y_1, \ldots, y_m)$ whenever $x_1 \wedge \cdots \wedge x_m = ay_1 \wedge \cdots \wedge y_m \in \bigwedge^m \mathbb{R}^\infty$ for some $a \in \mathbb{R}^*$. If one replaces the latter condition by the requirement that $a > 0$, then the result is a new equivalence relation $\sim'$ for which one easily establishes $(\mathbb{R}^\infty)^{m*}/\sim' = \tilde{G}^m(\mathbb{R}^\infty)$. Similarly, one can describe the total space $\tilde{E}^\infty$ of the universal oriented m-plane bundle $\tilde{\gamma}^m$ as a quotient $\mathbb{R}^{m*} \times \mathbb{R}^m/\approx'$ in which

$$\left((x_1, \ldots, x_m), \begin{pmatrix} s^1 \\ \vdots \\ s^m \end{pmatrix}\right) \approx' \left((y_1, \ldots, y_m), \begin{pmatrix} t^1 \\ \vdots \\ t^m \end{pmatrix}\right)$$

if and only if $x_1 \wedge \cdots \wedge x_m = ay_1 \wedge \cdots \wedge y_m \in \bigwedge^m \mathbb{R}^\infty$, for some $a > 0$, and $s^1x_1 + \cdots + s^mx_m = t^1y_1 + \cdots + t^my_m \in \mathbb{R}^\infty$. The subspace $\tilde{E}^{\infty *}$ of nonzero fibers in $\tilde{E}^\infty$ is then the quotient $(\mathbb{R}^\infty)^{m*} \times \mathbb{R}^{m*}/\approx'$, for the same equivalence relation $\approx'$.

10.9 Proposition: *For any $m > 1$ let $\tilde{E}^\infty \xrightarrow{\tilde{\pi}^\infty} \tilde{G}^m(\mathbb{R}^\infty)$ represent the universal oriented m-plane bundle $\tilde{\gamma}^m$, and let $\tilde{E}^{\infty *}$ be the space of nonzero fibers in $\tilde{E}^\infty$. Then there is a homotopy equivalence $\tilde{G}^{m-1}(\mathbb{R}^\infty) \xrightarrow{\tilde{h}} \tilde{E}^{\infty *}$ such that the composition $\tilde{G}^{m-1}(\mathbb{R}^\infty) \xrightarrow{\tilde{h}} \tilde{E}^{\infty *} \xrightarrow{\tilde{\pi}^\infty} \tilde{G}^m(\mathbb{R}^\infty)$ classifies the oriented Whitney sum $\tilde{\gamma}^{m-1} \oplus \varepsilon^1$ over $\tilde{G}^{m-1}(\mathbb{R}^\infty)$.*

PROOF: The proof is virtually identical to that of Proposition 8.10, using the preceding description of $\tilde{E}^\infty \xrightarrow{\tilde{\pi}^\infty} \tilde{G}^m(\mathbb{R}^\infty)$ in place of the corresponding description of $E^\infty \xrightarrow{\pi^\infty} G^m(\mathbb{R}^\infty)$. (One also uses the property that if m is even, then $\varepsilon^1 \oplus \tilde{\gamma}^{m-1}$ and $\tilde{\gamma}^{m-1} \oplus \varepsilon^1$ have opposite orientations.)

Let ξ be any real m-plane bundle over any $X \in \mathscr{B}$, as before. The *canonical involution* of the double covering $O(\xi) \xrightarrow{\pi(\xi)} X$ is the well-defined homeomorphism $O(\xi) \xrightarrow{\tau} O(\xi)$ that interchanges the two points of the fiber of $o(\xi)$ over each $x \in X$. If ξ is orientable, then for any orientation $X \xrightarrow{\sigma} O(\xi)$ the *opposite orientation* is the composition $X \xrightarrow{\tau \circ \sigma} O(\xi)$; the resulting *oppositely oriented* m-plane bundle is usually denoted $-\xi$.

Since $\tilde{G}^m(\mathbb{R}^\infty) = O(\gamma^m)$, there is a canonical involution of $\tilde{G}^m(\mathbb{R}^\infty)$.

10.10 Proposition: *If m is odd the canonical involution $\tilde{G}^m(\mathbb{R}^\infty) \xrightarrow{\tau} \tilde{G}^m(\mathbb{R}^\infty)$ is homotopic to the identity.*

PROOF: Let $(\ldots; x_r, x_{r+1}; \ldots)$ denote any element $(x_0, x_1; x_2, x_3; \ldots; x_r, x_{r+1}; \ldots) \in \mathbb{R}^\infty$, where r is even, and let $\mathbb{R}^\infty \times [0, 1] \xrightarrow{f} \mathbb{R}^\infty$ be the map that carries any $((\ldots; x_r, x_{r+1}; \ldots), t)$ into

$$(\ldots; x_r \cos \pi t - x_{r+1} \sin \pi t, x_r \sin \pi t + x_{r-1} \cos \pi t; \ldots).$$

The restriction $f|\mathbb{R}^\infty \times \{t\}$ is a linear isomorphism for each $t \in [0, 1]$, the restriction $f|\mathbb{R}^\infty \times \{0\}$ is the identity, and the restriction $f|\mathbb{R}^\infty \times \{1\}$ carries $(\ldots; x_r, x_{r+1}; \ldots)$ into $(\ldots; -x_r, -x_{r+1}; \ldots)$. Hence if $E \xrightarrow{\pi} G^m(\mathbb{R}^\infty)$ represents the universal bundle γ^m, then there is an induced homotopy $E \times [0, 1] \xrightarrow{F} E$ from the identity $E \to E$ to the map $E \xrightarrow{-1} E$ that carries each $e \in E_{\pi(e)}$ into $-e \in E_{\pi(-e)}$, where $\pi(-e) = \pi(e)$. If m is odd and $E' \xrightarrow{\pi'} G^m(\mathbb{R}^\infty)$ represents the mth exterior power $\bigwedge^m \gamma^m$ over $G^m(\mathbb{R}^\infty)$, F induces a homotopy F' from the identity $E' \to E'$ to the map $E' \xrightarrow{-1} E'$. The restriction of F' to any fiber is an isomorphism for each $t \in [0, 1]$, so that there is a corresponding homotopy of maps of the total space of the bundle $(\bigwedge^m \gamma^m)^*$ (Definition 10.2), which induces the desired homotopy of maps of the total space $\tilde{G}^m(\mathbb{R}^\infty)$ of $o(\gamma^m)$.

We now replace $\mathbb{R}^\infty$ by $\mathbb{R}^{m+n}$ for $m + n < \infty$.

10.11 Definition: For any $m > 0$ and $n > 0$ let $\tilde{G}^m(\mathbb{R}^{m+n})$ be the total space $O(\gamma_n^m)$ of the orientation bundle $o(\gamma_n^m)$ of the canonical real m-plane bundle γ_n^m over the Grassmann manifold $G^m(\mathbb{R}^{m+n})$; that is, $\tilde{G}^m(\mathbb{R}^{m+n}) \xrightarrow{\pi(\gamma_n^m)} G^m(\mathbb{R}^{m+n})$ represents $o(\gamma_n^m)$. The pullback $\pi(\gamma_n^m)^! \gamma_n^m$ over $\tilde{G}^m(\mathbb{R}^{m+n})$ is the *canonical oriented m-plane bundle* $\tilde{\gamma}_n^m$.

Since $G^m(\mathbb{R}^{m+n})$ is a smooth closed mn-dimensional manifold, by Proposition I.7.3, the same is true of the double covering $\tilde{G}^m(\mathbb{R}^{m+n})$. Furthermore, the bundle $\tilde{\gamma}_n^m$ over $\tilde{G}^m(\mathbb{R}^{m+n})$ is canonically oriented by Proposition 10.6.

10.12 Proposition: *Let $X \in \mathscr{B}$ be homotopy equivalent either to a compact space or to a finite-dimensional simplicial space, and let ξ be any real m-plane bundle over X. Then ξ is orientable if and only if $\xi = \tilde{f}^! \tilde{\gamma}_n^m$ for some finite $n > 0$ and some map $X \xrightarrow{\tilde{f}} \tilde{G}^m(\mathbb{R}^{m+n})$. If X is a smooth manifold, then ξ is orientable if and only if a smooth such $\tilde{f}$ exists.*

PROOF: By Propositions 8.12, 9.1, or 9.3 one has $\xi = f^! \gamma_n^m$ for an appropriate map $X \xrightarrow{f} G^m(\mathbb{R}^{m+n})$, and one substitutes $\mathbb{R}^{m+n}$ for $\mathbb{R}^\infty$ throughout the proof of Proposition 10.8.

10.13 Proposition: *If m and n are both odd natural numbers, the canonical involution $\tilde{G}^m(\mathbb{R}^{m+n}) \xrightarrow{\tau} \tilde{G}^m(\mathbb{R}^{m+n})$ is homotopic to the identity.*

PROOF: Since $m + n$ is even, one can substitute $\mathbb{R}^{m+n}$ for $\mathbb{R}^\infty$ in the proof of Proposition 10.10.

10.14 Corollary: *For any odd m and any $n > 0$ the canonical real m-plane bundle γ_n^m over $G^m(\mathbb{R}^{m+n})$ is nonorientable.*

PROOF: The inclusion $\mathbb{R}^{m+1} \to \mathbb{R}^{m+n}$ induces a map $G^m(\mathbb{R}^{m+1}) \xrightarrow{f} G^m(\mathbb{R}^{m+n})$ for which $\gamma_1^m = f^! \gamma_n^m$, so that nonorientability of γ_1^m implies nonorientability of γ_n^m, by Proposition 10.5. Hence it suffices to consider only the case that n is the odd number 1, for which Proposition 10.13 guarantees that $\tilde{G}^m(\mathbb{R}^{m+1}) \xrightarrow{\tau} \tilde{G}^m(\mathbb{R}^{m+1})$ is homotopic to the identity. If γ_1^m were orientable, then since the double covering $\tilde{G}^m(\mathbb{R}^{m+1})$ of $G^m(\mathbb{R}^{m+1})$ is the total space $O(\gamma_1^m)$ of the orientation bundle $o(\gamma_1^m)$, it would consist of two disjoint copies of $G^m(\mathbb{R}^{m+1})$, and τ would interchange the two copies of $\tilde{G}^m(\mathbb{R}^{m+1})$; however, such a map τ would *not* be homotopic to the identity.

Since trivial vector bundles are clearly orientable, Corollary 10.14 incidentally guarantees that the bundles γ_n^m are not trivial when m is odd. The simplest case is the canonical real line bundle γ_1^1 over RP^1, the base space RP^1 being diffeomorphic to the circle S^1.

We now develop a more concrete characterization of orientability. Let S^0 be the 0-sphere $\{+1, -1\}$, and let the orthogonal group $O(1)$ act on S^0 as usual, via multiplication by $+1$ or -1. Any coordinate bundle $E \xrightarrow{\pi} X$ with fiber S^0 and structure group $O(1)$ over a space $X \in \mathscr{B}$ is a *double covering* of X in the sense defined earlier; in fact, since $O(1) \subset GL(1, \mathbb{R})$, one easily constructs a real line bundle λ over X such that $E \xrightarrow{\pi} X$ is precisely the double covering $O(\lambda) \xrightarrow{\pi(\lambda)} X$.

10.15 Lemma: *Let $E \xrightarrow{\pi} X$ be a double covering of a space $X \in \mathscr{B}$, a coordinate bundle with respect to some open covering $\{U_i | i \in I\}$ of X. Then $E \xrightarrow{\pi} X$ is trivial if and only if there is a family of transition functions $U_i \cap U_j \xrightarrow{\psi_i^j} O(1)$ each of which has the constant value $1 \in O(1)$.*

PROOF: Clearly $E \xrightarrow{\pi} X$ is trivial if and only if there is a section $X \xrightarrow{\sigma} E$. Let $\{\Psi_i | i \in I\}$ be a family of local trivializations $E|U_i \xrightarrow{\Psi_i} U_i \times S^0$. If a section σ exists, then each composition $U_i \xrightarrow{\sigma|U_i} E|U_i \xrightarrow{\Psi_i} U_i \times S^0$ maps each $x \in U_i$ into $(x, \pm 1) \in U_i \times S^0$, and one can alter the local trivializations in such a way that $(\Psi_i \circ \sigma | U_i)(x) = (x, +1)$ for each $i \in I$ and each $x \in U_i$. The transition functions Ψ_i^j are defined by the requirement that the compositions $U_i \cap U_j \times S^0 \xrightarrow{\Psi_j \circ \Psi_i^{-1}} U_i \cap U_j \times S^0$ maps $(x, \pm 1)$ into $(x, \psi_i^j(x)(\pm 1))$; hence, using the altered local trivialization Ψ_i, one has $\psi_i^j(x) = 1 \in O(1)$ for $x \in U_i \cap U_j$, as required. Conversely, if the latter conditions are satisfied, then there is a unique section $X \xrightarrow{\sigma} E$ such that for each $i \in I$ the composition $\Psi_i \circ \sigma | U_i$ carries each $x \in U_i$ into $(x, +1) \in U_i \times S^0$.

The following property of orientable vector bundles is frequently used as the definition of orientability, where $GL^+(m, \mathbb{R}) \subset GL(m, \mathbb{R})$ is the subgroup of elements with positive determinants, as usual.

10.16 Proposition: *A real m-plane bundle ξ over any $X \in \mathscr{B}$ is orientable if and only if the structure group $GL(m, \mathbb{R})$ can be reduced to the subgroup $GL^+(m, \mathbb{R}) \subset GL(m, \mathbb{R})$.*

PROOF: Let $E \xrightarrow{\pi} X$ be a coordinate bundle representing ξ, with respect to some open covering $\{U_i | i \in I\}$ of X. Then according to the discussion following Definition II.5.1 we must show that ξ is orientable if and only if there is a family of transition functions $U_i \cap U_j \xrightarrow{\psi_i^j} GL(m, \mathbb{R})$ whose values all lie in the subgroup $GL^+(m, \mathbb{R}) \subset GL(m, \mathbb{R})$. The coordinate bundle $E \xrightarrow{\pi} X$ induces a double covering $E' \xrightarrow{\pi'} X$ that represents the orientation bundle $o(\xi)$ with respect to the same covering $\{U_i | i \in I\}$. Specifically, for each ψ_i^j the corresponding transition function $U_i \cap U_j \xrightarrow{\psi_i'^j} O(1)$ for $E' \xrightarrow{\pi'} X$ is given by setting $\psi_i'^j = \det \psi_i^j / |\det \psi_i^j|$. By Lemma 10.15 the orientation bundle $o(\xi)$ is trivial if and only if there are transition functions $\psi_i'^j$ for $E' \xrightarrow{\pi'} X$ such that $\psi_i' j(x) = 1 \in O(1)$ for each $x \in U_i \cap U_j$. Hence the bundle ξ is orientable if and only if there are transition functions ψ_i^j for $E \xrightarrow{\pi} X$ such that $\det \psi_i^j(x) > 0$ for each $x \in U_i \cap U_j$, as claimed.

10.17 Corollary: *A real m-plane bundle ξ over any $X \in \mathscr{B}$ is orientable if and only if the structure group can be reduced to the rotation subgroup $O^+(m) \subset GL(m, \mathbb{R})$.*

PROOF: By the linear reduction theorem (Theorem II.6.13) the structure group $GL^+(m, \mathbb{R})$ can always be reduced to the subgroup $O^+(m) \subset GL^+(m, \mathbb{R})$.

10.18 Corollary: *Let ξ be any real m-plane bundle over any base space $X \in \mathscr{B}$; then the Whitney sum $\xi \oplus \xi$ is orientable.*

PROOF: If ξ is represented by a coordinate bundle using some open covering $\{U_i | i \in I\}$ of X and transition functions $U_i \cap U_j \xrightarrow{\psi_i^j} GL(m, \mathbb{R})$, then $\xi \oplus \xi$ is also represented by a coordinate bundle using the same covering $\{U_i | i \in I\}$ and transition functions $U_i \cap U_j \xrightarrow{\psi_i^j \oplus \psi_i^j} GL(2m, \mathbb{R})$; however, $\det(\psi_i^j \oplus \psi_i^j) = (\det \psi_i^j)^2 > 0$ over $U_i \cap U_j$.

One can easily strengthen Corollary 10.18: $\xi \oplus \xi$ has a *natural orientation*, which will appear just before Proposition 12.8.

10.19 Corollary: *Let ξ and ξ' be real vector bundles over any base space $X \in \mathscr{B}$, and suppose that ξ is orientable; then ξ' is orientable if and only if $\xi \oplus \xi'$ is orientable.*

PROOF: If ξ and ξ' are represented by coordinate bundles using open coverings $\{U_i | i \in I\}$ and $\{U_j | j \in J\}$ of X, respectively, then they can both be represented by coordinate bundles using the common open covering $\{U_i \cap U_j | (i, j) \in I \times J\}$ of X; the latter covering will be denoted $\{U_i | i \in I\}$ for convenience. By hypothesis, the transition functions $U_i \cap U_j \xrightarrow{\psi_i^j} GL(m, \mathbb{R})$ for the representation of ξ can be chosen in such a way that $\det \psi_i^j > 0$ over $U_i \cap U_j$. If $U_i \cap U_j \xrightarrow{\psi_i'^j} GL(n, \mathbb{R})$ are transition functions for the representation of ξ', then $U_i \cap U_j \xrightarrow{\psi_i^j \oplus \psi_i'^j} GL(m + n, \mathbb{R})$ are transition functions for a representation of $\xi \oplus \xi'$; however, $\det(\psi_i^j \oplus \psi_i'^j) = (\det \psi_i^j)(\det \psi_i'^j)$, where $\det \psi_i^j > 0$, over $U_i \cap U_j$.

The Möbius band was introduced for motivation at the beginning of this section. We now sketch the proof that its tangent bundle is indeed nonorientable.

First, recall that the Möbius band was described in §II.0 as the total space E of a coordinate bundle $E \xrightarrow{\pi} S^1$ $(= RP^1)$ whose fiber is the closed interval $[-1, +1] \subset \mathbb{R}$ and whose structure group is $\mathbb{Z}/2$, acting on $[-1, +1]$ via multiplication by $+1$ or -1; that is, $\mathbb{Z}/2 = O(1) \subset GL(1, \mathbb{R})$. It is clear that $E \xrightarrow{\pi} RP^1$ is merely a restriction of a coordinate bundle $E' \xrightarrow{\pi'} RP^1$ representing the canonical line bundle γ_1^1 over RP^1. In any event, one can pull the tangent bundle $\tau(E)$ or $\tau(E')$ back along the zero-section $RP^1 \to E \subset E'$ to obtain a 2-plane bundle $\sigma^!\tau(E)$ over RP^1, and a direct computation shows that $\sigma^!\tau(E) = \gamma_1^1 \oplus \varepsilon^1$ for the trivial line bundle ε^1. However, γ_1^1 is nonorientable by Corollary 10.14, so that $\sigma^!\tau(E)$ is nonorientable by Corollary 10.19, so that $\tau(E)$ is nonorientable by Proposition 10.5, as claimed.

11. Complex Vector Bundles

One can replace the real field $\mathbb{R}$ by the complex field $\mathbb{C}$ throughout the development of vector bundles, with virtually no other changes. The replacement is more than an idle exercise, however, even if one is only interested in real vector bundles. For example, "complexifications" of real vector bundles will be used in Volume 2 to compute the real cohomology rings $H^*(G^m(\mathbb{R}^\infty); \mathbb{R})$ of real Grassmann manifolds; Theorem 8.9 suggests the real importance of such cohomology rings. Complex vector bundles are also used in the very definition of K-theory, in Volume 3, which is used to solve the classical problem of real vector fields on spheres. However, complex vector bundles are of geometric interest in their own right, especially if one studies complex manifolds by looking at their tangent bundles.

As always, the base space of a fib*er* bundle belongs to the category $\mathscr{B}$ of base spaces.

11.1 Definition: A *complex vector bundle of rank n*, or simply a *complex n-plane bundle*, is any fiber bundle whose fiber is the complex vector space $\mathbb{C}^n$ and whose structure group is the general linear group $GL(n, \mathbb{C})$ of invertible $n \times n$ matrices, acting in the usual way on $\mathbb{C}^n$. A *complex line bundle* is a complex vector bundle of rank 1.

The obvious complex analog of Definition 2.4 or Proposition 2.5 provides the *Whitney sum* $\zeta \oplus \zeta'$ and *product* $\zeta \otimes \zeta'$ of complex vector bundles ζ and ζ' over the same base space, with the associative, commutative, and distributive properties described for real vector bundles in Proposition 2.6.

If one substitutes hermitian inner products $\mathbb{C}^n \times \mathbb{C}^n \xrightarrow{\langle\,,\,\rangle} \mathbb{C}$ for real inner products $\mathbb{R}^m \times \mathbb{R}^m \xrightarrow{\langle\,,\,\rangle} \mathbb{R}$ in Definition 3.1, then the result is a *hermitian metric* for a given complex n-plane bundle ζ. The obvious complex analog of Proposition 3.4 then guarantees that any complex vector bundle (over a base space $X \in \mathscr{B}$, as always) has a hermitian metric. Consequently there is a complex analog of Proposition 3.6: for any complex *subbundle* ζ^1 of a given complex vector bundle ζ there is another subbundle ζ^2 of ζ such that $\zeta = \zeta^1 \oplus \zeta^2$.

One can also use hermitian metrics, as in the real case, to show that the structure group $GL(n, \mathbb{C})$ of any complex n-plane bundle over any $X \in \mathscr{B}$ can be reduced to the unitary subgroup $U(n) \subset GL(n, \mathbb{C})$; however, since this is a special case of the linear reduction theorem (Theorem II.6.13), the details will be left as an exercise (Exercise 13.19).

Definitions 7.2 and 8.2 have obvious complex analogs, which provide the *canonical complex m-plane bundle* γ_n^m over the Grassmann manifold $G^m(\mathbb{C}^{m+n})$ and the *universal complex m-plane bundle* γ^m over the Grassmann manifold $G^m(\mathbb{C}^\infty)$, respectively; in particular there is a *canonical complex line bundle* γ_n^1 over the projective space CP^n, and a *universal complex line bundle* γ^1 over the projective space CP^∞. We already know from Proposition I.7.5 that $G^m(\mathbb{C}^{m+n}) \in \mathscr{B}$ and $G^m(\mathbb{C}^\infty) \in \mathscr{B}$.

11.2 Theorem (Homotopy Classification Theorem): *Any complex m-plane bundle ζ over a base space $X \in \mathscr{B}$ is a pullback $f^!\gamma^m$ of the universal complex* *m-plane bundle* γ^m *over* $G^m(\mathbb{C}^\infty) \in \mathscr{B}$, *along a map* $X \xrightarrow{f} G^m(\mathbb{C}^\infty)$ *that is unique up to homotopy.*

PROOF: Substitute $\mathbb{C}$ for $\mathbb{R}$ throughout the proof of Theorem 8.9.

11.3 Proposition: *For any* $n > 1$ *let* $E^\infty \xrightarrow{\pi^\infty} G^n(\mathbb{C}^\infty)$ *represent the universal complex n-plane bundle* γ^n, *and let* $E^{\infty *}$ *be the space of nonzero fibers in* E^∞. *Then there is a homotopy equivalence* $G^{n-1}(\mathbb{C}^\infty) \xrightarrow{h} E^{\infty *}$ *such that the composition* $G^{n-1}(\mathbb{C}^\infty) \xrightarrow{h} E^{\infty *} \xrightarrow{\pi^\infty} G^n(\mathbb{C}^\infty)$ *classifies the Whitney sum* $\gamma^{n-1} \oplus \varepsilon^1$ *over* $G^{n-1}(\mathbb{C}^\infty)$.

PROOF: Substitute $\mathbb{C}$ for $\mathbb{R}$ (and n for m) throughout the proof of Proposition 8.10.

11.4 Proposition: *If* $E^\infty \xrightarrow{\pi^\infty} CP^\infty$ *represents the universal complex line bundle* γ^1, *then the space* $E^{\infty *}$ *of nonzero fibers in* E^∞ *is homeomorphic to the contractible space* $\mathbb{C}^{\infty *}$. *If* $E \xrightarrow{\pi} CP^n$ *represents the canonical complex line bundle* γ_n^1, *for a given* $n > 0$, *then the space* E^* *of nonzero fibers in E is homeomorphic to* $\mathbb{C}^{(n+1)*}$, *trivially homotopy equivalent to the* $(2n + 1)$*-sphere* S^{2n+1}.

PROOF: As in Proposition 8.11, $E^{\infty *}$ is a quotient $\mathbb{C}^{\infty *} \times \mathbb{C}^*/\approx$, with $(x, s) \approx (y, t)$ if and only if $sx = ty \in \mathbb{C}^{\infty *}$; similarly E^* is a quotient $\mathbb{C}^{(n+1)*} \times \mathbb{C}^*/\approx$, with $(x, s) \approx (y, t)$ if and only if $sx = ty \in \mathbb{C}^{(n+1)*}$. The proof that $\mathbb{C}^{\infty *}$ is contractible follows the pattern of the corresponding proof for $\mathbb{R}^{\infty *}$, given in Proposition 8.11.

The homotopy classification theorem for complex vector bundles has finite analogs, just as in the real case. Here are some corresponding complex uniqueness and existence results, in that order.

11.5 Proposition (Ersatz Homotopy Uniqueness Theorem): *Let* ζ *be a complex m-plane bundle over* $X \in \mathscr{B}$ *such that* $\zeta = f^!_0\gamma^m_n$ *and* $\zeta = f^!_1\gamma^{m'}_n$ *for maps* $X \xrightarrow{f_0} G^m(\mathbb{C}^{m+n})$ *and* $X \xrightarrow{f_1} G^m(\mathbb{C}^{m+n'})$, *where* γ^m_n *and* $\gamma^m_{n'}$ *are canonical complex m-plane bundles. Then there is an* $n'' \geqq \max(n, n')$ *with finite classifying extensions* $G^m(\mathbb{C}^{m+n}) \xrightarrow{g_{n,n''}} G^m(\mathbb{C}^{m+n''})$ *and* $G^m(\mathbb{C}^{m+n'}) \xrightarrow{g_{n',n''}} G^m(\mathbb{C}^{m+n''})$ *such that* $g_{n,n''} \circ f_0$ *and* $g_{n',n''} \circ f_1$ *are homotopic maps from X to* $G^m(\mathbb{C}^{m+n''})$, *for which*

$$(g_{n,n''} \circ f_0)^! \gamma^m_{n''} = \zeta = (g_{n',n''} \circ f_1)^! \gamma^m_{n''}.$$

PROOF: Substitute $\mathbb{C}$ for $\mathbb{R}$ throughout the proof of Theorem 8.14.

11.6 Proposition: *Let* $X \in \mathscr{B}$ *be a compact space, a finite-dimensional simplicial space, or a smooth manifold, and let* ζ *be a complex m-plane bundle* *over X. Then there is a natural number $n > 0$ such that ζ is a pullback $f^! \gamma_n^m$ of the canonical complex m-plane bundle γ_n^m over $G^m(\mathbb{C}^{m+n})$ along a map $X \xrightarrow{f} G^m(\mathbb{C}^{m+n})$.*

PROOF: Substitute $\mathbb{C}$ for $\mathbb{R}$ throughout the proofs of Propositions 8.12, 9.1, or 9.3, respectively.

11.7 Corollary: *Let $X \in \mathscr{B}$ be a compact space, a finite-dimensional simplicial space, or a smooth manifold, and let ζ be a complex m-plane bundle over X. Then for some $n > 0$ there is a complex n-plane bundle ζ' over X such that the Whitney sum $\zeta \oplus \zeta'$ is the trivial complex bundle ε^{m+n} over X.*

PROOF: This is the complex analog of Corollaries 8.13, 9.2, or 9.4, respectively.

11.8 Theorem: *Any complex m-plane bundle ζ over a smooth manifold X can be represented by a smooth coordinate bundle $E \xrightarrow{\pi} X$; that is, any complex vector bundle over X is itself smooth.*

PROOF: By Proposition 11.6, ζ is a pullback $f^! \gamma_n^m$ of the canonical complex m-plane bundle γ_n^m along some map $X \xrightarrow{f} G^m(\mathbb{C}^{m+n})$, and by Proposition I.6.19 one can choose f to be a smooth map. One then substitutes $\mathbb{C}$ for $\mathbb{R}$ throughout the proofs of Lemma 7.1 and Proposition 5.3 to conclude that γ_n^m and $f^! \gamma_n^m$ are smooth.

12. Realifications and Complexifications

For any complex n-plane bundle ζ over a base space $X \in \mathscr{B}$ there is a corresponding oriented $2n$-plane bundle $\zeta_{\mathbb{R}}$ over X; one can also ignore the orientation of $\zeta_{\mathbb{R}}$ and simply regard it as a real $2n$-plane bundle over X. Similarly, for any real m-plane bundle ξ over a base space $X \in \mathscr{B}$ there is a corresponding complex m-plane bundle $\xi_{\mathbb{C}}$ over X. The constructions $(\ )_{\mathbb{R}}$ and $(\ )_{\mathbb{C}}$ are studied in this section.

Recall from Definition II.2.6 that a morphism from a transformation group $G \times F \to F$ to a transformation group $G' \times F' \to F'$ is a pair (Γ, Φ) of maps $G \xrightarrow{\Gamma} G'$ and $F \xrightarrow{\Phi} F'$ such that Γ is a group homomorphism and the obvious diagram commutes. For any fiber bundle ζ with structure group G and fiber F over a space X, the morphism (Γ, Φ) induces a new fiber bundle ζ' with structure group G' and fiber F' over the same space X, as in Proposition II.2.7 and Definition II.2.8.

Let $G = GL(n, \mathbb{C})$, and let Γ assign to any $(b_p^q + ic_p^q) \in GL(n, \mathbb{C})$ the real $2n \times 2n$ matrix consisting of 2×2 blocks

$$\begin{pmatrix} b_p^q & -c_p^q \\ c_p^q & b_p^q \end{pmatrix},$$

where b_p^q and c_p^q are real and $i = \sqrt{-1}$. Similarly let $F = \mathbb{C}^n$, and let $\mathbb{C}^n \xrightarrow{\Phi} \mathbb{R}^{2n}$ carry column vectors with pth entry $x^p + iy^p$ into column vectors whose $(2p - 1)$th and $(2p)$th entries are x^p and y^p, respectively, where $p = 1, \ldots, n$. One easily verifies that (Γ, Φ) is a morphism from the transformation group $GL(n, \mathbb{C}) \times \mathbb{C}^n \to \mathbb{C}^n$ to the transformation group $GL(2n, \mathbb{R}) \times \mathbb{R}^{2n} \to \mathbb{R}^{2n}$.

12.1 Definition: For any complex n-plane bundle ζ over a base space $X \in \mathscr{B}$ the *realification* $\zeta_{\mathbb{R}}$ is the real $2n$-plane bundle over X induced by the preceding morphism (Γ, Φ).

Let $I \in GL(n, \mathbb{C})$ be the identity element, so that $iI \in GL(n, \mathbb{C})$ is scalar multiplication by $i = \sqrt{-1} \in \mathbb{C}$. Since $(iI)^2 = -I \in GL(n, \mathbb{C})$, the element $J = \Gamma(iI) \in GL(2n, \mathbb{R})$ satisfies $J^2 = -I \in GL(2n, \mathbb{R})$; furthermore J commutes with every element in the image of $GL(n, \mathbb{C}) \xrightarrow{\Gamma} GL(2n, \mathbb{R})$.

12.2 Definition: Let $E \xrightarrow{\pi} X$ be a coordinate bundle that represents a real m-plane bundle ξ over $X \in \mathscr{B}$. A vector bundle morphism

$$\begin{array}{ccc} E & \xrightarrow{\quad J \quad} & E \\ & {\scriptstyle \pi}\searrow \quad \swarrow{\scriptstyle \pi} & \\ & X & \end{array}$$

is a *complex structure* in $E \xrightarrow{\pi} X$ whenever the restriction $E_x \xrightarrow{J_x} E_x$ over each $x \in X$ satisfies $J_x^2 = -I$ for the identity element $I \in GL(m, \mathbb{R})$.

Clearly a complex structure in one coordinate bundle representing ξ induces a complex structure in any other coordinate bundle representing ξ, so that one can regard a complex structure as a structure in ξ itself. Equally clearly, the endomorphism $J = \Gamma(iI) \in GL(2n, \mathbb{R})$ induces a complex structure in the realification $\zeta_{\mathbb{R}}$ of the complex n-plane bundle ζ. We shall show that a real vector bundle has a complex structure if and only if it is of the form $\zeta_{\mathbb{R}}$ for a complex vector bundle ζ; furthermore, ζ is unique.

12.3 Lemma: *If $J \in GL(m, \mathbb{R})$ satisfies $J^2 = -I \in GL(m, \mathbb{R})$, then m is an even number $2n$ and there is a basis of $\mathbb{R}^{2n}$ of the form $(e_1, Je_1; \ldots; e_n, Je_n)$.*

PROOF: For any nonzero $e_1 \in \mathbb{R}^m$ suppose there were a linear relation $Je_1 = \lambda e_1$ over $\mathbb{R}$; then $(\lambda^2 + 1)e_1 = (J^2 + I)e_1 = 0$, so that $\lambda^2 + 1 = 0$, which is not possible in $\mathbb{R}$. Hence (e_1, Je_1) spans a 2-dimensional subspace $V \subset \mathbb{R}^m$, and J induces an endomorphism J of $\mathbb{R}^m/V$ such that $J^2 = -I$. The induction on dimension is clear.

It follows from Lemma 12.3 that up to a change of basis in $\mathbb{R}^{2n}$ one has $J = \Gamma(iI)$ for the homomorphism $GL(n, \mathbb{C}) \xrightarrow{\Gamma} GL(2n, \mathbb{R})$. We shall henceforth use the basis of $\mathbb{R}^{2n}$ described in the proof of Lemma 12.3.

12.4 Lemma: *If $A \in GL(2n, \mathbb{R})$ satisfies $AJ = JA$ for $J = \Gamma(iI)$, then A lies in the image of $GL(n, \mathbb{C}) \xrightarrow{\Gamma} GL(2n, \mathbb{R})$.*

PROOF: For the basis $(e_1, Je_1; \ldots; e_n, Je_n)$ of $\mathbb{R}^{2n}$ there are unique real $n \times n$ matrices $B = (b_p^q)$ and $C = (c_p^q)$ such that $Ae_p = \sum_{q=1}^n (b_p^q e_q + c_p^q Je_q)$ for $p = 1, \ldots, n$, and since $AJ = JA$, one also has $A(Je_p) = JAe_p = \sum_{q=1}^n (-c_p^q e_q + b_p^q Je_q)$; hence $A = \Gamma(B + iC)$.

12.5 Proposition: *A real vector bundle ξ has a complex structure J if and only if it is the realification $\zeta_{\mathbb{R}}$ of a complex vector bundle ζ; furthermore ζ is unique.*

PROOF: If $\xi = \zeta_{\mathbb{R}}$, then $\Gamma(iI)$ is a complex structure J in ξ. Conversely, suppose that J is a complex structure in ξ, and let $\{U_i | i \in I\}$ be an open covering of the base space $X \in \mathscr{B}$ of ξ for which there are trivializations $E|U_i \xrightarrow{\Psi_i} U_i \times \mathbb{R}^m$ of a representation $E \xrightarrow{\pi} X$ of ξ. The restriction of J to $E|U_i$ induces a map $U_i \xrightarrow{J_i} GL(m, \mathbb{R})$ such that

$$(\Psi_i \circ J \circ \Psi_i^{-1})(x, e) = (x, J_i(x)e) \in U_i \times \mathbb{R}^m$$

for every $(x, e) \in U_i \times \mathbb{R}^m$, and since $J^2 = -I$, one has $J_i(x)^2 = -I \in GL(m, \mathbb{R})$ for every $x \in U_i$. It follows as in Lemma 12.3 that m is an even number $2n$ and that there is a basis $(e_1, J_i e_1; \ldots; e_n, J_i e_n)$ of sections of $E|U_i \xrightarrow{\pi} U_i$, for each $i \in I$. With respect to any such basis the matrix representation $(J_i(x)) \in GL(2n, \mathbb{R})$ consists of n blocks

$$\begin{pmatrix} 0 & -1 \\ 1 & 0 \end{pmatrix}$$

down the main diagonal, with zeros elsewhere; in particular, $(J_i) = (J_j)$ over any nonvoid intersection $U_i \cap U_j$. The transition function $U_i \cap U_j \xrightarrow{\psi_i^j} GL(2n, \mathbb{R})$ is given by

$$(\Psi_j \circ \Psi_i^{-1})(x, e) = (x, \psi_i^j(x)e) \qquad \text{for} \quad (x, e) \in (U_i \cap U_j) \times \mathbb{R}^{2n},$$

so that the identity

$$(\Psi_j \circ \Psi_i^{-1}) \circ (\Psi_i \circ J \circ \Psi_i^{-1}) = \Psi_j \circ J \circ \Psi_i^{-1} = (\Psi_j \circ J \circ \Psi_j^{-1}) \circ (\Psi_j \circ \Psi_i^{-1})$$

implies $(\psi_i^j)(J_i) = (J_j)(\psi_i^j) \in GL(2m, \mathbb{R})$. Since $(J_i) = (J_j)$ and $(J_i)^2 = -(I)$, Lemma 12.4 then implies that (ψ_i^j) lies in the image of $GL(n, \mathbb{C}) \xrightarrow{\Gamma} GL(2n, \mathbb{R})$. The uniqueness assertion is an easy exercise.

Briefly, Proposition 12.5 asserts that any complex vector bundle ζ can equally well be regarded as a real vector bundle ξ with a complex structure J.

By Proposition II.6.19, $GL(n, \mathbb{C})$ and $GL(2n, \mathbb{R})$ consist of one and two components, respectively, and since the homomorphism $GL(n, \mathbb{C}) \xrightarrow{\Gamma} GL(2n, \mathbb{R})$ necessarily carries the neutral element into the neutral element, it follows that the image of Γ lies in the component $GL^+(2n, \mathbb{R}) \subset GL(2n, \mathbb{R})$. Hence Proposition 10.16 guarantees that any realification $\zeta_{\mathbb{R}}$ is orientable. In fact, there is a *natural orientation* $X \xrightarrow{\sigma} O(\zeta_{\mathbb{R}})$ given by letting the image of each $x \in X$ be the equivalence class of $e_1 \wedge Je_1 \wedge e_2 \wedge Je_2 \wedge \cdots \wedge e_n \wedge Je_n \in (\bigwedge^{2n} E_x)^*$ for any representation $E \xrightarrow{\pi} X$ of $\zeta_{\mathbb{R}}$ and any linearly independent elements $e_1, \ldots, e_n \in E_x$.

According to the discussion preceding Proposition 10.10, if a real m-plane bundle ξ over $X \in \mathscr{B}$ has an orientation $X \xrightarrow{\sigma} O(\xi)$, then the *oppositely oriented* m-plane bundle $-\xi$ has the orientation $X \xrightarrow{\tau \circ \sigma} O(\xi)$, where τ is the canonical involution of $O(\xi)$. We have just defined one natural orientation of the realification $\zeta_{\mathbb{R}}$ of any complex n-plane bundle ζ over any $X \in \mathscr{B}$. The *other natural orientation* $X \to O(\zeta_{\mathbb{R}})$ of $\zeta_{\mathbb{R}}$ is given by letting the image of each $x \in X$ be the equivalence class of $e_1 \wedge e_2 \wedge \cdots \wedge e_n \wedge Je_1 \wedge Je_2 \wedge \cdots \wedge Je_n \in (\bigwedge^{2n} E_x)^*$. One easily verifies that the two natural orientations of the realification $\zeta_{\mathbb{R}}$ of any complex n-plane bundle ζ differ by a factor $(-1)^{n(n-1)/2}$.

The preceding discussion suggests the converse question: is a given oriented $2n$-plane bundle ξ the realification $\zeta_{\mathbb{R}}$ of a complex n-plane bundle ζ? Here is the case $n = 1$.

12.6 Proposition: *Any oriented 2-plane bundle ξ is the realification of a unique complex line bundle λ.*

PROOF: Let $E \xrightarrow{\pi} X$ represent ξ, with trivializations $E|U_i \xrightarrow{\Psi_i} U_i \times \mathbb{R}^2$ over the sets U_i in some open covering $\{U_i | i \in I\}$ of the base space X of ξ. For each $i \in I$ the trivialization Ψ_i provides a basis $\{s_i, t_i\}$ of the sections $U_i \to E|U_i$. According to Corollary 10.17 one can reduce the structure group of ξ to the rotation group $O^+(2)$, which consists of real 2×2 matrices

$$\begin{pmatrix} \cos\theta & -\sin\theta \\ \sin\theta & \cos\theta \end{pmatrix},$$

so that the bases $\{s_i, t_i\}$ can be chosen in such a way that $s_j = s_i \cos\theta + t_i \sin\theta$ and $t_j = -s_i \sin\theta + t_i \cos\theta$ over any nonvoid $U_i \cap U_j \subset X$, for some map $U_i \cap U_j \xrightarrow{\theta = \theta_i^j} S^1$. For each $i \in I$ one defines $E|U_i \xrightarrow{J_i} E|U_i$ by setting

$J_i s_i = t_i$ and $J_i t_i = -s_i$, so that J_i^2 is multiplication by $-1 \in \mathbb{R}$. Over any nonvoid $U_i \cap U_j \subset X$ one has

$$J_i s_j = J_i(s_i \cos\theta + t_i \sin\theta) = -s_i \sin\theta + t_i \cos\theta = t_j = J_j s_j$$

and

$$J_i t_j = J_i(-s_i \sin\theta + t_i \cos\theta) = -s_i \cos\theta - t_i \sin\theta = -s_j = J_j t_j.$$

Thus $J_i | U_i \cap U_j = J_j | U_i \cap U_j$ for any nonvoid $U_i \cap U_j \subset X$, so that J_i and J_j are the restrictions $J | U_i$ and $J | U_j$ of a globally defined complex structure J. It follows from Proposition 12.5 that $\xi = \lambda_R$ for a unique complex line bundle λ over X, as asserted.

The preceding result occurs only in dimension 2, a reflection of the fact that the obvious homomorphism $U(1) \to O^+(2)$ is an isomorphism. For $n > 1$ one has monomorphisms $U(n) \to O^+(2n)$, but not isomorphisms. Furthermore, for $n > 1$ one can always find an oriented $2n$-plane bundle that is not a realification. For example, we shall show in Volume 2 that the tangent bundle $\tau(S^4)$ of the 4-sphere is not a realification.

The usual inclusion $\mathbb{R} \to \mathbb{C}$ of the real field into the complex field provides an obvious morphism (Γ, Φ) of the transformation group $GL(m, \mathbb{R}) \times \mathbb{R}^m \to \mathbb{R}^m$ into the transformation group $GL(m, \mathbb{C}) \times \mathbb{C}^m \to \mathbb{C}^m$.

12.7 Definition: For any real m-plane bundle ξ over a base space $X \in \mathscr{B}$ the *complexification* $\xi_{\mathbb{C}}$ is the complex m-plane bundle over X induced by the preceding morphism.

Recall from Corollary 10.18 that if ξ is a real m-plane bundle over $X \in \mathscr{B}$, then the Whitney sum $\xi \oplus \xi$ is orientable. In fact there is a *natural orientation* $X \to O(\xi \oplus \xi)$ in which the image of any $x \in X$ is the equivalence class of $e_1 \wedge \cdots \wedge e_m \wedge e'_1 \wedge \cdots \wedge e'_m \in (\bigwedge^{2m} E_x)^*$, where $(e_1, \ldots, e_m)$ and $(e'_1, \ldots, e'_m)$ are corresponding bases of the fibers over x of the two summands in $\xi \oplus \xi$.

12.8 Proposition: *For any real m-plane bundle ξ one has*

$$\xi_{\mathbb{C}\mathbb{R}} = (-1)^{m(m-1)/2}\, \xi \oplus \xi,$$

for the natural orientations of $\xi_{\mathbb{C}\mathbb{R}}$ and $\xi \oplus \xi$, respectively.

PROOF: This is an exercise in definitions, using the *other* natural orientation of $\xi_{\mathbb{C}\mathbb{R}}$.

Let $GL(n, \mathbb{C}) \xrightarrow{\Gamma} GL(n, \mathbb{C})$ and $\mathbb{C}^n \xrightarrow{\Phi} \mathbb{C}^n$ consist of complex conjugations, so that (Γ, Φ) is a morphism of transformation groups, as in Definition II.2.6. For any complex n-plane bundle ζ the *complex conjugate bundle* $\bar{\zeta}$ is the bundle induced by applying (Γ, Φ) to ζ, as in Definition II.2.8. Any coordinate bundle $E \xrightarrow{\pi} X$ representing ζ also represents $\bar{\zeta}$, except that scalar multiplication in each fiber by any complex number $x + iy$ is redefined to be scalar multiplication by the complex conjugate $x - iy$.

12.9 Proposition: *Let $\zeta_{\mathbb{R}}$ and $\bar{\zeta}_{\mathbb{R}}$ be the naturally oriented 2n-plane bundles associated to a complex n-plane bundle ζ and its conjugate $\bar{\zeta}$, respectively; then $\bar{\zeta}_{\mathbb{R}} = (-1)^n \zeta_{\mathbb{R}}$.*

PROOF: The bundles ζ and $\bar{\zeta}$ trivially have the same realifications, and it remains to verify that the natural orientations differ by the factor $(-1)^n$. Let $E \xrightarrow{\pi} X$ represent $\zeta_{\mathbb{R}}$, and let $X \xrightarrow{\sigma} O(\zeta_{\mathbb{R}})$ be its natural orientation. If $e_1, \ldots, e_n$ are any linearly independent elements of E_x, for any $x \in X$, the value of $\sigma(x)$ is the equivalence class of

$$e_1 \wedge Je_1 \wedge e_2 \wedge Je_2 \wedge \cdots \wedge e_n \wedge Je_n \in (\textstyle\bigwedge^{2n} E_x)^*,$$

by definition of σ. Similarly, if $X \xrightarrow{\tau} O(\xi_R)$ is the natural orientation of $\bar{\xi}_{\mathbb{R}}$, the value of $\tau(x)$ is the equivalence class of

$$e_1 \wedge (-Je_1) \wedge e_2 \wedge (-Je_2) \wedge \cdots \wedge e_n \wedge (-Je_n) \in (\textstyle\bigwedge^{2n} E_x)^*.$$

The factor $(-1)^n$ is clear.

In the following proof we use a fixed product $PQ \in GL(2n, \mathbb{C})$, for which (Γ', Φ') is defined as a morphism of transformation groups with $\Gamma' A = (PQ)A(PQ)^{-1} \in GL(2n, \mathbb{C})$ and $\Phi' e = (PQ)e \in \mathbb{C}^{2n}$ for any $A \in GL(2n, \mathbb{C})$ and $e \in \mathbb{C}^{2n}$. If one applies (Γ', Φ') to a given $2n$-plane bundle ζ', then the result is again ζ' itself: one uses PQ to change the basis of $\mathbb{C}^n$ in every trivialization of any coordinate bundle representing ζ'. (See Proposition II.2.9.)

12.10 Proposition: *$\zeta_{\mathbb{RC}} = \zeta \oplus \bar{\zeta}$ for any complex vector bundle ζ.*

PROOF: Let (Γ, Φ) be the composition of the morphisms used in Definitions 12.1 and 12.7 to define realification and complexification, respectively; for example, the image under Γ of $(b_p^q + ic_p^q) \in GL(n, \mathbb{C})$ is the matrix in $GL(2n, \mathbb{C})$ consisting of 2×2 blocks

$$\begin{pmatrix} b_p^q & -c_p^q \\ c_p^q & b_p^q \end{pmatrix},$$

where b_p^q and c_p^q are complex numbers whose imaginary parts happen to vanish. Then $\zeta_{\mathbb{RC}}$ is obtained by applying (Γ, Φ) to ζ. Let $P \in GL(2n, \mathbb{C})$ satisfy $Pe_{2p-1} = e_p$ and $Pe_{2p} = e_{n+p}$ for the usual basis $(e_1, \ldots, e_{2n})$ of $\mathbb{C}^{2n}$, where $p = 1, \ldots, n$, and let $Q \in GL(2n, \mathbb{C})$ consist of 2×2 blocks

$$\begin{pmatrix} 1 & i \\ 1 & -i \end{pmatrix}$$

down the main diagonal, with zeros elsewhere. If (Γ', Φ') is the morphism carrying any $A \in GL(2n, \mathbb{C})$ and $e \in \mathbb{C}^{2n}$ into $(PQ)A(PQ)^{-1} \in GL(2n, \mathbb{C})$ and $(PQ)e \in \mathbb{C}^{2n}$, respectively, then by the preceding remark the result of applying the composed morphism $(\Gamma', \Phi') \circ (\Gamma, \Phi)$ to ζ is still just $\zeta_{\mathbb{R}\mathbb{C}}$. However, for any $A \in GL(n, \mathbb{C})$ and $e \in \mathbb{C}^n$ one has

$$(\Gamma' \circ \Gamma)A = \begin{pmatrix} A & 0 \\ 0 & \bar{A} \end{pmatrix} \in GL(2n, \mathbb{C})$$

and $(\Phi' \circ \Phi)e = e \oplus \bar{e} \in \mathbb{C}^{2n}$, so that the bundle induced by applying the composition $(\Gamma', \Phi') \circ (\Gamma, \Phi)$ to ζ is also $\zeta \oplus \bar{\zeta}$.

12.11 Proposition: $\overline{\xi_{\mathbb{C}}} = \xi_{\mathbb{C}}$ *for any real vector bundle* ξ.

PROOF: Suppose that ξ is a real m-plane bundle over $X \in \mathscr{B}$, represented by a coordinate bundle $E \xrightarrow{\pi} X$ with trivializations $\{\Psi_i | i \in I\}$ defined over the sets of an open covering $\{U_i | i \in I\}$ of X. Let $E' \xrightarrow{\pi'} X$ and $E'' \xrightarrow{\pi''} X$ be the corresponding representations of the realifications $\xi_{\mathbb{C}\mathbb{R}}$ and $(\bar{\xi}_{\mathbb{C}})_{\mathbb{R}}$, respectively. If $\{e_1, \ldots, e_m\}$ is any basis of the sections $U_i \to E | U_i$ arising from a trivialization $E | U_i \xrightarrow{\Psi_i} U_i \times \mathbb{R}^m$, and if J is the complex structure of $\xi_{\mathbb{C}\mathbb{R}}$, then there are corresponding bases $\{e_1, \ldots, e_m, Je_1, \ldots, Je_m\}$ and $\{e_1, \ldots, e_m, -Je_1, \ldots, -Je_m\}$ for $E' | U_i$ and $E'' | U_i$. It is clear that the real linear isomorphism $E' | U_i \xrightarrow{\mathbf{f}_i} E'' | U_i$ satisfying $\mathbf{f}_i(e_j) = e_j$ and $\mathbf{f}_i(Je_j) = -Je_j$ for $j = 1, \ldots, m$ preserves the complex structure, and that there is a globally defined real linear isomorphism $E' \xrightarrow{\mathbf{f}} E''$ with $\mathbf{f}_i = \mathbf{f} | U_i$ for each $i \in I$, which also preserves the complex structure. Hence $\xi_{\mathbb{C}\mathbb{R}}$ and $(\overline{\xi_{\mathbb{C}}})_{\mathbb{R}}$ are isomorphic via an isomorphism that preserves the complex structure, so that $\xi_{\mathbb{C}} = \overline{\xi_{\mathbb{C}}}$ by Proposition 12.5.

13. Remarks and Exercises

13.1 Remark: Introductory accounts of vector bundles can be found in Atiyah [2, Chapter I], Dupont [1, Chapter I], M. W. Hirsch [4, Chapter 4], Husemoller [1, Chapter 3], Kahn [1, Chapter 4], Karoubi [2, Chapter I], Lang [1, Chapter 3], and Milnor and Stasheff [1, Section 2], for example.

13.2 Remark: Introductory accounts of tangent bundles in particular can be found in Auslander and MacKenzie [1], [2, Chapter 4], M. W. Hirsch [4, Chapter 1], Lang [1, Chapter 3], and Lashof [1].

13.3 Remark: According to the linear reduction theorem (Theorem II.6.13) the structure group $GL(m, \mathbb{R})$ of any real m-plane bundle can be reduced to the orthogonal group $O(m) \subset GL(m, \mathbb{R})$, which maps the $(m-1)$-sphere $S^{m-1} \subset \mathbb{R}^m$ into itself; moreover, the behavior of $O(m)$ on $\mathbb{R}^m$ is entirely determined by its behavior on S^{m-1}. With this observation in the background, vector bundles were first considered by Whitney [2, 4, 5, 6] as *sphere bundles.* In particular, tangent bundles to n-dimensional manifolds were introduced as $(n-1)$-sphere bundles, and Whitney's first consideration of Grassmann manifolds appeared within the same framework.

One can also consider more general *nonlinear sphere bundles* that are smooth fiber bundles with fiber S^{m-1} (over smooth manifolds), whose structure group is the group $\text{Diff}^0\, S^{m-1}$ of all orientation-preserving diffeomorphisms $S^{m-1} \to S^{m-1}$. S. P. Novikov [2] and Taniguchi [1] contain examples of such sphere bundles which are genuinely nonlinear: the structure group $\text{Diff}^0\, S^{m-1}$ cannot be reduced to the rotation group $O^+(m)$ $(\subset O(m))$.

Pontrjagin preferred the linear fiber $\mathbb{R}^m$ (and structure group $GL(m, \mathbb{R})$) in his approach to tangent bundles, in Pontrjagin [1, 3, 4, 5].

13.4 Remark: Both Whitney and Pontrjagin had the germ of the homotopy classification theorem (Theorem 8.9) in their very first constructions. The first explicit recognition of a homotopy classification theorem (actually an "ersatz homotopy classification theorem") occurs in Steenrod [2], and later in Chern [2, 3] and in Wu [5].

According to the linear reduction theorem (Theorem II.6.13) one can reduce the structure groups $GL(m, \mathbb{R})$ and $GL(m, \mathbb{C})$ of real and complex m-plane bundles to the orthogonal group $O(m) \subset GL(m, \mathbb{R})$ and the unitary group $U(m) \subset GL(m, \mathbb{C})$, respectively. Since the homotopy classification of such vector bundles is equivalent to the homotopy classification of the associated principal bundles, it follows that the classifying spaces $G^m(\mathbb{R}^\infty)$ and $G^m(\mathbb{C}^\infty)$ of Theorems 8.9 and 11.2 are classifying spaces BG in the sense of Remark II.8.18, for $G = O(m)$ and $G = U(m)$. For this reason, one frequently writes $BO(m)$ and $BU(m)$ in place of $G^m(\mathbb{R}^\infty)$ and $G^m(\mathbb{C}^\infty)$.

13.5 Remark: Other early definitions of tangent bundles appear in Steenrod [1] and in Ehresmann [5]. Stiefel [1] does not contain any vector bundles per se, in spite of its importance; instead, Stiefel works entirely with vector fields, which were later regarded as sections of tangent bundles.

13.6 Remark: The more general identification of vector bundles with corresponding modules of sections first occurs in Serre [4] and in Swan [1], with the observation that the "locally free" modules $\mathscr{F}$ of Theorem 4.6 and Proposition 5.6 are precisely the projective modules over the rings $C^0(X)$ and $C^\infty(X)$, respectively. In the case of tangent bundles, direct descriptions of corresponding modules of differentials and their dual modules of vector fields can be found in Osborn [2, 4] and in Sikorski [1]. One can even describe the underlying rings of smooth functions themselves algebraically, as in S. B. Myers [1] and in Nachbin [1], for example.

Kandelaki [1] gives a sweeping generalization of the results of Serre and Swan, and Lønsted [1] shows that if X is a finite CW space, then one can replace the ring $C^0(X)$ in Theorem 4.6 by a subring $A \subset C^0(X)$ that is Noetherian, with Krull dimension equal to the dimension of X: vector bundles over finite CW spaces are "algebraic" in a reasonable sense.

13.7 Remark: If $E \xrightarrow{\pi} X$ represents the tangent bundle $\tau(X)$ of a smooth manifold X, then for each $x \in X$ the fiber E_x is the corresponding *tangent space*. One can describe E_x as the vector space of derivations of the ring of germs of smooth functions at x into the real field $\mathbb{R}$, an observation due to Chevalley and Bohnenblust. (See Chevalley [1, pp. 76–78] for the analytic case, and Flanders [1, pp. 313–314] for the smooth case.) However, if X is merely a C^r-manifold for $0 < r < \infty$, then the identification no longer works properly. (See Papy [1, 2], Newns and Walker [1], and Osborn [1].) One can impose restrictions on the derivations for which one *does* recover E_x in the latter cases, as in Osborn [1], Sanchez Giralda [1], L. E. Taylor [1], and Ellis [1]; however, the simplest procedure is to always let "smooth structure" mean "C^∞ structure." (Incidentally, "universal" derivations in the sense of Kähler provide further surprises, not only in the cases $0 < r < \infty$ but also in the cases $r = 0$ and $r = \infty$; see Osborn [3, 5], for example.)

13.8 Remark: Tangent bundles are also more sensitive than one might expect to the global properties of the underlying manifolds: one cannot reasonably ignore second countability (or paracompactness). For example, the tangent bundle of the "long line" is a real line bundle, as one expects; however, it is *not* the trivial real line bundle. (See Morrow [1].)

13.9 Remark: In spite of the preceding cautionary remarks, even *topological manifolds* have tangent bundles in a certain sense. The first such definition was given by Nash [1], who successfully generalized constructions we shall describe for the smooth case in Chapter VI. Nash's definition can also be used to generalize various classical results, as in Fadell [4], R. F. Brown [1], and Brown and Fadell [1]; the smooth versions of these results will appear elsewhere in this work.

Milnor later introduced (tangent) *microbundles*, in Milnor [11, 14], which were not intended to be fiber bundles in the sense of Chapter II. However, results of Kister [1, 2] show that the tangent microbundle of an n-dimensional topological manifold *is* a fiber bundle with fiber $\mathbb{R}^n$; the structure group is not $GL(n, \mathbb{R})$, of course, but the group of *all* homeomorphisms $\mathbb{R}^n \to \mathbb{R}^n$ that leave the origin fixed. More generally, a fiber bundle with fiber $\mathbb{R}^m$ is a *topological* $\mathbb{R}^m$ *bundle* whenever the structure group consists of all homeomorphisms $\mathbb{R}^m \to \mathbb{R}^m$ that leave the origin fixed; thus the results of Milnor and Kister assign a tangent topological $\mathbb{R}^n$ bundle to any n-dimensional topological manifold. Further properties of such tangent bundles appear in Lashof and Rothenberg [1], Kuiper and Lashof [1], Kister [3], Derwent [2], and Kurogi [1], for example. Lashof [1] provides an excellent survey of tangent bundles in general.

13.10 Remark: Vector bundle sums can be constructed by other means than the one used in Definition 2.1. Let $\mathbb{R}^\infty \times \mathbb{R}^\infty \to \mathbb{R}^\infty$ map any pair $((x_0, x_1, \ldots), (y_0, y_1, \ldots)) \in \mathbb{R}^\infty \times \mathbb{R}^\infty$ into $(x_0, y_0, x_1, y_1, \ldots) \in \mathbb{R}^\infty$, and observe that if $U \subset \mathbb{R}^\infty$ and $V \subset \mathbb{R}^\infty$ are linear subspaces of dimensions m and n, respectively, then the image of $U \times V \subset \mathbb{R}^\infty \times \mathbb{R}^\infty$ is a linear subspace $W \subset \mathbb{R}^\infty$ of dimension $m + n$; hence there is an induced map $G^m(\mathbb{R}^\infty) \times G^n(\mathbb{R}^\infty) \xrightarrow{+} G^{m+n}(\mathbb{R}^\infty)$ of Grassmann manifolds. If ξ and ξ' are vector bundles of ranks m and n over $X \in \mathscr{B}$ and $X' \in \mathscr{B}$, classified by maps $X \xrightarrow{f} G^m(\mathbb{R}^\infty)$ and $X' \xrightarrow{f'} G^n(\mathbb{R}^\infty)$, respectively, then the composition

$$X \times X' \xrightarrow{f \times f'} G^m(\mathbb{R}^\infty) \times G^n(\mathbb{R}^\infty) \xrightarrow{+} G^{m+n}(\mathbb{R}^\infty)$$

classifies a unique $(m + n)$-plane bundle $\xi + \xi'$ over $X \times X' \in \mathscr{B}$.

13.11 Exercise: Verify that the preceding $(m + n)$-plane bundle $\xi + \xi'$ is the $(m + n)$-plane bundle $\xi + \xi'$ of Definition 2.1.

13.12 Remark: Vector bundle products can be constructed by other means than the one used in Definition 2.2. Let $\mathbb{N}$ be the set $\{0, 1, 2, \ldots\}$ of natural numbers, and let $\mathbb{N} \times \mathbb{N} \xrightarrow{e} \mathbb{N}$ be any bijection. Let $\mathbb{R}^\infty \times \mathbb{R}^\infty \to \mathbb{R}^\infty$ map any pair $((x_0, x_1, \ldots), (y_0, y_1, \ldots)) \in \mathbb{R}^\infty \times \mathbb{R}^\infty$ into that element of $\mathbb{R}^\infty$ whose $e(i, j)$th entry is $x_i y_j \in \mathbb{R}$ for each $(i, j) \in \mathbb{N} \times \mathbb{N}$, and observe that if $U \subset \mathbb{R}^\infty$ and $V \subset \mathbb{R}^\infty$ are linear subspaces of dimensions m and n, respectively, then the image of $U \times V \subset \mathbb{R}^\infty \times \mathbb{R}^\infty$ is a linear subspace $W \subset \mathbb{R}^\infty$ of dimension mn; hence there is an induced *Segre map* $G^m(\mathbb{R}^\infty) \times G^n(\mathbb{R}^\infty) \xrightarrow{\times} G^{mn}(\mathbb{R}^\infty)$ of Grassmann manifolds. If ξ and ξ' are vector bundles of ranks m and n over $X \in \mathscr{B}$ and $X' \in \mathscr{B}$, classified by maps $X \xrightarrow{f} G^m(\mathbb{R}^\infty)$ and $X' \xrightarrow{f'} G^n(\mathbb{R}^\infty)$, respectively, then the composition

$$X \times X' \xrightarrow{f \times f'} G^m(\mathbb{R}^\infty) \times G^n(\mathbb{R}^\infty) \xrightarrow{\times} G^{mn}(\mathbb{R}^\infty)$$

classifies a unique (mn)-plane bundle $\xi \times \xi'$ over $X \times X'$.

13.13 Exercise: Verify that the preceding (mn)-plane bundle $\xi \times \xi'$ is the (mn)-plane bundle $\xi \times \xi'$ of Definition 2.2.

13.14 Remark: Exercises 13.11 and 13.13 provide alternative constructions of the Whitney sum $\xi \oplus \xi'$ and product $\xi \otimes \xi'$ of two vector bundles ξ and ξ' over the same base space $X \in \mathscr{B}$. If $X \overset{\Delta}{\to} X \times X$ is the diagonal map, then one simply sets $\xi \oplus \xi' = \Delta^{!}(\xi + \xi')$ and $\xi \otimes \xi' = \Delta^{!}(\xi \times \xi')$ for the bundles $\xi + \xi'$ and $\xi \times \xi'$ over $X \times X \in \mathscr{B}$, as described in Remarks 13.10 and 13.12.

13.15 Remark: Direct sums $U \times V \mapsto U \oplus V$ and tensor products $U \times V \mapsto U \otimes V$ are *continuous functors* in the sense that there are induced maps

$$\mathrm{Hom}(U, U') \times \mathrm{Hom}(V, V') \to \mathrm{Hom}(U \oplus V, U' \oplus V')$$

and

$$\mathrm{Hom}(U, U') \times \mathrm{Hom}(V, V') \to \mathrm{Hom}(U \otimes V, U' \otimes V'),$$

which are themselves continuous, for any vector spaces U, V, U', V'. One can therefore use the method in Atiyah [2, pp. 6–9], for example, to construct Whitney sums $\xi \oplus \xi'$ and products $\xi \otimes \xi'$ directly, without introducing $\xi + \xi'$ and $\xi \times \xi'$. This construction also appears in Husemoller [1, pp. 65–67], for example. The commutative diagram

$$\begin{array}{ccc} G \times F & \longrightarrow & F \\ {\scriptstyle \Gamma \times \Phi}\big\downarrow & & \big\downarrow{\scriptstyle \Phi} \\ G' \times F' & \longrightarrow & F' \end{array}$$

used in Definition II.2.6 to define a morphism (Γ, Φ) of transformation groups is a generalized version of the condition describing continuous functors; in particular, the morphisms (Γ, Φ) of transformation groups used to construct $\xi + \xi'$ and $\xi \times \xi'$ in Definitions 2.1 and 2.2 are rephrased versions of the continuous functors $U \times V \mapsto U \oplus V$ and $U \times V \mapsto U \otimes V$, respectively.

13.16 Exercise: Verify that the Whitney sum $\xi \oplus \xi'$ and product $\xi \otimes \xi'$ of Definition 2.4 (or Proposition 2.5) coincide with the Whitney sum $\xi \oplus \xi'$ and product $\xi \otimes \xi'$ described via continuous functors in Atiyah [2, pp. 6–9].

13.17 Remark: The linear reduction theorem (Theorem II.6.13) implies that the structure group $GL(m, \mathbb{R})$ of any real m-plane bundle ξ over any base space $X \in \mathscr{B}$ can be reduced to the orthogonal subgroup $O(m) \subset GL(m, \mathbb{R})$. The existence of a riemannian metric on ξ (Proposition 3.4) leads to an alternative proof of the same result, as follows.

Let $E \xrightarrow{\pi} X$ represent ξ, let $\{U_i | i \in I\}$ be any open covering of X with local trivializations $E|U_i \xrightarrow{\Psi_i} U_i \times \mathbb{R}^m$, and let $U_i \times (\mathbb{R}^m \times \mathbb{R}^m) \xrightarrow{\langle,\rangle_i} \mathbb{R}$ be the restriction to U_i of a fixed riemannian metric on ξ. Let $(e_1, \ldots, e_m)$ be the usual orthonormal basis of $\mathbb{R}^m$ with respect to the usual inner product $\mathbb{R}^m \times \mathbb{R}^m \xrightarrow{\langle,\rangle} \mathbb{R}$, and for each $p = 1, \ldots, m$ and each $x \in U_i$ let $s_p(x) = \Psi_i^{-1}(x, e_p) \in E_x$. Then $(s_1(x), \ldots, s_m(x))$ is a basis of E_x, and one applies the Gram–Schmidt process to obtain a basis $(s'_1(x), \ldots, s'_m(x))$ that is orthonormal with respect to the inner product $\{x\} \times (\mathbb{R}^m \times \mathbb{R}^m) \xrightarrow{\langle,\rangle_i} \mathbb{R}$. Thus a given basis $(s_1, \ldots, s_m)$ of local sections $U_i \to E$ provides a new basis $(s'_1, \ldots, s'_m)$ of local sections $U_i \to E$, with $s'_p = \sum_{q=1}^m \lambda_p^q s_q$ for a map $U_i \xrightarrow{\lambda_i} GL(m, \mathbb{R})$ carrying each $x \in U_i$ into a triangular matrix in $GL(m, \mathbb{R})$. If $U_i \times \mathbb{R}^m \xrightarrow{\Lambda_i} U_i \times \mathbb{R}^m$ carries each $(x, e) \in U_i \times \mathbb{R}^m$ into $(x, \lambda_i(x)e) \in U_i \times \mathbb{R}^m$, the composition $E|U_i \xrightarrow{\Psi_i} U_i \times \mathbb{R}^m \xrightarrow{\Lambda_i^{-1}} U_i \times \mathbb{R}^m$ is a new local trivialization $E|U_i \xrightarrow{\Psi'_i} U_i \times \mathbb{R}^m$. There are then new transition functions $U_i \cap U_j \xrightarrow{\psi_i'^j} GL(m, \mathbb{R})$, defined by requiring $(\Psi'_j \circ \Psi_i'^{-1})(x, e) = (x, \psi_i'^j(x)e)$ for every $(x, e) \in U_i \cap U_j \times \mathbb{R}^m$, and one easily verifies that $\psi_i'^j(x) \in O(m)$ for every $x \in U_i \cap U_j$. Hence the structure group $GL(m, \mathbb{R})$ of ξ can be reduced to the subgroup $O(m) \subset GL(m, \mathbb{R})$, as claimed.

13.18 Exercise: Carry out the verification required to complete the proof of the preceding reduction theorem.

13.19 Exercise: The linear reduction theorem (Theorem II.6.13) implies that the structure group $GL(n, \mathbb{C})$ of any complex n-plane bundle ξ over any base space $X \in \mathscr{B}$ can be reduced to the unitary group $U(n) \subset GL(n, \mathbb{C})$. Prove the same result by imposing a hermitian metric on ζ and following the pattern of Remark 13.17 and Exercise 13.18.

13.20 Remark: The $C^\infty(X)$-module $\mathscr{E}(X)$ of differentials on a smooth manifold X was described in Definition 6.11 as the dual of the $C^\infty(X)$-module $\mathscr{E}^*(X)$ of smooth vector fields on X: however, one can also construct $\mathscr{E}(X)$ directly. The ring $C^\infty(X \times X)$ of smooth functions $X \times X \xrightarrow{F} \mathbb{R}$ is an algebra over the ring $C^\infty(X)$ of smooth functions $X \xrightarrow{f} \mathbb{R}$, with $fF \in C^\infty(X \times X)$ defined by setting $(fF)(x, y) = f(x)F(x, y) \in \mathbb{R}$ for any $(x, y) \in X \times X$. Let $J \subset C^\infty(X \times X)$ be the ideal of those $F \in C^\infty(X \times X)$ that vanish on the diagonal $\Delta(X) \subset X \times X$, and let $\mathscr{E}(X)$ be the quotient $C^\infty(X)$-module J/J^2. There is a real linear map $C^\infty(X) \to J$ carrying any $f \in C^\infty(X)$ into the function with value $f(y) - f(x) \in \mathbb{R}$ on $(x, y) \in X \times X$, and there is an induced real linear map $C^\infty(X) \xrightarrow{d} \mathscr{E}(X)$ that satisfies the classical product rule $d(fg) = f\,dg + g\,df$.

A smooth map $Y \xrightarrow{\Phi} X$ induces an algebra homomorphism $C^\infty(X \times X) \to C^\infty(Y \times Y)$, which in turn induces a module homomorphism $\mathscr{E}(X) \xrightarrow{\Phi^*} \mathscr{E}(Y)$ over the induced ring homomorphism $C^\infty(X) \xrightarrow{\Phi^*} C^\infty(Y)$, where $\Phi^* f = f \circ \Phi$ for any $f \in C^\infty(X)$, so that $\Phi^* df = d(\Phi^* f)$.

13.21 Exercise: Verify that the $C^\infty(X)$-module $\mathscr{E}(X)$ of Remark 13.20 is canonically isomorphic to the $C^\infty(X)$-module $\mathscr{E}(X)$ of Definition 6.11.

13.22 Exercise: Verify for any smooth map $Y \xrightarrow{\Phi} X$ that the induced homomorphism $\mathscr{E}(X) \xrightarrow{\Phi^*} \mathscr{E}(Y)$ of Remark 13.20 agrees up to canonical isomorphism with the homomorphism $\mathscr{E}(X) \xrightarrow{\Phi^*} \mathscr{E}(Y)$ constructed following Proposition 6.14.

13.23 Remark: There is a severe penalty for replacing $C^\infty(X \times X)$ by the subalgebra $C^\infty(X) \otimes C^\infty(X) \subset C^\infty(X \times X)$ in the constructions of Remark 13.20. Some of the resulting pathology is described in Osborn [3].

13.24 Exercise: Let $E \xrightarrow{\pi} X$ and $E' \xrightarrow{\pi'} X$ represent real m-plane bundles over the same $X \in \mathscr{B}$, and suppose that they are described by transition functions $U_i \cap U_j \to GL(m, \mathbb{R})$ whose values are transposed inverses of each other. Show that $E \xrightarrow{\pi} X$ and $E' \xrightarrow{\pi'} X$ represent the same real vector bundle.

13.25 Exercise: Replace "real" by "complex" in the preceding exercise, using *conjugate* transposed inverses. Show that the given coordinate bundles represent complex conjugate m-plane bundles.

13.26 Remark: A smooth n-dimensional manifold X is *parallelizable* whenever the tangent bundle $\tau(X)$ is the trivial bundle ε^n over X. For example, any Lie group G is parallelizable: the left-invariant vector fields provide a basis of sections of $\tau(G)$. The spheres S^1, S^3, and S^7 are also parallelizable; in fact S^1 and S^3 are the underlying manifolds of the Lie groups $U(1)$ and $U(2)$, respectively. However, there are no other parallelizable spheres, a result of Bott and Milnor [1] and Milnor [4], which was later simplified by Atiyah and Hirzebruch [3]; a proof will be given in Volume 3.

Products of spheres are better behaved. One result of Kervaire [1] is that $S^p \times S^q$ is parallelizable whenever at least one of the numbers $p > 0$ or $q > 0$ is odd. Staples [1] gives a simpler proof of the same result.

A classical result of Stiefel [1] asserts that every orientable 3-dimensional manifold is parallelizable; more generally, Dupont [3] shows that every orientable $(4k + 3)$-dimensional manifold has at least three linearly independent vector fields. Dediu [1, 2, 3] obtains similar results for $(4k+3)$-dimensional lens spaces, $k \geqq 0$; lens spaces are defined in Remark I.10.20.

Since parallelizability of the real projective space RP^n would imply parallelizability of the corresponding sphere S^n, it follows that RP^1, RP^3, and RP^7 are the only parallelizable real projective spaces. The question concerning more general real Grassmann manifolds is apparently still open.

13.27 Exercise: Recall the identity $\tau(RP^n) \oplus \varepsilon^1 = (n+1)\gamma_n^1$ of Proposition 7.4, where RP^n is the Grassmann manifold $G^1(\mathbb{R}^{n+1})$ and γ_n^1 is the canonical real line bundle over $G^1(\mathbb{R}^{n+1})$. Show more generally that

$$\tau(G^m(\mathbb{R}^{m+n})) \oplus (\gamma_n^m \otimes \gamma_n^m) = (m+n)\gamma_n^m,$$

where γ_n^m is the canonical m-plane bundle over the Grassmann manifold $G^m(\mathbb{R}^{m+n})$.

13.28 Remark: The preceding exercise is not entirely trivial; its solution can be found in Hsiang and Szczarba [1], along with corresponding results for the complex and quaternionic cases. In the latter cases the summand $\gamma_n^m \otimes \gamma_n^m$ is replaced by $\gamma_n^m \otimes \bar{\gamma}_n^m$, for the conjugate bundle $\bar{\gamma}_n^m$. Different generalizations of these results are given in Borel and Hirzebruch [1] and in Lam [4].

13.29 Remark: Whitney sums $m\gamma_n^1$ of m copies of the canonical real line bundle γ_n^1 over RP^n serve other useful purposes. For example, the immersion problem for real projective spaces RP^n is equivalent to the problem of finding the largest numbers of linearly independent sections of $m\gamma_n^1$ for all $m > 0$ and $n > 0$. The immersion problem for real projective spaces is approached from this point of view in Lam [3] and Yoshida [1, 2]. (Recent catalogs of best-possible immersions $RP^n \to \mathbb{R}^{2n-k}$ can be found in Gitler [1], James [1], and Berrick [1].)

13.30 Remark: If a *complex* vector bundle ζ over a polyhedron $|K|$ has a finite structure group $G \subset GL(n, \mathbb{C})$, then there is a finite covering $X \xrightarrow{f} |K|$ such that the pullback $f^!\zeta$ over X is trivial; see Deligne and Sullivan [1]. The corresponding statement is false in the real case; see Milnor [5].

13.31 Remark: If $Y \in \mathscr{B}$ is a closed subspace of $X \in \mathscr{B}$, then any trivial bundle ε^n over Y is the restriction to Y of a corresponding trivial bundle ε^n over X. In general, however, a vector bundle over $Y \subset X$ is *not* the restriction to Y of a vector bundle over X, even when X and Y are real projective spaces or lens spaces. Counterexamples can be found in Horrocks [1],

Schwarzenberger [3], Maki [1], and Kobayashi, Maki, and Yoshida [1], for example. There is an especially easy complex counterexample in Schwarzenberger [4, pp. 65–66].

13.32 Exercise: According to Definition 3.1, a riemannian metric on a real m-plane bundle ξ over $X \in \mathscr{B}$ restricts to an inner product $\langle\ ,\ \rangle_x$ for each fiber E_x; in particular, there is a matrix representation of $\langle\ ,\ \rangle_x$ as a diagonal matrix with m positive entries. One can equally well consider nondegenerate *metrics of type* (p, q): the latter requirement is replaced by requiring the existence of a diagonal representation of each $\langle\ ,\ \rangle_x$ with p positive entries and q negative entries, where $p + q = m$. Show that ξ possesses such a metric of type (p, q) if and only if it is the Whitney sum of a real p-plane bundle and a real q-plane bundle.

13.33 Remark: According to Theorem 9.5, any (continuous) real vector bundle over a smooth manifold can be represented by a smooth coordinate bundle. Can a (continuous) complex vector bundle over a complex manifold X be represented by a *holomorphic* coordinate bundle? For complex line bundles the answer is affirmative, and for complex 2-plane bundles the answer is affirmative in the special cases $X = CP^2$ (Schwarzenberger [1, 2]) and $X = CP^3$ (Atiyah and Rees [1]). These results, and others, are presented in detail on pages 111–138 of Okonek, Schneider, and Spindler [1], where it is shown for $n \leqq 3$ that *every* complex vector bundle over CP^n can be represented by a holomorphic coordinate bundle. However, a construction of Rees [1] strongly suggests that for each $n > 4$ there is a complex 2-plane bundle over CP^n that *cannot* be represented by a holomorphic coordinate bundle.

Here is an apparently narrower problem: can a (continuous) complex vector bundle over a complex algebraic variety be represented by a coordinate bundle which is *algebraic* in the obvious sense? A basic result of Serre [3] implies that this question reduces to the preceding one: a complex vector bundle over a complex algebraic variety is algebraic if and only if it is holomorphic.

13.34 Remark: If $E \xrightarrow{\pi} X$ represents a real m-plane bundle ξ, then π is itself a homotopy equivalence; hence the fiber homotopy equivalence relation of Remark II.8.21 is of little direct interest. *However*, if $E_s \xrightarrow{\pi_s} X$ represents the corresponding $(m - 1)$-sphere bundle ξ_s, described in Remark 13.3, the projection π_s is no longer a homotopy equivalence. Accordingly, two real vector bundles ξ and η over the same base space are *defined* to be *fiber homotopy equivalent* (or of the same *fiber homotopy type*) whenever the statement is true of the corresponding sphere bundles ξ_s and η_s: fiber homotopy equivalence of complex vector bundles is described in terms of their realifications. It is clear that one can equally well consider fiber homotopy equivalence of vector bundles over different base spaces, provided the base spaces themselves are homotopy equivalent in the usual sense.

The first major application of fiber homotopy equivalence coincided with its very definition in Thom [4]: *the fiber homotopy type of the tangent bundle* $\tau(X)$ *of a smooth closed manifold* X *is independent of the smooth structure assigned to* X. A stronger version of Thom's result occurs in Benlian and Wagoner [1]: *if two smooth closed manifolds are homotopy equivalent, then their tangent bundles are fiber homotopy equivalent*; a simplified proof of this statement is given in Dupont [2]. Incidentally, Benlian and Wagoner also show that if the given homotopy equivalent manifolds are n-dimensional, and if one of the manifolds has k linearly independent vector fields for some $k \leqq (n-1)/2$, then the other manifold also has k linearly independent vector fields.

Fiber homotopy equivalence provides a natural setting for other results. Recall that parallelizability of Lie groups was easily established in Remark 13.26, and that H-spaces are natural generalizations of Lie groups. According to Kaminker [1], the tangent bundle of any (smooth) H-space is fiber homotopy equivalent to a trivial bundle; this is clearly the natural generalization of parallelizability.

13.35 Remark: In general one *cannot* strengthen the preceding remark to conclude that a homotopy equivalence $X \xrightarrow{f} X'$ pulls the tangent bundle $\tau(X')$ back to the tangent bundle $\tau(X)$. However, one *does* have $f^!\tau(X') = \tau(X)$ in certain special cases considered in Shiraiwa [1] and Ishimoto [1].

13.36 Remark: Two vector bundles ξ and η over the same base space are *stably equivalent* if $\xi \oplus \varepsilon^p = \eta \oplus \varepsilon^q$ for trivial vector bundles ε^p and ε^q. For example, Proposition 7.4 asserts that the $(n+1)$-fold Whitney sum $(n+1)\gamma_n^1$ of the real canonical line bundle γ_n^1 over RP^n is stably equivalent to the tangent bundle $\tau(RP^n)$. In Shiraiwa [2] one learns that if $X \xrightarrow{f} X'$ is a homotopy equivalence of closed even-dimensional smooth manifolds such that $f^!\tau(X')$ is stably equivalent to $\tau(X)$, then the stronger conclusion $f^!\tau(X') = \tau(X)$ is valid.

There are natural homotopy classification theorems for stable equivalence classes of vector bundles. For each $m > 0$, let $G^m(\mathbb{R}^\infty) \to G^{m+1}(\mathbb{R}^\infty)$ classify the Whitney sum $\gamma^m \oplus \varepsilon^1$ of the universal real m-plane bundle γ^m and the trivial line bundle ε^1 over $G^m(\mathbb{R}^\infty)$. In the notation of Remark 13.4, this is a map $BO(m) \to BO(m+1)$, and one can form the inductive limit

$BO = \lim_m BO(m)$. Clearly the stable equivalence classes of real vector bundles over any $X \in \mathscr{B}$ are classified by homotopy classes of maps $X \to BO$. Similarly, the stable equivalence classes of complex vector bundles over any $X \in \mathscr{B}$ are classified by homotopy classes of maps $X \to BU = \lim_m BU(m)$, where $BU(m)$ is also defined in Remark 13.4.

Stable equivalence classes will be studied in more detail in Volume 3: they are the very essence of K-theory. However, representations of such classes are of independent interest. For example, Fossum [1] shows that every real or complex vector bundle over any sphere S^n is stably equivalent to a vector bundle represented by an *algebraic* coordinate bundle. In the special case $n = 4k > 16$ Barratt and Mahowald [1] show that any real vector bundle over S^{4k} is either stably trivial or stably equivalent to a bundle of rank $2k + 1$ that is *irreducible* in the sense that it is not a Whitney sum of nontrivial bundles of lower rank; an alternative proof of the same result appears in Mahowald [1]. Glover, Homer, and Stong [1] show for any $k > 0$ that the tangent bundle $\tau(CP^{2k})$ of the complex projective space CP^{2k} is similarly irreducible, as a complex vector bundle; this partially strengthens a result of Tango [1], that for *any* $n > 2$, there is an irreducible complex vector bundle of rank $n - 1$ over CP^n.

13.37 Remark: Surfaces have special properties as base spaces. Cavenaugh [1] shows that every nontrivial real 2-plane bundle over any orientable surface is irreducible in the sense of the preceding remark. Moore [1] sharpens the result of Fossum [1] as follows: every vector bundle over the 2-sphere S^2 can itself be represented by an algebraic coordinate bundle. (The latter result is clearly related to the result of Lønsted [1], cited in Remark 13.6, that any vector bundle over any finite CW space is "algebraic" in a reasonable sense.)

13.38 Exercise: Let X be an m-dimensional CW space as in §I.5, so that X has no n-cells for any $n > m$, and for some fixed $n > m$ let ξ be a real n-plane bundle over X. Use induction on $p = 1, \ldots, m$ and the fact that every map $S^{p-1} \to S^{n-1}$ is homotopic to a constant map for $p < n$ to show that ξ has a nowhere-vanishing section. (This is an easy exercise, which is carried out in Husemoller [1, p. 99], for example; it will also be done in detail in Volume 2 of the present work, for any m-dimensional weak simplicial space X.)

13.39 Exercise: Use Exercise 13.38 and Proposition 4.8 to conclude that if $n > m$, then any real n-plane bundle ξ over an m-dimensional CW space X is of the form $\eta \oplus \varepsilon^{n-m}$ for some real m-plane bundle η over X.

13.40 Remark: The *geometric dimension* of a real vector bundle ξ is the least integer n such that ξ is stably equivalent to a real n-plane bundle η; in case ξ is stably trivial its geometric dimension is 0. For example, according to Exercise 13.39 the geometric dimension of any vector bundle over an m-dimensional CW space is at most m. However, stronger results are possible in some cases: Remark 13.36 asserts for $k > 4$ that the geometric dimension of any real vector bundle over the sphere S^{4k} is either 0 or $2k + 1$. There are other base spaces over which all real vector bundles have severely limited geometric dimensions; several examples can be found in Sjerve [1], Hill [1], and Davis and Mahowald [1].

13.41 Remark: Given an m-plane bundle ξ, a *stable inverse* of ξ is any n-plane bundle η, over the same base space and for some $n \geqq 0$, such that $\xi \oplus \eta = \varepsilon^{m+n}$. The existence of stable inverses of real m-plane bundles over certain kinds of base spaces was established in Corollaries 8.13, 9.2, and 9.4, and there were indications of upper bounds on n. In the following situation one can take $n = m$.

For any $X \in \mathscr{B}$ and any point $* \in X$ the (*reduced*) *suspension* ΣX is the quotient of the product $X \times [0, 1]$ by the subspace $X \times \{0\} \cup \{*\} \times [0, 1] \cup X \times \{1\}$, in the quotient topology. If $X \in \mathscr{B}$ is a compact hausdorff space, and if ξ is a complex m-plane bundle over ΣX, then there is a complex m-plane bundle η over ΣX such that $\xi \oplus \eta = \varepsilon^{2m}$. A proof is given in Chan and Hoffman [1].

13.42 Remark: The equivalence relations of Remarks 13.34 and 13.36 lead to a useful weaker equivalence relation. Two vector bundles ξ and η over the same base space are *J-equivalent* (or *stably fiber homotopy equivalent*) if there are trivial bundles ε^p and ε^q such that $\xi \oplus \varepsilon^p$ and $\eta \oplus \varepsilon^q$ are fiber homotopy equivalent in the sense of Remark 13.34. For example, the theorems of Benlian and Wagoner [1] described in Remark 13.35 first appeared as J-equivalence theorems in Atiyah [1] and Sutherland [1], respectively. The following J-equivalence theorem is due to Atiyah and Todd [1] and to Adams and Walker [1] with a later simplification in Lam [2]: a necessary and sufficient condition that the m-fold Whitney sum $m\gamma_n^1$ of the canonical complex line bundle γ_n^1 over CP^n be J-equivalent to a trivial bundle is that m be divisible by an integer $i(m, n)$ defined in Atiyah and Todd [1]. An application of this result will be indicated in Remark VI.9.22.

13.43 Exercise: Let $Y \xrightarrow{\sigma} E$ be the zero-section of a smooth coordinate bundle $E \xrightarrow{\pi} Y$ representing a real vector bundle η over a smooth manifold Y. Show that $\sigma^! \tau(E) = \eta \oplus \tau(Y)$ for the homotopy equivalence σ.

13.44 Remark: Let $X \xrightarrow{f} Y$ be the restriction of a smooth embedding $X \to \mathbb{R}^n$ of a smooth manifold X to an open tubular neighborhood $Y \subset \mathbb{R}^n$ of $f(X) \subset \mathbb{R}^n$, so that f is a homotopy equivalence. Then any vector bundle ξ over X is of the form $f^!\eta$ for a vector bundle η over Y, and since $\tau(Y) = \varepsilon^n$ over Y the preceding exercise implies $(\sigma \circ f)^!\tau(E) = \xi \oplus \varepsilon^n$ over X. Thus, up to the homotopy equivalence $\sigma \circ f$, any vector bundle ξ over X is stably equivalent to the tangent bundle of a smooth manifold E.

CHAPTER IV

$\mathbb{Z}/2$ Euler Classes

0. Introduction

Let $H^*(X; \mathbb{Z}/2)$ be the singular $\mathbb{Z}/2$ cohomology ring of a base space $X \in \mathscr{B}$. In this chapter we assign a cohomology class $e(\xi) \in H^n(X; \mathbb{Z}/2)$ to any real n-plane bundle ξ over X. Such classes will be used in Chapter V to obtain further $\mathbb{Z}/2$ classes $w(\xi) = 1 + w_1(\xi) + \cdots + w_n(\xi) \in H^*(X; \mathbb{Z}/2)$ for ξ, with $w_n(\xi) = e(\xi)$, which have many useful geometric applications.

Let $E \xrightarrow{\pi} X$ represent ξ, and let $E^* \subset E$ consist of the nonzero elements in E. The Mayer–Vietoris technique is used in §1 to construct a relative cohomology class $U_\xi \in H^n(E, E^*; \mathbb{Z}/2)$, whose properties are further described in §2. The zero-section $X \xrightarrow{\sigma} E$ and inclusion $E \xrightarrow{j} E, E^*$ then provide a composition

$$H^*(E, E^*; \mathbb{Z}/2) \xrightarrow{j^*} H^*(E; \mathbb{Z}/2) \xrightarrow{\sigma^*} H^*(X; \mathbb{Z}/2),$$

hence a class $e(\xi) = \sigma^* j^* U_\xi \in H^n(X; \mathbb{Z}/2)$ as desired.

For the universal real line bundle γ^1 over RP^∞ (Definition III.8.2) the class $e(\gamma^1) \in H^1(RP^\infty; \mathbb{Z}/2)$ plays a privileged role: the cohomology ring $H^*(RP^\infty; \mathbb{Z}/2)$ is the polynomial ring $\mathbb{Z}/2[e(\gamma^1)]$ over $\mathbb{Z}/2$ generated by $e(\gamma^1) \in H^1(RP^\infty; \mathbb{Z}/2)$. Similarly, for the canonical real line bundle γ_1^1 over RP^1 (Definition III.7.2) the class $e(\gamma_1^1) \in H^1(RP^1; \mathbb{Z}/2)$ generates $H^*(RP^1; \mathbb{Z}/2)$, subject to the condition $e(\gamma_1^1) \cup e(\gamma_1^1) = 0$. These properties of $e(\gamma^1)$ and $e(\gamma_1^1)$ lead to an axiomatic characterization of all the classes $e(\xi)$, given in §6.

The notations $H^*(-,-)$ and $H^*(-)$ are used consistently in this chapter, and in the remainder of this volume, to denote singular cohomology

$H^*(-,-;\mathbb{Z}/2)$ and $H^*(-;\mathbb{Z}/2)$ with $\mathbb{Z}/2$ coefficients. Other coefficient rings will appear in Volumes 2 and 3.

1. The $\mathbb{Z}/2$ Thom Class $U_\xi \in H^n(E, E^*)$

Let $E \xrightarrow{\pi} X$ represent a real n-plane bundle ξ over a base space $X \in \mathscr{B}$, and let E^* consist of the nonzero elements in E. We use the local behavior of $H^*(E, E^*)$ to construct a class $U_\xi \in H^n(E, E^*)$. Specifically, for each $x \in X$ let $E_x = \pi^{-1}(\{x\})$ and $E_x^* = E_x \cap E^*$, and let $E_x, E_x^* \xrightarrow{j_x} E, E^*$ be the inclusion of pairs; then U_ξ will be defined by the behavior of the images of the homomorphisms $H^n(E, E^*) \xrightarrow{j_x^*} H^n(E_x, E_x^*)$. An obvious first step is to describe the $\mathbb{Z}/2$ cohomology modules $H^*(E_x, E_x^*)$, each of which is isomorphic to $H^*(\mathbb{R}^n, \mathbb{R}^{n*})$.

1.1 Lemma: *If $n > 0$, the $\mathbb{Z}/2$-module $H^*(\mathbb{R}^n, \mathbb{R}^{n*})$ vanishes except for a single generator in $H^n(\mathbb{R}^n, \mathbb{R}^{n*})$.*

PROOF: The excision and homotopy axioms for cohomology permit one to replace the pair $\mathbb{R}^n$, $\mathbb{R}^{n*}$ by the pair D^n, S^{n-1}, where $D^n \subset \mathbb{R}^n$ is the unit n-disk and its boundary $S^{n-1} \subset \mathbb{R}^{n*}$ is the unit $(n-1)$-sphere. For the inclusions $S^{n-1} \xrightarrow{i} D^n$ and $D^n \xrightarrow{j} D^n, S^{n-1}$ the usual exact cohomology sequence

$$\cdots \xrightarrow{j^*} H^{q-1}(D^n) \xrightarrow{i^*} H^{q-1}(S^{n-1})$$
$$\xrightarrow{\delta} H^q(D^n, S^{n-1}) \xrightarrow{j^*} H^q(D^n) \xrightarrow{i^*} \cdots$$

contains many 0's; in particular, all terms with $q < 0$ vanish. In case $q > 1$ the result follows from the portion

$$0 \longrightarrow H^{q-1}(S^{n-1}) \xrightarrow{\delta} H^q(D^n, S^{n-1}) \longrightarrow 0$$

of the preceding exact sequence and the known behavior of $H^{q-1}(S^{n-1})$. For smaller values of q the exact sequence is

$$0 \xrightarrow{\delta} H^0(D^n, S^{n-1}) \xrightarrow{j^*} H^0(D^n) \xrightarrow{i^*} H^0(S^{n-1})$$
$$\xrightarrow{\delta} H^1(D^n, S^{n-1}) \xrightarrow{j^*} 0.$$

If $n > 1$, then $H^0(D^n) \xrightarrow{i^*} H^0(S^{n-1})$ is an isomorphism, so that $H^0(D^n, S^{n-1}) = 0 = H^1(D^n, S^{n-1})$. If $n = 1$, then one has $H^0(D^1) = \mathbb{Z}/2$, $H^0(S^0) = \mathbb{Z}/2 \oplus \mathbb{Z}/2$, and $i^*(a) = a \oplus (-a)$ for any $a \in H^0(D^1)$; hence necessarily $H^1(D^1, S^0) = \mathbb{Z}/2$ and $\delta(a \oplus b) = a + b$ for any $a \oplus b \in H^0(S^0)$.

Let $E \xrightarrow{\pi} X$ represent a real n-plane bundle over $X \in \mathscr{B}$ and define $E^* \subset E$ as before.

1.2 Lemma: *$H^p(E, E^*) = 0$ for $p < n$.*

PROOF: Since $H^p(E, E^*)$ depends only on the homotopy type of the base space $X \in \mathscr{B}$, one can assume that X is of finite type as in Definition I.1.1. (This normalization is described in detail in the proof of Theorem II.7.2, the absolute Leray–Hirsch theorem.) Let $\mathcal{O}(X)$ denote the category of open sets $U \subset X$, and let $E|U = \pi^{-1}(U)$ and $E^*|U = \pi^{-1}(U) \cap E^*$ for any $U \in \mathcal{O}(X)$. There is then a Mayer–Vietoris functor $\{h^q | q \in \mathbb{Z}\}$ on $\mathcal{O}(X)$ to $\mathbb{Z}/2$-modules, with $h^q(U) = H^q(E|U, E^*|U)$ for any $U \in \mathcal{O}(X)$; the connecting homomorphisms are the classical ones for relative cohomology, where $E^*|U$ is open in $E|U$. There is also a trivial Mayer–Vietoris functor $\{k^q | q \in \mathbb{Z}\}$ on $\mathcal{O}(X)$ to $\mathbb{Z}/2$-modules, with $k^q(U) = 0$ for any $U \in \mathcal{O}(X)$. Finally there is a unique natural transformation θ from $\{h^q | q \in \mathbb{Z}\}$ to $\{k^q | q \in \mathbb{Z}\}$ which annihilates everything. If $U \in \mathcal{O}(X)$ is contractible to a point $x \in U$, then $h^q(U) = H^q(E_x, E_x^*)$ by the homotopy axiom for cohomology, where $H^q(E_x, E_x^*) = H^q(\mathbb{R}^n, \mathbb{R}^{n*}) = 0$ for $q < n$ by Lemma 1.1. Thus $h^q(U) \xrightarrow{\theta_U} k^q(U) = 0$ is an isomorphism whenever U is contractible and $q < n$, so that $H^q(E, E^*) = h^q(X) \xrightarrow{\theta_X} k^q(X) = 0$ is also an isomorphism whenever $q < n$, by Corollary I.9.5.

The main result of this section is obtained by refining the trivial Mayer–Vietoris functor $\{k^q | q \in \mathbb{Z}\}$ of the previous proof. For any $U \in \mathcal{O}(X)$ let $k^q(U) = 0$ as before *except* in the case $q = n$, and let $k^n(U)$ be the $\mathbb{Z}/2$-module of functions $U \to \mathbb{Z}/2$ which are continuous in the discrete topology of $\mathbb{Z}/2$. For any ordered pair $(U, V) \in \mathcal{O}(X) \times \mathcal{O}(X)$ the sequence

$$0 \longrightarrow k^n(U \cup V) \xrightarrow{i^*_{U,V}} k^n(U) \oplus k^n(V) \xrightarrow{j^*_{U,V}} k^n(U \cap V) \longrightarrow 0$$

is trivially exact for the Mayer–Vietoris homomorphisms $i^*_{U,V}$ and $j^*_{U,V}$ of §I.9, so that $\{k^q | q \in \mathbb{Z}\}$ is still a Mayer–Vietoris functor. There is a natural transformation θ from the old $\{h^q | q \in \mathbb{Z}\}$ to the new $\{k^q | q \in \mathbb{Z}\}$ which annihilates $h^q(U)$ for every $U \in \mathcal{O}(X)$ and every $q \neq n$. To define $h^n(U) \xrightarrow{\theta_U} k^n(U)$ let $E_x, E_x^* \xrightarrow{j_x} E|U, E^*|U$ be the inclusion of pairs for any $x \in U$, where $H^n(E_x, E_x^*) = \mathbb{Z}/2$ by Lemma 1.1, and for any $\alpha \in h^n(U) = H^n(E|U, E^*|U)$ let $\theta_U\alpha$ be that function $U \to \mathbb{Z}/2$ in $k^n(U)$ with value $j_x^*\alpha \in \mathbb{Z}/2$ on $x \in U$.

1.3 Proposition: *If $E \xrightarrow{\pi} X$ represents a real n-plane bundle ξ over a base space $X \in \mathscr{B}$, then there is a unique class $U_\xi \in H^n(E, E^*)$ such that $j_x^* U_\xi \in H^n(E_x, E_x^*)$ is the generator of the $\mathbb{Z}/2$-module $H^*(E_x, E_x^*)$ for each $x \in X$.*

PROOF: As in Lemma 1.2 one can assume that X is of finite type, and for any contractible $U \in \mathcal{O}(X)$ one has isomorphisms $h^q(U) \xrightarrow{\theta_U} k^q(U)$ whenever $q \leqq n$, including the case $q = n$; hence Corollary I.9.5 provides isomorphisms

$h^q(X) \xrightarrow{\theta_X} k^q(X)$ for $q \leqq n$, including the case $q = n$. In particular the isomorphism $h^n(X) \xrightarrow{\theta_X} k^n(X)$ carries a unique cohomology class $U_\xi \in h^n(X) = H^n(E, E^*)$ into the constant function $X \to \mathbb{Z}/2$ with value $1 \in \mathbb{Z}/2$.

1.4 Definition: If $E \xrightarrow{\pi} X$ represents a real n-plane bundle ξ over a base space $X \in \mathscr{B}$, then the $\mathbb{Z}/2$ *Thom class* of ξ is the unique class $U_\xi \in H^n(E, E^*)$ such that $j_x^* U_\xi \in H^n(E_x, E_x^*)$ is the generator of $H^*(E_x, E_x^*)$ for each $x \in X$.

The coordinate bundle $E \xrightarrow{\pi} X$ of Definition 1.4 induces an isomorphism $H^*(X) \xrightarrow{\pi^*} H^*(E)$ of cohomology rings, which one can combine with the relative cup product $H^*(E) \otimes H^*(E, E^*) \xrightarrow{\cup} H^*(E, E^*)$ to regard $H^*(E, E^*)$ as an $H^*(X)$-module. Specifically, the product of any $\alpha \in H^*(E, E^*)$ by any scalar $\beta \in H^*(X)$ is $\pi^*\beta \cup \alpha \in H^*(E, E^*)$.

1.5 Proposition (A Relative Leray–Hirsch Theorem): *If $E \xrightarrow{\pi} X$ represents a real n-plane bundle ξ over $X \in \mathscr{B}$, then the $H^*(X)$-module $H^*(E, E^*)$ is the free $H^*(X)$-module generated by the Thom class $U_\xi \in H^n(E, E^*)$.*

PROOF: For each $x \in X$ the class $j_x^* U_\xi \in H^n(E_x, E_x^*)$ generates the $\mathbb{Z}/2$-module $H^*(E_x, E_x^*)$, by Proposition 1.3, and the remainder of the proof is identical to that of Theorem II.7.2 (the absolute Leray–Hirsch theorem) with the singleton set $\{U_\xi\}$ in place of $\{\alpha_1, \ldots, \alpha_r\}$.

2. Properties of $\mathbb{Z}/2$ Thom Classes

We now establish two properties of $\mathbb{Z}/2$ Thom classes. In order to formulate the first property suppose that $X' \xrightarrow{f} X$ is a map in the category $\mathscr{B}$ of base spaces and that $E \xrightarrow{\pi} X$ represents a real n-plane bundle ξ over X. If $E' \xrightarrow{\pi'} X'$ represents the pullback $f^!\xi$ over X', there is then a map $E' \xrightarrow{\mathbf{f}} E$ which induces a linear isomorphism on each fiber, and for which the diagram

$$\begin{array}{ccc} E' & \xrightarrow{\mathbf{f}} & E \\ {\scriptstyle \pi'}\downarrow & & \downarrow{\scriptstyle \pi} \\ X' & \xrightarrow{f} & X \end{array}$$

commutes as in Lemma II.1.3. In particular, $\mathbf{f}$ carries the set $E'^* \subset E'$ of nonzero elements into the corresponding set $E^* \subset E$, so that there is an induced map $E', E'^* \xrightarrow{\mathbf{f}} E, E^*$ of pairs. Consequently there is also an induced map $H^n(E, E^*) \xrightarrow{\mathbf{f}^*} H^n(E', E'^*)$ in $\mathbb{Z}/2$ singular cohomology.

2.1 Proposition: *Let $E \xrightarrow{\pi} X$ represent a real n-plane bundle ξ over a base space $X \in \mathscr{B}$, let $X' \xrightarrow{f} X$ be any map in $\mathscr{B}$, and let $E' \xrightarrow{\pi'} X'$ represent the pullback $f^!\xi$ over $X' \in \mathscr{B}$. Then $U_{f^!\xi} = \mathbf{f}^* U_\xi \in H^n(E', E'^*)$, for the $\mathbb{Z}/2$ Thom classes $U_\xi \in H^n(E, E^*)$ and $U_{f^!\xi} \in H^n(E', E'^*)$.*

PROOF: For any $y \in X'$ there is a commutative diagram

$$\begin{array}{ccc} E, E^* & \xleftarrow{\ \mathbf{f}\ } & E', E'^* \\ \uparrow{\scriptstyle j_{f(y)}} & & \uparrow{\scriptstyle j_y} \\ E_{f(y)}, E^*_{f(y)} & \xleftarrow{\ \mathbf{f}_y\ } & E'_y, E'^*_y, \end{array}$$

where $\mathbf{f}_y$ is the restriction of $\mathbf{f}$ to the fiber pair over y, hence a commutative diagram

$$\begin{array}{ccc} H^n(E, E^*) & \xrightarrow{\ \mathbf{f}^*\ } & H^n(E', E'^*) \\ \downarrow{\scriptstyle j^*_{f(y)}} & & \downarrow{\scriptstyle j^*_y} \\ H^n(E_{f(y)}, E^*_{f(y)}) & \xrightarrow{\ \mathbf{f}^*_y\ } & H^n(E'_y, E'^*_y). \end{array}$$

Since $\mathbf{f}_y$ is a homeomorphism of the pair E_y, E^*_y, the induced isomorphism $\mathbf{f}^*_y$ carries the generator $j^*_{f(y)} U_\xi$ of $H^n(E_{f(y)}, E^*_{f(y)})$ into the generator of $H^n(E'_y, E'^*_y)$; thus $j^*_y(\mathbf{f}^* U_\xi)$ is the generator of $H^n(E'_y, E'^*_y)$ for each $y \in X'$. According to Proposition 1.3, however, the Thom class $U_{f^!\xi} \in H^n(E', E'^*)$ is the unique class with this property of $\mathbf{f}^* U_\xi \in H^n(E', E'^*)$.

In order to formulate the second property of $\mathbb{Z}/2$ Thom classes we recall that if $X \in \mathscr{B}$ and $X' \in \mathscr{B}$, then $X \times X' \in \mathscr{B}$, by Proposition I.1.4. We also recall from Definition III.2.1 that if $E \xrightarrow{\pi} X$ and $E' \xrightarrow{\pi'} X'$ represent real vector bundles ξ and ξ', then one applies a morphism (Γ, Φ) of transformation groups to the product $E \times E' \xrightarrow{\pi \times \pi'} X \times X'$ to obtain a coordinate bundle representing a real vector bundle $\xi + \xi'$ over $X \times X'$; Γ and Φ are the obvious maps $GL(n, \mathbb{R}) \times GL(n', \mathbb{R}) \to GL(n, \mathbb{R}) \oplus GL(n', \mathbb{R}) \subset GL(n + n', \mathbb{R})$ and $\mathbb{R}^n \times \mathbb{R}^{n'} \to \mathbb{R}^{n+n'}$, respectively, which can conveniently be ignored. If E^* and E'^* are the sets of nonzero elements of E and E', respectively, then $E^* \times E' \cup E \times E'^*$ is the set of nonzero elements of $E \times E'$, so that

$$(E, E^*) \times (E', E'^*) = (E \times E', E^* \times E' \cup E \times E'^*) = (E \times E', (E \times E')^*).$$

2.2 Proposition: *If $E \xrightarrow{\pi} X$ and $E' \xrightarrow{\pi'} X'$ represent real vector bundles ξ and ξ' over $X \in \mathscr{B}$ and $X' \in \mathscr{B}$, respectively, then the cross product of the $\mathbb{Z}/2$ Thom classes $U_\xi \in H^n(E, E^*)$ and $U_{\xi'} \in H^{n'}(E', E'^*)$ satisfies*

$$U_\xi \times U_{\xi'} = U_{\xi+\xi'} \in H^{n+n'}(E \times E', (E \times E')^*).$$

PROOF: For any point $(x, x') \in X \times X'$ let $E_x, E_x^* \xrightarrow{j_x} E, E^*$, $E'_{x'}, E'^*_{x'} \xrightarrow{j_{x'}} E', E'^*$, and $E''_{(x,x')}, E''^*_{(x,x')} \xrightarrow{j_{(x,x')}} E'', E''^*$ be inclusions of fiber pairs, where $E'' = E \times E'$. Then $j_{(x,x')} = j_x \times j_{x'}$, so that

$$\begin{aligned} j^*_{(x,x')}(U_\xi \times U_{\xi'}) &= (j_x \times j_{x'})^*(U_\xi \times U_{\xi'}) \\ &= ((j_x \times j_{x'})^* \circ \times)(U_\xi \otimes U_{\xi'}) \\ &= (\times \circ (j_x^* \otimes j_{x'}^*))(U_\xi \otimes U_{\xi'}) \\ &= j_x^* U_\xi \times j_{x'}^* U_{\xi'}, \end{aligned}$$

by naturality of the cross product. Since $j_x^* U_\xi \in H^n(E_x, E_x^*)$ and $j_{x'}^* U_{\xi'} \in H^{n'}(E'_{x'}, E'^*_{x'})$ are generators, by definition of Thom classes, and since

$$H^*(E_x, E_x^*) \otimes H^*(E'_{x'}, E'^*_{x'}) \xrightarrow{\times} H^*(E''_{(x,x')}, E''^*_{(x,x')})$$

is an isomorphism by the Künneth theorem, it follows for each $(x, x') \in X \times X'$ that $j^*_{(x,x')}(U_\xi \times U_{\xi'})$ is the generator of $H^{n+n'}(E''_{(x,x')}, E''_{(x,x')})$. However, $U_{\xi+\xi'} \in H^{n+n'}(E'', E''^*)$ is uniquely characterized by this property of $U_\xi \times U_{\xi'}$, so that $U_\xi \times U_{\xi'} = U_{\xi+\xi'}$, as claimed.

3. $\mathbb{Z}/2$ Euler Classes

We now construct the most accessible characteristic classes of real vector bundles, establish two of their algebraic properties, and apply those properties to obtain a necessary condition for the existence of nowhere-vanishing sections of real vector bundles.

3.1 Definition: Let $E \xrightarrow{\pi} X$ represent a real n-plane bundle ξ over $X \in \mathscr{B}$, with $\mathbb{Z}/2$ Thom class $U_\xi \in H^n(E, E^*)$, let $X \xrightarrow{\sigma} E$ be the zero section, and let $E \xrightarrow{j} E, E^*$ be the inclusion. Then the $\mathbb{Z}/2$ *Euler class* $e(\xi) \in H^n(X)$ is given by setting $e(\xi) = \sigma^* j^* U_\xi$.

The trivial property that $e(\xi)$ belongs to $H^n(X) \subset H^*(X)$ will play a role in the axiomatic characterization of $\mathbb{Z}/2$ Euler classes, given later. Here is the first nontrivial property of $\mathbb{Z}/2$ Euler classes.

3.2 Proposition (Naturality): *Let $X' \xrightarrow{f} X$ be any map in the category $\mathscr{B}$ of base spaces, let ξ be a real n-plane bundle over X, and let $f^!\xi$ be its pullback. Then $e(f^!\xi) = f^*e(\xi)$, for the $\mathbb{Z}/2$ Euler classes $e(\xi) \in H^n(X)$ and $e(f^!\xi) \in H^n(X')$.*

PROOF: If $E' \xrightarrow{\pi'} X'$ represents $f^!\xi$, then as in §2 there is a map $E' \to E$ which is a linear isomorphism on each fiber, inducing a map $E', E'^* \xrightarrow{\mathbf{f}} E, E^*$ of pairs such that the diagram

$$\begin{array}{ccc} E, E^* & \xleftarrow{\mathbf{f}} & E', E'^* \\ \uparrow{\scriptstyle j} & & \uparrow{\scriptstyle j'} \\ E & & E' \\ \uparrow{\scriptstyle \sigma} & & \uparrow{\scriptstyle \sigma'} \\ X & \xleftarrow{f} & X' \end{array}$$

commutes, for the zero section σ' and inclusion j' associated to $f^!\xi$. Since $U_{f^!\xi} = \mathbf{f}^*U_\xi \in H^n(E', E'^*)$ by Proposition 2.1, it follows that $e(f^!\xi) = \sigma'^*j'^*U_{f^!\xi} = \sigma'^*j'^*\mathbf{f}^*U_\xi = f^*\sigma^*j^*U_\xi = f^*e(\xi)$, as claimed.

One easily shows the equivalence of the following two properties of $\mathbb{Z}/2$ Euler classes, and we identify both of them by the name "Whitney product formula." There will be many further "Whitney product formulas" throughout this work.

3.3 Proposition (Whitney Product Formula): *Let $E \xrightarrow{\pi} X$ and $E' \xrightarrow{\pi'} X'$ represent real vector bundles ξ and ξ' of ranks n and n' over base spaces $X \in \mathscr{B}$ and $X' \in \mathscr{B}$, respectively, so that $E \times E' \xrightarrow{\pi \times \pi'} X \times X'$ represents the bundle $\xi + \xi'$ of rank $n + n'$ over $X \times X' \in \mathscr{B}$. Then the cross product of the $\mathbb{Z}/2$ Euler classes $e(\xi) \in H^n(X)$ and $e(\xi') \in H^{n'}(X')$ satisfies*

$$e(\xi) \times e(\xi') = e(\xi + \xi') \in H^{n+n'}(X \times X').$$

PROOF: Let $X \xrightarrow{\sigma} E$ and $X' \xrightarrow{\sigma'} E'$ be the zero sections, and let $E \xrightarrow{j} E, E^*$ and $E' \xrightarrow{j'} E', E'^*$ be the inclusions, associated to ξ and to ξ', respectively. Since $(E, E^*) \times (E', E'^*) = (E \times E'), (E \times E')^*$, as in Proposition 2.2, it follows that $X \times X' \xrightarrow{\sigma \times \sigma'} E \times E'$ and $E \times E' \xrightarrow{j \times j'} (E, E^*) \times (E', E'^*)$ are the zero section and inclusion associated to $\xi + \xi'$. According to Proposition 2.2 the $\mathbb{Z}/2$ Thom classes satisfy

$$U_\xi \times U_{\xi'} = U_{\xi + \xi'} \in H^n(E \times E', (E \times E')^*),$$

so that by naturality of the cross-product one has

$$\begin{aligned} e(\xi) \times e(\xi') &= \sigma^* j^* U_\xi \times \sigma'^* j'^* U_{\xi'} \\ &= (\times \circ (\sigma^* \otimes \sigma'^*) \circ (j^* \otimes j'^*))(U_\xi \otimes U_{\xi'}) \\ &= ((\sigma \times \sigma')^* \circ (j \times j')^* \circ \times)(U_\xi \otimes U_{\xi'}) \\ &= (\sigma \times \sigma')^*(j \times j')^*(U_\xi \times U_{\xi'}) \\ &= (\sigma \times \sigma')^*(j \times j')^* U_{\xi + \xi'} \\ &= e(\xi + \xi'), \end{aligned}$$

as claimed.

Here is an equivalent formulation of Proposition 3.3.

3.4 Proposition (Whitney Product Formula): *Let ξ and ξ' be real vector bundles over the same base space $X \in \mathscr{B}$, and let $\xi \oplus \xi'$ be their Whitney sum. Then the cup product of the $\mathbb{Z}/2$ Euler classes $e(\xi) \in H^n(X)$ and $e(\xi') \in H^{n'}(X)$ satisfies $e(\xi) \cup e(\xi') = e(\xi \oplus \xi') \in H^{n+n'}(X)$.*

PROOF: If $X \xrightarrow{\Delta} X \times X$ is the diagonal map, then

$$\begin{aligned} e(\xi) \cup e(\xi') &= \Delta^*(e(\xi) \times e(\xi')) = \Delta^* e(\xi + \xi') \\ &= e(\Delta^!(\xi + \xi')) = e(\xi \oplus \xi'), \end{aligned}$$

by the definition of the cup product, Proposition 3.3, Proposition 3.2, and the definition $\xi \oplus \xi' = \Delta^!(\xi + \xi')$, respectively.

The following lemma will be used to establish a necessary condition for the existence of nowhere-vanishing sections of real vector bundles.

3.5 Lemma: *If ε^n is the trivial real n-plane bundle over a base space $X \in \mathscr{B}$, for any $n > 0$, then $e(\varepsilon^n) = 0 \in H^n(X)$.*

PROOF: ε^n can be classified by a constant map into $G^n(\mathbb{R}^\infty)$, which can be expressed as the composition $X \xrightarrow{f} \{*\} \xrightarrow{g} G^n(\mathbb{R}^\infty)$, where $* \in G^n(\mathbb{R}^\infty)$ is the constant value and g is the inclusion; that is, $\varepsilon^n = f^! g^! \gamma^n$ for the universal n-plane bundle γ^n over $G^n(\mathbb{R}^\infty)$. Since $H^n(\{*\}) = 0$ one has $e(g^! \gamma^n) = 0$, hence $e(\varepsilon^n) = e(f^! g^! \gamma^n) = f^* e(g^! \gamma^n) = 0$ by Proposition 3.2.

3.6 Proposition: *If a real n-plane bundle ξ over a base space $X \in \mathscr{B}$ admits a nowhere-vanishing section, then the $\mathbb{Z}/2$ Euler class vanishes: $e(\xi) = 0 \in H^n(X)$.*

PROOF: By Proposition III.4.8, $\xi = \varepsilon^1 \oplus \eta$ for the trivial line bundle ε^1 and some $(n - 1)$-plane bundle η; hence Proposition 3.4 and Lemma 3.5 yield $e(\xi) = e(\varepsilon^1 \oplus \eta) = e(\varepsilon^1) \cup e(\eta) = 0$.

According to Proposition 1.5, if $E \xrightarrow{\pi} X$ represents a real n-plane bundle ξ over $X \in \mathscr{B}$, then the $H^*(X)$-module $H^*(E, E^*)$ is free of rank one, the unique generator being the $\mathbb{Z}/2$ Thom class $U_\xi \in H^n(E, E^*)$. It follows that cup product by U_ξ induces an $H^*(X)$-module isomorphism $H^*(X) \to H^{*+n}(E, E^*)$.

3.7 Definition: Let $U_\xi \in H^n(E, E^*)$ be the $\mathbb{Z}/2$ Thom class of a coordinate bundle $E \xrightarrow{\pi} X$ representing a real n-plane bundle ξ over a base space $X \in \mathscr{B}$. The $\mathbb{Z}/2$ *Thom isomorphism* $H^*(X) \xrightarrow{\Phi_\xi} H^{*+n}(E, E^*)$ maps any $\beta \in H^*(X)$ into $\pi^*\beta \cup U_\xi \in H^{*+n}(E, E^*)$.

The $\mathbb{Z}/2$ Thom isomorphism provides a new characterization of $\mathbb{Z}/2$ Euler classes, as follows.

3.8 Proposition: *If $E \xrightarrow{\pi} X$ represents a real n-plane bundle ξ over $X \in \mathscr{B}$, with $\mathbb{Z}/2$ Thom class $U_\xi \in H^n(E, E^*)$ and $\mathbb{Z}/2$ Thom isomorphism $H^*(X) \xrightarrow{\Phi_\xi} H^{*+n}(E, E^*)$, then the $\mathbb{Z}/2$ Euler class $e(\xi) \in H^n(X)$ is given by*

$$e(\xi) = \Phi_\xi^{-1}(U_\xi \cup U_\xi)$$

for the cup product $U_\xi \cup U_\xi \in H^{2n}(E, E^*)$.

PROOF: Since $E \xrightarrow{\pi} X \xrightarrow{\sigma} E$ is homotopic to the identity, one can apply Definition 3.7 to $e(\xi) = \sigma^* j^* U_\xi \in H^n(X)$ to compute

$$\Phi_\xi e(\xi) = \pi^*(\sigma^* j^* U_\xi) \cup U_\xi = j^* U_\xi \cup U_\xi = U_\xi \cup U_\xi.$$

4. Gysin Sequences and $H^*(RP^\infty; \mathbb{Z}/2)$

The goal of this section is to compute the $\mathbb{Z}/2$ cohomology of real projective spaces, including RP^∞.

4.1 Definition: Let $E \xrightarrow{\pi} X$ represent a real n-plane bundle ξ over $X \in \mathscr{B}$, let E^* be the nonzero portion of E, let $H^*(E^*) \xrightarrow{\delta} H^{*+1}(E, E^*)$ be the connecting homomorphism in the $\mathbb{Z}/2$ cohomology exact sequence of the pair E, E^*, and let $H^*(X) \xrightarrow{\Phi_\xi} H^{*+n}(E, E^*)$ be the $\mathbb{Z}/2$ Thom isomorphism of Definition 3.7. The *Gysin map* $H^{*+n}(E^*) \xrightarrow{\Psi_\xi} H^{*+1}(X)$ is the composition $\Phi_\xi^{-1}\delta$:

$$H^{*+n}(E^*) \xrightarrow{\delta} H^{*+n+1}(E, E^*) \xleftarrow{\Phi_\xi} H^{*+1}(X).$$

Trivially the Gysin map Ψ_ξ is a $\mathbb{Z}/2$-module homomorphism of degree $1 - n$.

4.2 Proposition (The Gysin Sequence): *Let $E \xrightarrow{\pi} X$ represent a real n-plane bundle ξ over $X \in \mathscr{B}$, let $E^* \xrightarrow{\bar{\pi}} X$ be the restriction of π to the nonzero portion of E, let Ψ_ξ be the Gysin map, and let $H^*(X) \xrightarrow{\cup e(\xi)} H^{*+n}(X)$ be cup product by the $\mathbb{Z}/2$ Euler class $e(\xi) \in H^n(X)$. Then the sequence*

$$\cdots \to H^{q+n-1}(E^*) \xrightarrow{\Psi_\xi} H^q(X) \xrightarrow{\cup e(\xi)} H^{q+n}(X) \xrightarrow{\bar{\pi}^*} H^{q+n}(E^*) \to \cdots$$

is exact.

PROOF: The bottom row of the accompanying diagram is the $\mathbb{Z}/2$ cohomology exact sequence of the pair E, E^*,

$$\begin{array}{ccccccccccc}
\cdots \longrightarrow & H^{q+n-1}(E^*) & \xrightarrow{\Psi_\xi} & H^q(X) & \xrightarrow{\cup e(\xi)} & H^{q+n}(X) & \xrightarrow{\bar{\pi}^*} & H^{q+n}(E^*) & \longrightarrow \cdots \\
& \downarrow \text{id} & & \downarrow \Phi_\xi & & \downarrow \pi^* & & \downarrow \text{id} & \\
\cdots \longrightarrow & H^{q+n-1}(E^*) & \xrightarrow{\delta} & H^{q+n}(E, E^*) & \xrightarrow{j^*} & H^{q+n}(E) & \xrightarrow{i^*} & H^{q+n}(E^*) & \longrightarrow \cdots
\end{array}$$

and the vertical maps are all isomorphisms. The left-hand square commutes by definition of Ψ_ξ, and the right-hand square commutes because $E^* \xrightarrow{\bar{\pi}} X$ is the composition of the inclusion $E^* \xrightarrow{i} E$ and the projection $E \xrightarrow{\pi} X$. To show that the middle square commutes, observe that if $X \xrightarrow{\sigma} E$ is the zero section, then $E \xrightarrow{\sigma \circ \pi} E$ is homotopic to the identity, so that the endomorphism $H^n(E) \xrightarrow{\pi^*\sigma^*} H^n(E)$ is the identity; hence for any $\beta \in H^q(X)$ one has

$$\begin{aligned}
\pi^*(\beta \cup e(\xi)) &= \pi^*\beta \cup \pi^*(\sigma^* j^* U_\xi) = \pi^*\beta \cup j^* U_\xi \\
&= j^*(\pi^*\beta \cup U_\xi) = j^*\Phi_\xi \beta,
\end{aligned}$$

as required.

4.3 Proposition: *Let γ^1 be the universal real line bundle over RP^∞, with $\mathbb{Z}/2$ Euler class $e(\gamma^1) \in H^1(RP^\infty)$. Then the $\mathbb{Z}/2$ cohomology ring $H^*(RP^\infty)$ is a polynomial ring in a single variable over $\mathbb{Z}/2$, the generator being $e(\gamma^1)$.*

PROOF: If $E \xrightarrow{\pi} RP^\infty$ represents γ^1, then $E^* = \mathbb{R}^{\infty *}$, which is contractible, as in Proposition III.8.11. The Gysin sequence (for $n = 1$) is then of the form

$$H^q(\mathbb{R}^{\infty *}) \xrightarrow{\Psi_{\gamma^1}} H^q(RP^\infty) \xrightarrow{\cup e(\gamma^1)} H^{q+1}(RP^\infty) \xrightarrow{\bar{\pi}^*} H^{q+1}(\mathbb{R}^\infty),$$

where $H^{q+1}(\mathbb{R}^{\infty *}) = 0$ for $q \geqq 0$, and where $H^q(\mathbb{R}^{\infty *}) = 0$ for $q > 0$; in particular, $\cup e(\gamma^1)$ is an isomorphism for $q > 0$. Furthermore, the beginning of the Gysin sequence is

$$H^0(RP^\infty) \xrightarrow{\bar{\pi}^*} H^0(\mathbb{R}^{\infty *}) \xrightarrow{\Psi_\xi} H^0(RP^\infty) \xrightarrow{\cup e(\gamma^1)} H^1(RP^\infty) \to 0,$$

where the initial homomorphism $\bar{\pi}^*$ is the trivial isomorphism $\mathbb{Z}/2 \to \mathbb{Z}/2$, so that $\cup e(\gamma^1)$ is also an isomorphism for $q = 0$. Since $H^0(RP^\infty) = \mathbb{Z}/2$, this completes the proof.

A similar method applies to the $\mathbb{Z}/2$ cohomology ring $H^*(RP^n)$ for any finite $n > 0$.

4.4 Proposition: *For any $n > 0$ let γ_n^1 be the canonical line bundle over RP^n, with $\mathbb{Z}/2$ Euler class $e(\gamma_n^1) \in H^1(RP^n)$. Then the $\mathbb{Z}/2$ cohomology ring $H^*(RP^n)$ is a truncated polynomial ring in a single variable over $\mathbb{Z}/2$, the generator being $e(\gamma_n^1)$ and the relation being $e(\gamma_n^1)^{n+1} = 0$.*

PROOF: If $E \xrightarrow{\pi} RP^n$ represents γ_n^1, then, as in Proposition III.8.11, E^* is a quotient $\mathbb{R}^{(n+1)*} \times \mathbb{R}^*/\approx$, with $(x, s) \approx (y, t)$ if and only if $sx = ty \in \mathbb{R}^{(n+1)*}$; that is, $E^* = \mathbb{R}^{(n+1)*}$, which is trivially homotopy equivalent to the sphere S^n. Since $H^q(S^n) = 0$ for $0 < q < n$, the argument used for Proposition 4.3 shows that $H^0(RP^n) = \mathbb{Z}/2$, that $H^q(RP^n) \xrightarrow{\cup e(\gamma_n^1)} H^{q+1}(RP^n)$ is an isomorphism for $0 \leqq q < n - 1$, and that $H^{n-1}(RP^n) \xrightarrow{\cup e(\gamma_n^1)} H^n(RP^n)$ is a monomorphism; the latter conclusion implies that $H^n(RP^n)$ contains $\mathbb{Z}/2$ as a submodule. Since RP^n is an n-dimensional CW space, by Proposition I.5.5, one has $H^q(RP^n) = 0$ for $q > n$. In particular, since $H^{n+1}(RP^n) = 0$, the Gysin sequence terminates in the fashion $H^n(S^n) \xrightarrow{\Psi_{\gamma_n^1}} H^n(RP^n) \to 0$, and since $H^n(S^n) = \mathbb{Z}/2$, the monomorphism $H^{n-1}(RP^n) \xrightarrow{\cup e(\gamma_n^1)} H^n(RP^n)$ is in fact an isomorphism, which completes the proof.

4.5 Corollary: *For any $n > 0$ let $RP^n \xrightarrow{f} RP^\infty$ classify the canonical line bundle γ_n^1 over RP^n; then for any $p \leqq n$ the induced $\mathbb{Z}/2$-module homomorphism $H^p(RP^\infty) \xrightarrow{f^*} H^p(RP^n)$ is an isomorphism carrying $e(\gamma^1)^p$ into $e(\gamma_n^1)^p$.*

PROOF: $H^p(RP^\infty)$ is free on the single generator $e(\gamma^1)^p$, and $H^p(RP^n)$ is free on the single generator $e(\gamma_n^1)^p$. Since $\gamma_n^1 = f^!\gamma^1$, it follows from the naturality of $\mathbb{Z}/2$ Euler classes (Proposition 3.2) that $e(\gamma_n^1) = f^*e(\gamma^1)$, hence that $e(\gamma_n^1)^p = f^*e(\gamma^1)^p$, as required.

Observe that Propositions 4.3 and 4.4 establish the existence of nonzero $\mathbb{Z}/2$ Euler classes, hence that the condition of Proposition 3.6 for the existence of nowhere-vanishing sections is nonvacuous. In particular, γ^1 and γ_n^1 admit no nowhere-vanishing sections; a fortiori, they are not trivial line bundles.

5. The Splitting Principle

In general, an n-plane bundle ξ over a base space $X \in \mathscr{B}$ is *not* a Whitney sum of n line bundles. Nevertheless, one can always find a map $X' \xrightarrow{g} X$ in the category $\mathscr{B}$ such that the pullback $g^!\xi$ over $X' \in \mathscr{B}$ *is* a Whitney sum of n line bundles, and such that the induced homomorphism $H^*(X) \xrightarrow{g^*} H^*(X')$ of $\mathbb{Z}/2$ cohomology rings is a monomorphism. This result will be used often to verify properties of cohomology classes assigned to arbitrary vector bundles: one need only consider sums of line bundles. The construction of g will itself be used in the next chapter to define further $\mathbb{Z}/2$ characteristic classes.

For any n-plane bundle ξ over a base space X there is a corresponding fiber bundle P_ξ over X with fiber RP^{n-1}, the group being the projective group. To construct P_ξ let $E \xrightarrow{\pi} X$ represent ξ and let $E^* \subset E$ consist of nonzero points, as usual. For each $x \in X$ one identifies two points of $E_x^* \subset E_x$ if and only if one point is a real (nonzero) multiple of the other; the total space of P_ξ is the quotient of E^* by this equivalence relation, in the quotient topology. There is an obvious induced projection onto X, each fiber is homeomorphic to RP^{n-1}, and the action of $GL(n, \mathbb{R})$ on the fibers in E induces an action of the corresponding projective group on the fibers in P_ξ.

5.1 Definition: Given any real n-plane bundle ξ over $X \in \mathscr{B}$, the preceding bundle P_ξ over X is the *projective bundle* associated to ξ.

For convenience we use the symbol P_ξ ambiguously to denote either the projective bundle associated to ξ or to denote its total space; in most of what follows P_ξ will represent the total space. Since the fiber RP^{n-1} of the projective bundle P_ξ is a finite CW space, by Proposition I.5.5, one easily shows that the total space P_ξ itself belongs to the category $\mathscr{B}$ of base spaces.

Let $P_\xi \xrightarrow{f} X$ represent the preceding projective bundle over the base space X. One can pull the original n-plane bundle ξ back along f to obtain an n-plane bundle $f^!\xi$ over P_ξ. One can also identify an especially attractive line subbundle λ_ξ of $f^!\xi$, as follows. The elements of P_ξ can be regarded as 1-dimensional subspaces L of the fibers $E_x \subset E$, and the total space of $f^!\xi$ consists of those pairs $(L, e) \in P_\xi \times E$ such that $L \subset E_x$ and $e \in E_x$; the total space of the line subbundle λ_ξ is then defined to consist of those points $(L, e) \in P_\xi \times E$ such that $e \in L$, in the relative topology of the total space of $f^!\xi$.

5.2 Definition: For any real n-plane bundle ξ over a base space $X \in \mathscr{B}$, the preceding line sub-bundle λ_ξ of $f^!\xi$ over the base space $P_\xi \in \mathscr{B}$ is the *splitting bundle* of ξ; the $\mathbb{Z}/2$ Euler class $e(\lambda_\xi) \in H^1(P_\xi)$ is the *splitting class* of ξ.

The $\mathbb{Z}/2$-module $H^*(P_\xi)$ becomes an $H^*(X)$-module with respect to the product $\beta \cdot \alpha = f^*\beta \cup \alpha \in H^*(P_\xi)$ of $\beta \in H^*(X)$ and $\alpha \in H^*(P_\xi)$.

5.3 Proposition: *For any real n-plane bundle ξ over a base space $X \in \mathscr{B}$, the $H^*(X)$-module $H^*(P_\xi)$ is free on the basis $\{1, e(\lambda_\xi), \ldots, e(\lambda_\xi)^{n-1}\}$, where $e(\lambda_\xi) \in H^1(P_\xi)$ is the splitting class of ξ.*

PROOF: For each $x \in X$ let $P_{\xi,x} \xrightarrow{j_x} P_\xi$ be the inclusion of the fiber of the projective bundle over X. Then if $E \xrightarrow{\pi} P_\xi$ represents the splitting bundle λ_ξ

over P_ξ, there is a pullback diagram

$$\begin{array}{ccc} E_x & \xrightarrow{j_x} & E \\ \downarrow{\scriptstyle \pi_x} & & \downarrow{\scriptstyle \pi} \\ P_{\xi,x} & \xrightarrow{j_x} & P_\xi \end{array}$$

in which $P_{\xi,x}$ is homeomorphic to RP^{n-1}; up to this homeomorphism the map $E_x \xrightarrow{\pi_x} P_{\xi,x}$ represents a bundle $j_x^! \lambda_\xi$ which corresponds to the canonical line bundle γ_{n-1}^1 over RP^{n-1}. Since $H^*(RP^{n-1})$ is a free $\mathbb{Z}/2$-module on the basis $\{1, e(\gamma_{n-1}^1), \ldots, e(\gamma_{n-1}^1)^{n-1}\}$, by Proposition 4.4, it follows that $H^*(P_{\xi,x})$ is a free $\mathbb{Z}/2$-module on the basis $\{1, e(j_x^! \lambda_\xi), \ldots, e(j_x^! \lambda_\xi)^{n-1}\}$. Since $e(j_x^! \lambda_\xi) = j_x^* e(\lambda_\xi)$ by naturality of $\mathbb{Z}/2$ Euler classes (Proposition 3.2), we have therefore shown for each $x \in X$ that the $\mathbb{Z}/2$-module $H^*(P_{\xi,x})$ is free, with basis $\{j_x^* 1, j_x^* e(\lambda_\xi), \ldots, j_x^* e(\lambda_\xi)^{n-1}\}$. Hence the absolute Leray–Hirsch theorem (Theorem II.7.2) guarantees that $H^*(P_\xi)$ is a free $H^*(X)$-module, with basis $\{1, e(\lambda_\xi), \ldots, e(\lambda_\xi)^{n-1}\}$, as claimed.

5.4 Corollary: *Let $P_\xi \xrightarrow{f} X$ represent the projective bundle associated to a real n-plane bundle ξ over a base space $X \in \mathscr{B}$; then the induced homomorphism $H^*(X) \xrightarrow{f} H^*(P_\xi)$ of $\mathbb{Z}/2$ cohomology modules is a monomorphism.*

PROOF: The image of f^* is precisely the free $H^*(X)$-submodule of $H^*(P_\xi)$ spanned by the element 1 in the basis $\{1, e(\lambda_\xi), \ldots, e(\lambda_\xi)^{n-1}\}$ of $H^*(P_\xi)$.

We have reached the goal of this section.

5.5 Proposition (The Splitting Principle): *Let $\xi_1, \ldots, \xi_r$ be finitely many real vector bundles over the same base space $X \in \mathscr{B}$. Then there is a map $X' \xrightarrow{g} X$ in $\mathscr{B}$ such that each of the pullbacks $g^! \xi_1, \ldots, g^! \xi_r$ over $X' \in \mathscr{B}$ is a Whitney sum of line bundles, and such that $H^*(X) \xrightarrow{g^*} H^*(X')$ is a monomorphism of $\mathbb{Z}/2$ cohomology rings.*

PROOF: Since pullbacks of sums of line bundles are sums of line bundles, and since compositions of monomorphisms are monomorphisms, it suffices to consider just the case $r = 1$. Let ξ be an n-plane bundle over X and let $P_\xi \xrightarrow{f} X$ represent the corresponding projective bundle. Then the splitting bundle λ_ξ is a line subbundle of the pullback $f^! \xi$ over $P_\xi \in \mathscr{B}$; by Proposition III.3.6 there is an $(n-1)$-plane subbundle η of $f^! \xi$ such that $f^! \xi = \lambda_\xi \oplus \eta$, and $H^*(X) \xrightarrow{f^*} H^*(P_\xi)$ is a monomorphism by Corollary 5.4. One applies the same construction to η, splitting off a line subbundle by a map $P_\eta \to P_\xi$ which induces a $\mathbb{Z}/2$ cohomology monomorphism $H^*(P_\xi) \to H^*(P_\eta)$, and the induction is clear.

5.6 Definition: A *splitting map* for finitely many real vector bundles $\xi_1,\ldots,\xi_r$ over the same $X \in \mathscr{B}$ is any map $X' \xrightarrow{g} X$ in the category $\mathscr{B}$ such that each of $g^!\xi_1, \ldots, g^!\xi_r$ is a Whitney sum of line bundles, and such that $H^*(X) \xrightarrow{g^*} H^*(X')$ is a monomorphism of $\mathbb{Z}/2$ cohomology rings.

Proposition 5.5 asserts that splitting maps always exist, for real vector bundles and $\mathbb{Z}/2$ cohomology. They are not necessarily unique, however, nor need their construction be the one given in Proposition 5.5. In particular, for any $n > 0$ the universal real n-plane bundle γ^n over $G^n(\mathbb{R}^\infty)$ admits an especially nice splitting map, as follows.

Let γ^1 be the universal real line bundle over RP^∞, let $(RP^\infty)^n$ denote the n-fold product $RP^\infty \times \cdots \times RP^\infty$, and let $\gamma^1 + \cdots + \gamma^1$ be the corresponding sum of n copies of γ^1 over $(RP^\infty)^n$, as in Definition III.2.1. Then if $(RP^\infty)^n \xrightarrow{\mathrm{pr}_j} RP^\infty$ denotes the jth projection map, for $j = 1, \ldots, n$, one has $\gamma^1 + \cdots + \gamma^1 = \mathrm{pr}_1^!\gamma^1 \oplus \cdots \oplus \mathrm{pr}_n^!\gamma^1$, so that $\gamma^1 + \cdots + \gamma^1$ is itself a Whitney sum of n real line bundles over $(RP^\infty)^n$, hence an n-plane bundle. Consequently the homotopy classification theorem (Theorem III.8.9) provides a classifying map $(RP^\infty)^n \xrightarrow{h} G^n(R^\infty)$ for $\gamma^1 + \cdots + \gamma^1$; that is, $\gamma^1 + \cdots + \gamma^1 = h^!\gamma^n$ for the universal real n-plane bundle γ^n over $G^n(R^\infty)$.

5.7 Proposition: *Any classifying map $(RP^\infty)^n \xrightarrow{h} G^n(R^\infty)$ for the real n-plane bundle $\gamma^1 + \cdots + \gamma^1$ over $(RP^\infty)^n$ is also a splitting map for the universal real n-plane bundle γ^n over $G^n(\mathbb{R}^\infty)$.*

Proof: Since $h^!\gamma^n$ is already a Whitney sum $\mathrm{pr}_1^!\gamma^1 \oplus \cdots \oplus \mathrm{pr}_n^!\gamma^1$ of n line bundles, it remains only to show that $H^*(G^n(\mathbb{R}^\infty)) \xrightarrow{h^*} H^*((RP^\infty)^n)$ is monic. Let $X \xrightarrow{f} G^n(\mathbb{R}^\infty)$ be any splitting map for γ^n, as in Proposition 5.5, so that $f^!\gamma^n$ is a sum $\lambda_1 \oplus \cdots \oplus \lambda_n$ of line bundles $\lambda_1, \ldots, \lambda_n$ over X and $H^*(G^n(\mathbb{R}^\infty)) \xrightarrow{f^*} H^*(X)$ is monic; trivially f classifies $\lambda_1 \oplus \cdots \oplus \lambda_n$. If the maps $X \xrightarrow{g_1} RP^\infty, \ldots, X \xrightarrow{g_n} RP^\infty$ classify the line bundles $\lambda_1, \ldots, \lambda_n$, and if g denotes the map $X \xrightarrow{(g_1, \ldots, g_n)} (RP^\infty)^n$, then the composition $X \xrightarrow{g} (RP^\infty)^n \xrightarrow{h} G^n(\mathbb{R}^\infty)$ also classifies $\lambda_1 \oplus \cdots \oplus \lambda_n$. By the homotopy classification theorem (Theorem III.8.9) the classifying maps f and $h \circ g$ for $\lambda_1 \oplus \cdots \oplus \lambda_n$ are necessarily homotopic, and consequently there is a commutative triangle

$$\begin{array}{ccc} H^*(G^n(\mathbb{R}^\infty)) & \xrightarrow{\quad f^* \quad} & H^*(X) \\ & {\scriptstyle h^*}\searrow \qquad \nearrow{\scriptstyle g^*} & \\ & H^*((RP^\infty)^n). & \end{array}$$

Since f^* is monic, it follows that h^* is also monic, as required.

A finite-dimensional version of Proposition 5.7 will be established in Proposition V.4.6.

6. Axioms for $\mathbb{Z}/2$ Euler Classes

The following description of $\mathbb{Z}/2$ Euler classes will serve as a model for characterizing many characteristic classes of many categories of vector bundles.

6.1 Theorem (Axioms for $\mathbb{Z}/2$ Euler Classes): *For real vector bundles ξ over base spaces $X \in \mathscr{B}$ there are unique homogeneous $\mathbb{Z}/2$ singular cohomology classes $e(\xi) \in H^*(X)$ which satisfy the following axioms:*

(1) **Naturality:** *If ξ is a real vector bundle over $X \in \mathscr{B}$, and if $X' \xrightarrow{f} X$ is a map in the category $\mathscr{B}$, then $e(f^!\xi) = f^*e(\xi) \in H^*(X')$.*

(2) **Whitney product formula:** *If ξ and η are real vector bundles over the same $X \in \mathscr{B}$, with Whitney sum $\xi \oplus \eta$ over X, then $e(\xi) \cup e(\eta) = e(\xi \oplus \eta) \in H^*(X)$ for the cup product $e(\xi) \cup e(\eta)$.*

(3) **Normalization:** *If γ_1^1 is the canonical real line bundle over $RP^1(=S^1)$, then $e(\gamma_1^1)$ is the generator of $H^1(RP^1)$.*

Proof: The Euler classes of Definition 3.1 satisfy Axioms 1, 2, and 3 by virtue of Propositions 3.2, 3.4, and 4.4, respectively. Conversely, suppose that $e(\ )$ satisfies Axioms 1, 2, and 3, let γ^1 be the universal real line bundle over RP^∞, and let $RP^1 \xrightarrow{f} RP^\infty$ classify the canonical line bundle γ_1^1 over RP^1. Then $H^1(RP^\infty) \xrightarrow{f^*} H^1(RP^1)$ is an isomorphism by Corollary 4.5, and $f^*e(\gamma^1) = e(f^!\gamma^1) = e(\gamma_1^1)$ by Axiom 1, so that $e(\gamma^1) \in H^1(RP^\infty)$ is necessarily the generator of $H^*(RP^\infty)$, by Axiom 3. Now let γ^n be the universal real n-plane bundle over $G^n(\mathbb{R}^\infty)$ and let $(RP^\infty)^n \xrightarrow{h} G^n(\mathbb{R}^\infty)$ be the splitting map for γ^n described in Proposition 5.7, with

$$h^!\gamma^n = \gamma^1 + \cdots + \gamma^1 = \mathrm{pr}_1^!\gamma^1 \oplus \cdots \oplus \mathrm{pr}_n^!\gamma^1$$

for the n projections $(RP^\infty)^n \xrightarrow{\mathrm{pr}_j} RP^\infty$. Then by Axioms 1 and 2 one has

$$\begin{aligned} h^*e(\gamma^n) &= e(h^!\gamma^n) = e(\mathrm{pr}_1^!\gamma^1 \oplus \cdots \oplus \mathrm{pr}_n^!\gamma^1) \\ &= e(\mathrm{pr}_1^!\gamma^1) \cup \cdots \cup e(\mathrm{pr}_n^!\gamma^1) = \mathrm{pr}_1^*e(\gamma^1) \cup \cdots \cup \mathrm{pr}_n^*e(\gamma^1), \end{aligned}$$

which is a unique element of $H^n((RP^\infty)^n)$ since $e(\gamma^1)$ is uniquely defined, as we have just learned. Since h^* is monic by Proposition 5.7, it follows that $e(\gamma^n)$ is a unique element of $H^n(G^n(\mathbb{R}^\infty))$. By one final appeal to the homotopy classification theorem (Theorem III.8.9) *any* real n-plane bundle ξ over *any*

$X \in \mathscr{B}$ can be classified by a map $X \xrightarrow{g} G^n(\mathbb{R}^\infty)$ which is unique up to homotopy, so that by Axiom 1 one has $e(\xi) = e(g^! \gamma^n) = g^* e(\gamma^n) \in H^n(X)$ for a unique homomorphism $H^n(G^n(\mathbb{R}^\infty)) \xrightarrow{g^*} H^n(X)$ and the unique element $e(\gamma^n) \in H^n(G^n(\mathbb{R}^\infty))$.

7. $\mathbb{Z}/2$ Euler Classes of Product Line Bundles

If λ and μ are real line bundles over the same base space $X \in \mathscr{B}$, then the product $\lambda \otimes \mu$ is also a real line bundle over X, and the real line bundles over X form an abelian group with respect to this product. Furthermore, the map carrying the real line bundle λ over X into the $\mathbb{Z}/2$ Euler class $e(\lambda) \in H^1(X)$ is an isomorphism. In this section we verify only that the latter map is a monomorphism; proofs of the stronger result are indicated later.

7.1 Proposition: *For any base space $X \in \mathscr{B}$ let $\Gamma(X)$ denote the set of real line bundles over X, with the product $\Gamma(X) \times \Gamma(X) \to \Gamma(X)$ carrying $(\lambda, \mu) \in \Gamma(X) \times \Gamma(X)$ into $\lambda \otimes \mu \in \Gamma(X)$. Then $\Gamma(X)$ is an abelian group in which every element is of order* 2.

PROOF: The associative and commutative properties

$$(\lambda \otimes \mu) \otimes \nu = \lambda \otimes (\mu \otimes \nu) \qquad \text{and} \qquad \lambda \otimes \mu = \mu \otimes \lambda$$

were observed in Proposition III.2.6, and the trivial line bundle $\varepsilon^1 \in \Gamma(X)$ is the neutral element: $\lambda \otimes \varepsilon^1 = \lambda$ for every $\lambda \in \Gamma(X)$. By the linear reduction theorem (Theorem II.6.13) the structure group $GL(\mathbb{R}, 1)$ of any real line bundle can be reduced to the orthogonal group $O(1)$, which consists of two elements, in the discrete topology, and since the square of each of these elements is the neutral element, it follows for any real line bundle λ that $\lambda \otimes \lambda = \varepsilon^1$. Thus each $\lambda \in \Gamma(X)$ is of order 2, so that each $\lambda \in \Gamma(X)$ is its own inverse.

7.2 Definition: The abelian group $\Gamma(X)$ of Proposition 7.1 is the *real line bundle group* of the base space $X \in \mathscr{B}$.

The following trivial example will be used later.

7.3 Proposition: $\Gamma(S^1) = \mathbb{Z}/2$.

PROOF: There is a covering $\{U_0, U_1\}$ of the circle S^1 by two slightly lengthened open half-circles U_0 and U_1 such that the intersection $U_0 \cap U_1$ consists of two connected components V_0 and V_1. Since the structure group $GL(1, \mathbb{R})$ of any real line bundle λ over S^1 can be reduced to $O(1) \subset GL(1, \mathbb{R})$

as before, where $O(1) = \{+1, -1\}$, any transition functions describing λ assume only one of the values $+1$ or -1 on each of the components V_0 and V_1. If the values agree, then the result is $\lambda = \varepsilon^1$, and if the values differ, then the result is $\lambda = \gamma_1^1$ (over the base space $S^1 = RP^1$).

For any real vector bundles ξ and η over the same base space $X \in \mathscr{B}$ the product bundle $\xi \times \eta$ over $X \times X$ was described in Definition III.2.2, the product $\xi \otimes \eta$ over X being the pullback $\Delta^!(\xi \times \eta)$ along the diagonal map $X \xrightarrow{\Delta} X \times X$, as in Definition III.2.4. We use the former product in the following lemma.

7.4 Lemma: *Let γ^1 be the universal real line bundle over RP^∞, and let $RP^\infty \times RP^\infty \xrightarrow{h} RP^\infty$ classify the real line bundle $\gamma^1 \times \gamma^1$ over $RP^\infty \times RP^\infty$. If $RP^\infty \times RP^\infty \xrightarrow{\pi_1} RP^\infty$ and $RP^\infty \times RP^\infty \xrightarrow{\pi_2} RP^\infty$ are first and second projections, respectively, and if α_1, α_2, and α denote the generator $e(\gamma^1)$ in each of the three copies of $H^1(RP^\infty)$, it follows that $h^*\alpha = \pi_1^*\alpha_1 + \pi_2^*\alpha_2 \in H^1(RP^\infty \times RP^\infty)$.*

PROOF: By the Künneth theorem $H^1(RP^\infty \times RP^\infty)$ is isomorphic to $H^0(RP^\infty) \otimes H^1(RP^\infty) + H^1(RP^\infty) \otimes H^0(RP^\infty)$, where $H^0(RP^\infty)$ can be regarded as the ground ring $\mathbb{Z}/2$ in each summand. Since α_1 and α_2 generate the initial two copies of $H^1(RP^\infty)$ it follows that $h^*\alpha = b_1\pi_1^*\alpha_1 + b_2\pi_2^*\alpha_2$ for unique elements $b_1, b_2 \in \mathbb{Z}/2$, and it remains to show that $b_1 = b_2 = 1$. Choose a base point $* \in RP^\infty$ and let $RP^\infty \xrightarrow{i_1} RP^\infty \times RP^\infty$ carry $x_1 \in RP^\infty$ into $(x_1, *) \in RP^\infty \times RP^\infty$. Then $\pi_1 i_1$ is an identity map and $\pi_2 i_1$ is a constant map, so that

$$i_1^*\pi_1^*\alpha_1 = (\pi_1 i_1)^* = \alpha_1 \qquad \text{and} \qquad i_1^*\pi_2^*\alpha_2 = (\pi_2 i_1)^*\alpha_2 = 0.$$

Furthermore, the pullback $i_1^!(\gamma^1 \times \gamma^1)$ is clearly the bundle γ^1 over the first copy of RP^∞, so that

$$\begin{aligned} i_1^*h^*\alpha &= i_1^*h^*e(\gamma^1) = e(i_1^!h^!\gamma^1) \\ &= e(i_1^!(\gamma^1 \times \gamma^1)) = e(\gamma^1) = \alpha_1. \end{aligned}$$

Hence

$$\alpha_1 = i_1^*h^*\alpha = b_1 i_1^*\pi_1^*\alpha_1 + b_2 i_1^*\pi_2^*\alpha_2 = b_1\alpha_1,$$

so that $b_1 = 1$. A similar argument shows that $b_2 = 1$.

7.5 Proposition: *For any real line bundles λ, μ over a base space $X \in \mathscr{B}$ one has $e(\lambda \otimes \mu) = e(\lambda) + e(\mu) \in H^1(X)$; that is, the computation of $\mathbb{Z}/2$ Euler classes of real line bundles over X induces a homomorphism $\Gamma(X) \to H^1(X)$ of abelian groups.*

PROOF: Let $X \xrightarrow{f} RP^\infty$ and $X \xrightarrow{g} RP^\infty$ classify λ and μ, respectively, let $RP^\infty \times RP^\infty \xrightarrow{h} RP^\infty$ classify $\gamma^1 \times \gamma^1$, and let $X \xrightarrow{\Delta} X \times X$ be the diagonal map; then the composition

$$X \xrightarrow{\Delta} X \times X \xrightarrow{f \times g} RP^\infty \times RP^\infty \xrightarrow{h} RP^\infty$$

classifies $\lambda \otimes \mu$. Since $\pi_1(f \times g)\Delta = f$ and $\pi_2(f \times g)\Delta = g$, the preceding lemma implies

$$\begin{aligned} e(\lambda \otimes \mu) &= \Delta^*(f \times g)^* h^* \alpha \\ &= \Delta^*(f \times g)^*(\pi_1^*\alpha_1 + \pi_2^*\alpha_2) \\ &= (\pi_1(f \times g)\Delta)^*\alpha_1 + (\pi_2(f \times g)\Delta)^*\alpha_2 \\ &= f^*\alpha_1 + g^*\alpha_2 = e(\lambda) + e(\mu), \end{aligned}$$

as claimed.

The homomorphism $\Gamma(X) \to H^1(X)$ is in fact an isomorphism, for any $X \in \mathscr{B}$. However, for the moment we prove only that it is a monomorphism; proofs that it is an isomorphism are indicated in Remarks 9.7 and 9.8.

7.6 Lemma: *$\Gamma(S^1) \to H^1(S^1)$ is an isomorphism for the circle $S^1 \in \mathscr{B}$.*

PROOF: By Proposition 7.3, $\Gamma(S^1) = \mathbb{Z}/2$, the nonneutral element being the canonical real line bundle γ_1^1 over RP^1 $(=S^1)$; but $e(\gamma_1^1) \neq 0 \in H^1(RP^1)$ by Proposition 4.4.

7.7 Proposition: *$\Gamma(X) \to H^1(X)$ is a monomorphism for any $X \in \mathscr{B}$.*

PROOF: If $e(\lambda) = 0$ for a real line bundle λ over X, and if $S^1 \xrightarrow{f} X$ is any map, then by naturality of $\mathbb{Z}/2$ Euler classes (Proposition 3.2) one has $e(f^!\lambda) = f^*e(\lambda) = 0 \in H^1(S^1)$; hence by Lemma 7.6, $f^!\lambda = \varepsilon^1$ over S^1. However, since each connected component of X is pathwise connected (Exercise I.10.22), λ is trivial if (and only if) $f^!\lambda$ is trivial for every map $S^1 \xrightarrow{f} X$.

In Chapter V we shall derive a necessary and sufficient condition for any real n-plane bundle ξ to be orientable, formulated in terms of a $\mathbb{Z}/2$ cohomology class. The proof will depend upon the special case that ξ is a Whitney sum of real line bundles, which we now consider.

7.8 Lemma: *Let ξ be the Whitney sum $\lambda_1 \oplus \cdots \oplus \lambda_n$ of n real line bundles $\lambda_1, \ldots, \lambda_n$ over $X \in \mathscr{B}$, let $\lambda_1 \otimes \cdots \otimes \lambda_n \in \Gamma(X)$ be the product of the same line bundles, and let $\bigwedge^n \xi \in \Gamma(X)$ be the nth exterior power of ξ, as in Definition* III.10.1; *then $\bigwedge^n \xi = \lambda_1 \otimes \cdots \otimes \lambda_n$.*

PROOF: Let S be the linear subspace of the real tensor product $\bigotimes^n \mathbb{R}^n$ spanned by elements $x_1 \otimes \cdots \otimes x_n$ such that $x_1, \ldots, x_n$ are linearly dependent in $\mathbb{R}^n$, and let T be the 1-dimensional linear subspace of $\bigwedge^n \mathbb{R}^n$ spanned by elements $e_1 \otimes \cdots \otimes e_n$ such that each $e_i \in R^n$ vanishes outside the ith summand of $\mathbb{R}^n$. The classical computation

$$(e_i \otimes e_j + e_j \otimes e_i) + e_i \otimes e_i + e_j \otimes e_j = (e_i + e_j) \otimes (e_i + e_j)$$

shows that $e_i \otimes e_j + e_j \otimes e_i \in S$, hence that $\bigotimes^n \mathbb{R}^n = S \oplus T$, so that $\bigwedge^n \mathbb{R}^n = \bigotimes^n \mathbb{R}^n / S \cong T$, by definition of $\bigwedge^n \mathbb{R}^n$.

7.9 Proposition: *A Whitney sum $\lambda_1 \oplus \cdots \oplus \lambda_n$ of real line bundles $\lambda_1, \ldots, \lambda_n$ over a base space $X \in \mathscr{B}$ is orientable if and only if the $\mathbb{Z}/2$ Euler classes satisfy*

$$e(\lambda_1) + \cdots + e(\lambda_n) = 0 \in H^1(X).$$

PROOF: Let $\xi = \lambda_1 \oplus \cdots \oplus \lambda_n$. Then according to Definition III.10.4, ξ is orientable if and only if the nth exterior power $\bigwedge^n \xi$ is the trivial line bundle ε^1 over X. Hence by Lemma 7.8, ξ is orientable if and only if $\lambda_1 \otimes \cdots \otimes \lambda_n = \varepsilon^1 \in \Gamma(X)$. By Proposition 7.7, $\Gamma(X) \to H^1(X)$ is a monomorphism, so that ξ is orientable if and only if $e(\lambda_1 \otimes \cdots \otimes \lambda_n) = 0 \in H^1(X)$, and by Proposition 7.5 one has $e(\lambda_1 \otimes \cdots \otimes \lambda_n) = e(\lambda_1) + \cdots + e(\lambda_n)$.

The following computation will play a crucial role in the construction of other $\mathbb{Z}/2$ characteristic classes, in Chapter V.

7.10 Proposition: *If λ is a line subbundle of a real n-plane bundle ξ over a base space $X \in \mathscr{B}$, then $e(\lambda \otimes \xi) = 0 \in H^n(X)$.*

PROOF: By Proposition III.3.6, $\xi = \lambda \oplus \eta$ for some $(n-1)$-plane bundle η, so that $\lambda \otimes \xi = (\lambda \otimes \lambda) \oplus (\lambda \otimes \eta) = \varepsilon^1 \oplus \zeta$ for the $(n-1)$-plane bundle $\zeta = \lambda \otimes \eta$. Hence $e(\lambda \otimes \xi) = e(\varepsilon^1 \oplus \zeta) = e(\varepsilon^1) \cup e(\zeta) = 0$ as in Proposition 3.6.

8. The $\mathbb{Z}/2$ Thom Class $V_\xi \in H^n(P_{\xi\oplus 1}, P_\xi)$

There are many constructions of $\mathbb{Z}/2$ Euler classes other than the ones given in Definition 3.1 and Proposition 3.8. We consider an alternative in this section.

Let $E \xrightarrow{\pi} X$ represent a real n-plane bundle ξ over a base space X, let 1 denote the trivial bundle ε^1 over X, and let $\xi \oplus 1$ be the Whitney sum of ξ and 1. Then the total space $P_{\xi\oplus 1}$ of the projective bundle associated to $\xi \oplus 1$ consists of equivalence classes $[e \oplus s1_{\pi(e)}]$ of nonzero elements $e \oplus s1_{\pi(e)}$ in the total space of $\xi \oplus 1$, where $e \in E$ projects to $\pi(e) \in X$, and where $s1_{\pi(e)}$ is a real multiple in the fiber over $\pi(e)$ of a fixed nowhere-vanishing section of 1. The subset of those $[e \oplus s1_{\pi(e)}] \in P_{\xi\oplus 1}$ with $s = 0$ is clearly the total space P_ξ of the projective bundle associated to ξ, so that there is an inclusion $P_\xi \xrightarrow{i} P_{\xi\oplus 1}$ and an inclusion $P_{\xi\oplus 1} \xrightarrow{j} P_{\xi\oplus 1}, P_\xi$.

8.1 Definition: The *zero-section* $X \xrightarrow{\tau} P_{\xi\oplus 1}$ carries each $x \in X$ into the point $[0 \oplus 1_x] \in P_{\xi\oplus 1}$.

If $P_{\xi\oplus 1} \xrightarrow{\bar{\pi}} X$ represents the preceding projective bundle associated to $\xi \oplus 1$, then $H^*(P_{\xi\oplus 1}, P_\xi)$ is clearly an $H^*(X)$-module with respect to the product $\beta \cdot \alpha = \bar{\pi}^*\beta \cup \alpha \in H^*(P_{\xi\oplus 1}, P_\xi)$ of $\beta \in H^*(X)$ and $\alpha \in H^*(P_{\xi\oplus 1}, P_\xi)$. We shall show that $H^*(P_{\xi\oplus 1}, P_\xi)$ is isomorphic to the $H^*(X)$-module $H^*(E, E^*)$, and that if $V_\xi \in H^n(P_{\xi\oplus 1}, P_\xi)$ corresponds to the Thom class $U_\xi \in H^n(E, E^*)$, then $\tau^* j^* V_\xi$ is the $\mathbb{Z}/2$ Euler class $e(\xi) \in H^n(X)$.

In order to prove the preceding assertions, let $P^*_{\xi\oplus 1} \subset P_{\xi\oplus 1}$ consist of all points of the form $[e \oplus s1_{\pi(e)}]$ for $e \neq 0$. In the following lemma the letters i and j each represent three distinct (but obvious) inclusions. Let $E \xrightarrow{f_1} P_{\xi\oplus 1}$ carry e into $[e \oplus 1_{\pi(e)}]$, with restriction $E^* \xrightarrow{f_2} P^*_{\xi\oplus 1}$, let $P_{\xi\oplus 1} \xrightarrow{g_1} P_{\xi\oplus 1}$ be the identity map, let $P_\xi \xrightarrow{g_2} P^*_{\xi\oplus 1}$ be an inclusion of subspaces of $P_{\xi\oplus 1}$, and set $f_0 = (f_1, f_2)$ and $g_0 = (g_1, g_2)$.

8.2 Lemma: *There is a commutative diagram*

$$\begin{array}{ccccccccc}
\cdots & \xrightarrow{\delta} & H^*(E, E^*) & \xrightarrow{j^*} & H^*(E) & \xrightarrow{i^*} & H^*(E^*) & \xrightarrow{\delta} & \cdots \\
 & & \uparrow{\scriptstyle f_0^*}\ \approx & & \uparrow{\scriptstyle f_1^*} & & \uparrow{\scriptstyle f_2^*} & & \\
\cdots & \xrightarrow{\delta} & H^*(P_{\xi\oplus 1}, P^*_{\xi\oplus 1}) & \xrightarrow{j^*} & H^*(P_{\xi\oplus 1}) & \xrightarrow{i^*} & H^*(P^*_{\xi\oplus 1}) & \xrightarrow{\delta} & \cdots \\
 & & \downarrow{\scriptstyle g_0^*}\ \approx & & \downarrow{\scriptstyle g_1^*}\ = & & \downarrow{\scriptstyle g_2^*}\ \approx & & \\
\cdots & \xrightarrow{\delta} & H^*(P_{\xi\oplus 1}, P_\xi) & \xrightarrow{j^*} & H^*(P_{\xi\oplus 1}) & \xrightarrow{i^*} & H^*(P_\xi) & \xrightarrow{\delta} & \cdots
\end{array}$$

in which the four homomorphisms f_0^*, g_0^*, g_1^*, g_2^* are isomorphisms.

PROOF: Commutativity is immediate since f_2 and g_2 are restrictions of f_1 and g_1. Since f_1 and f_2 are injections of E and E^* into $P_{\xi\oplus 1} \backslash P_\xi$ and $P^*_{\xi\oplus 1} \backslash P_\xi$, respectively, f_0^* is an isomorphism by the excision axiom, and g_1^* is trivially an isomorphism. To show that g_2^* is an isomorphism define $P^*_{\xi\oplus 1} \xrightarrow{h_2} P_\xi$ by setting $h_2([e \oplus s1_{\pi(e)}]) = [e]$, so that $h_2 g_2$ is the identity on P_ξ. Then $g_2 h_2$ carries $[e \oplus s1_{\pi(e)}]$ into $[e \oplus 0\,1_{\pi(e)}]$ (for $e \neq 0$), a map which is trivially homotopic to the identity. Thus g_2 is a homotopy equivalence, so that g_2^* is an isomorphism. Finally, since g_1^* and g_2^* are isomorphisms the 5-lemma implies that g_0^* is also an isomorphism.

The isomorphisms f_0^* and g_0^* of Lemma 8.2 provide new *Thom classes* $W_\xi = f_0^{*-1}U_\xi \in H^n(P_{\xi\oplus 1}, P_{\xi\oplus 1}^*)$ and $V_\xi = g_0^* W_\xi \in H^n(P_{\xi\oplus 1}, P_\xi)$. We are primarily interested in V_ξ.

8.3 Proposition: *For any real n-plane bundle ξ over a base space X the $H^*(X)$-module $H^*(P_{\xi\oplus 1}, P_\xi)$ is free on a single generator $V_\xi \in H^n(P_{\xi\oplus 1}, P_\xi)$, and $\tau^* j^* V_\xi$ is the Euler class $e(\xi) \in H^n(X)$.*

PROOF: The first statement is just Proposition 1.5, combined with the isomorphism $H^*(E, E^*) \xrightarrow{g_0^* f_0^{*-1}} H^*(P_{\xi\oplus 1}, P_\xi)$ of Lemma 8.2. To prove the second statement one applies $H^n(-)$ to the commutative diagram

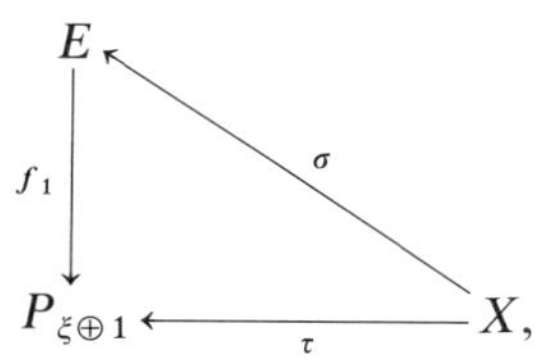

and one combines the result with the j^* portion of Lemma 8.2 to conclude that the diagram

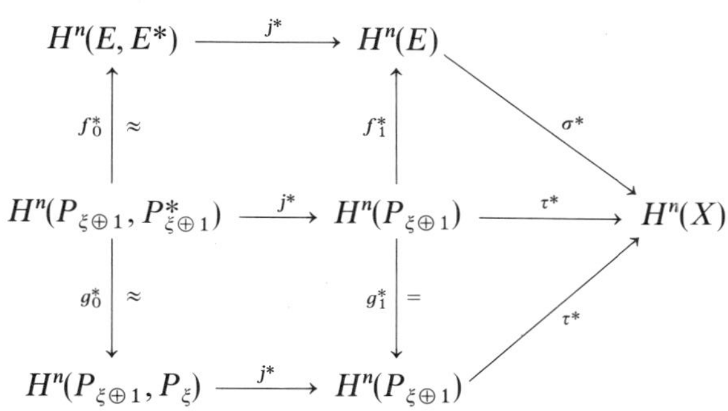

also commutes; hence

$$\tau^* j^* V_\xi = \tau^* j^* g_0^* W_\xi = \sigma^* j^* f_0^* W_\xi = \sigma^* j^* U_\xi = e(\xi),$$

as claimed.

9. Remarks and Exercises

9.1 Remark: Thom isomorphisms and Thom classes (of sphere bundles over polyhedral base spaces) first appeared in Thom [1], with sketches of elementary proofs; the constructions were extended in Thom [4], using sheaf theory. Both Thom [1] and [4] contain the identity $\Phi_\xi e(\xi) = U_\xi \cup U_\xi$ of Proposition 3.8, of which special cases had appeared in Whitney [6, p. 119] and in Gysin [1]. Thom [4] also contains the Gysin sequence in more or less the form reported in Proposition 4.2. An independent development of the Thom isomorphism and the Gysin sequence was given by Chern and Spanier [1].

The Mayer–Vietoris argument for the existence of Thom classes can be found in Milnor [3] and in Milnor and Stasheff [1]. An entirely different approach to the Thom isomorphism theorem is given in Cockroft [1], and a generalization of Thom classes to nonspherical fibrations appears in Schultz [4]. Also, see Bott and Tu [1, pp. 63–64].

9.2 Remark: If $E \to X$ represents a vector bundle, the pair E, E^* is the corresponding *Thom complex*. The same terminology is sometimes applied to the one-point compactification of E itself, which is more often called the *Thom space* of the given bundle: the cohomology of the Thom complex and the Thom space coincide except in degree 0. (The original construction of Thom [1, 4], applied to sphere bundles rather than vector bundles, presents yet a third variant of Thom complexes and Thom spaces.)

Thom spaces were used by Atiyah [1] to show that the J-equivalence type of the tangent bundle $\tau(X)$ of a smooth manifold X depends only on the homotopy type of X. (This result was strengthened by Benlian and Wagoner [1] and Dupont [2], as noted in Remarks III.13.34 and III.13.42.) Atiyah's result (applied to normal bundles rather than tangent bundles) is also valid for arbitrary closed topological manifolds: see Horvath [1]. Other properties of Thom spaces and their relations to J-equivalence can be found in Held and Sjerve [1–3], for example.

The most important application of Thom spaces occurs in the computation of cobordism rings, considered later in their work.

9.3 Exercise: Show for any real vector bundle ξ over a base space $X \in \mathscr{B}$ that the total space P_ξ of the corresponding projective bundle also belongs to $\mathscr{B}$, as claimed following Definition 5.1. (If one replaces the category $\mathscr{B}$ by the category $\mathscr{W} \subset \mathscr{B}$ of Definition I.3.8 and Corollary I.5.3 one obtains a stronger result, which can be found in Stasheff [1] and in Schön [1]: If $E \xrightarrow{\pi} X$ is *any* fibration with base space $X \in \mathscr{W}$ and fiber $F \in \mathscr{W}$, then $E \in \mathscr{W}$.)

9.4 Remark: The cohomology ring $H^*(RP^n; \mathbb{Z}/2)$ was computed in Proposition 4.4. This computation and other computations of the $\mathbb{Z}/2$ homology and cohomology of RP^n can be found in Artin and Braun [2, pp. 178–181], Feder [1], Gray [1, pp. 194–195, 285–286], Greenberg

[1, p. 141], Hu [4, pp. 133–138], Lam [1], Massey [6, pp. 241–242], Maunder [1, pp. 140, 169–170, 348–349], Spanier [4, pp. 264–265], and Ta [1], for example.

9.5 Remark: The splitting principle of Proposition 5.5 was introduced in Borel [1]. It was used in Grothendieck [1] to provide axioms for certain characteristic classes we shall study later; Grothendieck's axioms are similar to those of Theorem 6.1. We shall use Grothendieck's procedure often, in other circumstances described in later chapters.

9.6 Exercise: Let γ_n^1 be the canonical real line bundle over the real projective space RP^n, and let $\tau(RP^n)$ be the tangent bundle of RP^n. Show that

$$e(\tau(RP^n)) = \begin{cases} e(\gamma_n^1)^n & \text{if } n \text{ is even} \\ 0 & \text{if } n \text{ is odd.} \end{cases}$$

9.7 Remark: We constructed a $\mathbb{Z}/2$-module homomorphism $\Gamma(X) \to H^1(X; \mathbb{Z}/2)$ for any $X \in \mathscr{B}$ in Proposition 7.5, and we showed in Proposition 7.7 that it is a monomorphism. A stronger result is valid: *the $\mathbb{Z}/2$-module homomorphism $\Gamma(X) \to H^1(X; \mathbb{Z}/2)$ is an isomorphism.*

To start the proof recall that any $X \in \mathscr{B}$ is homotopy equivalent to a metric space of finite type (Definition I.1.2), a fortiori homotopy equivalent to a paracompact space. Lemma II.4.5 implies that $\Gamma(X)$ is invariant under homotopy equivalence, and since $H^1(X; \mathbb{Z}/2)$ is also homotopy invariant, one can therefore assume at the outset that X is paracompact. Clearly one may as well further assume that X is connected, hence pathwise connected by Exercise I.10.22.

Now let $\pi_1(X)$ be the fundamental group of X, let $\pi_1(X) \to H_1(X; \mathbb{Z})$ be the Hurewicz map, and let $H_1(X; \mathbb{Z}) \to H_1(X; \mathbb{Z}/2)$ be the coefficient homomorphism. Both homomorphisms are surjective, so that if $L \subset \pi_1(X)$ is the kernel of the composition $\pi_1(X) \to H_1(X; \mathbb{Z}) \to H_1(X; \mathbb{Z}/2)$ the quotient group $\pi_1(X)/L$ is a $\mathbb{Z}/2$-module isomorphic to $H_1(X; \mathbb{Z}/2)$. Since $\mathbb{Z}/2$ is a field, $\pi_1(X)/L$ is free, and since X is pathwise connected, one can represent each element of a basis B of $\pi_1(X)/L$ by a map $S^1 \to X$ carrying a fixed base point in S^1 to a fixed base point in X. It follows that there is a *bouquet* $\bigvee_B S^1$ of circles attached at a common base point, and a base-point-preserving map $\bigvee_B S^1 \xrightarrow{f} X$ such that (i) the singular homology homomorphism $H_1(\bigvee_B S^1; \mathbb{Z}/2) \xrightarrow{f_*} H_1(X; \mathbb{Z}/2)$ is an isomorphism, and (ii) the homomorphism $\Gamma(X) \xrightarrow{f^!} \Gamma(\bigvee_B S^1)$ induced by pulling line bundles back along f is an epimorphism; the second property is an easy consequence of the paracompactness of X.

Since $\mathbb{Z}/2$ is a field, one has $H^1(X;\mathbb{Z}/2) = \mathrm{Hom}(H_1(X;\mathbb{Z}/2),\mathbb{Z}/2)$ and $H^1(\bigvee_B S^1) = \mathrm{Hom}(H_1(\bigvee_B S^1;\ \mathbb{Z}/2),\ \mathbb{Z}/2) = \mathrm{Hom}(\coprod_B H_1(S^1;\ \mathbb{Z}/2),\ \mathbb{Z}/2) = \prod_B H^1(S^1;\ \mathbb{Z}/2)$, for direct sums $\coprod_B$ and direct products $\prod_B$. Clearly $\Gamma(\bigvee_B S^1) = \prod_B \Gamma(S^1)$, so that naturality of $\mathbb{Z}/2$ Euler classes (Proposition 3.2) provides a commutative diagram

$$\begin{array}{ccc} \Gamma(X) & \longrightarrow & H^1(X;\mathbb{Z}/2) \\ {\scriptstyle f^!}\downarrow & & \downarrow{\scriptstyle \approx\ f^*} \\ \prod_B \Gamma(S^1) & \xrightarrow{\approx} & \prod_B H^1(S^1;\mathbb{Z}/2) \end{array}$$

with an epimorphism $f^!$ and an isomorphism f^*. The bottom map is also an isomorphism, by Lemma 7.6, so that $\Gamma(X) \to H^1(X;\mathbb{Z}/2)$ is an epimorphism. Since the latter map is also a monomorphism, by Proposition 7.7, this completes the proof.

9.8 Remark: Another proof of the preceding result is briefly sketched in Husemoller [1, 2nd ed., pp. 235–236].

9.9 Remark: If one substitutes *complex* line bundles for real line bundles in Definition 7.2, the resulting analog of $\Gamma(X)$ is the *Picard group* of X, which is isomorphic to $H^2(X;\mathbb{Z})$; details will appear in Chapter X. Picard groups of complex algebraic varieties are frequently defined by other means; see Griffiths and Harris [1, pp. 133–135], for example.

CHAPTER V

Stiefel–Whitney Classes

0. Introduction

Let $\mathbb{Z}/2[t]$ be the polynomial ring in a single variable t over the field $\mathbb{Z}/2$, and let $f(t) \in \mathbb{Z}/2[t]$ be a polynomial $1 + a_1 t + \cdots + a_p t^p$ with leading term $1 \in \mathbb{Z}/2$. Then for any real m-plane bundle ξ over a base space $X \in \mathscr{B}$ the splitting class $e(\lambda_\xi) \in H^1(P_\xi; \mathbb{Z}/2)$ leads to an inhomogeneous characteristic class $u_f(\xi) \in H^*(X; \mathbb{Z}/2)$. The case $f(t) = 1 + t$ provides the *total Stiefel–Whitney class* $w(\xi) \in H^*(X; \mathbb{Z}/2)$, which can be used to compute all the other classes $u_f(\xi)$. More generally, if $\mathbb{Z}/2[[t]]$ is the formal power series ring in a single variable t over $\mathbb{Z}/2$, and if $f(t) \in \mathbb{Z}/2[[t]]$ is a formal power series $1 + a_1 t + a_2 t^2 + \cdots$ with leading term $1 \in \mathbb{Z}/2$, there is a corresponding inhomogeneous characteristic class $u_f(\xi)$ in the direct product $H^{**}(X; \mathbb{Z}/2)$. The classes $u_f(\xi)$ and their properties are developed in §1 and §2.

For each $q \geqq 0$ the qth summand of the total Stiefel–Whitney class $w(\xi) \in H^*(X; \mathbb{Z}/2)$ of a real m-plane bundle ξ over $X \in \mathscr{B}$ is the qth *Stiefel–Whitney class* $w_q(\xi) \in H^q(X; \mathbb{Z}/2)$. One easily shows that $w_0(\xi) = 1 \in H^0(X; \mathbb{Z}/2)$, that $w_m(\xi) \in H^m(X; \mathbb{Z}/2)$ is the $\mathbb{Z}/2$ Euler class $e(\xi)$ of Definition IV.3.1, and that $w_q(\xi) = 0 \in H^q(X; \mathbb{Z}/2)$ for $q > m$. In particular, it is shown in §4 that if γ^m is the universal real m-plane bundle over the Grassmann manifold $G^m(\mathbb{R}^\infty)$, then $H^*(G^m(\mathbb{R}^\infty); \mathbb{Z}/2)$ is the polynomial ring $\mathbb{Z}/2[w_1(\gamma^m), \ldots, w_m(\gamma^m)]$ over $\mathbb{Z}/2$ generated by the Stiefel–Whitney classes $w_1(\gamma^m), \ldots, w_m(\gamma^m)$. This result, combined with the homotopy classification theorem, Theorem III.8.9, provides an alternative characterization of all $\mathbb{Z}/2$ characteristic classes of real vector bundles.

Many Stiefel–Whitney classes have a direct geometric interpretation. For example, it is shown in §3 that the first Stiefel–Whitney class $w_1(\xi) \in H^1(X; \mathbb{Z}/2)$ of any real vector bundle ξ over any $X \in \mathscr{B}$ vanishes if and only if ξ is orientable. A formal result proved in §6 will lead in the next chapter to several geometric properties of smooth manifolds X, which are proved by computing Stiefel–Whitney classes of their tangent bundles $\tau(X)$.

As in the previous chapter, the notations $H^*(-,-)$ and $H^*(-)$ are used consistently to indicate singular cohomology $H^*(-,-;\mathbb{Z}/2)$ and $H^*(-;\mathbb{Z}/2)$ with $\mathbb{Z}/2$ coefficients.

1. Multiplicative $\mathbb{Z}/2$ Classes

Let ξ be a real m-plane bundle over a base space $X \in \mathscr{B}$, let $P_\xi \in \mathscr{B}$ be the total space of the corresponding projective bundle, as in Definition IV.5.1, and let λ_ξ be the splitting bundle and $e(\lambda_\xi) \in H^1(P_\xi)$ the $\mathbb{Z}/2$ splitting class, as in Definition IV.5.2. According to Proposition IV.5.3 the $H^*(X)$-module $H^*(P_\xi)$ is free on the basis $\{1, e(\lambda_\xi), \ldots, e(\lambda_\xi)^{m-1}\}$. It follows that the operation $\cup e(\lambda_\xi)$ of cup product by the splitting class is a cyclic $H^*(X)$-module endomorphism $H^*(P_\xi) \xrightarrow{\cup e(\lambda_\xi)} H^*(P_\xi)$ of degree $+1$. More generally, for any polynomial $f(t) \in \mathbb{Z}/2[t]$ the operation $\cup f(e(\lambda_\xi))$ of cup product by the class $f(e(\lambda_\xi)) \in H^*(P_\xi)$ is also an $H^*(X)$-module endomorphism

$$H^*(P_\xi) \xrightarrow{\cup f(e(\lambda_\xi)) \,=\, f(\cup e(\lambda_\xi))} H^*(P_\xi),$$

which is in general inhomogeneous. Since $H^*(P_\xi)$ is a free $H^*(X)$-module of rank $m < \infty$ one can form the determinant of the endomorphism $f(\cup e(\lambda_\xi))$, with value in $H^*(X)$.

1.1 Definition: Let $f(t)$ be any polynomial $1 + a_1 t + \cdots + a_p t^p$ over $\mathbb{Z}/2$, with leading term $1 \in \mathbb{Z}/2$, and let ξ be any real m-plane bundle over a base space $X \in \mathscr{B}$. The *multiplicative* $\mathbb{Z}/2$ *class* $u_f(\xi) \in H^*(X)$ is given by $u_f(\xi) = \det f(\cup e(\lambda_\xi))$. In the special case $f(t) = 1 + t$ the class $u_f(\xi)$ is the *total Stiefel–Whitney class*: $w(\xi) = \det(\cup(1 + e(\lambda_\xi))) \in H^*(X)$.

Clearly if $f(t)$ is of degree p over $\mathbb{Z}/2$, then the class $u_f(\xi) \in H^*(X)$ is of the form $1 + u_{f,1}(\xi) + \cdots + u_{f,mp}(\xi)$, where $u_{f,q}(\xi) \in H^q(X)$ for $q = 1, \ldots, mp$; that is, one has $u_{f,0}(\xi) = 1 \in H^0(X)$ and $u_{f,q}(\xi) = 0 \in H^q(X)$ for $q > mp$. In particular, the total Stiefel–Whitney class $w(\xi) \in H^*(X)$ of the real m-plane bundle ξ over the base space $X \in \mathscr{B}$ is of the form $1 + w_1(\xi) + \cdots + w_m(\xi)$; for each $q = 1, \ldots, m$ the class $w_q(\xi) \in H^q(X)$ is the qth *Stiefel–Whitney class* of ξ.

1.2 Proposition: *Let μ be a real line bundle over a base space $X \in \mathscr{B}$, with $\mathbb{Z}/2$ Euler class $e(\mu) \in H^1(X)$. Then for any polynomial $f(t) = 1 + a_1 t + \cdots + a_p t^p \in \mathbb{Z}/2[t]$ with leading term 1 the multiplicative $\mathbb{Z}/2$ class $u_f(\mu) \in H^*(X)$ satisfies*

$$u_f(\mu) = f(e(\mu)) = 1 + a_1 e(\mu) + \cdots + a_p e(\mu)^p;$$

in particular, the total Stiefel–Whitney class $w(\mu) \in H^*(X)$ is given by $w(\mu) = 1 + e(\mu)$.

Proof: Since μ is of rank one, any representation $P_\mu \xrightarrow{\pi} X$ of the corresponding projective bundle is a homeomorphism, for which $\pi^!\mu$ is the splitting bundle λ_μ over P_μ. Hence $e(\lambda_\mu) = e(\pi^!\mu) = \pi^* e(\mu)$, and since $\beta \cdot \alpha = \pi^*\beta \cup \alpha$ for $\beta \in H^*(X)$ and $\alpha \in H^*(P_\mu)$, by definition of scalar multiplication in the $H^*(X)$-module $H^*(P_\mu)$, it follows that the endomorphism $f(\cup e(\lambda_\mu))$ of $H^*(P_\mu)$ is scalar multiplication by $f(e(\mu))$. Since $H^*(P_\mu)$ is of rank one over $H^*(X)$, this implies that $u_f(\mu) = \det f(\cup e(\lambda_\mu)) = f(e(\mu))$, as claimed.

Now let $X' \xrightarrow{g} X$ be any map of base spaces, and let ξ be any real vector bundle over X. The pullback diagram for the bundle $g^!\xi$ over X' induces a corresponding pullback diagram

$$\begin{array}{ccc} P_{g^!\xi} & \xrightarrow{\quad g \quad} & P_\xi \\ \downarrow & & \downarrow \\ X' & \xrightarrow{\quad g \quad} & X \end{array}$$

of projective bundles, over which one can place the diagram for the pullback $g^!\lambda_\xi$ of the splitting bundle λ_ξ, as indicated:

$$\begin{array}{ccc} \bar{E}' & \longrightarrow & \bar{E} \\ \downarrow & & \downarrow \\ P_{g^!\xi} & \xrightarrow{\quad g \quad} & P_\xi \\ \downarrow & & \downarrow \\ X' & \xrightarrow{\quad g \quad} & X. \end{array}$$

If E denotes the total space of ξ, then $\bar{E}$ consists of pairs (l_x, e_x) for each $x \in X$, where $l_x \in P_\xi$ can be regarded as a 1-dimensional subspace of the fiber E_x, and where $e_x \in l_x$. There is a corresponding description of the total space $\bar{E}'$

of the splitting bundle $\lambda_{g^!\xi}$, for which uniqueness of the preceding pullback diagram guarantees that $\lambda_{g^!\xi} = \mathbf{g}^!\lambda_\xi$.

1.3 Proposition (Naturality): *Let $X' \xrightarrow{g} X$ be a map in the category $\mathscr{B}$ of base spaces, let ξ be a real m-plane bundle over X, and let $g^!\xi$ be the pullback over X'. Then for any polynomial $f(t) \in \mathbb{Z}/2[t]$ with leading term $1 \in \mathbb{Z}/2$ the multiplicative $\mathbb{Z}/2$ class $u_f(g^!\xi) \in H^*(X')$ satisfies $u_f(g^!\xi) = g^*u_f(\xi)$; in particular, the total Stiefel–Whitney class $w(g^!\xi) \in H^*(X')$ satisfies $w(g^!\xi) = g^*w(\xi)$.*

PROOF: The lower portion of the preceding diagram guarantees that $H^*(P_\xi) \xrightarrow{\mathbf{g}^*} H^*(P_{g^!\xi})$ is a module homomorphism over the ground ring homomorphism $H^*(X) \xrightarrow{g^*} H^*(X')$; that is, $\mathbf{g}^*(\beta \cdot \alpha) = (g^*\beta) \cdot (\mathbf{g}^*\alpha)$ for $\beta \in H^*(X)$ and $\alpha \in H^*(P_\xi)$. Furthermore, since $\lambda_{g^!\xi} = \mathbf{g}^!\lambda_\xi$, it follows from the naturality of $\mathbb{Z}/2$ Euler classes (Proposition IV.3.2) that $e(\lambda_{g^!\xi}) = \mathbf{g}^*e(\lambda_\xi)$, hence that $\mathbf{g}^*$ carries the ordered basis $(1, e(\lambda_\xi), \ldots, e(\lambda_\xi)^{m-1})$ of $H^*(P_\xi)$ into the ordered basis $(1, e(\lambda_{g^!\xi}), \ldots, e(\lambda_{g^!\xi})^{m-1})$ of $H^*(P_{g^!\xi})$. Since the endomorphism $H^*(P_\xi) \xrightarrow{\cup e(\lambda_\xi)} H^*(P_\xi)$ is cyclic, it has a matrix representation

$$\begin{pmatrix} 0 & \cdots & 0 & \beta_m \\ 1 & & 0 & \beta_{m-1} \\ & \ddots & & \vdots \\ 0 & & 1 & \beta_1 \end{pmatrix}$$

with respect to the former basis, where $\beta_1, \ldots, \beta_m \in H^*(X)$. It follows that $H^*(P_{g^!\xi}) \xrightarrow{\cup e(\lambda_{g^!\xi})} H^*(P_{g^!\xi})$ is represented by

$$\begin{pmatrix} 0 & \cdots & 0 & g^*\beta_m \\ 1 & & 0 & g^*\beta_{m-1} \\ & \ddots & & \vdots \\ 0 & & 1 & g^*\beta_1 \end{pmatrix}$$

with respect to the latter basis, hence that

$$u_f(g^!\xi) = \det f(\cup e(\lambda_{g^!\xi})) = g^* \det f(\cup e(\lambda_\xi)) = g^*u_f(\xi),$$

as claimed.

We briefly reexamine the preceding matrix representation of the endomorphism $\cup e(\lambda_\xi)$ with respect to the basis $(1, e(\lambda_\xi), \ldots, e(\lambda_\xi)^{m-1})$. By definition, $\cup e(\lambda_\xi)$ carries $e(\lambda_\xi)^{q-1}$ into $e(\lambda_\xi)^q$ for $q < m$, and $\cup e(\lambda_\xi)$ carries $e(\lambda_\xi)^{m-1}$ into $\beta_m \cdot 1 + \beta_{m-1} \cdot e(\lambda_\xi) + \cdots + \beta_1 \cdot e(\lambda_\xi)^{m-1}$, representing $e(\lambda_\xi)^m \in H^m(P_\xi)$. It follows that $\beta_1 \in H^1(X), \ldots, \beta_m \in H^m(X)$, and in particular that $\beta_1, \ldots, \beta_m$ are of strictly positive degree in $H^*(X)$. Hence, if

$$f(t) = 1 + a_1 t + \cdots + a_p t^p$$

and

$$u_f(\xi) = \det f(\cup e(\lambda_\xi)) = 1 + u_{f,1}(\xi) + \cdots + u_{f,mp}(\xi),$$

then for each $q \leqq mp$ the class $u_{f,q}(\xi) \in H^q(X)$ depends only on the coefficients $1, a_1, \ldots, a_q$ of $f(t)$. Thus if one replaces polynomials $f(t) \in \mathbb{Z}/2[t]$ by formal power series $f(t) = 1 + a_1 t + a_2 t^2 + \cdots \in \mathbb{Z}/2[[t]]$, one can compute analogous classes $u_f(\xi)$ in the direct product $H^{**}(X)$ of the modules $H^q(X)$.

1.4 Definition: Let $f(t)$ be any formal power series $1 + a_1 t + a_2 t^2 + \cdots \in \mathbb{Z}/2[[t]]$ over $\mathbb{Z}/2$, with leading term $1 \in \mathbb{Z}/2$, and let ξ be any real m-plane bundle over a base space $X \in \mathscr{B}$. Then the *multiplicative* $\mathbb{Z}/2$ *class* $u_f(\xi) \in H^{**}(X)$ is given by $u_f(\xi) = \det f(\cup e(\lambda_\xi))$, for the endomorphism $f(\cup e(\lambda_\xi))$ of the free $H^{**}(X)$-module $H^{**}(P_\xi)$.

One can extend Proposition 1.3 from polynomials $f(t) \in \mathbb{Z}/2[t]$ to formal power series $f(t) \in \mathbb{Z}/2[[t]]$.

1.5 Proposition (Naturality): *Let $X' \xrightarrow{g} X$ be a map in the category $\mathscr{B}$ of base spaces, let ξ be a real m-plane bundle over X, and let $g^!\xi$ be the pullback over X'. Then for any formal power series $f(t) \in \mathbb{Z}/2[[t]]$ with leading term $1 \in \mathbb{Z}/2$ the multiplicative $\mathbb{Z}/2$ class $u_f(g^!\xi) \in H^{**}(X')$ satisfies $u_f(g^!\xi) = g^* u_f(\xi)$.*

PROOF: The proof of Proposition 1.3 remains valid if one substitutes "formal power series" throughout for "polynomial."

The advantage of using formal power series $f(t) \in \mathbb{Z}/2[[t]]$ rather than polynomials, and the rationale for insisting that the leading term of $f(t)$ be $1 \in \mathbb{Z}/2$, lies in the existence of multiplicative inverses $1/f(t) \in \mathbb{Z}/2[[t]]$, which one can compute directly from $f(t)$. For example, the multiplicative inverse of $1 + t$ $(= 1 - t)$, as an element of $\mathbb{Z}/2[[t]]$, is the formal geometric series $1 + t + t^2 + \cdots \in \mathbb{Z}/2[[t]]$. The relation between the classes $u_f(\xi)$ and $u_{1/f}(\xi)$ will be developed later.

In most applications the cohomology $H^*(X)$ of the base space $X \in \mathscr{B}$ will itself be finite, so that $H^*(X) = H^{**}(X)$. However, if $X = RP^\infty$, for example, then $H^*(RP^\infty)$ is indeed a proper subring of $H^{**}(RP^\infty)$.

2. Whitney Product Formulas

We show for any formal power series $f(t) \in \mathbb{Z}/2[[t]]$ with leading term $1 \in \mathbb{Z}/2$, and for any real vector bundles ξ and η over the same base space $X \in \mathscr{B}$, with Whitney sum $\xi \oplus \eta$ over X, that $u_f(\xi \oplus \eta) = u_f(\xi) \cup u_f(\eta) \in H^{**}(X)$.

Let $H^*(X)$ be the singular $\mathbb{Z}/2$ cohomology ring of a base space $X \in \mathscr{B}$, and let $H^*(X)[t]$ be the polynomial ring in a variable t over $H^*(X)$. If ξ is a real vector bundle over X, then the free $H^*(X)$-module $H^*(P_\xi)$ becomes a free $H^*(X)[t]$-module $H^*(P_\xi)[t]$ in the obvious way, and one can construct endomorphisms of $H^*(P_\xi)[t]$.

2.1 Definition: For any real vector bundle ξ over a base space $X \in \mathscr{B}$, the *characteristic polynomial* $\sigma_\xi(t) \in H^*(X)[t]$ is the determinant of the endomorphism $\cup(t \cdot 1 - e(\lambda_\xi))$ of the free $H^*(X)[t]$-module $H^*(P_\xi)[t]$, cup product by $t \cdot 1 - e(\lambda_\xi)$.

The Cayley–Hamilton theorem guarantees that the endomorphism $H^*(P_\xi) \xrightarrow{\sigma_\xi(\cup e(\lambda_\xi))} H^*(P_\xi)$ vanishes, and since $\cup e(\lambda_\xi)$ is a cyclic endomorphism, the kernel of the epimorphism $H^*(X)[t] \to H^*(P_\xi)$ carrying t into $e(\lambda_\xi)$ is precisely the principal ideal $(\sigma_\xi(t))$. Hence one can regard the $H^*(X)$-module $H^*(P_\xi)$ as the quotient $H^*(X)[t]/(\sigma_\xi(t))$.

Now suppose that $E \to X$ represents a real m-plane bundle ξ over the base space $X \in \mathscr{B}$, and that $E' \to X$ represents the Whitney sum $\xi \oplus \eta$ of ξ and a real n-plane bundle η over X, for any $m > 0$ and any $n > 0$. The inclusion diagram

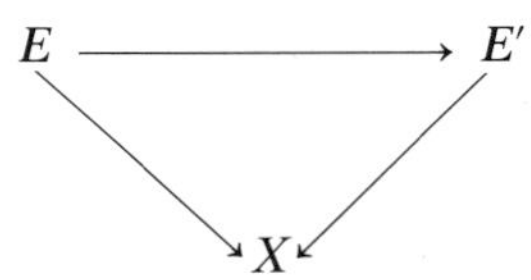

induces an inclusion diagram

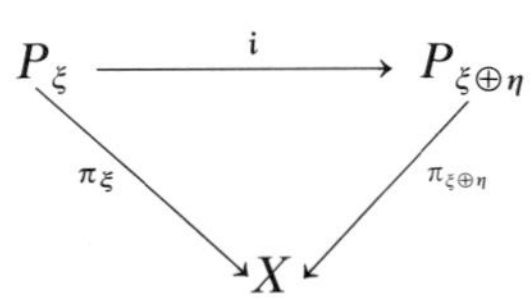

of corresponding projective bundles, and there is an induced $H^*(X)$-module homomorphism $H^*(P_{\xi\oplus\eta}) \xrightarrow{i^*} H^*(P_\xi)$.

2.2 Lemma: *The kernel of i^* is the principal ideal*

$$(\sigma_\xi(e(\lambda_{\xi\oplus\eta}))) \subset H^*(P_{\xi\oplus\eta}).$$

PROOF: Since the splitting bundle λ_ξ is the restriction of the splitting bundle $\lambda_{\xi\oplus\eta}$ to $P_\xi \subset P_{\xi\oplus\eta}$, it follows that $\lambda_\xi = i^!\lambda_{\xi\oplus\eta}$, hence that $e(\lambda_\xi) = i^*e(\lambda_{\xi\oplus\eta})$. Since $H^*(P_\xi)$ is isomorphic to the quotient $H^*(X)[t]/(\sigma_\xi(t))$, this completes the proof.

Observe that the splitting bundle $\lambda_{\xi\oplus\eta}$ of $\xi \oplus \eta$ and the pullback $\pi^!_{\xi\oplus\eta}\xi$ of the bundle ξ are both real vector bundles over the base space $P_{\xi\oplus\eta} \in \mathscr{B}$.

2.3 Lemma: $e(\lambda_{\xi\oplus\eta} \otimes \pi^!_{\xi\oplus\eta}\xi) = \sigma_\xi(e(\lambda_{\xi\oplus\eta}))$.

PROOF: Since λ_ξ is a line subbundle of the bundle $\pi^!_\xi\xi$ over P_ξ, it follows from Proposition IV.7.10 that

$$i^*e(\lambda_{\xi\oplus\eta} \otimes \pi^!_{\xi\oplus\eta}\xi) = e(i^!\lambda_{\xi\oplus\eta} \otimes i^!\pi^!_{\xi\oplus\eta}\xi) = e(\lambda_\xi \otimes \pi^!_\xi\xi) = 0.$$

Hence $e(\lambda_{\xi\oplus\eta} \otimes \pi^!_{\xi\oplus\eta}\xi) \in \ker i^*$, so that Lemma 2.2 implies that

$$e(\lambda_{\xi\oplus\eta} \otimes \pi^!_{\xi\oplus\eta}\xi) = \alpha\sigma_\xi(e(\lambda_{\xi\oplus\eta}))$$

for some $\alpha \in H^*(P_{\xi\oplus\eta})$. If ξ is of rank m, then both $e(\lambda_{\xi\oplus\eta} \otimes \pi^!_{\xi\oplus\eta}\xi)$ and $\sigma_\xi(e(\lambda_{\xi\oplus\eta}))$ lie in $H^m(P_{\xi\oplus\eta})$, so that $\alpha \in H^0(P_{\xi\oplus\eta})$. Hence α is of the form $\pi^*_{\xi\oplus\eta}\beta$ for a unique $\beta \in H^0(X)$; that is, $e(\lambda_{\xi\oplus\eta} \otimes \pi^!_{\xi\oplus\eta}\xi) = \beta \cdot \sigma_\xi(e(\lambda_{\xi\oplus\eta}))$.

In order to verify that $\beta = 1$, observe that for any $x \in X$ there is an inclusion diagram

$$\begin{array}{ccc} P_{\xi\oplus\eta,\,x} & \xrightarrow{k_x} & P_{\xi\oplus\eta} \\ \downarrow{\scriptstyle \pi_{\xi\oplus\eta,\,x}} & & \downarrow{\scriptstyle \pi_{\xi\oplus\eta}} \\ \{x\} & \xrightarrow{j_x} & X, \end{array}$$

and since any bundle over $\{x\}$ is trivial it follows that $k^!_x\pi^!_{\xi\oplus\eta}\xi$ is the trivial bundle ε^m over $P_{\xi\oplus\eta,\,x}$, hence that

$$\begin{aligned} k^*_x e(\lambda_{\xi\oplus\eta} \otimes \pi^!_{\xi\oplus\eta}\xi) &= e(k^!_x\lambda_{\xi\oplus\eta} \otimes k^!_x\pi^!_{\xi\oplus\eta}\xi) \\ &= e(k^!_x\lambda_{\xi\oplus\eta} \otimes \varepsilon^m) = e(mk^!_x\lambda_{\xi\oplus\eta}) \\ &= e(k^!_x\lambda_{\xi\oplus\eta})^m = k^*_x e(\lambda_{\xi\oplus\eta})^m. \end{aligned}$$

The same diagram shows that

$$H^*(P_{\xi\oplus\eta}) \xrightarrow{k^*_x} H^*(P_{\xi\oplus\eta,\,x})$$

is a module homomorphism over the ring homomorphism $H^*(X) \xrightarrow{j^*_x} H^*(\{x\}) \approx \mathbb{Z}/2$, and since j^*_x clearly annihilates everything outside of $H^0(X) \subset H^*(X)$, it follows that k^*_x annihilates all coefficients of $\sigma_\xi(t)$ lying outside of $H^0(X) \subset H^*(X)$, so that $k^*_x\sigma_\xi(e(\lambda_{\xi\oplus\eta})) = k^*_x e(\lambda_{\xi\ominus\eta})^m$. Thus the identity $e(\lambda_{\xi\oplus\eta} \otimes \pi^!_{\xi\oplus\eta}\xi) = \beta \cdot \sigma_\xi(e(\lambda_{\xi\oplus\eta}))$ has the consequence $k^*_x e(\lambda_{\xi\oplus\eta})^m = (j^*_x\beta) \cdot k^*_x e(\lambda_{\xi\oplus\eta})^m$. However, $H^*(P_{\xi\oplus n,\,x})$ is the free $H^*(\{x\})$-module on the basis $\{1, k^*_x e(\lambda_{\xi\oplus\eta}), \ldots, k^*_x e(\lambda_{\xi\oplus\eta})^{m+n-1}\}$, where $m+n-1 \geq n$ and $H^*(\{x\}) \approx \mathbb{Z}/2$, so that $j^*_x\beta = 1$. Hence β has the value 1 on each path component of X, which means that $\beta = 1 \in H^0(X)$, as required.

2.4 Proposition: *Let $\sigma_\xi(t) \in H^*(X)[t]$ and $\sigma_\eta(t) \in H^*(X)[t]$ be the characteristic polynomials of real vector bundles ξ and η over the same base space $X \in \mathscr{B}$; then $\sigma_\xi(t) \cdot \sigma_\eta(t) = \sigma_{\xi\oplus\eta}(t)$.*

Proof: Two applications of Lemma 2.3, combined with the Whitney product formula for $\mathbb{Z}/2$ Euler classes (Proposition IV.3.4) give

$$\begin{aligned}\sigma_\xi(e(\lambda_{\xi\oplus\eta})) \cup \sigma_\eta(e(\lambda_{\xi\oplus\eta})) &= e(\lambda_{\xi\oplus\eta} \otimes \pi^!_{\xi\oplus\eta}\xi) \cup e(\lambda_{\xi\oplus\eta} \otimes \pi^!_{\xi\oplus\eta}\eta)\\ &= e(\lambda_{\xi\oplus\eta} \otimes \pi^!_{\xi\oplus\eta}(\xi \oplus \eta)),\end{aligned}$$

and since $\lambda_{\xi\oplus\eta}$ is a line subbundle of $\pi^!_{\xi\oplus\eta}(\xi \oplus \eta)$, Proposition IV.7.10 implies that the latter expression vanishes. Thus the map $t \mapsto e(\lambda_{\xi\oplus\eta})$ annihilates the polynomial $\sigma_\xi(t) \cdot \sigma_\eta(t)$, which has highest-order term t^{m+n}; but $\sigma_{\xi\oplus\eta}(t)$ is the unique polynomial with this property.

2.5 Proposition (Whitney Product Formula): *For any polynomial $f(t) \in \mathbb{Z}/2[t]$ with leading term $1 \in \mathbb{Z}/2$, or for any formal power series $f(t) \in \mathbb{Z}/2[[t]]$ with leading term $1 \in \mathbb{Z}/2$, and for any real vector bundles ξ and η over the same base space $X \in \mathscr{B}$, with Whitney sum $\xi \oplus \eta$ over X, it follows that*

$$u_f(\xi \oplus \eta) = u_f(\xi) \cup u_f(\eta) \in H^*(X),$$

or that

$$u_f(\xi \oplus \eta) = u_f(\xi) \cup u_f(\eta) \in H^{**}(X);$$

in particular, for $f(t) = 1 + t \in \mathbb{Z}/2[t]$, one has $w(\xi \oplus \eta) = w(\xi) \cup w(\eta) \in H^*(X)$.

Proof: Suppose that the characteristic polynomials of ξ and η are given by $\sigma_\xi(t) = t^m - \alpha_1 t^{m-1} - \cdots - \alpha_m \cdot 1$ and $\sigma_\eta(t) = t^n - \beta_1 t^{n-1} - \cdots - \beta_n \cdot 1$, respectively, so that the cyclic endomorphisms $\cup e(\lambda_\xi)$ and $\cup e(\lambda_\eta)$ have matrix representations

$$A = \left(\begin{array}{ccc|c} 0 & \cdots & 0 & \alpha_m \\ \hline 1 & & 0 & \alpha_{m-1} \\ & \ddots & & \vdots \\ 0 & & 1 & \alpha_1 \end{array}\right) \quad\text{and}\quad B = \left(\begin{array}{ccc|c} 0 & \cdots & 0 & \beta_n \\ \hline 1 & & 0 & \beta_{n-1} \\ & \ddots & & \vdots \\ 0 & & 1 & \beta_1 \end{array}\right)$$

with respect to the ordered bases $(1, e(\lambda_\xi), \ldots, e(\lambda_\xi)^{m-1})$ and $(1, e(\lambda_\eta), \ldots, e(\lambda_\eta)^{n-1})$, respectively. The cyclic endomorphism $\cup e(\lambda_{e\oplus\eta})$ has a corresponding matrix representation with respect to the ordered bases $(1, e, \ldots, e^{m+n-1})$, where $e = e(\lambda_{\xi\oplus\eta})$; however, in this case the basis $(1, e, \ldots, e^{m-1}; \sigma_\xi(e), \sigma_\xi(e)e, \ldots, \sigma_\xi(e)e^{n-1})$ is of more interest. To compute the matrix representation of $\cup e(\lambda_{\xi\oplus\eta})$ with respect to the latter basis observe that $(\cup e(\lambda_{\xi\oplus\eta}))e^{p-1} = e^p$ for $1 \leqq p < m$, and that

$$(\cup e(\lambda_{\xi\oplus\eta}))e^{m-1} = e^m = \alpha_m 1 + \cdots + \alpha_1 e^{m-1} + \sigma_\xi(e).$$

Similarly $(\cup e(\lambda_{\xi\oplus\eta}))\sigma_\xi(e)e^{q-1} = \sigma_\xi(e)e^q$ for $1 \leqq q < n$, and

$$(\cup e(\lambda_{\xi\oplus\eta}))\sigma_\xi(e)e^{n-1} = \beta_n\sigma_\xi(e) + \cdots + \beta_1\sigma_\xi(e)e^{n-1} + \sigma_\xi(e)\sigma_\eta(e).$$

However, in the second computation one has $\sigma_\xi(t)\sigma_\eta(t) = \sigma_{\xi\oplus\eta}(t)$ by Proposition 2.4, so that $\sigma_\xi(e)\sigma_\eta(e) = \sigma_{\xi\oplus\eta}(e) = \sigma_{\xi\oplus\eta}(\cup e(\lambda_{\xi\oplus\eta}))1 = 0$ by the Cayley–Hamilton theorem; thus

$$(\cup e(\lambda_{\xi\oplus\eta}))\sigma_\xi(e)e^{n-1} = \beta_n\sigma_\xi(e) + \cdots + \beta_1\sigma_\xi(e)e^{n-1}.$$

It follows that the matrix representation C of $\cup e(\lambda_{\xi\oplus\eta})$ with respect to the basis $(1, e, \ldots, e^{m-1}; \sigma_\xi(e), \sigma_\xi(e)e, \ldots, \sigma_\xi(e)e^{n-1})$ for $e = e(\lambda_{\xi\oplus\eta})$ is given by

$$C = \begin{pmatrix} A & 0 \\ D & B \end{pmatrix},$$

where D consists of 0's except for a single $1 \in \mathbb{Z}/2$ in the upper right-hand corner. Consequently, for the given $f \in \mathbb{Z}/2[[t]]$ with leading term $1 \in \mathbb{Z}/2$ there is a matrix D_f with formal power series entries for which

$$f(C) = \begin{pmatrix} f(A) & 0 \\ D_f & f(B) \end{pmatrix},$$

so that

$$u_f(\xi \oplus \eta) = \det f(C) = \det f(A) \cup \det f(B) = u_f(\xi) \cup u_f(\eta),$$

as claimed.

2.6 Corollary: *Let ε^m be the trivial m-plane bundle over a base space $X \in \mathscr{B}$. Then for any formal power series $f(t) \in \mathbb{Z}/2[[t]]$ with leading term $1 \in \mathbb{Z}/2$ it follows that $u_f(\varepsilon^m) = 1 \in H^{**}(X)$.*

PROOF: Let $f(t) = 1 + a_1t + a_2t^2 + \cdots$. Then as in Proposition 1.2 one has

$$u_f(\varepsilon^1) = f(e(\varepsilon^1)) = 1 + a_1e(\varepsilon^1) + a_2e(\varepsilon^1)^2 + \cdots$$

for the trivial line bundle ε^1 over X, and since $e(\varepsilon^1) = 0$ by Lemma IV.3.5, one has $u_f(\varepsilon^1) = 1$. The Whitney product formula then gives

$$u_f(\varepsilon^m) = u_f(\varepsilon^1 \oplus \cdots \oplus \varepsilon^1) = u_f(\varepsilon^1) \cup \cdots \cup u_f(\varepsilon^1) = 1,$$

as claimed.

In the following result we use the elementary symmetric functions $\sigma_1(t_1, \ldots, t_m) = t_1 + \cdots + t_m, \ldots, \sigma_m(t_1, \ldots, t_m) = t_1 \cdots t_m$ in the polynomial ring $\mathbb{Z}/2[t_1, \ldots, t_m]$ in m variables $t_1, \ldots, t_m$. The elementary symmetric functions are characterized by the condition

$$(1 + t_1) \cdots (1 + t_m) = 1 + \sigma_1(t_1, \ldots, t_m) + \cdots + \sigma_m(t_1, \ldots, t_m).$$

2.7 Corollary: *Let $\lambda_1, \ldots, \lambda_m$ be m real line bundles over the same base space $X \in \mathscr{B}$, and let $\lambda_1 \oplus \cdots \oplus \lambda_m$ be their Whitney sum. Then for each $q = 1, \ldots, m$ the qth Stiefel–Whitney class $w_q(\lambda_1 \oplus \cdots \oplus \lambda_m) \in H^q(X)$ is the qth elementary symmetric function $\sigma_q(e(\lambda_1), \ldots, e(\lambda_m))$ in the $\mathbb{Z}/2$ Euler classes $e(\lambda_1), \ldots, e(\lambda_m)$.*

PROOF: By the Whitney product formula (Proposition 2.5) and Proposition 1.2 one has

$$\begin{aligned} 1 + w_1(\lambda_1 \oplus \cdots \oplus \lambda_m) + \cdots + w_m(\lambda_1 \oplus \cdots \oplus \lambda_m) \\ = w(\lambda_1 \oplus \cdots \oplus \lambda_m) = w(\lambda_1) \cup \cdots \cup w(\lambda_m) \\ = (1 + e(\lambda_1)) \cup \cdots \cup (1 + e(\lambda_m)). \end{aligned}$$

2.8 Corollary: *For any real m-plane bundle ξ over any base space $X \in \mathscr{B}$ one has $w_m(\xi) = e(\xi) \in H^m(X)$.*

PROOF: For any Whitney sum $\lambda_1 \oplus \cdots \oplus \lambda_m$ of m real line bundles Corollary 2.7 and the Whitney product formula for $\mathbb{Z}/2$ Euler classes (Proposition IV.3.4) give

$$w_m(\lambda_1 \oplus \cdots \oplus \lambda_m) = e(\lambda_1) \cup \cdots \cup e(\lambda_m) = e(\lambda_1 \oplus \cdots \oplus \lambda_m).$$

For any real m-plane bundle ξ over $X \in \mathscr{B}$ one applies the splitting principle (Proposition IV.5.5) to find a map $X' \xrightarrow{g} X$ in $\mathscr{B}$ such that $g^!\xi$ is a Whitney sum $\lambda_1 \oplus \cdots \oplus \lambda_m$ of line bundles over $X' \in \mathscr{B}$ and $H^*(X) \xrightarrow{g^*} H^*(X')$ is a monomorphism. By naturality of Stiefel–Whitney classes (Proposition 1.3), the preceding special case, and naturality of $\mathbb{Z}/2$ Euler classes (Proposition IV.3.2), it then follows that

$$\begin{aligned} g^*w_m(\xi) &= w_m(g^!\xi) = w_m(\lambda_1 \oplus \cdots \oplus \lambda_m) \\ &= e(\lambda_1 \oplus \cdots \oplus \lambda_m) = e(g^!\xi) = g^*e(\xi) \end{aligned}$$

for the monomorphism g^*.

For any formal power series $f(t) \in \mathbb{Z}/2[[t]]$ with leading term $1 \in \mathbb{Z}/2$ one can use the Whitney product formula of Proposition 2.5 to compute the multiplicative $\mathbb{Z}/2$ class $u_f(\xi) \in H^{**}(X)$ of any real m-plane bundle ξ over any $X \in \mathscr{B}$ in terms of the Stiefel–Whitney classes $w_1(\xi), \ldots, w_m(\xi)$. The computation depends upon a simple algebraic device.

Let $\mathbb{Z}/2[[u_1, u_2, \ldots]]$ denote the graded formal power series algebra in denumerably many variables $u_1, u_2, \ldots$, in which u_p is assigned degree p for each natural number $p > 0$; for example, any monomial $(u_1)^{n_1} \cdots (u_p)^{n_p}$ is of degree $n_1 + 2n_2 + \cdots + pn_p$. If $n > p$, then no polynomial of degree p contains u_n, so that every element of $\mathbb{Z}/2[[u_1, u_2, \ldots]]$ is uniquely of the form $\sum_{n \geqq 0} P_n(u_1, \ldots, u_n)$, where $P_n(u_1, \ldots, u_n)$ is a degree n polynomial in $u_1, \ldots, u_n$. In what follows we suppose that $P_0 = 1 \in \mathbb{Z}/2$.

For any natural number $p > 0$ let $\mathbb{Z}/2[[t_1, \ldots, t_p]]$ denote the graded formal power series algebra in which each of $t_1, \ldots, t_p$ is assigned degree 1; for example, any monomial $(t_1)^{n_1} \cdots (t_p)^{n_p}$ is of degree $n_1 + \cdots + n_p$. Then if $f(t) \in \mathbb{Z}/2[[t]]$ has leading term $1 \in \mathbb{Z}/2$, the product $f(t_1) \cdots f(t_p) \in \mathbb{Z}/2[[t_1, \ldots, t_p]]$ is symmetric in $t_1, \ldots, t_p$. It follows that if $u'_1, \ldots, u'_p$ are the elementary symmetric functions $\sigma_1(t_1, \ldots, t_p) = t_1 + \cdots + t_p, \ldots, \sigma_p(t_1, \ldots, t_p) = t_1 \cdots t_p$ in $t_1, \ldots, t_p$, then there are unique polynomials $P_n(u_1, \ldots, u_n)$ of degree n in $\mathbb{Z}/2[[u_1, \ldots, u_n]]$ such that

$$f(t_1) \cdots f(t_p) = \sum_{0 \leqq n \leqq p} P_n(u'_1, \ldots, u'_n) + \sum_{n > p} P_n(u'_1, \ldots, u'_p).$$

Suppose that $u''_1, \ldots, u''_{p-1}$ are the elementary symmetric functions $\sigma_1(t_1, \ldots, t_{p-1}), \ldots, \sigma_{p-1}(t_1, \ldots, t_{p-1})$ in $t_1, \ldots, t_{p-1}$ obtained from $u'_1, \ldots, u'_p$ by setting $t_p = 0$; clearly $u''_p = t_1 \cdots t_{p-1} \cdot 0 = 0$. Since $f(0) = 1$, one has

$$f(t_1) \cdots f(t_{p-1}) = f(t_1) \cdots f(t_{p-1}) f(0) = \sum_{0 \leqq n \leqq p-1} P_n(u''_1, \ldots, u''_n)$$
$$+ \sum_{n > p-1} P_n(u''_1, \ldots, u''_{p-1}, 0).$$

Since there are no polynomial relations among the elementary symmetric functions, it follows for each $n \geqq 0$ that there is a unique $P_n(u_1, \ldots, u_n) \in \mathbb{Z}/2[[u_1, \ldots, u_n]]$ of degree n such that $P_n(u'_1, \ldots, u'_n)$ is the term of degree n in $f(t_1) \cdots f(t_p)$, for any $p \geqq n$.

2.9 Definition: For any formal power series $f(t) \in \mathbb{Z}/2[[t]]$ with leading term $1 \in \mathbb{Z}/2$ the corresponding *multiplicative sequence* $\sum_{n \geqq 0} P_n(u_1, \ldots, u_n) \in \mathbb{Z}/2[[u_1, u_2, \ldots]]$ is determined by requiring the nth degree term of $f(t_1) \cdots f(t_n) \in \mathbb{Z}/2[[t_1, \ldots, t_n]]$ to be $P_n(u'_1, \ldots, u'_n)$, where $u'_1, \ldots, u'_n$ are the elementary symmetric functions in $t_1, \ldots, t_n$.

For example, if $f(t) = 1 + t$ then one clearly has $P_n(u_1, \ldots, u_n) = u_n$ for each $n > 0$.

2.10 Proposition: *Let $\sum_{n \geqq 0} P_n(u_1, \ldots, u_n) \in \mathbb{Z}/2[[u_1, u_2, \ldots]]$ be the multiplicative sequence of any formal power series $f(t) \in \mathbb{Z}/2[[t]]$ with leading term $1 \in \mathbb{Z}/2$, and let ξ be any real m-plane bundle over a base space $X \in \mathscr{B}$. Then the $\mathbb{Z}/2$ singular cohomology class $u_f(\xi) \in H^{**}(X)$ is given by*

$$u_f(\xi) = \sum_{n \geqq 0} u_{f,n}(\xi) \qquad \textit{for} \quad u_{f,n}(\xi) = P_n(w_1(\xi), \ldots, w_n(\xi)) \in H^n(X),$$

where $w_n(\xi) \in H^n(X)$ is the nth Stiefel–Whitney class of ξ for $n \leqq m$, and where $w_n(\xi) = 0$ for $n > m$.

PROOF: By applying a splitting map $X' \to X$, if necessary, one may as well assume that ξ is a Whitney sum $\lambda_1 \oplus \cdots \oplus \lambda_m$ of line bundles. Then

$$u_f(\lambda_1 \oplus \cdots \oplus \lambda_m) = u_f(\lambda_1) \cup \cdots \cup u_f(\lambda_m) = f(e(\lambda_1)) \cup \cdots \cup f(e(\lambda_m))$$

by Proposition 2.5 and the obvious extension of Proposition 1.2 to formal power series, so that $u_f(\xi)$ is of the form $\sum_{n \geqq 0} P_n(u_1(\xi), \ldots, u_n(\xi))$, where $u_n(\xi)$ is the nth elementary symmetric function in $e(\lambda_1), \ldots, e(\lambda_m)$ for $n \leqq m$, and where $u_n(\xi) = 0$ for $n > m$. However, Corollary 2.7 guarantees that $w_n(\xi)$ is the nth elementary symmetric function in $e(\lambda_1), \ldots, e(\lambda_m)$ for $n \leqq m$.

There is another useful version of the Whitney product formula, expressed in terms of cross products rather than cup products. Recall that if $X_1 \times X_2 \xrightarrow{\text{pr}_1} X_1$ and $X_1 \times X_2 \xrightarrow{\text{pr}_2} X_2$ are the first and second projections of a product $X_1 \times X_2$ of topological spaces X_1 and X_2, the cross product of any $\alpha \in H^r(X_1)$ and any $\beta \in H^s(X_2)$ is a cup product: $\alpha \times \beta = \text{pr}_1^*\alpha \cup \text{pr}_2^*\beta \in H^{r+s}(X_1 \times X_2)$. Recall also that the category $\mathscr{B}$ is closed with respect to products (Proposition I.1.4), and that if ξ_1 and ξ_2 are real vector bundles over $X_1 \in \mathscr{B}$ and $X_2 \in \mathscr{B}$, respectively, then the bundle $\xi_1 + \xi_2$ over $X_1 \times X_2 \in \mathscr{B}$ is the Whitney sum $\text{pr}_1^!\xi_1 \oplus \text{pr}_2^!\xi_2$ of the pullbacks $\text{pr}_1^!\xi_1$ and $\text{pr}_2^!\xi_2$ over $X_1 \times X_2$.

2.11 Proposition (Whitney Product Formula): *For any formal power series $f(t) \in \mathbb{Z}/2[[t]]$ with leading term $1 \in \mathbb{Z}/2$, and for any real vector bundles ξ_1 and ξ_2 over $X_1 \in \mathscr{B}$ and $X_2 \in \mathscr{B}$, respectively, the sum $\xi_1 + \xi_2$ over $X_1 \times X_2 \in \mathscr{B}$ satisfies*

$$u_f(\xi_1 + \xi_2) = u_f(\xi_1) \times u_f(\xi_2) \in H^{**}(X_1 \times X_2).$$

PROOF: Replace ξ, ξ' and X in Proposition 2.5 by $\text{pr}_1^!\xi_1$, $\text{pr}_2^!\xi_2$, and $X_1 \times X_2$, respectively. Then Propositions 2.5 and 1.5 imply

$$\begin{aligned} u_f(\xi_1 + \xi_2) &= u_f(\text{pr}_1^!\xi_1 \oplus pr_2^!\xi_2) \\ &= u_f(\text{pr}_1^!\xi_1) \cup u_f(\text{pr}_2^!\xi_2) \\ &= \text{pr}_1^* u_f(\xi_1) \cup \text{pr}_2^* u_f(\xi_2) \\ &= u_f(\xi_1) \times u_f(\xi_2), \end{aligned}$$

as claimed.

3. Orientability

First Stiefel–Whitney classes provide a criterion for the orientability of arbitrary real vector bundles; it is essentially a generalization of Proposition IV.7.9.

3.1 Lemma: *Let $\bigwedge^m \xi$ be the mth exterior power of a real m-plane bundle ξ over a base space $X \in \mathscr{B}$; then $e(\bigwedge^m \xi) = w_1(\xi)$ for the $\mathbb{Z}/2$ Euler class $e(\bigwedge^m \xi) \in H^1(X)$ and the first Stiefel–Whitney class $w_1(\xi) \in H^1(X)$.*

PROOF: Let $X' \xrightarrow{f} X$ be a splitting map for ξ, so that $f^!\xi$ is a Whitney sum $\lambda_1 \oplus \cdots \oplus \lambda_m$ of real line bundles over X' and $H^*(X) \xrightarrow{f^*} H^*(X')$ is a monomorphism. It suffices to show that $f^*e(\bigwedge^m \xi) = f^*w_1(\xi) \in H^1(X')$. On the one hand, $f^*e(\bigwedge^m \xi) = e(f^! \bigwedge^m \xi) = e(\bigwedge^m f^!\xi)$ by naturality of $\mathbb{Z}/2$ Euler classes and naturality of exterior powers, and since $f^!\xi = \lambda_1 \oplus \cdots \oplus \lambda_m$, Lemma IV.7.8 and Proposition IV.7.5 give

$$\begin{aligned} f^*e(\textstyle\bigwedge^m \xi) &= e(\textstyle\bigwedge^m(f^!\xi)) = e(\lambda_1 \otimes \cdots \otimes \lambda_m) \\ &= e(\lambda_1) + \cdots + e(\lambda_m) \in H^1(X'). \end{aligned}$$

On the other hand,

$$\begin{aligned} f^*w(\xi) &= w(f^!\xi) = w(\lambda_1 \oplus \cdots \oplus \lambda_m) = w(\lambda_1) \cup \cdots \cup w(\lambda_m) \\ &= (1 + e(\lambda_1)) \cup \cdots \cup (1 + e(\lambda_m)) \in H^*(X') \end{aligned}$$

by naturality and the Whitney product formula for (total) Stiefel–Whitney classes, and by Proposition 1.2. In $H^1(X')$ this implies $f^*w_1(\xi) = e(\lambda_1) + \cdots + e(\lambda_m) = f^*e(\bigwedge^m\xi)$, and since f^* is a monomorphism one has $w_1(\xi) = e(\bigwedge^m \xi) \in H^1(X)$, as claimed.

3.2 Proposition: *A real m-plane bundle ξ over a base space $X \in \mathscr{B}$ is orientable if and only if its first Stiefel–Whitney class satisfies $w_1(\xi) = 0 \in H^1(X)$.*

PROOF: By Definition III.10.4, ξ is orientable if and only if $\bigwedge^m \xi$ is the trivial line bundle over X, and by Proposition IV.7.7 one has $\bigwedge^m \xi = \varepsilon^1 \in \Gamma(X)$ if and only if the $\mathbb{Z}/2$ Euler class satisfies $e(\bigwedge^m \xi) = 0 \in H^1(X)$. Hence the result follows from Lemma 3.1.

4. The Rings $H^*(G^m(\mathbb{R}^\infty); \mathbb{Z}/2)$

We now show that the $\mathbb{Z}/2$ cohomology ring $H^*(G^m(\mathbb{R}^\infty))$ of the Grassmann manifold $G^m(\mathbb{R}^\infty)$ is the polynomial algebra generated over $\mathbb{Z}/2$ by the Stiefel–Whitney classes $w_1(\gamma^m), \ldots, w_m(\gamma^m)$ of the universal bundle γ^m over $G^m(\mathbb{R}^\infty)$; similarly, for any $q > 0$ and sufficiently large $n > 0$, the portion of the ring $H^*(G^m(\mathbb{R}^{m+n}))$ in degrees $0, \ldots, q$ is isomorphic to the corresponding portion of $\mathbb{Z}/2[w_1(\gamma_n^m), \ldots, w_m(\gamma_n^m)]$.

In the next lemmas $(RP^\infty)^m$ is the m-fold product $RP^\infty \times \cdots \times RP^\infty$ of copies of the real projective space RP^∞, and $\gamma^1 + \cdots + \gamma^1$ is the corresponding sum of m copies of the universal real line bundle over RP^∞, as in Definition III.2.1; that is, $\gamma^1 + \cdots + \gamma^1$ is the Whitney sum $\mathrm{pr}_1^!\gamma^1 \oplus \cdots \oplus \mathrm{pr}_m^!\gamma^1$, where $(RP^\infty)^m \xrightarrow{\mathrm{pr}_j} RP^\infty$ is the jth projection map, for $j = 1, \ldots, m$. We recall from Proposition IV.5.7 that if $(RP^\infty)^m \xrightarrow{h} G^m(\mathbb{R}^\infty)$ classifies $\gamma^1 + \cdots + \gamma^1$ then h is also a splitting map for the universal real m-plane bundle γ^m over $G^m(\mathbb{R}^\infty)$.

According to Proposition IV.4.3 the $\mathbb{Z}/2$ cohomology ring $H^*(RP^\infty)$ is the polynomial ring over $\mathbb{Z}/2$ generated by the $\mathbb{Z}/2$ Euler class $e(\gamma^1) \in H^1(RP^\infty)$. Hence the Künneth theorem guarantees that the cross product $H^*(RP^\infty) \otimes \cdots \otimes H^*(RP^\infty) \xrightarrow{\times} H^*((RP^\infty)^m)$ is an isomorphism, so that $H^*((RP^\infty)^m)$ is the polynomial ring over $\mathbb{Z}/2$ generated by the elements $p_1^*e(\gamma^1), \ldots, p_m^*e(\gamma^1) \in H^1((RP^\infty)^m)$.

4.1 Lemma: *Let $(RP^\infty)^m \xrightarrow{h} G^m(\mathbb{R}^\infty)$ classify the bundle $\gamma^1 + \cdots + \gamma^1$ over $(RP^\infty)^m$; then the image of $H^*(G^m(\mathbb{R}^\infty)) \xrightarrow{h^*} H^*((RP^\infty)^m)$ lies in the ring of symmetric functions in the elements* $\mathrm{pr}_1^*e(\gamma^1), \ldots, \mathrm{pr}_m^*e(\gamma^1) \in H^1((RP^\infty)^m)$.

PROOF: Let $(RP^\infty)^m \xrightarrow{\pi} (RP^\infty)^m$ be any permutation of the m factors in $RP^\infty \times \cdots \times RP^\infty$. Then the composition $(RP^\infty)^m \xrightarrow{\pi} (RP^\infty)^m \xrightarrow{h} G^m(\mathbb{R}^\infty)$ also classifies $\gamma^1 + \cdots + \gamma^1$, so that the homotopy classification theorem provides a commutative triangle

$$\begin{array}{ccc} H^*(G^m(\mathbb{R}^\infty)) & \xrightarrow{\quad h^* \quad} & H^*((RP^\infty)^m) \\ & {\scriptstyle h^*}\searrow \qquad \nearrow{\scriptstyle \pi^*} & \\ & H^*((RP^\infty)^m). & \end{array}$$

In particular, the image of h^* is invariant under π^*.

4.2 Lemma: *Let $(RP^\infty)^m \xrightarrow{h} G^m(\mathbb{R}^\infty)$ classify the bundle $\gamma^1 + \cdots + \gamma^1$ over $(RP^\infty)^m$; then the image of $H^*(G^m(\mathbb{R}^\infty)) \xrightarrow{h^*} H^*((RP^\infty)^m)$ contains the ring of symmetric functions in the elements* $\mathrm{pr}_1^*e(\gamma^1), \ldots, \mathrm{pr}_m^*e(\gamma^1) \in H^1((RP^\infty)^m)$.

PROOF: For each $i = 1, \ldots, m$ let $w_i(\gamma^m) \in H^i(G^m(\mathbb{R}^\infty))$ be the ith Stiefel–Whitney class of the universal real m-plane bundle γ^m. Then

$$\begin{aligned} h^*w_i(\gamma^m) &= w_i(h^!\gamma^m) = w_i(\gamma^1 + \cdots + \gamma^1) \\ &= w_i(\mathrm{pr}_1^!\gamma^1 \oplus \cdots \oplus \mathrm{pr}_m^!\gamma^1) \end{aligned}$$

by naturality of Stiefel–Whitney classes, where $w_i(\mathrm{pr}_1^!\gamma^1 \oplus \cdots \oplus \mathrm{pr}_m^!\gamma^1)$ is the ith elementary symmetric function $\sigma_i(e(\mathrm{pr}_1^!\gamma^1), \ldots, e(\mathrm{pr}_m^!\gamma^1))$ by Corollary 2.7, and where $e(\mathrm{pr}_j^!\gamma^1) = \mathrm{pr}_j^*e(\gamma^1)$ for $j = 1, \ldots, m$ by naturality of $\mathbb{Z}/2$ Euler classes. It remains to recall that the ring of symmetric functions in a polynomial ring in m variables is generated by the m elementary symmetric functions.

4.3 Proposition: *For any $m > 0$ the $\mathbb{Z}/2$ cohomology ring $H^*(G^m(\mathbb{R}^\infty))$ of the Grassmann manifold $G^m(\mathbb{R}^\infty)$ is the polynomial ring $\mathbb{Z}/2[w_1(\gamma^m), \ldots, w_m(\gamma^m)]$ generated over $\mathbb{Z}/2$ by the Stiefel–Whitney classes $w_i(\gamma^m) \in H^i(G^m(\mathbb{R}^\infty))$ of the universal m-plane bundle γ^m over $G^m(\mathbb{R}^\infty)$ for $i = 1, \ldots, m$.*

PROOF: According to Proposition IV.5.7, if $(RP^\infty)^m \xrightarrow{h} G^m(\mathbb{R}^\infty)$ classifies $\gamma^1 + \cdots + \gamma^1$ then h is also a splitting map for γ^m. In particular, one can use the induced monomorphism $H^*(G^m(\mathbb{R}^\infty)) \xrightarrow{h^*} H^*((RP^\infty)^m)$ to identify $H^*(G^m(\mathbb{R}^\infty))$ as a subring $H^*_{\text{inv}}((RP^\infty)^m) \subset H^*((RP^\infty)^m)$, and Lemmas 4.1 and 4.2 imply that $H^*_{\text{inv}}((RP^\infty)^m)$ is precisely the subring generated by the elementary symmetric functions $\sigma_i(\text{pr}_1^* e(\gamma^1), \ldots, \text{pr}_m^* e(\gamma^1))$, for $i = 1, \ldots, m$. However, $\sigma_i(\text{pr}_1^* e(\gamma^1), \ldots, \text{pr}_m^* e(\gamma^1)) = h^* w_i(\gamma^m)$, as in the proof of Lemma 4.2, and since h^* is a monomorphism, $H^*(G^m(\mathbb{R}^\infty))$ is therefore the polynomial ring over $\mathbb{Z}/2$ generated by $w_1(\gamma^m), \ldots, w_m(\gamma^m)$, as claimed.

Proposition 4.3 provides an elegant alternative characterization of the Stiefel–Whitney classes of any real vector bundle whatsoever. One temporarily ignores the provenance of the generators $w_1(\gamma^m), \ldots, w_m(\gamma^m)$ of $H^*(G^m(\mathbb{R}^\infty))$, merely identifying $H^*(G^m(\mathbb{R}^\infty))$ as a polynomial ring $\mathbb{Z}/2[w_1, \ldots, w_m]$ over $\mathbb{Z}/2$ with one generator $w_i \in H^i(G^m(\mathbb{R}^\infty))$ in each degree $i = 1, \ldots, m$. Then if $X \xrightarrow{f} G^m(\mathbb{R}^\infty)$ classifies a given m-plane bundle ξ over a base space $X \in \mathscr{B}$, naturality of Stiefel–Whitney classes yields $w_i(\xi) = w_i(f^!\gamma^m) = f^* w_i(\gamma^m) = f^* w_i$. The abbreviated result $w_i(\xi) = f^* w_i$ can be regarded as a *definition* of the ith Stiefel–Whitney class $w_i(\xi) \in H^i(X)$ of ξ.

The computation $H^*(G^m(\mathbb{R}^\infty)) = \mathbb{Z}/2[w_1, \ldots, w_m]$ of Proposition 4.3 has further importance: it shows that *every* $\mathbb{Z}/2$ characteristic class of *any* m-plane bundle ξ over *any* $X \in \mathscr{B}$ can be computed from the Stiefel–Whitney classes $w_1(\xi), \ldots, w_m(\xi)$. For if a characteristic class $u_q(\xi) \in H^q(X)$ satisfies the naturality condition, and if $X \xrightarrow{f} G^m(\mathbb{R}^\infty)$ classifies ξ, then $u_q(\xi) = u_q(f^!\gamma^m) = f^* u_q(\gamma^m)$, where $u_q(\gamma^m) \in H^q(G^m(\mathbb{R}^\infty))$ is necessarily a polynomial $v_q(w_1, \ldots, w_m)$ in $w_1, \ldots, w_m$; hence

$$\begin{aligned} u_q(\xi) &= f^* v_q(w_1, \ldots, w_m) = v_q(f^* w_1, \ldots, f^* w_m) \\ &= v_q(w_1(f^!\gamma^m), \ldots, w_n(f^!\gamma^m)) = v_q(w_1(\xi), \ldots, w_m(\xi)). \end{aligned}$$

For example, we already know from Corollary 2.8 that the $\mathbb{Z}/2$ Euler class is given by $e(\xi) = w_m(\xi) \in H^m(X)$, and Proposition 2.10 gives a specific prescription $u_f(\xi) = \sum_n P_n(w_1(\xi), \ldots, w_n(\xi)) \in H^{**}(X)$ for any $f(t) \in \mathbb{Z}/2[[t]]$ with leading term $1 \in \mathbb{Z}/2$, where $w_n(\xi) = 0$ for $n > m$.

Proposition 4.3 can also be proved in other ways. One of the nicest alternatives is an induction which uses $H^*(G^{m-1}(\mathbb{R}^\infty))$ to compute $H^*(G^m(\mathbb{R}^\infty))$. The case $m = 1$ is the identity $H^*(G^1(\mathbb{R}^\infty)) = H^*(RP^\infty) = \mathbb{Z}/2[e(\gamma^1)]$ of Proposition IV.4.3, which was obtained from a Gysin sequence. The following key to the inductive step is also obtained from a Gysin sequence.

4.4 Proposition: *For any $m > 1$ let γ^{m-1} and γ^m be the universal real vector bundles over $G^{m-1}(\mathbb{R}^\infty)$ and $G^m(\mathbb{R}^\infty)$, respectively, let $\cup e(\gamma^m)$ be cup product by the $\mathbb{Z}/2$ Euler class $e(\gamma^m) \in H(G^m(\mathbb{R}^\infty))$, and let $G^{m-1}(\mathbb{R}^\infty) \xrightarrow{f} G^m(\mathbb{R}^\infty)$ classify the real m-plane bundle $\gamma^{m-1} \oplus \varepsilon^1$ over $G^{m-1}(\mathbb{R}^\infty)$. Then there are $\mathbb{Z}/2$-module homomorphisms g for which there is an exact sequence*

$$\cdots \xrightarrow{g} H^q(G^m(\mathbb{R}^\infty)) \xrightarrow{\cup e(\gamma^m)} H^{q+m}(G^m(\mathbb{R}^\infty))$$
$$\xrightarrow{f^*} H^{q+m}(G^{m-1}(\mathbb{R}^\infty)) \xrightarrow{g} H^{q+1}(G^m(\mathbb{R}^\infty)) \to \cdots$$

of $\mathbb{Z}/2$-modules.

PROOF: According to Proposition III.8.10, if $E^\infty \xrightarrow{\pi^\infty} G^m(\mathbb{R}^\infty)$ represents γ^m, then there is a homotopy equivalence $G^{m-1}(\mathbb{R}^\infty) \xrightarrow{h} E^{\infty *}$ such that the composition $G^{m-1}(\mathbb{R}^\infty) \xrightarrow{h} E^{\infty *} \xrightarrow{\pi^\infty} G^m(\mathbb{R}^\infty)$ is a classifying map for $\gamma^{m-1} \oplus \varepsilon^1$. Since classifying maps are unique up to homotopy, we automatically have $f^* = h^* \circ \pi^{\infty *}$ in the following diagram, whose top line is the Gysin sequence for γ^m, as in Proposition IV.4.2. Since h^* is an isomorphism, one can set $g = \Psi_{\gamma^m} \circ (h^*)^{-1}$ to complete the proof:

$$\cdots \xrightarrow{\cup e(\gamma^m)} H^{q+m}(G^m(\mathbb{R}^\infty)) \xrightarrow{\pi^{\infty *}} H^{q+m}(E^{\infty *}) \xrightarrow{\Psi_{\gamma^m}} H^{q+1}(G^m(\mathbb{R}^\infty)) \longrightarrow \cdots$$
$$f^* \searrow \qquad h^* \downarrow \approx \qquad \nearrow g$$
$$H^{q+m}(G^{m-1}(\mathbb{R}^\infty)).$$

We omit the application of Proposition 4.4 as an inductive step in the computation of the rings $H^*(G^m(\mathbb{R}^\infty))$; it is an easy exercise, which appears with a brief hint as Exercise 7.4.

Here is a finite-dimensional version of Proposition 4.3.

4.5 Proposition: *For any $m > 0$ and any $q > 0$ there is an $N(m, q) > 0$ such that for any $n \geqq N(m, q)$ the $\mathbb{Z}/2$ cohomology ring $H^*(G^m(\mathbb{R}^{m+n}))$ agrees in dimensions $0, 1, \ldots, q$ with the polynomial ring $\mathbb{Z}/2[w_1(\gamma_n^m), \ldots, w_m(\gamma_n^m)]$ generated over $\mathbb{Z}/2$ by the Stiefel–Whitney classes $w_r(\gamma_n^m) \in H^r(G^m(\mathbb{R}^{m+n}))$ of the canonical real m-plane bundle γ_n^m over $G^m(\mathbb{R}^{m+n})$.*

PROOF: Let γ_q^1 be the canonical real line bundle over the projective space RP^q, and let $\gamma_q^1 + \cdots + \gamma_q^1$ be the sum of m copies of γ_q^1 over the m-fold product

$RP^q \times \cdots \times RP^q = (RP^q)^m$; that is, $\gamma_q^1 + \cdots + \gamma_q^1 = \mathrm{pr}_1^! \gamma_q^1 \oplus \cdots \oplus \mathrm{pr}_m^! \gamma_q^1$ for the m projections $(RP^q)^m \xrightarrow{\mathrm{pr}_j} RP^q$, as usual. Since $(RP^q)^m$ is a smooth qm-dimensional manifold, by Proposition I.7.3, it follows from Proposition III.9.3 that there is an $N(m, q) > 0$ for which there is a "finite classifying map" $(RP^q)^m \xrightarrow{k} G^m(\mathbb{R}^{m+n})$ for $\gamma_q^1 + \cdots + \gamma_q^1$ whenever $n \geqq N(m, q)$; that is, $\gamma_q^1 + \cdots + \gamma_q^1 = k^! \gamma_n^m$ for $n \geqq N(m, q)$. If $G^m(\mathbb{R}^{m+n}) \xrightarrow{j} G^m(\mathbb{R}^\infty)$ classifies γ_n^m (in the usual sense), then the composition $j \circ k$ classifies $\gamma_q^1 + \cdots + \gamma_q^1$. Now let $RP^q \xrightarrow{i} RP^\infty$ classify the real line bundle γ_q^1 over RP^q, so that the m-fold product $(RP^q)^m \xrightarrow{i^m} (RP^\infty)^m$ satisfies $\gamma_q^1 + \cdots + \gamma_q^1 = (i^m)^!(\gamma^1 + \cdots + \gamma^1)$; if $(RP^\infty)^m \xrightarrow{h} G^m(\mathbb{R}^\infty)$ classifies $\gamma^1 + \cdots + \gamma^1$, then the composition $h \circ i^m$ is another classifying map for $\gamma_q^1 + \cdots + \gamma_q^1$. By the homotopy classification theorem (Theorem III.8.9) the two classifying maps $j \circ k$ and $h \circ i^m$ for $\gamma_q^1 + \cdots + \gamma_q^1$ are homotopic, so that there is a commutative diagram

$$\begin{array}{ccc} H^*(G^m(R^\infty)) & \xrightarrow{j^*} & H^*(G^m(R^{m+n})) \\ \downarrow{\scriptstyle h^*} & & \downarrow{\scriptstyle k^*} \\ H^*((RP^\infty)^m) & \xrightarrow{(i^m)^*} & H^*((RP^q)^m) \end{array}$$

of $\mathbb{Z}/2$-modules. We already know for Lemmas 4.1 and 4.2 that h^* is a monomorphism whose image $H^*_{\mathrm{inv}}((RP^\infty)^m) \subset H^*((RP^\infty)^m)$ consists of those elements invariant under the automorphisms induced by permutations of the factors RP^∞, as in the proof of Proposition 4.3. By Corollary IV.4.5 $H^p(RP^\infty) \xrightarrow{i^*} H^p(RP^q)$ is an isomorphism for $p \leqq q$, so that $(i^m)^*$ induces isomorphisms $H^p_{\mathrm{inv}}((RP^\infty)^m) \xrightarrow{(i^m)^*} H^p_{\mathrm{inv}}((RP^q)^m)$ for $p \leqq q$, for the corresponding sub-ring $H^*_{\mathrm{inv}}((RP^q)^m) \subset H^*((RP^q)^m)$. Hence for each $p \leqq q$ there is a commutative diagram

$$\begin{array}{ccc} H^p(G^m(R^\infty)) & \xrightarrow{j^*} & H^p(G^m(R^{m+n})) \\ \downarrow{\scriptstyle h^*\ \approx} & & \downarrow{\scriptstyle k^*} \\ H^p_{\mathrm{inv}}((RP^\infty)^m) & \underset{\approx}{\xrightarrow{(i^m)^*}} & H^p_{\mathrm{inv}}((RP^q)^m) \end{array}$$

for any $n \geqq N(m, q)$, with isomorphisms h^* and $(i^m)^*$. Thus j^* is a monomorphism (and k^* is an epimorphism).

Now let $X \xrightarrow{f} G^m(\mathbb{R}^{m+n})$ be any splitting map for γ_n^m, so that $H^*(G(\mathbb{R}^{m+n})) \xrightarrow{f^*} H^*(X)$ is a monomorphism and $f^! \gamma_n^m$ is a sum $\lambda_1 \oplus \cdots \oplus \lambda_m$ of line bundles over X; the composition $j \circ f$ classifies $\lambda_1 \oplus \cdots \oplus \lambda_m$. If $X \xrightarrow{l_1} RP^\infty, \ldots, X \xrightarrow{l_m} RP^\infty$ classify $\lambda_1, \ldots, \lambda_m$, respectively, then the map

$X \xrightarrow{l=(l_1,\ldots,l_m)} (RP^\infty)^m$ induces a homomorphism $H^*((RP^\infty)^m) \xrightarrow{l^*} H^*(X)$; the composition $h \circ l$ also classifies $\lambda_1 \oplus \cdots \oplus \lambda_m$. Since $j \circ f$ and $h \circ l$ both classify $\lambda_1 \oplus \cdots \oplus \lambda_m$ the homotopy classification theorem (Theorem III.8.9) provides a homotopy commutative diagram

$$\begin{array}{ccc} G^m(\mathbb{R}^\infty) & \xleftarrow{\quad j \quad} & G^m(\mathbb{R}^{m+n}) \\ \uparrow{\scriptstyle h} & & \uparrow{\scriptstyle f} \\ (RP^\infty)^m & \xleftarrow{\quad l \quad} & X, \end{array}$$

hence a commutative diagram

$$\begin{array}{ccc} H^p(G^m(\mathbb{R}^\infty)) & \xrightarrow{\quad j^* \quad} & H^p(G^m(\mathbb{R}^{m+n})) \\ \downarrow{\scriptstyle h^*} & & \downarrow{\scriptstyle f^*} \\ H^p((RP^\infty)^m) & \xrightarrow{\quad l^* \quad} & H^p(X), \end{array}$$

whenever $p \leqq q$ and $n \geqq N(m, q)$. We already know that h^* is a monomorphism with image $H^*_{\text{inv}}((RP^\infty)^m)$, and that j^* and f^* are monomorphisms. Consequently the preceding diagram reduces to

$$\begin{array}{ccc} H^p(G^m(\mathbb{R}^\infty)) & \xrightarrow[\text{mono}]{\quad j^* \quad} & H^p(G^m(\mathbb{R}^{m+n})) \\ \downarrow{\scriptstyle h^*\,\approx} & & {\scriptstyle \text{mono}}\downarrow{\scriptstyle f^*} \\ H^p_{\text{inv}}((RP^\infty)^m) & \xrightarrow{\quad l^* \quad} & H^p(X), \end{array}$$

where h^* is an isomorphism and the composition $f^* \circ j^*$ is a monomorphism. It follows that l^* is a monomorphism, which becomes an isomorphism in the further reduced diagram

$$\begin{array}{ccc} H^p(G^m(\mathbb{R}^\infty)) & \xrightarrow{\quad j^* \quad} & H^p(G^m(\mathbb{R}^{m+n})) \\ \downarrow{\scriptstyle h^*\,\approx} & & {\scriptstyle \text{mono}}\downarrow{\scriptstyle f^*} \\ H^p_{\text{inv}}((RP^\infty)^m) & \xrightarrow[\approx]{\quad l^* \quad} & \operatorname{Im} l^*. \end{array}$$

Commutativity of the latter diagram guarantees that the monomorphism f^* is also an epimorphism, hence an isomorphism; consequently j^* is also an isomorphism, assuming as always that $p \leqq q$ and $n \geqq N(m,q)$. Since $j^*w_r(\gamma^m) = w_r(j^!\gamma^m) = w_r(\gamma_n^m) \in H^r(G^m(\mathbb{R}^{m+n}))$ for $r = 1, \ldots, m$, Proposition 4.3 implies the desired result.

In the preceding proof we appealed to Proposition III.9.3 for the existence of a "finite classifying map" $(RP^q)^m \xrightarrow{k} G^m(\mathbb{R}^{m+n})$ for the bundle $\gamma_q^1 + \cdots + \gamma_q^1$ over $(RP^q)^m$, for sufficiently large $n > 0$. The following property of such maps is a finite-dimensional version of Proposition IV.5.7.

4.6 Proposition: *Given $q > 0$, any "finite classifying map" $(RP^q)^m \xrightarrow{k} G^m(\mathbb{R}^{m+n})$ for the real m-plane bundle $\gamma_q^1 + \cdots + \gamma_q^1$ over $(RP^q)^m$ is also a splitting map for the canonical real m-plane bundle γ_n^m over $G^m(\mathbb{R}^{m+n})$, in the restricted sense that $H^p(G^m(\mathbb{R}^{m+n})) \xrightarrow{k^*} H^p((RP^q)^m)$ is monic for $p \leqq q$ and $n \geqq N(m,q)$.*

PROOF: The preceding proof contains a commutative diagram

$$\begin{array}{ccc} H^p(G^m(\mathbb{R}^\infty)) & \xrightarrow{j^*} & H^p(G^m(\mathbb{R}^{m+n})) \\ {\scriptstyle h^*}\downarrow{\scriptstyle \approx} & & \downarrow{\scriptstyle k^*} \\ H^p_{\text{inv}}((RP^\infty)^m) & \underset{\approx}{\xrightarrow{(i^m)^*}} & H^p_{\text{inv}}((RP^q)^m) \end{array}$$

with isomorphisms h^* and $(i^m)^*$, and we later verified that j^* is also an isomorphism. Hence k^* is an isomorphism onto the submodule $H^p_{\text{inv}}((RP^q)^m) \subset H^p((RP^q)^m)$, for $p \leqq q$ and $n \geqq N(m,q)$.

5. Axioms for Stiefel–Whitney Classes

We now establish an axiomatic characterization of Stiefel–Whitney classes of real vector bundles over arbitrary base spaces $X \in \mathscr{B}$, following the pattern used in Theorem IV.6.1 for $\mathbb{Z}/2$ Euler classes. Even if one considers only smooth real vector bundles over smooth manifolds, the corresponding axioms uniquely describe Stiefel–Whitney classes, as we also show.

5.1 Theorem (Axioms for Stiefel–Whitney Classes, for the Category $\mathscr{B}$): *For real vector bundles ξ over base spaces $X \in \mathscr{B}$, there are unique inhomogeneous $\mathbb{Z}/2$ cohomology classes $w(\xi) \in H^*(X)$ which satisfy the following axioms:*

(0) **Dimension:** *If ξ is a real m-plane bundle over $X \in \mathscr{B}$, then $w(\xi) \in H^0(X) \oplus \cdots \oplus H^m(X) \subset H^*(X)$.*

(1) **Naturality:** *If* $X' \xrightarrow{g} X$ *is a map in* $\mathscr{B}$, *and if* ξ *is a real vector bundle over* X, *then* $w(g^!\xi) = g^*w(\xi) \in H^*(X')$.

(2) **Whitney product formula:** *If* ξ *and* η *are real vector bundles over the same* $X \in \mathscr{B}$, *with Whitney sum* $\xi \oplus \eta$ *over* X, *then* $w(\xi) \cup w(\eta) = w(\xi \oplus \eta) \in H^*(X)$, *for the cup product* $w(\xi) \cup w(\eta)$.

(3) **Normalization:** *If* γ_1^1 *is the canonical real line bundle over* RP^1 $(=S^1)$, *then* $w(\gamma_1^1) = 1 + e(\gamma_1^1) \in H^0(RP^1) \oplus H^1(RP^1)$, *where* $e(\gamma_1^1)$ *is the generator of* $H^*(RP^1)$.

PROOF: The total Stiefel–Whitney classes of Definition 1.1 trivially satisfy Axiom (0), and they satisfy Axioms (1)–(3) by virtue of Propositions 1.3, 2.5, and 1.2, respectively. Conversely, suppose that $w(\;)$ satisfies Axioms (0)–(3), let γ^1 be the universal real line bundle over RP^∞, and let $RP^1 \xrightarrow{f} RP^\infty$ classify the canonical line bundle γ_1^1 over RP^1. Then $w(\gamma_1^1) \in H^0(RP^1) \oplus H^1(RP^1)$ by Axiom (0), and $H^0(RP^\infty) \oplus H^1(RP^\infty) \xrightarrow{f^*} H^0(RP^1) \oplus H^1(RP^1)$ is an isomorphism such that $f^*(1 + e(\gamma^1)) = 1 + e(\gamma_1^1)$ for the generator $e(\gamma^1) \in H^1(RP^\infty)$, by Corollary IV.4.5. Since one has $f^*w(\gamma^1) = w(f^!\gamma^1) = w(\gamma_1^1) = 1 + e(\gamma_1^1)$ by Axioms (1) and (3), for the same isomorphism f^*, it follows that $w(\gamma^1) = 1 + e(\gamma^1)$. Now let γ^m be the universal real m-plane bundle over $G^m(\mathbb{R}^\infty)$ and let $(RP^\infty)^m \xrightarrow{h} G^m(\mathbb{R}^\infty)$ be the splitting map for γ^m described in Proposition IV.5.7, with

$$h^!\gamma^m = \gamma^1 + \cdots + \gamma^1 = \mathrm{pr}_1^!\gamma^1 \oplus \cdots \oplus \mathrm{pr}_m^!\gamma^1$$

for the m projections $(RP^\infty)^m \xrightarrow{\mathrm{pr}_j} RP^\infty$. Then by Axioms (1) and (2) one has

$$\begin{aligned} h^*w(\gamma^m) &= w(h^!\gamma^m) = w(\mathrm{pr}_1^!\gamma^1 \oplus \cdots \oplus \mathrm{pr}_m^!\gamma^1) \\ &= w(\mathrm{pr}_1^!\gamma^1) \cup \cdots \cup w(\mathrm{pr}_m^!\gamma^1) \\ &= \mathrm{pr}_1^*w(\gamma^1) \cup \cdots \cup \mathrm{pr}_m^*w(\gamma^1), \end{aligned}$$

which is a unique element of $H^*((RP^\infty)^m)$ since $w(\gamma^1)$ is uniquely defined, as we have just learned. Since h^* is monic, by Proposition IV.5.7, it follows that $w(\gamma^m)$ is a unique element of $H^*(G^m(\mathbb{R}^\infty))$. By one final appeal to the homotopy classification theorem (Theorem III.8.9) *any* real m-plane bundle ξ over *any* $X \in \mathscr{B}$ can be classified by a map $X \xrightarrow{g} G^m(\mathbb{R}^\infty)$ which is unique up to homotopy, so that by Axiom (1) one has $w(\xi) = w(g^!\gamma^m) = g^*w(\gamma^m) \in H^*(X)$ for a unique homomorphism $H^*(G^m(\mathbb{R}^\infty)) \xrightarrow{g^*} H^*(X)$ and the unique element $w(\gamma^m) \in H^*(G^m(\mathbb{R}^\infty))$.

Now let $\mathscr{M}$ denote the category of smooth manifolds and smooth maps, as before. We know from Corollary I.6.8 that $\mathscr{M} \subset \mathscr{B}$, and we know from Theorem III.9.5 that every real vector bundle ξ over any $X \in \mathscr{M}$ is itself smooth, in the sense that it can be represented by a smooth coordinate bundle $E \xrightarrow{\pi} X$. Since many applications of characteristic classes concern only smooth manifolds $X \in \mathscr{M}$ and their tangent bundles $\tau(X)$, for example, it is of interest to know that the preceding characterization of Stiefel–Whitney classes applies equally well to the subcategory $\mathscr{M} \subset \mathscr{B}$.

5.2 Theorem (Axioms for Stiefel–Whitney Classes, for the Category $\mathscr{M}$): *For smooth real vector bundles ξ over smooth manifolds $X \in \mathscr{M}$, there are unique inhomogeneous $\mathbb{Z}/2$ cohomology classes $w(\xi) \in H^*(X)$ which satisfy the following axioms:*

(0) **Dimension:** *If ξ is a smooth real m-plane bundle over $X \in \mathscr{M}$, then*

$$w(\xi) \in H^0(X) \oplus \cdots \oplus H^m(X) \subset H^*(X).$$

(1) **Naturality:** *If $X' \xrightarrow{g} X$ is a smooth map, and if ξ is a smooth real vector bundle over X, then $w(g^!\xi) = g^*w(\xi) \in H^*(X')$.*

(2) **Whitney product formula:** *If ξ and η are smooth real vector bundles over the same smooth manifold $X \in \mathscr{M}$, with Whitney sum $\xi \oplus \eta$ over X, then $w(\xi) \cup w(\eta) = w(\xi \oplus \eta) \in H^*(X)$, for the cup product $w(\xi) \cup w(\eta)$.*

(3) **Normalization:** *If γ_1^1 is the canonical real line bundle over RP^1 $(=S^1)$, then $w(\gamma_1^1) = 1 + e(\gamma_1^1) \in H^0(RP^1) \oplus H^1(RP^1)$, where $e(\gamma_1^1)$ is the generator of $H^*(RP^1)$.*

PROOF: The existence of such classes follows automatically from the inclusion $\mathscr{M} \subset \mathscr{B}$ and Theorem 5.1. Conversely, suppose that $w(\ )$ satisfies Axioms (0)–(3). For any $q > 0$ one easily obtains $w(\gamma_q^1) = 1 + e(\gamma_q^1) \in H^0(RP^q) \oplus H^1(RP^q)$ from Corollary IV.4.5 and Axioms (0), (1), and (3), exactly as in the proof of Theorem 5.1, where γ_q^1 is the canonical real line bundle over RP^q, and where $e(\gamma_q^1) \in H^1(RP^q)$ is the uniquely defined generator of $H^*(RP^q)$. Now, for any $m > 0$, let $q = m$, so that there is a finite classifying map $(RP^m)^m \xrightarrow{k} G^m(\mathbb{R}^{m+n})$ for the real m-plane bundle $\gamma_m^1 + \cdots + \gamma_m^1$ over $(RP^m)^m$ whenever $n \geqq N(m, q) = N(m, m)$, as in Proposition 4.5; that is,

$$k^!\gamma_n^m = \gamma_m^1 + \cdots + \gamma_m^1 = \mathrm{pr}_1^!\gamma_m^1 \oplus \cdots \oplus \mathrm{pr}_m^!\gamma_m^1$$

for the canonical real m-plane bundle γ_n^m over $G^m(\mathbb{R}^{m+n})$. By Axioms (1) and (2) one has

$$\begin{aligned} k^*w(\gamma_n^m) &= w(k^!\gamma_n^m) = w(\mathrm{pr}_1^!\gamma_m^1 \oplus \cdots \oplus \mathrm{pr}_m^!\gamma_m^1) \\ &= w(\mathrm{pr}_1^!\gamma_m^1) \cup \cdots \cup w(\mathrm{pr}_m^!\gamma_m^1) \\ &= \mathrm{pr}_1^*w(\gamma_m^1) \cup \cdots \cup \mathrm{pr}_m^*w(\gamma_m^1), \end{aligned}$$

and since we already know that $w(\gamma_m^1) = 1 + e(\gamma_m^1)$, it follows that $k^*w(\gamma_n^m)$ is a uniquely defined element of $H^*((RP^m)^m)$. However, Axiom (0) guarantees that $w(\gamma_n^m)$ vanishes in degrees above m, and since $q = m$, Proposition 4.6

guarantees that $H^p(G^m(\mathbb{R}^{m+n})) \xrightarrow{k^*} H^p((RP^m)^m)$ is monic for $p \leqq m$; hence $w(\gamma_n^m)$ is uniquely defined in $H^*(G^m(\mathbb{R}^{m+n}))$, for any $n \geqq N(m,m)$. Finally, if ξ is any smooth real m-plane bundle over any $X \in \mathcal{M}$, then Proposition III.9.3 provides a finite classifying map $X \xrightarrow{f} G^m(\mathbb{R}^{m+n})$ for sufficiently large n. By Theorem I.6.19 one can assume that f is itself smooth; furthermore, one may as well suppose $n \geqq N(m,m)$, so that Axiom (1) implies $w(\xi) = w(f^!\gamma_n^m) = f^*w(\gamma_n^m) \in H^*(X)$ for the preceding class $w(\gamma_n^m) \in H^*(G^m(\mathbb{R}^{m+n}))$.

The final step of the preceding proof clearly does not depend on the choice of the finite classifying map f; for if $w(\ )$ and $\tilde{w}(\ )$ both satisfy the axioms, the earlier steps of the proof give $w(\gamma_n^m) = \tilde{w}(\gamma_n^m) \in H^*(G^m(\mathbb{R}^{m+n}))$, hence

$$w(\xi) = w(f^!\gamma_n^m) = f^*w(\gamma_n^m) = f^*\tilde{w}(\gamma_n^m) = \tilde{w}(f^!\gamma_n^m) = \tilde{w}(\xi) \in H^*(X).$$

However, there is also an explicit demonstration that the choice of f is immaterial. Suppose that $X \xrightarrow{g} G^m(\mathbb{R}^{m+n'})$ is another smooth finite classifying map for the smooth real m-plane bundle ξ over $X \in \mathcal{M}$, for some $n' \geqq N(m,m)$. The ersatz homotopy uniqueness theorem (Proposition III.8.14) then provides (smooth) finite classifying extensions $g_{n,n''}$ and $g_{n',n''}$ for which the compositions

$$X \xrightarrow{f} G^m(\mathbb{R}^{m+n}) \xrightarrow{g_{n,n''}} G^m(\mathbb{R}^{m+n''}) \quad \text{and} \quad X \xrightarrow{g} G^m(\mathbb{R}^{m+n'}) \xrightarrow{g_{n',n''}} G^m(\mathbb{R}^{m+n''})$$

are homotopic. The axioms then imply

$$\begin{aligned} f^*w(\gamma_n^m) &= f^*w(g_{n,n''}^!\gamma_{n''}^m) = f^*g_{n,n''}^*w(\gamma_{n''}^m) \\ &= (g_{n,n''} \circ f)^*w(\gamma_{n''}^m) = (g_{n',n''} \circ g)^*w(\gamma_{n''}^m) \\ &= g^*g_{n',n''}^*w(\gamma_{n''}^m) = g^*w(g_{n',n''}^!\gamma_{n''}^m) \\ &= g^*w(\gamma_{n'}^m) \end{aligned}$$

as expected.

Since the Grassmann manifolds $G^m(\mathbb{R}^{m+n})$ are both closed and smooth (Proposition I.7.3), they are a fortiori compact and triangulable. One therefore has analogs of Theorem 5.2 in which the category $\mathcal{M}$ of smooth manifolds is replaced by any one of several categories of topological spaces. For example, Theorem 5.2 is valid for real vector bundles over spaces in the category of smooth *closed* manifolds. Theorem 5.2 is also valid if one replaces $\mathcal{M}$ by the category of those compact spaces which happen to lie in $\mathcal{B}$, or by the category of finite-dimensional simplicial spaces (which automatically lie in $\mathcal{B}$ by Corollary I.2.2 and Proposition I.4.6). In each case one of Propositions III.8.12, III.9.1, or III.9.3 furnishes the required finite classifying maps.

6. Dual Classes

Since $\mathbb{Z}/2$ is a field, and since we work exclusively with formal power series $f(t) \in \mathbb{Z}/2[[t]]$ with leading term $1 \in \mathbb{Z}/2$, it follows for each such $f(t)$ that there is a unique formal power series $(1/f)(t) \in \mathbb{Z}/2[[t]]$ with leading term $1 \in \mathbb{Z}/2$ for which $f(t) \cdot (1/f)(t) = 1 \in \mathbb{Z}/2[[t]]$. For example, if $f(t) = 1 + t$, then $(1/f)(t) = 1 + t + t^2 + \cdots$.

6.1 Definition: For any formal power series $f(t) \in \mathbb{Z}/2[[t]]$ with leading term $1 \in \mathbb{Z}/2$, and for any real vector bundle ξ over a base space $X \in \mathscr{B}$, the multiplicative $\mathbb{Z}/2$ class $u_{1/f}(\xi) \in H^{**}(X)$ is the *dual class* of the multiplicative $\mathbb{Z}/2$ class $u_f(\xi) \in H^{**}(X)$. In particular, for $f(t) = 1 + t$ and $(1/f)(t) = 1 + t + t^2 + \cdots$, the class $u_{1/f}(\xi) \in H^{**}(X)$ is the *dual Stiefel–Whitney class* of ξ, denoted $\bar{w}(\xi)$.

6.2 Lemma: *For any real line bundle μ over a base space $X \in \mathscr{B}$, and for any formal power series $f(t) \in \mathbb{Z}/2[[t]]$ with leading term $1 \in \mathbb{Z}/2$, it follows that $u_f(\mu) \cup u_{1/f}(\mu) = 1 \in H^{**}(X)$.*

PROOF: There is a unique multiplicative homomorphism $\mathbb{Z}/2[[t]] \to H^{**}(X)$ carrying t into $e(\mu)$, and since μ is a line bundle, the relation $f(t) \cdot (1/f)(t) = 1$ defining $(1/f)(t)$ and Proposition 1.2 imply that

$$u_f(\mu) \cup u_{1/f}(\mu) = f(e(\mu)) \cup (1/f)(e(\mu)) = 1 \in H^{**}(X).$$

6.3 Lemma: *For any real vector bundle ξ over a base space $X \in \mathscr{B}$, and for any formal power series $f(t) \in \mathbb{Z}/2[[t]]$ with leading term $1 \in \mathbb{Z}/2$, it follows that $u_{1/f}(\xi)$ is the unique class such that $u_f(\xi) \cup u_{1/f}(\xi) = 1 \in H^{**}(X)$.*

PROOF: Any splitting map $X' \xrightarrow{g} X$ provides a sum $g^!\xi = \mu_1 \oplus \cdots \oplus \mu_n$ of line bundles $\mu_1, \ldots, \mu_n$ over X' and a monomorphism $H^{**}(X) \xrightarrow{g^{**}} H^{**}(X')$, and one uses Lemma 6.2 to complete the proof.

6.4 Proposition: *Suppose that ξ and η are real vector bundles over the same base space $X \in \mathscr{B}$, and suppose that there are trivial bundles ε^p and ε^q over X such that $\xi \oplus \eta \oplus \varepsilon^p = \varepsilon^q$. Then for any formal power series $f(t) \in \mathbb{Z}/2[[t]]$ with leading term $1 \in \mathbb{Z}/2$ it follows that $u_{1/f}(\xi) = u_f(\eta) \in H^{**}(X)$.*

PROOF: Since the leading coefficient of $f(t)$ is $1 \in \mathbb{Z}/2$, it follows that the coefficient of $u_f(\xi)$ in $H^0(X)$ is also 1, and since $\mathbb{Z}/2$ is a field, there is a unique cohomology class $\alpha \in H^{**}(X)$ with $u_f(\xi) \cup \alpha = 1$. However, $u_f(\xi) \cup u_{1/f}(\xi) = 1$ by Lemma 6.3, so that $u_{1/f}(\xi) = \alpha$. On the other hand, according to Corollary 2.6 one has $u_f(\varepsilon^p) = 1$ and $u_f(\varepsilon^q) = 1$, so that one can apply the Whitney product formula to the identity $\xi \oplus \eta \oplus \varepsilon^p = \varepsilon^q$ to conclude that $u_f(\xi) \cup u_f(\eta) = 1$, so that $u_f(\eta) = \alpha$. Hence $u_{1/f}(\xi) = \alpha = u_f(\eta)$, as claimed.

6.5 Corollary: *Suppose that ξ and η are real vector bundles over the same base space $X \in \mathscr{B}$, and suppose that there are trivial bundles ε^p and ε^q such that $\xi \oplus \eta \oplus \varepsilon^p = \varepsilon^q$. Then the Stiefel–Whitney class $w(\eta) \in H^*(X)$ and dual Stiefel–Whitney class $\bar{w}(\xi) \in H^{**}(X)$ satisfy $w(\eta) = \bar{w}(\xi) \in H^{**}(X)$; a fortiori $\bar{w}(\xi) \in H^*(X)$.*

We recall once more that in most applications the base space X will have a finite-dimensional cohomology ring $H^*(X)$, so that $H^*(X) = H^{**}(X)$.

7. Remarks and Exercises

7.1 Remark: Stiefel–Whitney classes were first constructed in Stiefel [1] and in Whitney [2], independently written papers of 1935.

Stiefel considered the existence of k linearly independent vector fields on a smooth n-dimensional manifold X. Suppose for convenience that X is oriented and that it has a fixed triangulation K; the $(n-k)$-*skeleton* $|K_{n-k}| \subset |K| = X$ consists of all geometric p-simplexes in $|K|$ with $p \leqq n-k$. There are always k linearly independent vector fields on $|K_{n-k}|$. However, Stiefel found an *obstruction class* in $H^{n-k+1}(X)$ whose vanishing is a condition for the existence of k linearly independent vector fields on $|K_{n-k+1}|$, where the coefficient group of $H^{n-k+1}(X)$ is $\mathbb{Z}$ or $\mathbb{Z}/2$ according as $n-k$ is even or odd. Since vector fields can equally well be regarded as sections of the tangent bundle $\tau(X)$, Stiefel's obstruction classes can be regarded as characteristic classes of $\tau(X)$; furthermore, if one reduces the coefficient ring $\mathbb{Z}$ to $\mathbb{Z}/2$, then Stiefel's obstruction classes are precisely the Stiefel–Whitney classes $w_{n-k+1}(\tau(X)) \in H^{n-k+1}(X; \mathbb{Z}/2)$.

We already noted in Remark III.13.3 that the first definition of fiber bundles is in Whitney [2]; the very same paper contains a sketch which generalizes Stiefel's construction to arbitrary oriented sphere bundles over a polyhedron $|K|$. In either paper one can ignore the orientation by reducing the coefficient ring $\mathbb{Z}$ to $\mathbb{Z}/2$; however, one then loses part of the obstruction information.

A more recent account of Stiefel [1] and Whitney [2] is given in Steenrod [4, pp. 190–199], and a summary of the same material is given in E. Thomas [2, pp. 159–161]; the homotopy computation leading to the coefficient groups $\mathbb{Z}$ and $\mathbb{Z}/2$ is presented in modern dress on pages 202–203 of G. W. Whitehead [1] for example. The special case $k = 1$ will be treated in detail in Chapter VIII of the present work, and the cases $1 < k \leqq n$ will be summarized in the Remarks at the end of that chapter.

There are many other constructions of Stiefel–Whitney classes, in addition to that of Definition 1.1. One of these (suggested by Proposition 4.3) is discussed in the following remark, an entirely different one of Thom [1, 4] and Wu [3] is presented in Exercise 7.17 and Remark 7.18, and de Carvalho [1] contains yet another construction, for example. However, the original methods of Stiefel [1] and Whitney [2] are still of current interest: Stern [1, 2] assigns Stiefel–Whitney classes to the topological $\mathbb{R}^m$ bundles of Remark III.13.9, with an interpretation as obstruction classes in the sense of Stiefel [1] and Whitney [2] (over suitably restricted base spaces).

7.2 Remark: Following Proposition 4.3 it was observed that the Stiefel–Whitney classes $w_1(\xi), \ldots, w_m(\xi)$ of a real m-plane bundle ξ over any $X \in \mathscr{B}$ could be *defined* by setting $w_i(\xi) = f^*w_i$ for the generators $w_1, \ldots, w_m \in H^*(G^m(\mathbb{R}^\infty); \mathbb{Z}/2)$; over appropriate base spaces one can replace $G^m(\mathbb{R}^\infty)$ by $G^m(\mathbb{R}^{m+n})$ for suitably large n and appeal to the ersatz homotopy classification theorem for a similar result. This technique was first introduced in Pontrjagin [1], the details appearing in Pontrjagin [5]. The technique also appears in Chern [2, 3], in Wu [2, 5], in Takizawa [1], and in Borel and Hirzebruch [1, pp. 483–487], for example; a group-theoretic interpretation of the construction is given on pp. 496–497 of the latter paper. Hodge [3] contains one of the many early complex analogs of the same technique.

7.3 Remark: The preceding characterization of Stiefel–Whitney classes requires the structure of the ring $H^*(G^m(\mathbb{R}^\infty); \mathbb{Z}/2)$, which we computed in Proposition 4.3 in terms of the particular Stiefel–Whitney classes $w_1(\gamma^m), \ldots, w_m(\gamma^m)$. However, the study of $H^*(G^m(\mathbb{R}^\infty); \mathbb{Z}/2)$ and $H^*(G^m(\mathbb{R}^{m+n}); \mathbb{Z}/2)$ antedates the very definition of cohomology. In order to trace the early history of $H^*(G^m(\mathbb{R}^{m+n}); \mathbb{Z}/2)$ we begin with the work which led to the first computation of the complex analog, $H^*(G^m(\mathbb{C}^{m+n}); \mathbb{Z})$, considered in Volume 2.

During the 1870's Hermann Cäsar Hannibal Schubert developed a technique for solving certain geometric intersection problems in $\mathbb{R}^3$, the *enumerative calculus* of Schubert [1]. The technique was applied during 1886 to similar problems in (complex) vector spaces of arbitrary finite dimension, for which the Grassmann manifolds $G^m(\mathbb{C}^{m+n})$ provided a natural setting. In particular, *Schubert varieties* were constructed as submanifolds of $G^m(\mathbb{C}^{m+n})$ in Schubert [2], and it was shown (with the rigor of 1886) that certain Schubert varieties represent cycles whose dual cohomology classes form a convenient basis of the cohomology modules $H^*(G^m(\mathbb{C}^{m+n}); \mathbb{Z})$. Intersections of such *Schubert cycles* were computed in Pieri [1] and in Giambelli [1], furnishing 1894 and 1902 versions, respectively, of the cup product structure of $H^*(G^m(\mathbb{C}^{m+n}); \mathbb{Z})$. Todd [1] later used the original methods of Schubert [2] to derive Giambelli's results from those of Pieri. In 1934 Ehresmann [1] used *topological* intersection theory to give the first rigorous statement and proof of the Schubert "basis theorem" for $G^m(\mathbb{C}^{m+n})$. Schubert varieties and Schubert cycles are constructed for $G^m(\mathbb{R}^{m+n})$ as they are for $G^m(\mathbb{C}^{m+n})$, and in 1937 Ehresmann [2] used *topological* intersection theory to establish an analogous "basis theorem" for $G^m(\mathbb{R}^{m+n})$. In 1942 (and 1947) Pontrjagin [1, 5] used Ehresmann's methods to give a corresponding "basis theorem" for the oriented double covering $\tilde{G}^m(\mathbb{R}^{m+n})$ of $G^m(\mathbb{R}^{m+n})$. (The cohomology of $G^m(\mathbb{C}^{m+n})$ and $\tilde{G}^m(\mathbb{R}^{m+n})$ will be considered in Volume 2.)

Meanwhile, Hodge [1] developed a rigorous and entirely *algebraic* justification for the Schubert "basis theorem", in 1941, followed by a further investigation of intersections of Schubert cycles in Hodge [2]. The Schubert calculus computation of $H^*(G^m(\mathbb{C}^{m+n}); \mathbb{Z})$ is presented in Hodge and Pedoe [1, Chap. XIV] and in Griffiths and Harris [1, pp. 193–206]; as in all the work just cited, there is no explicit mention of cohomology. General expositions of Schubert calculus are given in Zeuthen [1], Pieri and Zeuthen [1], Severi [1, 2], Kleiman and Laksov [1], and Kleiman [1].

In 1947 Chern observed that the closest real analog of $H^*(G^m(\mathbb{C}^{m+n}); \mathbb{Z})$ is $H^*(G^m(\mathbb{R}^{m+n}); \mathbb{Z}/2)$, with the coefficient ring $\mathbb{Z}/2$, and he used Schubert calculus in Chern [2, 3] to provide the first proof of Proposition 4.5. Chern's results inspired entirely different computations of $H^*(G^m(\mathbb{R}^\infty); \mathbb{Z}/2)$ and $H^*(G^m(\mathbb{R}^{m+n}); \mathbb{Z}/2)$, such as the ones used for Propositions 4.3 and 4.5, respectively. Other computations of $H^*(G^m(\mathbb{R}^\infty); \mathbb{Z}/2)$ are sketched in Exercise 7.4 and Remark 7.5, and similar computations of $H^*(G^m(\mathbb{R}^\infty); \mathbb{Z}/2)$ are given in Milnor [3, pp. 26–31] and in Milnor and Stasheff [1, pp. 83–88]; an alternative using spectral sequences is given in Liulevicius [2, 117–122].

7.4 Exercise: Use Proposition 4.4 to compute $H^*(G^m(\mathbb{R}^\infty); \mathbb{Z}/2)$ by induction on m. (*Hint*: The initial step is in Proposition IV.4.3. To establish the inductive step observe that the homomorphisms f^* in the exact sequence of Proposition 4.4 are surjective, hence that there are short exact sequences

$$0 \to H^q(G^m(\mathbb{R}^\infty)) \xrightarrow{\smile e(\gamma^m)} H^{q+m}(G^m(\mathbb{R}^\infty)) \xrightarrow{f^*} H^{q+m}(G^{m-1}(\mathbb{R}^\infty)) \to 0.)$$

7.5 Remark: Since $H^*(RP^\infty; \mathbb{Z}/2)$ is the polynomial ring generated over $\mathbb{Z}/2$ by the $\mathbb{Z}/2$ Euler class $e(\gamma^1) = w_1(\gamma^1) \in H^1(RP^1)$, one can use Remark 7.2 to define the Stiefel–Whitney class $w_1(\lambda) \in H^1(X; \mathbb{Z}/2)$ of any line bundle λ over any $X \in \mathscr{B}$, hence the total Stiefel–Whitney class $1 + w_1(\lambda) \in H^*(X; \mathbb{Z}/2)$. The Whitney product formula (Proposition 2.5) and the splitting principle (Proposition IV.5.5) then provide yet another characterization of Stiefel–Whitney classes, essentially the uniqueness theorem proved earlier. This construction originated in Grothendieck [1], in a slightly different setting; it can also be found in Braemer [1].

7.6 Exercise: Some of the preceding remarks indicate alternative constructions of Stiefel–Whitney classes, and there are more to come. For example, show that the special case $f(t) = 1 + t$ of Definition 1.1 is equivalent to the following *formula of Hirsch* for any real m-plane bundle ξ over any $X \in \mathscr{B}$:

$$e(\lambda_\xi)^m + w_1(\xi)e(\lambda_\xi)^{m-1} + \cdots + w_{m-1}(\xi)e(\lambda_\xi) + w_m(\xi)1 = 0$$

in the free $H^*(X)$-module $H^*(P_\xi)$, where λ_ξ is the splitting bundle over $P_\xi \in \mathscr{B}$. This identity was apparently known to Chern and Wu before its publication (in a slightly different setting) in G. Hirsch [7]. The language "formula of Hirsch" appears in Hirzebruch [3, p. 75].

7.7 Remark: The axiomatic characterization of Stiefel–Whitney classes reported in Theorems 5.1 and 5.2 was first given in Hirzebruch [2]; it can be found in Hirzebruch [3, p. 73].

7.8 Remark: A special case of the Whitney product formula of Proposition 2.5 was first announced as a "duality theorem" in Whitney [6], without proof. The proof is implicit in Chern's computation of the ring structure of $H^*(G^m(\mathbb{R}^{m+n}); \mathbb{Z}/2)$ in Chern [2, 3], for example, and an explicit proof appears in Wu [1]. More recent product formulae in $H^*(G^m(\mathbb{R}^{m+n}); \mathbb{Z}/2)$ can be found in Oproiu [1].

7.9 Exercise: Use the identity $e(\lambda \otimes \mu) = e(\lambda) + e(\mu)$ of Proposition IV.7.5 to prove the following property of Stiefel–Whitney classes: For any natural numbers $m \geqq 0$ and $n \geqq 0$ there is a unique polynomial $P_{m,n}$ in $m + n$ variables over $\mathbb{Z}/2$ such that

$$w(\xi \otimes \eta) = P_{m,n}(w_1(\xi), \ldots, w_m(\xi); w_1(\eta), \ldots, w_n(\eta))$$

for any real vector bundles ξ and η of ranks m and n, respectively, over the same base space $X \in \mathscr{B}$. (This is a relatively easy exercise; however, alternative approaches to its solution can be found in Borel and Hirzebruch [1] and in E. Thomas [1].)

7.10 Remark: In Thom [4] one learns that the total Stiefel–Whitney class $w(\xi)$ of a vector bundle ξ depends only on the J-equivalence class of ξ. This suggests that Stiefel–Whitney classes should be assigned directly to J-equivalence classes; it also suggests that one should try to construct Stiefel–Whitney classes *directly* from less information than a vector bundle carries.

For example, since Thom [4] also shows that the J-equivalence class of the tangent bundle $\tau(X)$ of a smooth manifold X is independent of the smooth structure assigned to X, it follows that $w(\tau(X))$ is independent of the smooth structure of X. Nash [1] proves the latter result directly; in fact, Nash's construction assigns total Stiefel–Whitney classes to arbitrary *topological* manifolds X, the results agreeing with $w(\tau(X))$ for smooth manifolds X. (Nash's construction also leads to generalizations of several classical results about tangent bundles of smooth manifolds, as noted earlier in Remark III.13.9, in Fadell [4], R. F. Brown [1], and Brown and Fadell [1].)

Stiefel–Whitney classes of more general fibrations, including the topological $\mathbb{R}^m$ bundles of Remark III.13.9, have been constructed more recently in Teleman [1, 2, 3] and Bordoni [1]. According to Teleman [2], the specialization to the tangent topological $\mathbb{R}^n$ bundle of an n-dimensional manifold X provides an alternative computation of Nash's total Stiefel–Whitney class of X. It has already been observed in Remark 7.1 that the individual Stiefel–Whitney classes arising in such constructions can be interpreted as obstruction classes, as in Stern [1, 2].

An earlier construction of the Stiefel–Whitney classes of the preceding paragraph can be found in Vazquez [1], using generalizations of methods described in some of the following remarks; an application of the result appears in Vazquez [2].

7.11 Exercise: Let Λ be any commutative ring with unit, and let $P_0 = 1$, $P_1(u_1)$, $P_2(u_1, u_2), \ldots$ be a sequence of polynomials $P_n(u_1, \ldots, u_n) \in \Lambda[u_1, \ldots, u_n]$ such that if u_q is assigned degree q, then $P_n(u_1, \ldots, u_n)$ is of degree n. The sequence can be regarded as an element $\sum_{n \geqq 0} P_n(u_1, \ldots, u_n)$ of the formal power series ring $\Lambda[[u_1, u_2, \ldots]]$, as in Definition 2.9; for example, if $u_0 = 1$, then $\sum_{n \geqq 0} u_n$ is itself such an element. One can compute the formal product

$$\left(\sum_{n \geqq 0} u_n\right)\left(\sum_{n \geqq 0} v_n\right) = \sum_{n \geqq 0} w_n,$$

where $w_n = \sum_{i=0}^{n} u_i v_{n-i}$, and $\sum_{n \geqq 0} P_n(u_1, \ldots, u_n)$ is *multiplicative* whenever

$$\left(\sum_{n \geqq 0} P_n(u_1, \ldots, u_n)\right)\left(\sum_{n \geqq 0} P_n(v_1, \ldots, v_n)\right) = \sum_{n \geqq 0} P_n(w_1, \ldots, w_n).$$

Show that if $\sum_{n \geqq 0} P_n(u_1, \ldots, u_n)$ is the multiplicative sequence assigned to a formal power series $f(t) \in \Lambda[[t]]$ as in Definition 2.9, then it is multiplicative in the present sense.

7.12 Exercise: Conversely, let $\sum_{n \geqq 0} P_n(u_1, \ldots, u_n)$ be multiplicative in the sense of Exercise 7.11; show that it is the multiplicative sequence assigned as in Definition 2.9 to the formal power series

$$f(t) = \sum_{n \geqq 0} P_n(t, 0, 0, \ldots, 0) \in \Lambda[[t]].$$

7.13 Remark: Multiplicative sequences were introduced in Hirzebruch [1]. Other expositions can be found in of Hirzebruch [3, pp. 9–16], Milnor [3, pp. 111–116], and Milnor and Stasheff [1, pp. 219–222].

7.14 Remark: Thom provided a useful alternative construction of Stiefel–Whitney classes in Thom [1, 4]. In order to discuss it we first describe a classical $\mathbb{Z}/2$ cohomology operation which is not ordinarily presented in beginning algebraic topology courses. Let $H^*(-,-)$ (and $H^*(-)$) denote singular cohomology with $\mathbb{Z}/2$ coefficients, as before, and for each pair (X, Y) of topological spaces with $Y \subset X$ let Sq^i denote a $\mathbb{Z}/2$-module homomorphism $H^*(X, Y) \to H^*(X, Y)$ of degree $i \geqq 0$, carrying each summand $H^n(X, Y)$ into the summand $H^{n+i}(X, Y)$. The direct sum $\mathrm{Sq} = \mathrm{Sq}^0 \oplus \mathrm{Sq}^1 \oplus \mathrm{Sq}^2 \cdots$ is the *Steenrod square* if it satisfies the following axioms:

(0) **Dimension:** *If* $\alpha \in H^0(X, Y) \oplus \cdots \oplus H^n(X, Y)$, *then* $\mathrm{Sq}^i \alpha = 0$ *for* $i > n$.

(1) **Naturality:** *If f is any map from a pair* (X', Y') *to a pair* (X, Y), *then* $\mathrm{Sq}\, f^*\alpha = f^*\, \mathrm{Sq}\, \alpha \in H^*(X', Y')$ *for any* $\alpha \in H^*(X, Y)$.

(2) **Cartan formula:** *The cup product* $\alpha \cup \beta \in H^*(X, Y)$ *of* $\alpha \in H^*(X, Y)$ *and* $\beta \in H^*(X, Y)$ *satisfies*

$$\mathrm{Sq}(\alpha \cup \beta) = \mathrm{Sq}\, \alpha \cup \mathrm{Sq}\, \beta.$$

(3) **Normalization:** *If* $\alpha \in H^n(X, Y)$ *for any* $n \geqq 0$, *then*

$$\mathrm{Sq}^0\, \alpha = \alpha \qquad \textit{and} \qquad \mathrm{Sq}^n\, \alpha = \alpha \cup \alpha \in H^{2n}(X, Y).$$

The Steenrod square first appeared in Steenrod [3], and the preceding axiomatic characterization (including the Cartan formula) first appeared in Cartan [1]. Early general expositions of the Steenrod square can be found in Exposés 14 and 15 of Cartan [3], in Thom [5] and Wu [6], and in Steenrod [5]; the latter papers contain generalizations for other primes p than the prime $p = 2$, independently presented in Bott [1]. Somewhat more recent expositions of the Steenrod square (and its generalizations) can be found in Steenrod and Epstein [1, pp. 112–113, 124–132], Spanier [4, pp. 269–276], Mosher and Tangora [1, pp. 12–32], and Steenrod [9], for example. R. J. Milgram developed a simplified construction of the Steenrod square, first published in Gray [1, pp. 310–323].

7.15 Remark: Among the consequences of the preceding axioms are the *Adem relations*, which will figure in later remarks and exercises, and which can be found in most of the preceding references; they were discovered independently by Adem [1, 2] and Cartan [2].

7.16 Remark: The Cartan formula of Remark 7.14 can also be expressed in terms of cross products. Recall (from Dold [8, pp. 221–222], Greenberg [1, p. 201], or Spanier [4, p. 253], for example) that if X and Y are any topological spaces whatsoever, then the cross product $H^*(X) \otimes H^*(Y) \overset{\times}{\to} H^*(X \times Y)$ can be obtained as the composition

$$H^*(X) \otimes H^*(Y) \xrightarrow{\mathrm{pr}_1^* \otimes \mathrm{pr}_2^*} H^*(X \times Y) \otimes H^*(X \times Y) \overset{\cup}{\to} H^*(X \times Y)$$

for the projections $X \times Y \xrightarrow{\mathrm{pr}_1} X$ and $X \times Y \xrightarrow{\mathrm{pr}_2} Y$. Naturality and the preceding Cartan formula for Sq then imply $\mathrm{Sq}(\alpha \times \beta) = \mathrm{Sq}\,\alpha \times \mathrm{Sq}\,\beta \in H^*(X \times Y)$ for any $\alpha \in H^*(X)$ and $\beta \in H^*(Y)$. There is an obvious relative version of the same result, and both versions are regarded as Cartan formulae. Conversely, if $X \overset{\Delta}{\to} X \times X$ is the diagonal map one uses the composition

$$H^*(X) \otimes H^*(X) \overset{\times}{\to} H^*(X \times X) \xrightarrow{\Delta^*} H^*(X)$$

to define cup products in $H^*(X)$, so that naturality and the latter Cartan formula for Sq imply the absolute version of the Cartan formula of Remark 7.14; the relative version of the same construction is clear.

7.17 Exercise: Let ξ be a real n-plane bundle over $X \in \mathscr{B}$, with $\mathbb{Z}/2$ Thom class $U_\xi \in H^n(E, E^*)$ and $\mathbb{Z}/2$ Thom isomorphism $H^*(X) \xrightarrow{\Phi_\xi} H^{*+n}(E, E^*)$, and let Sq be the Steenrod square of Remark 7.14. Use the axiomatic characterization of Stiefel–Whitney classes (Theorem 5.1) to show that

$$w(\xi) = \Phi_\xi^{-1}(\mathrm{Sq}\, U_\xi) = \Phi_\xi^{-1}\, \mathrm{Sq}\, \Phi_\xi(1) \in H^*(X)$$

for the element $1 \in H^0(X)$.

7.18 Remark: The identity $w(\xi) = \Phi_\xi^{-1}\, \mathrm{Sq}\, \Phi_\xi(1)$ can also be written in the concrete form $\pi^* w(\xi) \cup U_\xi = \mathrm{Sq}\, U_\xi \in H^*(E, E^*)$. Since $H^*(E, E^*)$ is the free $H^*(X)$-module generated by the Thom class U_ξ, this clearly characterizes $w(\xi)$.

7.19 Remark: If ξ is a real n-plane bundle over $X \in \mathscr{B}$, then Corollary 2.8 and Proposition IV.3.8 provide an identity $w_n(\xi) = e(\xi) = \Phi_\xi^{-1}(U_\xi \cup U_\xi) \in H^n(X)$. Since $U_\xi \cup U_\xi = \mathrm{Sq}^n\, U_\xi$, this identity can be regarded as the dimension n portion of the identity $w(\xi) = \Phi_\xi^{-1}(\mathrm{Sq}\, U_\xi)$ of Exercise 7.17; in fact, one uses this observation as a part of Exercise 7.17.

7.20 Remark: Since $\mathbb{Z}/2$ is a field, one can regard the Steenrod square as an automorphism of the $\mathbb{Z}/2$ cohomology rings $H^*(X, Y)$, satisfying the obvious naturality condition; in particular, there is an inverse Steenrod square Sq^{-1} which is also natural. The same remarks apply if one substitutes the group of units in $H^{**}(X, Y)$ for $H^*(X, Y)$; that is, 0th entries are required to be $1 \in H^0(X, Y)$, and one considers products only. One can therefore define the *total Wu class* of any real vector bundle ξ over any base space $X \in \mathscr{B}$ by setting

$$\mathrm{Wu}(\xi) = \mathrm{Sq}^{-1} w(\xi) = \mathrm{Sq}^{-1} \Phi_\xi^{-1} \mathrm{Sq}\, \Phi_\xi(1) \in H^{**}(X)$$

for the total Stiefel–Whitney class $w(\xi) \in H^*(X)$ $(\subset H^{**}(X))$ and the Thom isomorphism Φ_ξ; that is, $\mathrm{Wu}(\xi) \in H^{**}(X)$ is the unique class such that $\mathrm{Sq}\,\mathrm{Wu}(\xi) = w(\xi) \in H^*(X)$.

7.21 Exercise: Show that there is a formal power series $f(t) \in \mathbb{Z}/2[[t]]$ with leading term $1 \in \mathbb{Z}/2$ such that $\mathrm{Wu}(\xi) = u_f(\xi) \in H^{**}(X)$, as in Definition 1.4, for any vector bundle ξ over any $X \in \mathscr{B}$.

7.22 Exercise: Show that the "Wu series" of Exercise 7.21 is given by $f(t) = 1 + \sum_{q \geqq 0} t^{2^q} \in \mathbb{Z}/2[[t]]$. Show furthermore that $f(t)^2 = f(t^2) = f(t) + t$, and that $(1/f)(t) = \sum_{q \geqq 0} t^{2^q - 1} \in \mathbb{Z}/2[[t]]$.

7.23 Remark: Wu classes of tangent bundles first appeared in Wu [3]. The definition was extended to arbitrary real vector bundles by Adams [3], who observed that Wu classes could be computed in terms of Stiefel–Whitney classes. The specific recipe $\mathrm{Wu}(\xi) = \mathrm{Sq}^{-1} \Phi_\xi^{-1} \mathrm{Sq}\, \Phi_\xi(1)$ was given by Atiyah and Hirzebruch [2]. Exercises 7.21 and 7.22 provide an alternative construction of Wu classes, *without* Steenrod squares.

7.24 Exercise: Let $\binom{a}{b}$ denote the $\mathbb{Z}/2$ image of the usual binomial coefficient $\binom{a}{b}$ whenever $a \geqq b \geqq 0$, let $\binom{-1}{0} = 1 \in \mathbb{Z}/2$, and let $\binom{a}{b} = 0 \in \mathbb{Z}/2$ otherwise. Show that the Stiefel–Whitney classes $w_0 = 1, w_1, w_2, \ldots$ of any real vector bundle satisfy the following *Wu relation*:

$$\mathrm{Sq}^j w_k = \sum_{i=0}^{j} \binom{k - j + i - 1}{i} w_{j-1} \cup w_{k+i}.$$

7.25 Remark: The Wu relation of Exercise 7.24 first appeared in Wu [4], the complete proof appearing in Wu [7]; a proof related to that of Wu [7] is given in Borel [2]. One can also use the Adem relations of Remark 7.15 to obtain the same result, as in Hsiang [1]. The best way to prove the Wu relations involves a technique developed in Brown and Peterson [1].

7.26 Remark: Any real m-plane bundle ξ over a contractible base space Y is the trivial bundle ε^m, by Proposition II.3.5, so that Corollary 2.6 gives

$u_f(\xi) = u_f(\varepsilon^m) = 1 \in H^{**}(Y)$ for any $f(t) \in \mathbb{Z}/2[[t]]$ with leading term $1 \in \mathbb{Z}/2$; equivalently, all the Stiefel–Whitney classes $w_1(\xi), \ldots, w_m(\xi)$ vanish. There are other base spaces Y with the latter property. For example, let X be any CW space and choose a point $* \in X$. As in Remark III.13.41 the *(reduced) suspension* ΣX is the quotient of the product $X \times [0,1]$ by the subspace $X \times \{0\} \cup \{*\} \times [0,1] \cup X \times \{1\}$, in the quotient topology; by iteration one obtains the *n-fold (reduced) suspension* $\Sigma^n X$ for any $n \geqq 0$. One of the consequences of Atiyah and Hirzebruch [3] is that all the Stiefel–Whitney classes of any real vector bundle over $\Sigma^n X$ vanish whenever $n \geqq 9$; that is, such suspensions have the preceding property of contractible base spaces Y. According to Sutherland [2], weaker results of a similar nature apply to the Stiefel–Whitney classes of real vector bundles over suspensions $\Sigma^2 X$ and $\Sigma^5 X$.

7.27 Remark: The Stiefel–Whitney classes of various specialized real vector bundles satisfy certain polynomial relations, some of which will be developed for geometric applications in the next chapter. For example, we shall learn in Corollary VI.8.6 that if $\tau(X)$ is the tangent bundle of any smooth closed n-dimensional manifold X, then the Wu class $\mathrm{Wu}(\tau(X)) = 1 + \mathrm{Wu}_1(\tau(X)) + \cdots + \mathrm{Wu}_n(\tau(X))$ satisfies $\mathrm{Wu}_p(\tau(X)) = 0 \in H^p(X)$ whenever $2p > n$; since the individual Wu classes are polynomials in the Stiefel–Whitney classes $w_1(\tau(X)), \ldots, w_n(\tau(X))$, the result can be regarded as a set of polynomial relations among the latter classes. In Brown and Peterson [1, 2] one finds all the polynomial relations satisfied by $w_1(\tau(X)), \ldots, w_n(\tau(X))$ for every smooth closed n-dimensional manifold X. Incidentally, E. H. Brown [2] and Stong [1] provide independent proofs that if $2p \leqq n$, then there are no such "universal" polynomial relations $P(w_1(\tau(X)), \ldots, w_n(\tau(X)) = 0 \in H^p(X)$ in degree p.

In sufficiently high degrees there are "universal" polynomial relations satisfied by the dual Stiefel–Whitney classes $\bar{w}_1(\tau(X)), \ldots, \bar{w}_n(\tau(X))$ of every smooth closed n-dimensional manifold X, where $1 + \bar{w}_1(\tau(X)) + \cdots + \bar{w}_n(\tau(X)) = \bar{w}(\tau(X)) = u_f(\tau(X))$ for $f(t) = 1 + t + t^2 + \cdots \in \mathbb{Z}/2[[t]]$. Such relations were investigated in Massey [1, 3] and in Massey and Peterson [1]; the latter paper shows that if $0 \leqq p < \alpha(n)$, where $\alpha(n)$ is the number of 1's in the dyadic expansion of n, then $\bar{w}_{n-p}(\tau(X)) = 0$ for every smooth closed n-dimensional manifold X. The set of *all* polynomial relations satisfied by $\bar{w}_1(\tau(X)), \ldots, \bar{w}_n(\tau(X))$ for every smooth closed n-dimensional manifold X is implicitly described in Brown and Peterson [1, 2]: one merely replaces w by $\bar{w}$ throughout their original argument (and result); related results are given in Bendersky [1] and in Papastavridis [2]. The result itself is an essential ingredient of our later Remark VI.9.14.

7.28 Remark: According to Proposition 3.2 one can reasonably define orientability of a smooth manifold X by the requirement that $w_1(\tau(X)) = 0 \in H^1(X)$. An orientable manifold which satisfies the additional relation $w_2(\tau(X)) = 0 \in H^2(X)$ is called a *spin manifold.* (Alternative characterizations of spin manifolds are given in Borel and Hirzebruch [2, p. 350], the end of Milnor [12], and at the beginning of Milnor [16].) The total Stiefel–Whitney class $w(\tau(X))$ of a spin manifold automatically satisfies further relations, some of which are discussed in Wilson [1], for example.

7.29 Remark: Since realifications of complex vector bundles are somewhat specialized, one expects their Stiefel–Whitney classes to satisfy universal polynomial relations. For example, the realification $\zeta_{\mathbb{R}}$ of a complex vector bundle ζ is naturally oriented, so that $w_1(\zeta_{\mathbb{R}}) = 0$ by Proposition 3.2; other such relations will appear in the second volume of this work. There are also universal polynomial relations satisfied by the Stiefel–Whitney classes of realifications of "quaternionic bundles," some of which appear in Marchiafava and Romani [1, 2, 3].

7.30 Remark: One of the classical results of differential topology is that S^1, S^3, and S^7 are the only standard spheres which are parallelizable in the sense of Remark III.13.26; that is, if S^n has n linearly independent vector fields, then $n = 1$, 3, or 7. This result was proved in Bott and Milnor [1] and Milnor [4], using the following property of Stiefel–Whitney classes: If S^n is the base space of a real vector bundle ξ with $w_n(\xi) \neq 0$, then $n = 1, 2, 4$, or 8. (Incidentally, the result of Barratt and Mahowald [1] reported in Remark III.13.36 instantly implies that if $n = 4k > 16$, then $w_n(\xi) = 0$ for any real vector bundle ξ over S^n.)

In the third volume of this work we shall formulate and prove a result of Adams [1, 2], one of whose principal corollaries is that S^1, S^3, and S^7 are the only parallelizable standard spheres. Adams's original proof, which also appears in Cartan [5], uses many of the techniques introduced in this chapter, including Steenrod squares; however, the proof given later will be based on an entirely different technique of Adams and Atiyah [1].

7.31 Remark: There is a more general result than the one just described: the maximum number of linearly independent vector fields on the standard sphere S^n is known for *any* $n > 0$. One of the early guidelines for the computation was provided by Stiefel [2], who used Stiefel–Whitney classes to show that if $n = 2^k u - 1$ for an odd number u, then the projective space RP^n cannot have 2^k linearly independent vector fields. Steenrod and Whitehead [1] used Steenrod squares to obtain the same result with S^n substituted for RP^n, and their work eventually led to the complete solution of the problem, in Adams [4]. (We shall not attempt to present a proof of this result. Adams's original proof is outlined in Eilenberg [1] and in Husemoller [1], and some simplifications by Karoubi are reported in Gordon [1], for example; Adams himself streamlined the proof in many ways, and a major simplification appeared in Woodward [1]. A complete proof is given in Karoubi [2], incorporating all improvements through 1978; a complete proof is also given in Mahammed *et al.* [1], incorporating all improvements through 1980.)

7.32 Remark: To any finite-dimensional real representation V of a finite group G one can associate a real vector bundle ξ_V over the classifying space BG of Remark II.8.18; the total Stiefel–Whitney class $w(\xi_V) \in H^*(BG)$ is then the *total Stiefel–Whitney class* $w(V)$ of the representation V. Segal and Stretch [1] present an alternative construction of $w(V)$, which leads to new results about group representations.

7.33 Remark: There is an algebraic construction of Delzant [1], which assigns Stiefel–Whitney classes to certain "quadratic modules" rather than to real vector bundles; the construction was further developed in Milnor [19]. A related construction exists for real vector bundles, given in Patterson [1]; however, this variant produces only the terms $1 + w_1(\xi) + w_2(\xi)$ of the total Stiefel–Whitney class $w(\xi)$ of a real vector bundle ξ. A recent exposition of the constructions of Delzant and Milnor is in Chapter 4 of Micali and Revoy [1].

7.34 Remark: Lemma 6.3 is a special case of a more general result, with essentially the same proof. Let $f(t)$ and $g(t)$ be any formal power series in $\mathbb{Z}/2[[t]]$ with leading terms $1 \in \mathbb{Z}/2$, so that the product $(fg)(t)$ is also such a formal power series. Then for any real vector bundle ξ over a base space $X \in \mathscr{B}$ one has $u_f(\xi) \cup u_g(\xi) = u_{fg}(\xi) \in H^{**}(X)$. Thus ξ induces a homomorphism from the multiplicative group of formal power series in $\mathbb{Z}/2[[t]]$ with leading terms $1 \in \mathbb{Z}/2$ to a corresponding multiplicative group in $H^{**}(X; \mathbb{Z}/2)$.

CHAPTER VI

Unoriented Manifolds

0. Introduction

Let $\tau(X)$ be the tangent bundle of a smooth manifold X that is not necessarily orientable. Several geometric properties of such manifolds X can be described in terms of the multiplicative $\mathbb{Z}/2$ classes $u_f(\tau(X)) \in H^*(X;\mathbb{Z}/2)$ of the bundles $\tau(X)$. For example, if X is n-dimensional one can use the dual Stiefel–Whitney class $\bar{w}(\tau(X)) \in H^*(X;\mathbb{Z}/2)$ to provide necessary conditions for the existence of immersions or embeddings $X \to \mathbb{R}^{2n-p}$ for a given $p > 0$. As another example, if X is a closed such manifold the Stiefel–Whitney classes $w_1(\tau(X)), \ldots, w_n(\tau(X))$ provide elements in $\mathbb{Z}/2$ whose vanishing is a necessary and sufficient condition for X to be the boundary of a smooth compact $(n+1)$-dimensional manifold Y.

The chapter begins with some standard properties of the $\mathbb{Z}/2$ homology module $H_*(X, \dot{X};\mathbb{Z}/2)$ and $\mathbb{Z}/2$ cohomology module $H^*(X;\mathbb{Z}/2)$ of any smooth compact n-dimensional manifold X with boundary $\dot{X}$. The best-known feature of $H_*(X, \dot{X};\mathbb{Z}/2)$ is the existence of a *fundamental* $\mathbb{Z}/2$ *homology class* $\mu_X \in H_n(X, \dot{X};\mathbb{Z}/2)$, which is used to provide the $\mathbb{Z}/2$ *Poincaré–Lefschetz duality isomorphisms* $H^q(X;\mathbb{Z}/2) \xrightarrow{\cap\mu_X} H_{n-q}(X, \dot{X};\mathbb{Z}/2)$ for any $q \in \mathbb{Z}$. The classes μ_X and isomorphisms $\cap\mu_X$ are introduced in §§1 and 2, respectively.

The classes μ_X and isomorphisms $\cap\mu_X$ are temporarily ignored in §§3 and 4, which treat the immersion and embedding problems mentioned earlier. Specifically, §3 consists entirely of the computation of the dual Stiefel–Whitney classes $\bar{w}(\tau(RP^n))$ of real projective spaces RP^n. Then immersibility and embeddability criteria for arbitrary smooth manifolds X are established in §4, in terms of the dual Stiefel–Whitney classes $\bar{w}(\tau(X))$; the computation of §3 provides explicit examples of n-dimensional manifolds X which can neither be embedded nor immersed in $\mathbb{R}^{2n-p}$ for $p \geqq 0$ unless p is sufficiently small.

If $1 + w_1(\tau(X)) + \cdots + w_n(\tau(X))$ is the total Stiefel–Whitney class $w(\tau(X)) \in H^*(X; \mathbb{Z}/2)$ of a smooth closed n-dimensional manifold X, then for any ordered n-tuple $(r_1, \ldots, r_n)$ of natural numbers such that $r_1 + 2r_2 + \cdots + nr_n = n$ the Kronecker product $\langle w_1(\tau(X))^{r^1} \cup \cdots \cup w_n(\tau(X))^{r_n}, \mu_X \rangle \in \mathbb{Z}/2$ is the $(r_1, \ldots, r_n)$th *Stiefel–Whitney number* of X. In §5 it is shown that if X is the boundary $\dot{Y}$ of a smooth compact $(n+1)$-dimensional manifold Y, then all the Stiefel–Whitney numbers of X vanish. The converse assertion is also true, although not proved here, and the combined results can be used to compute the *unoriented cobordism ring* $\mathfrak{N}$, whose elements are equivalence classes $[X]$ of smooth closed manifolds X.

The set of diffeomorphism classes of smooth closed manifolds X itself forms a semi-ring $\mathscr{U}$, and there is a natural epimorphism $\mathscr{U} \to \mathfrak{N}$ whose kernel corresponds to those manifolds which are boundaries. For any formal power series $f(t) \in \mathbb{Z}/2[[t]]$ with leading term $1 \in \mathbb{Z}/2$ there is a homomorphism $\mathscr{U} \xrightarrow{G(f)} \mathbb{Z}/2$ whose construction uses the multiplicative $\mathbb{Z}/2$ classes $u_f(\tau(X))$ of the tangent bundles $\tau(X)$ of manifolds representing elements of $\mathscr{U}$. The intersection $\bigcap_f \ker G(f)$ properly contains the kernel of the epimorphism $\mathscr{U} \to \mathfrak{N}$, so that the homomorphisms $G(f)$ can equally well be regarded as homomorphisms $\mathfrak{N} \xrightarrow{G(f)} \mathbb{Z}/2$. Such *Stiefel–Whitney genera* $G(f)$, in either interpretation, are considered in §6.

The boundary $\dot{X}$ of any smooth closed manifold X is void. Consequently if $X \in \mathscr{U}$ is n-dimensional its fundamental class μ_X is an element of $H_n(X; \mathbb{Z}/2)$, and the $\mathbb{Z}/2$ Poincaré–Lefschetz duality isomorphism $\cap \mu_X$ reduces to the *Poincaré duality isomorphism* $H^*(X; \mathbb{Z}/2) \xrightarrow{D_P = \cap \mu_X} H_*(X; \mathbb{Z}/2)$. Since $\mathbb{Z}/2$ is a field one has $H_*(X; \mathbb{Z}/2) = \mathrm{Hom}_{\mathbb{Z}/2}(H^*(X; \mathbb{Z}/2), \mathbb{Z}/2)$, so that D_p can equally well be treated as a nondegenerate bilinear form

$$H^*(X; \mathbb{Z}/2) \otimes H^*(X; \mathbb{Z}/2) \xrightarrow{(\,,\,)_P} \mathbb{Z}/2,$$

which happens to be symmetric. The dual $\mathbb{Z}/2$ *Thom form*

$$H_*(X; \mathbb{Z}/2) \otimes H_*(X; \mathbb{Z}/2) \xrightarrow{(\,,\,)_T} \mathbb{Z}/2$$

is constructed geometrically in §7 as an element

$$j^*T_X \in H^n(X \times X; \mathbb{Z}/2) \quad (\approx H^*(X; \mathbb{Z}/2) \otimes H^*(X; \mathbb{Z}/2)),$$

thereby providing a specific inverse $H_*(X; \mathbb{Z}/2) \xrightarrow{D_T} H^*(X; \mathbb{Z}/2)$ to the $\mathbb{Z}/2$ Poincaré duality isomorphism D_p. One of the most useful properties of j^*T_X is that if $X \xrightarrow{\Delta} X \times X$ is the diagonal map, then $\Delta^* j^* T_X \in H^n(X; \mathbb{Z}/2)$ is the $\mathbb{Z}/2$ Euler class $e(\tau(X))$ of the tangent bundle $\tau(X)$.

For any smooth closed manifold $X \in \mathcal{U}$, and for any formal power series $f(t) \in \mathbb{Z}/2[[t]]$ with leading term $1 \in \mathbb{Z}/2$, the $\mathbb{Z}/2$ multiplicative class $u_f(\tau(X)) \in H^*(X; \mathbb{Z}/2)$ depends only on the homotopy type of X; in particular, $u_f(\tau(X))$ is independent of the smooth structure assigned to X. The volume closes with a proof of this classical result, due to Thom and Wu.

As in the previous two chapters, the notations $H^*(-,-)$ and $H^*(-)$ are used to indicate singular cohomology $H^*(-,-; \mathbb{Z}/2)$ and $H^*(-; \mathbb{Z}/2)$ with $\mathbb{Z}/2$ coefficients. A corresponding convention applies to direct products $H^{**}(-)$ and to singular homology $H_*(-,-)$ and $H_*(-)$.

1. $\mathbb{Z}/2$ Fundamental Classes

Let X be a smooth compact n-dimensional manifold with boundary $\dot{X} = X \backslash \mathring{X}$, where $\mathring{X}$ is the interior of X. We shall show that the $\mathbb{Z}/2$ homology module $H_n(X, \dot{X})$ is free, with one basis element for each connected component of X; in case X is *closed*, in the sense that $\dot{X}$ is void, the result applies to $H_n(X)$.

For example, for any $n > 0$ let D^n be the n-disk, whose boundary $\dot{D}^n$ is the $(n-1)$-sphere S^{n-1}; in case $n = 1$, S^0 consists of just two points. A portion of the exact homology sequence of the pair D^n, S^{n-1} is

$$H_n(D^n) \xrightarrow{j_*} H_n(D^n, S^{n-1}) \xrightarrow{\partial} H_{n-1}(S^{n-1}) \xrightarrow{i_*} H_{n-1}(D^n),$$

where $H_n(D^n) = 0$. If $n > 1$, then $H_{n-1}(D^n) = 0$, so that $H_n(D^n, S^{n-1}) = H_{n-1}(S^{n-1}) = \mathbb{Z}/2$, and if $n = 1$, then the preceding exact sequence is

$$0 \xrightarrow{j_*} H_1(D^1, S^0) \xrightarrow{\partial} \mathbb{Z}/2 \oplus \mathbb{Z}/2 \xrightarrow{i_*} \mathbb{Z}/2,$$

which implies $H_1(D^1, S^0) = \mathbb{Z}/2$. Thus $H_n(D^n, S^{n-1})$ is the free $\mathbb{Z}/2$-module of rank 1 for every $n > 0$.

1.1 Lemma: *Let X be any n-dimensional manifold with interior $\mathring{X}$. Then for any $x \in \mathring{X}$ the $\mathbb{Z}/2$-module $H_n(X, X\backslash\{x\})$ is free of rank* 1, *and $H_q(X, X\backslash\{x\}) = 0$ for $q \neq n$.*

Proof: Since x lies in the interior $\mathring{X} \subset X$, it also lies in the interior of some n-cell $D^n \subset \mathring{X}$, and since the inclusion $D^n, D^n\backslash\{x\} \to X, X\backslash\{x\}$ is an excision, it suffices to consider just the case $X = D^n$. The boundary S^{n-1} of D^n is a strong deformation retract of $D^n\backslash\{x\}$, so that the 5-lemma provides isomorphisms $H_q(D^n, S^{n-1}) \xrightarrow{\approx} H_q(D^n, D^n\backslash\{x\})$ for all $q \in \mathbb{Z}$; however, we have just just observed that $H_n(D^n, S^{n-1}) = \mathbb{Z}/2$, so that $H_n(D^n, D^n\backslash\{x\}) = \mathbb{Z}/2$. A similar argument gives $H_q(D^n, D^n\backslash\{x\}) = 0$ for $q \neq n$.

One feature of Lemma 1.1 deserves special mention. Since there is one and only one nonzero element in $\mathbb{Z}/2$, there is also one and only one generator of the $\mathbb{Z}/2$-module $H_n(X, X\backslash\{x\})$ for any $x \in \mathring{X}$: there are no choices to make. In later chapters, where $H_*(-)$ will have coefficients other than $\mathbb{Z}/2$, corresponding results will require choices.

In Proposition I.8.4 we learned that if X is a smooth compact manifold with interior $\mathring{X}$, then there is a finite covering $\{U_1, \ldots, U_q\}$ of $\mathring{X}$ by open cells such that each nonvoid intersection $U_i \cap U_j$ is a cell U_k in the covering, and such that the closures in X satisfy $\overline{U_i \cap U_j} = \bar{U}_i \cap \bar{U}_j$. Then in Proposition I.9.7 we let $\mathscr{Q}(\mathring{X})$ be the category whose objects are unions of the sets $U_1, \ldots, U_q$, morphisms being inclusions, and we learned that there is a Mayer–Vietoris functor $\{h^q | q \in \mathbb{Z}\}$ on $\mathscr{Q}(\mathring{X})$ with $h^q(U) = H_{n-q}(X, X\backslash U)$ for every $U \in \mathscr{Q}(\mathring{X})$.

There is another Mayer–Vietoris functor $\{k^q | q \in \mathbb{Z}\}$ on $\mathscr{Q}(\mathring{X})$, defined as follows. For any $U \in \mathscr{Q}(\mathring{X})$ let $k^0(U)$ be the $\mathbb{Z}/2$-module of continuous functions $U \to \mathbb{Z}/2$, in the discrete topology of $\mathbb{Z}/2$, and let $k^q(U) = 0$ for $q \neq 0$. If $U \to V$ is a morphism (inclusion) in $\mathscr{Q}(\mathring{X})$, then the restriction to U of any element of $k^0(V)$ is clearly an element of $k^0(U)$, so that k^0 is indeed a contravariant functor from $\mathscr{Q}(\mathring{X})$ to the category $\mathfrak{M}$ of $\mathbb{Z}/2$-modules. The remainder of the verification that $\{k^q | q \in \mathbb{Z}\}$ is a Mayer–Vietoris functor on $\mathscr{Q}(\mathring{X})$ is trivial.

For any $U \in \mathscr{Q}(\mathring{X})$ and any $\alpha \in h^0(U)\,(=H_n(X, X\backslash U))$ let $\theta_U\alpha$ be the function $U \to \mathbb{Z}/2$ whose value on any $x \in U$ is the image of α under the homomorphism $H_n(X, X\backslash U) \to H_n(X, X\backslash\{x\}) = \mathbb{Z}/2$ induced by the inclusion $(X, X\backslash U) \to (X, X\backslash\{x\})$. A function $U \to \mathbb{Z}/2$ is continuous if and only if it is constant on each connected component of U, and since U is an open set in a manifold one can take "connected component" to mean "arcwise connected component." Now suppose that x and y are distinct points of the same arcwise connected component of U, so that they are end points of an embedded arc. One can fatten the arc into a closed n-cell $D^n \subset U$ containing both x and y in its interior $\mathring{D}^n$, for which the image of α under the homomorphism $H_n(X, X\backslash U) \to H_n(X, X\backslash\mathring{D}^n) = H_n(D^n, S^{n-1}) = \mathbb{Z}/2$ is the common value of $\theta_U\alpha(x)$ and $\theta_U\alpha(y)$. Hence $\theta_U\alpha$ is continuous, so that the map $\alpha \mapsto \theta_U\alpha$ is a $\mathbb{Z}/2$-module homomorphism $h^0(U) \xrightarrow{\theta_U} k^0(U)$. Naturality of the homomorphisms $\{\theta_U\}$ for $U \in \mathscr{Q}(\mathring{X})$ is trivial, and since $k^q(U) = 0$ for $q \neq 0$, the construction automatically extends to a natural transformation $h \xrightarrow{\theta} k$ of Mayer–Vietoris functors $\{h^q\}$ and $\{k^q\}$ on $\mathscr{Q}(\mathring{X})$.

1.2 Lemma: *For any smooth n-dimensional compact manifold X with interior $\mathring{X}$ the $\mathbb{Z}/2$-module homomorphisms $h^q(\mathring{X}) \xrightarrow{\theta_X} k^q(\mathring{X})$ are isomorphisms for $q \leqq 0$.*

PROOF: Let $\{\mathring{D}_i^n\}$ be the family of open n-cells $U_1, \ldots, U_q$ used to construct $\mathcal{Q}(\mathring{X})$. Then $h^q(\mathring{D}_i^n) = H_{n-q}(X, X\backslash\mathring{D}_i^n) = H_{n-q}(D_i^n, S_i^{n-1})$ for a closed n-cell D_i^n with boundary S_i^{n-1}, so that $h^0(\mathring{D}_i^n) = \mathbb{Z}/2$ and $h^q(\mathring{D}_i^n) = 0$ for $q < 0$. Since $k^q(\mathring{D}_i^n) = 0$ for $q < 0$, the homomorphisms $h^q(\mathring{D}_i^n) \xrightarrow{\theta_{\mathring{D}_i^n}} k^q(\mathring{D}_i^n)$ are the trivial isomorphisms $0 \to 0$ for $q < 0$. For any $\alpha \in h^0(\mathring{D}_i^n)$ $(=H_n(D_i^n, S_i^{n-1}))$ the function $\mathring{D}_i^n \xrightarrow{\theta_{\mathring{D}_i^n}\alpha} \mathbb{Z}/2$ carries $x \in \mathring{D}_i^n$ into the image of α under $H_n(D_i^n, S_i^{n-1}) \to H_n(D_i^n, D_i^n\backslash\{x\}) = \mathbb{Z}/2$, so that the homomorphisms $h^0(D_i^n) \xrightarrow{\theta_{\mathring{D}_i^n}} k^0(D_i^n)$ are also isomorphisms. Hence the Mayer–Vietoris comparison theorem (Theorem I.9.8) implies the desired result.

1.3 Proposition: *For any smooth n-dimensional compact manifold X with boundary $\dot{X} = X\backslash\mathring{X}$, the $\mathbb{Z}/2$-module $H_n(X, \dot{X})$ is free, with one basis element for each connected component of $\mathring{X}$.*

PROOF: For $q = 0$, Lemma 1.2 provides an isomorphism $H_n(X, \dot{X}) = H_n(X, X\backslash\mathring{X}) \xrightarrow{\theta_X} k^0(\mathring{X})$, where the $\mathbb{Z}/2$-module $k^0(\mathring{X})$ consists of the continuous functions $X \to \mathbb{Z}/2$. Clearly $k^0(X)$ is free, with one basis element for each connected component of X.

Since there is only one nonzero element in $\mathbb{Z}/2$, the basis elements assigned by Proposition 1.3 are unique, so that their sum is also unique.

1.4 Definition: For any smooth n-dimensional compact manifold X with boundary $\dot{X}$, the *fundamental $\mathbb{Z}/2$ homology class* $\mu_X \in H_n(X, \dot{X})$ is the sum of the basis elements described in Proposition 1.3.

1.5 Corollary: *Let X be any smooth n-dimensional compact manifold with interior $\mathring{X}$ and boundary $\dot{X} = X\backslash\mathring{X}$, and for any $x \in \mathring{X}$ let j_x be the inclusion $(X, \dot{X}) = (X, X\backslash\mathring{X}) \to (X, X\backslash\{x\})$. Then the fundamental class $\mu_X \in H_n(X, \dot{X})$ is uniquely characterized by the property that $(j_x)_*\mu_X \in H_n(X, X\backslash\{x\})$ generates for each $x \in X$.*

PROOF: This is just a reformulation of the case $q = 0$ of Lemma 1.2.

1.6 Corollary: *Let X be a smooth compact n-dimensional manifold, with interior $\mathring{X}$ and (smooth) boundary $\dot{X} = X\backslash\mathring{X}$ as usual, let $\mathcal{Q}(\mathring{X})$ be the category of open sets in $\mathring{X}$ described just before Proposition I.9.6, and for any $U \in \mathcal{Q}(\mathring{X})$ let $\bar{U}$ be the closure of U in X, with boundary $\dot{\bar{U}} = \bar{U}\backslash U$. Then there is a unique class $\mu_{\bar{U}} \in H_n(\bar{U}, \dot{\bar{U}})$ with the property that if $(\bar{U}, \dot{\bar{U}}) \xrightarrow{j_x} (\bar{U}, \bar{U}\backslash\{x\})$ is the inclusion for any $x \in U$ the element $(j_x)_*\mu_{\bar{U}}$ generates $H_n(\bar{U}, \bar{U}\backslash\{x\})$.*

PROOF: The method of Lemma 1.2 and Proposition 1.3 applies equally well to any $U \in \mathcal{Q}(\mathring{X})$, not just to $\mathring{X} \in \mathcal{Q}(\mathring{X})$ itself.

The class $\mu_{\bar{U}}$ is the *fundamental* $\mathbb{Z}/2$ *homology class of* $\bar{U}$, for any $U \in \mathscr{Q}(\mathring{X})$. It will appear along with the following constructions as part of a Mayer–Vietoris argument in the next section.

1.7 Corollary: *If X is a smooth compact manifold as in Corollary 1.6, then for any inclusion $U \xrightarrow{f} V$ of $U \in \mathscr{Q}(\mathring{X})$ into $V \in \mathscr{Q}(\mathring{X})$ there is an induced $\mathbb{Z}/2$-module homomorphism $H_{n-q}(\bar{V}, \dot{\bar{V}}) \xrightarrow{f^*} H_{n-q}(\bar{U}, \dot{\bar{U}})$ for every $q \in \mathbb{Z}$, such that for $q = 0$ one has $f^*\mu_{\bar{V}} = \mu_{\bar{U}} \in H_n(\bar{U}, \dot{\bar{U}})$.*

Proof: Let g and h be the inclusions $(\bar{V}, \dot{\bar{V}}) = (\bar{V}, \bar{V}\backslash V) \to (\bar{V}, \bar{V}\backslash U)$ and $(\bar{U}, \dot{\bar{U}}) = (\bar{U}, \bar{U}\backslash U) \to (\bar{V}, \bar{V}\backslash U)$, respectively. Then h is an excision, so that $H_{n-q}(\bar{U}, \dot{\bar{U}}) \xrightarrow{h_*} H_{n-q}(\bar{V}, \bar{V}\backslash U)$ is an isomorphism, and one can define f^* as the composition $(h_*)^{-1} \circ g_*$. The property $f^*\mu_{\bar{V}} = \mu_{\bar{U}}$ is a consequence of the characterization of fundamental classes given in Corollaries 1.5 and 1.6.

1.8 Corollary: *If X is a smooth compact n-dimensional manifold, then there is a Mayer–Vietoris functor $\{k^q \mid q \in \mathbb{Z}\}$ on $\mathscr{Q}(\mathring{X})$ with $k^q(U) = H_{n-q}(\bar{U}, \dot{\bar{U}})$ for every $U \in \mathscr{Q}(\mathring{X})$ and every $q \in \mathbb{Z}$, such that the homomorphisms $k^q(U \cup V) \xrightarrow{i^*_{U,V}} k^q(U) \oplus k^q(V)$ and $k^q(U) \oplus k^q(V) \xrightarrow{j^*_{U,V}} k^q(U \cap V)$ are obtained as in §I.9 by applying Corollary 1.7 to the inclusions $U \xrightarrow{i_U} U \cup V$, $V \xrightarrow{i_V} U \cup V$, $U \cap V \xrightarrow{j_U} U$, and $U \cap V \xrightarrow{j_V} V$.*

Proof: According to Proposition I.9.7 there is a Mayer–Vietoris functor $\{k^q \mid q \in \mathbb{Z}\}$ on $\mathscr{Q}(\mathring{X})$ with $k^q(U) = H_{n-q}(X, X\backslash U)$ for each $U \in \mathscr{Q}(\mathring{X})$. However, the excision $(\bar{U}, \dot{\bar{U}}) = (\bar{U}, \bar{U}\backslash U) \to (X, X\backslash U)$ induces isomorphism $H_{n-q}(\bar{U}, \dot{\bar{U}}) \xrightarrow{\approx} H_{n-q}(X, X\backslash U)$, and one uses further such excision isomorphisms to verify that the homomorphisms $i^*_{U,V}$ and $j^*_{U,V}$ implicit in Proposition I.9.7 correspond to the homomorphisms $i^*_{U,V}$ and $j^*_{U,V}$ obtained from Corollary 1.7.

In case X is *closed* in the usual sense that its boundary $\dot{X}$ is empty, then the fundamental class of Definition 1.4 is an element of $H_n(X)$.

In case X does have a nonempty boundary $\dot{X}$, then $\dot{X}$ is itself an $(n-1)$-dimensional manifold which is closed and smooth. Furthermore, the connecting homomorphism ∂ in the exact $\mathbb{Z}/2$ homology sequence

$$H_n(X) \xrightarrow{j_*} H_n(X, \dot{X}) \xrightarrow{\partial} H_{n-1}(\dot{X}) \xrightarrow{i_*} H_{n-1}(X)$$

carries the fundamental class $\mu_X \in H_n(X, \mathring{X})$ into a class $\partial\mu_X \in H_{n-1}(\dot{X})$.

1.9 Proposition: *Let X be a smooth compact n-dimensional manifold with a nonempty boundary $\dot{X}$, and with fundamental $\mathbb{Z}/2$ homology class $\mu_X \in H_n(X, \dot{X})$. Then the class $\partial\mu_X \in H_{n-1}(\dot{X})$ is the fundamental $\mathbb{Z}/2$ homology class $\mu_{\dot{X}} \in H_{n-1}(\dot{X})$ of the boundary $\dot{X}$.*

PROOF: As in Corollary 1.5 the class $\mu_{\dot{X}} \in H_{n-1}(\dot{X})$ is uniquely characterized by the property that if $\dot{X} \xrightarrow{j_y} \dot{X}, \dot{X}\backslash\{y\}$ is the inclusion induced by any $y \in \dot{X}$, then $(j_y)_*\mu_{\dot{X}}$ generates $H_{n-1}(\dot{X}, \dot{X}\backslash\{y\})$; we shall show that $\partial\mu_X$ has the same property. Let $x \in \mathring{X}$ lie in the same arcwise connected component of X as $y \in \mathring{X}$, let l be an arc connecting x and y, with interior in $\mathring{X}$, and let $(X, \dot{X}) \xrightarrow{j_x} (X, X\backslash\{x\})$ be the inclusion. Then l can be fattened to an n-cell D^n

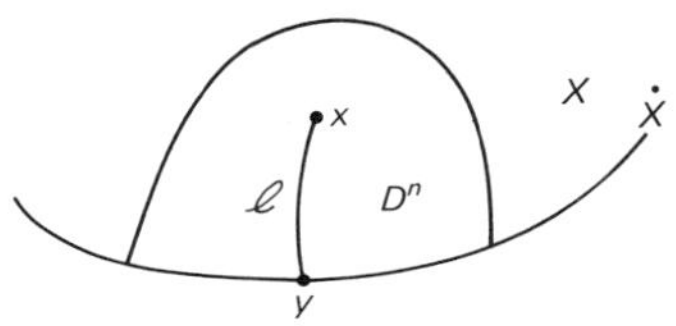

containing $l\backslash\{y\}$ in its interior (see the accompanying figure). One easily verifies that the diagram

$$\begin{array}{ccccc}
H_n(X, \dot{X}) & \xrightarrow{\partial} & H_{n-1}(\dot{X}) & \xrightarrow{(j_y)_*} & H_{n-1}(\dot{X}, \dot{X}\backslash\{y\}) \\
\downarrow{\scriptstyle (j_x)_*} & & \downarrow & & \downarrow{\scriptstyle \approx} \\
H_n(X, X\backslash\{x\}) & \xrightarrow{\partial} & H_{n-1}(X\backslash\{x\}) & \longrightarrow & H_{n-1}(X\backslash\{x\}, X\backslash l) \\
\uparrow{\scriptstyle \approx} & & \uparrow & & \uparrow{\scriptstyle \approx} \\
H_n(D^n, D^n\backslash\{x\}) & \xrightarrow[\approx]{\partial} & H_{n-1}(D^n\backslash\{x\}) & \xrightarrow[\approx]{} & H_{n-1}(D^n\backslash\{x\}, D^n\backslash l)
\end{array}$$

commutes, all maps except the three connecting homomorphisms ∂ arising from inclusions. Three of the vertical maps are excision isomorphisms, as indicated, and the two lower maps are isomorphisms because they are portions of long exact sequences

$$H_n(D^n) \longrightarrow H_n(D^n, D^n\backslash\{x\}) \xrightarrow[\approx]{\partial} H_{n-1}(D^n\backslash\{x\}) \longrightarrow H_{n-1}(D^n)$$

and

$$H_{n-1}(D^n\backslash l) \longrightarrow H_{n-1}(D^n\backslash\{x\}) \xrightarrow[\approx]{} H_{n-1}(D^n\backslash\{x\}, D^n\backslash l) \xrightarrow{\partial} H_{n-2}(D^n\backslash l)$$

whose outer terms vanish because D^n is an n-cell and $D^n\backslash l$ is homotopy equivalent to an n-cell. (For $n = 2$ the second sequence still produces the desired isomorphism because $H_0(D^n\backslash l) \to H_0(D^n\backslash\{x\})$ is an isomorphism; the adjustments needed for the degenerate case $n = 1$ are trivial.) Since $j_{x*}\mu_X$ generates $H_n(X, X\backslash\{x\})$, it follows that $j_{y*}\partial\mu_X$ generates $H_{n-1}(\dot{X}, \dot{X}\backslash\{y\})$, as required.

2. $\mathbb{Z}/2$ Poincaré–Lefschetz Duality

Let X be a smooth compact n-dimensional manifold with boundary $\dot{X}$, and let $\mu_X \in H_n(X, \dot{X})$ be the fundamental $\mathbb{Z}/2$ homology class of Definition 1.4. One of the classical products of singular homology and cohomology is the *cap product* $H^q(X) \otimes H_n(X, \dot{X}) \xrightarrow{\cap} H_{n-q}(X, \dot{X})$ for $q \in \mathbb{Z}$, some of whose properties will be recalled as needed. (Further details about cap products are readily available on pages 238–245 of Dold [8] and on pages 254–255 of Spanier [4], for example.) In particular, cap product $\cap\mu_X$ by the fundamental $\mathbb{Z}/2$ homology class μ_X is a $\mathbb{Z}/2$-module homomorphism $H^q(X) \xrightarrow{\cap\mu_X} H_{n-q}(X, \dot{X})$ for any $q \in \mathbb{Z}$. We show in this section that $\cap\mu_X$ is an isomorphism for any $q \in \mathbb{Z}$.

We continue to use the category $\mathscr{Q}(\mathring{X})$ of open sets described just before Proposition I.9.6. Specifically, $\mathscr{Q}(\mathring{X})$ consists of unions of certain open n-cells $U_1, \ldots, U_r$ covering $\mathring{X}$, such that each nonvoid intersection $U_i \cap U_j$ of two such open n-cells is itself one of the open n-cells U_k, and such that the closures in X satisfy $\overline{U_i \cap U_j} = \bar{U}_i \cap \bar{U}_j$ in addition to the usual identity $\overline{U_i \cup U_j} = \bar{U}_i \cup \bar{U}_j$. One Mayer–Vietoris functor $\{h^q \mid q \in \mathbb{Z}\}$ on $\mathscr{Q}(\mathring{X})$ is given in Proposition I.9.6 itself: one has $h^q(U) = H^q(\bar{U})$ for every $U \in \mathscr{Q}(\mathring{X})$. Another Mayer–Vietoris functor $\{k^q \mid q \in Z\}$ on $\mathscr{Q}(\mathring{X})$ is given in Proposition I.9.7, and modified via excision isomorphisms in Corollary 1.8: one has $k^q(U) = H_{n-q}(\bar{U}, \dot{\bar{U}})$ for every $U \in \mathscr{Q}(\mathring{X})$. The following lemma will be used to construct a natural transformation from $\{h^q \mid q \in \mathbb{Z}\}$ to $\{k^q \mid q \in \mathbb{Z}\}$.

2.1 Lemma: *Let $U \xrightarrow{f} V$ be a morphism (inclusion of one open set into another) in $\mathscr{Q}(\mathring{X})$, inducing an inclusion $\bar{U} \xrightarrow{f} \bar{V}$, hence a $\mathbb{Z}/2$-module homorphism $H^q(\bar{V}) \xrightarrow{f^*} H^q(\bar{U})$ for any $q \in \mathbb{Z}$, and let $H_{n-q}(\bar{V}, \dot{\bar{V}}) \xrightarrow{f^*} H_{n-q}(\bar{U}, \dot{\bar{U}})$ be the induced $\mathbb{Z}/2$-module homomorphism of Corollary 1.7. Then cap products by the $\mathbb{Z}/2$ fundamental classes $\mu_{\bar{V}} \in H_n(\bar{V}, \dot{\bar{V}})$ and $\mu_{\bar{U}} \in H_n(\bar{U}, \dot{\bar{U}})$ provide commutative diagrams*

$$
\begin{array}{ccc}
H^q(\bar{V}) & \xrightarrow{\quad f^* \quad} & H^q(\bar{U}) \\
\Big\downarrow {\scriptstyle \cap\mu_{\bar{V}}} & & \Big\downarrow {\scriptstyle \cap\mu_{\bar{U}}} \\
H_{n-q}(\bar{V}, \dot{\bar{V}}) & \xrightarrow{\quad f^* \quad} & H_{n-q}(\bar{U}, \dot{\bar{U}}).
\end{array}
$$

PROOF: The lower $\mathbb{Z}/2$-module homomorphism f^* was described in Corollary 1.7 as a composition, which appears in the bottom line of the following diagram:

$$\begin{array}{ccccc} H^q(\bar V) & \xleftarrow{\quad \text{id}^* \quad} & H^q(\bar V) & \xrightarrow{\quad f^* \quad} & H^q(\bar U) \\ \downarrow{\scriptstyle \cap\mu_{\bar V}} & & {\scriptstyle \cap g_*\mu_{\bar V}}\downarrow{\scriptstyle \cap h_*\mu_{\bar U}} & & \downarrow{\scriptstyle \cap\mu_{\bar U}} \\ H_{n-q}(\bar V, \bar V) & \xrightarrow{\ g_*\ } & H_{n-q}(\bar V, \bar V\backslash U) & \underset{\approx}{\xleftarrow{\ h_*\ }} & H_{n-q}(\bar U, \bar U). \end{array}$$

Each of the two squares commutes by naturality of the cap product (as on page 239 of Dold [8] or on page 254 of Spanier [4]). The ambiguity of the definition of the middle vertical arrow is quickly resolved: each of the classes $g_*\mu_{\bar V} \in H_n(\bar V, \bar V\backslash U)$ and $h_*\mu_{\bar U} \in H_n(\bar V, \bar V\backslash U)$ is uniquely defined by the property that if $(\bar V, \bar V\backslash U) \xrightarrow{j_x} (\bar V, \bar V\backslash\{x\})$ is the inclusion for any $x \in U$, then each of $(j_x)_* g_* \mu_{\bar V}$ and $(j_x)_* h_* \mu_{\bar U}$ is the unique generator of $H_n(\bar V, \bar V\backslash\{x\})$.

2.2 Lemma: *For any smooth compact n-dimensional manifold X with boundary $\dot X$ there is a natural transformation θ from the Mayer–Vietoris functor $\{h^q \mid q \in \mathbb{Z}\}$ on $\mathcal{Q}(\mathring{X})$ to the Mayer–Vietoris functor $\{k^q \mid q \in \mathbb{Z}\}$ on $\mathcal{Q}(\mathring{X})$, the homomorphisms $h^q(U) \xrightarrow{\theta_U} k^q(U)$ being cap products $H^q(\bar U) \xrightarrow{\cap\mu_{\bar U}} H_{n-q}(\bar U, \dot{\bar U})$ by $\mathbb{Z}/2$ fundamental classes $\mu_{\bar U} \in H_n(\bar U, \dot{\bar U})$, for each $U \in \mathcal{Q}(\mathring{X})$ and each $q \in \mathbb{Z}$.*

PROOF: One must show that the diagram

$$\begin{array}{ccccccc} h^q(U \cup V) & \xrightarrow{i^*_{U,V}} & h^q(U) \oplus h^q(V) & \xrightarrow{j^*_{U,V}} & h^q(U \cap V) & \xrightarrow{\delta_{U,V}} & h^{q+1}(U \cup V) \\ \downarrow{\scriptstyle \theta_{U\cup V}} & & \downarrow{\scriptstyle \theta_U \oplus \theta_V} & & \downarrow{\scriptstyle \theta_{U\cap V}} & & \downarrow{\scriptstyle \theta_{U\cup V}} \\ k^q(U \cup V) & \xrightarrow{i^*_{U,V}} & k^q(U) \oplus k^q(V) & \xrightarrow{j^*_{U,V}} & k^q(U \cap V) & \xrightarrow{\delta_{U,V}} & k^{q+1}(U \cup V) \end{array}$$

commutes for every $U \in \mathcal{Q}(\mathring{X})$, $V \in \mathcal{Q}(\mathring{X})$, and $q \in \mathbb{Z}$. However, both the left-hand and middle squares break into two copies of Lemma 2.1, using the inclusions $U \xrightarrow{i_U} U \cup V$ and $V \xrightarrow{i_V} U \cup V$ for the left-hand square, and using the inclusions $U \cap V \xrightarrow{j_U} U$ and $U \cap V \xrightarrow{j_V} V$ for the middle square. The right-hand square commutes by the stability property of the cap product (as on pages 239–240 of Dold [8]), of which a special case asserts that the diagram

$$\begin{array}{ccc} H^q(\overline{U \cap V}) & \xrightarrow{\qquad \delta \qquad} & H^{q+1}(\overline{U \cup V}) \\ \downarrow{\scriptstyle \cap\mu_{U\cap V}} & & \downarrow{\scriptstyle \cap\mu_{U\cup V}} \\ H_{n-q}(\overline{U \cap V}, \dot{\overline{U \cap V}}) & \xrightarrow{\ \partial\ } & H_{n-q-1}(\overline{U \cup V}, \dot{\overline{U \cup V}}) \end{array}$$

commutes, for the Mayer–Vietoris connecting homomorphisms δ and ∂ in cohomology and homology, respectively.

2.3 Theorem **($\mathbb{Z}/2$ Poincaré–Lefschetz Duality):** *Let X be a smooth compact n-dimensional manifold, with boundary $\dot{X}$ and $\mathbb{Z}/2$ fundamental class $\mu_X \in H_n(X, \dot{X})$; then for each $q \in \mathbb{Z}$ the cap product $H^q(X) \xrightarrow{\cap \mu_X} H_{n-q}(X, \dot{X})$ is a $\mathbb{Z}/2$-module isomorphism.*

PROOF: Let θ be the natural transformation of Lemma 2.2, from the Mayer–Vietoris functor $\{h^q \mid q \in \mathbb{Z}\}$ on $\mathscr{Q}(\mathring{X})$ to the Mayer–Vietoris functor $\{k^q \mid q \in \mathbb{Z}\}$ on $\mathscr{Q}(\mathring{X})$. Let $\mathring{D}^n$ be any one of the open n-cells $U_1, \ldots, U_r$ used to construct the category $\mathscr{Q}(\mathring{X})$, the closure of $\mathring{D}^n$ being an n-disk D^n with boundary S^{n-1}. Both $H^q(D^n) = 0$ and $H_{n-q}(D^n, S^{n-1}) = 0$ for $q \neq 0$, and since $1 \cap \mu_{D^n} = \mu_{D^n}$ for the generator $1 \in H^0(D^n)$ and the fundamental $\mathbb{Z}/2$ class $\mu_{D^n} \in H_n(D^n, S^{n-1})$ (as on page 239 of Dold [8] or on page 254 of Spanier [4]) it follows that each $h^q(U_i) \xrightarrow{\theta_{U_i}} k^q(U_i)$ is an isomorphism. Hence each $h^q(\mathring{X}) \xrightarrow{\theta_{\mathring{X}}} k^q(\mathring{X})$ is an isomorphism by the Mayer–Vietoris comparison theorem (Theorem I.9.8); that is, $H^q(X) \xrightarrow{\cap \mu_X} H_{n-q}(X, \mathring{X})$ is an isomorphism for any $q \in \mathbb{Z}$, as asserted.

Recall that a compact manifold is *closed* if its boundary is empty.

2.4 Corollary **($\mathbb{Z}/2$ Poincaré Duality):** *Let X be a smooth closed n-dimensional manifold with $\mathbb{Z}/2$ fundamental class $\mu_X \in H_n(X)$; then for any $q \in \mathbb{Z}$ the cap product $H^q(X) \xrightarrow{\cap \mu_X} H_{n-q}(X)$ is a $\mathbb{Z}/2$-module isomorphism.*

PROOF: This is the special case $\dot{X} = \varnothing$ of Theorem 2.3.

There are also $\mathbb{Z}/2$ Poincaré–Lefschetz duality isomorphisms $H^{n-q}(X, \dot{X}) \to H_q(X)$ that specialize to Poincaré duality isomorphisms; see Exercise 9.41. Explicit inverses to Poincaré duality isomorphisms will appear in Corollary 7.9.

3. Multiplicative $\mathbb{Z}/2$ Classes of RP^n

As indicated in the Introduction on this chapter, we now put $\mathbb{Z}/2$ fundamental classes μ_X and $\mathbb{Z}/2$ Poincaré–Lefschetz duality isomorphisms $\cap \mu_X$ aside, temporarily. The classes μ_X will reappear in §5, and the isomorphisms $\cap \mu_X$ will reappear in §7.

In this section we compute the Stiefel–Whitney classes $w(\tau(RP^n))$ and the dual Stiefel–Whitney class $\overline{w}(\tau(RP^n))$ of the (tangent bundle of the) real projective space RP^n. The former computation will show that $w_1(\tau(RP^n)) = 0$, hence that RP^n is orientable, if and only if n is odd. The latter computation will be used in the next section to obtain necessary conditions for the existence of immersions $RP^n \subseteq \mathbb{R}^{2n-p}$ and embeddings $RP^n \subset \mathbb{R}^{2n-q}$.

3.1 Proposition: *Let $\tau(RP^n)$ be the tangent bundle of RP^n, and let γ_n^1 be the canonical real line bundle over RP^n. Then*

$$w(\tau(RP^n)) = (1 + e(\gamma_n^1))^{n+1}$$

for the total Stiefel–Whitney class $w(\tau(RP^n)) \in H^*(RP^n)$ and Euler class $e(\gamma_n^1) \in H^1(RP^n)$.

PROOF: In Proposition III.7.4 we learned that $\tau(RP^n) \oplus \varepsilon^1 = (n+1)\gamma_n^1$, so that the Whitney product formula and the normalization of Stiefel–Whitney classes give

$$\begin{aligned} w(\tau(RP^n)) &= w(\tau(RP^n)) \cup w(\varepsilon^1) = w(\tau(RP^n) \oplus \varepsilon^1) \\ &= w((n+1)\gamma_n^1) = w(\gamma_n^1)^{n+1} = (1 + e(\gamma_n^1))^{n+1}, \end{aligned}$$

as claimed, since $w(\varepsilon^1) = 1$ by Corollary V.2.6.

In computing the product $(1 + e(\gamma_n^1))^{n+1}$ it is understood that the highest term $e(\gamma_n^1)^{n+1}$ vanishes, since the $(n+1)$st cohomology module $H^{n+1}(RP^n)$ of the n-dimensional CW space RP^n vanishes. It is also understood that if one expands $(1 + e(\gamma_n^1))^{n+1}$ by the binomial theorem, then the coefficient of $e(\gamma_n^1)^p$ is the $\mathbb{Z}/2$ image of the integral binomial coefficient $\binom{n+1}{p}$.

3.2 Corollary: *RP^n is orientable if n is odd; RP^n is nonorientable if n is even.*

PROOF: The $\mathbb{Z}/2$ image of $\binom{n+1}{1}$ is $0 \in \mathbb{Z}/2$ if n is odd, and $1 \in \mathbb{Z}/2$ if n is even. However, the product of this coefficient and $e(\gamma_n^1) \in H^1(RP^n)$ is the first Stiefel–Whitney class $w_1(\tau(RP^n))$, to which one then applies Proposition V.3.2.

We need the following lemma to compute the dual Stiefel–Whitney class $\bar{w}(\tau(RP^n)) \in H^*(RP^n)$.

3.3 Lemma: *For any element α of a $\mathbb{Z}/2$-algebra, and for any natural number $r \geqq 0$, one has $(1+\alpha)^{2^r} = 1 + \alpha^{2^r}$.*

PROOF: This is trivial for $r = 0$, and the computation

$$(1+\alpha)^{2^{r+1}} = ((1+\alpha)^{2^r})^2 = (1 + \alpha^{2^r})^2 = 1 + 2\alpha^{2^r} + \alpha^{2^{r+1}} = 1 + \alpha^{2^{r+1}}$$

is the inductive step.

3.4 Proposition: *If $2^r \leqq n < 2^{r+1}$ the dual Stiefel–Whitney class of RP^n is given by* $\overline{w}(\tau(RP^n)) = (1 + e(\gamma_n^1))^{2^{r+1}-n-1} \in H^*(RP^n)$, where $e(\gamma_n^1) \in H^1(RP^n)$ *is the $\mathbb{Z}/2$ Euler class of the canonical line bundle γ_n^1.*

PROOF: Since $H^{2^{r+1}}(RP^n) = 0$ by the hypothesis $n < 2^{r+1}$, Proposition 3.1 and Lemma 3.3 imply that

$$\begin{aligned} w(\tau(RP^n))(1 + e(\gamma_n^1))^{2^{r+1}-n-1} &= (1 + e(\gamma_n^1))^{n+1}(1 + e(\gamma_n^1))^{2^{r+1}-n-1} \\ &= (1 + e(\gamma_n^1))^{2^{r+1}} = 1, \end{aligned}$$

hence that $(1 + e(\gamma_n^1))^{2^{r+1}-n-1}$ is the unique class satisfying the conclusion of Lemma V.6.3.

4. Nonimmersions and Nonembeddings

According to the Whitney immersion theorem (Theorem I.6.9) there is at least one immersion $X \to S^{2n-1}$ of any smooth n-dimensional manifold X into the $(2n - 1)$-sphere S^{2n-1}; if $n > 1$, then there is at least one immersion $X \xrightarrow{f} \mathbb{R}^{2n-1}$ of X into the euclidean space $\mathbb{R}^{2n-1}$. Similarly, according to the Whitney embedding theorem (Theorem I.6.6) there is at least one proper embedding $X \xrightarrow{g} \mathbb{R}^{2n}$ of X into the euclidean space $\mathbb{R}^{2n}$. We shall show for some dimensions $n > 0$ that these results are best-possible: for certain $n > 0$ there are smooth n-dimensional manifolds which cannot be immersed in $\mathbb{R}^{2n-2}$ nor embedded in $\mathbb{R}^{2n-1}$. The proof gives a suggestion of best-possible results for any dimension $n > 0$.

In view of the Whitney immersion and embedding theorems, it is reasonable to try to find the largest natural number $p \geqq 0$ such that a given smooth n-dimensional manifold X immerses in $\mathbb{R}^{2n-p}$ or embeds in $\mathbb{R}^{2n-p}$. The tangent bundle $\tau(\mathbb{R}^{2n-p})$ is the trivial bundle ε^{2n-p} over $\mathbb{R}^{2n-p}$, and ε^{2n-p} has a riemannian metric induced by the usual inner product $\mathbb{R}^{2n-p} \times \mathbb{R}^{2n-p} \xrightarrow{\langle\,,\,\rangle} \mathbb{R}$. Then if $X \xrightarrow{f} \mathbb{R}^{2n-p}$ is an immersion, the pullback $f^!\tau(\mathbb{R}^{2n-p})$ is the trivial bundle ε^{2n-p} over X, and according to Lemma III.3.3 there is an induced riemannian metric $\langle\,,\,\rangle$ on $f^!\tau(\mathbb{R}^{2n-p})$. Since f is an immersion, the tangent bundle $\tau(X)$ of X itself is a subbundle of $f^!\tau(\mathbb{R}^{2n-p})$, of rank n, and according to Proposition III.3.6 the riemannian metric on $f^!\tau(\mathbb{R}^{2n-p})$ provides a well-defined subbundle ν_f of rank $n - p$ such that $\tau(X) \oplus \nu_f = f^!\tau(\mathbb{R}^{2n-p}) = \varepsilon^{2n-p}$ over X.

4.1 Definition: For any immersion $X \xrightarrow{f} \mathbb{R}^{2n-p}$ of a smooth n-dimensional manifold X into the euclidean space $\mathbb{R}^{2n-p}$, the preceding $(n - p)$-plane bundle ν_f over X is the *normal bundle* of the immersion f.

Since $\tau(X) \oplus \nu_f = \varepsilon^{2n-p}$, it follows from Corollary V.6.5 that the total Stiefel–Whitney class $w(\nu_f) \in H^*(X)$ of the normal bundle ν_f agress with the *dual* Stiefel–Whitney class $\bar{w}(\tau(X)) \in H^*(X)$ of the tangent bundle: $w(\nu_f) = \bar{w}(\tau(X))$. Furthermore, since $\bar{w}(\tau(X))$ is independent of f, one can use its properties to establish properties of *any* immersions $X \to \mathbb{R}^{2n-p}$, including their nonexistence for sufficiently large values of p. We now carry out such a program.

4.2 Lemma: *Let $\bar{w}(\tau(X))$ be the dual Stiefel–Whitney class of the tangent bundle of a smooth n-dimensional manifold X, and suppose that X immerses in $\mathbb{R}^{2n-p}$ for some $p > 0$; then $\bar{w}_q(\tau(X)) = 0 \in H^q(X)$ for any $q > n - p$.*

PROOF: The normal bundle ν_f of any immersion $X \xrightarrow{f} \mathbb{R}^{2n-p}$ is of rank $n - p$, so that $w_q(\nu_f) = 0$ for $q > n - p$ by the dimension axiom of Theorem V.5.2. However, we have just noted that $w(\nu_f) = \bar{w}(\tau(X))$, so that $\bar{w}_q(\tau(X)) = 0$ for $q > n - p$, as claimed.

Thus if $\bar{w}_q(\tau(X)) \neq 0$ for some $q > n - p$ then X cannot possibly immerse in $\mathbb{R}^{2n-p}$. Here is the simplest example of such a result.

4.3 Proposition: *If $n = 2^r$ for $r \geqq 0$, then the real projective space RP^n does not immerse in $\mathbb{R}^{2n-2}$.*

PROOF: According to Proposition 3.4 the dual Stiefel–Whitney class of RP^n is given by

$$\bar{w}(\tau(RP^n)) = (1 + e(\gamma_n^1))^{2^{r+1}-2^r-1} = (1 + e(\gamma_n^1))^{n-1},$$

so that $\bar{w}_{n-1}(\tau(RP^n))$ is the nonzero element $e(\gamma_n^1)^{n-1} \in H^{n-1}(RP^n)$.

There is a natural generalization of Proposition 4.3, which will be given in the following notation.

4.4 Definition: For any natural number $n > 0$ let $\alpha(n) > 0$ be the number of 1's in the dyadic expansion of n; that is, if $n = a_0 2^0 + a_1 2^1 + \cdots + a_r 2^r$, where the natural numbers $a_0, a_1, \ldots, a_r$ are either 0 or 1, then $\alpha(n) = \sum_{j=0}^r a_j$.

4.5 Proposition: *For any natural number $n > 0$ there is a smooth closed n-dimensional manifold X that does not immerse in $\mathbb{R}^{2n-\alpha(n)-1}$.*

PROOF: For the set J of those indices j with $a_j = 1$ in the expansion $n = a_0 2^0 + \cdots + a_r 2^r$, let X be the product $\prod_{j \in J} RP^{2^j}$. Then, for the projections $X \xrightarrow{\mathrm{pr}_j} RP^{2^j}$, the tangent bundle of X is a Whitney sum $\tau(X) = \bigoplus_{j \in J} \mathrm{pr}_j^! \tau(RP^{2^j})$, so that the dual Stiefel–Whitney class of $\tau(X)$ is a cup product $\bar{w}(\tau(X)) = \prod_{j \in J} \bar{w}(\mathrm{pr}_j^!\tau(RP^{2^j})) = \prod_{j \in J} \mathrm{pr}_j^* \bar{w}(\tau(RP^{2^j}))$. The highest-order nonvanishing class in $\bar{w}(\tau(RP^{2^j}))$ is $\bar{w}_{2^j-1}(\tau(RP^{2^j})) = e(\gamma_{2^j}^1)^{2^j-1} \in H^{2^j-1}(RP^{2^j})$, as in Proposition 4.3. Since the Künneth isomorphism $\bigotimes_{j \in J} H^*(RP^{2^j}) \xrightarrow[\approx]{\times} H^*(X)$ carries the highest-order nonvanishing tensor product $\bigotimes_{j \in J} \bar{w}_{2^j-1}(\tau(RP^{2^j}))$ into the cup product $\prod_{j \in J} \mathrm{pr}_j^* \bar{w}_{2^j-1}(\tau(RP^{2^j}))$, and since $\sum_{j \in J}(2^j - 1) = n - \alpha(n)$, it follows that $\bar{w}_{n-\alpha(n)}(\tau(X)) \neq 0 \in H^{n-\alpha(n)}(X)$. Hence by Lemma 4.2 the product X cannot immerse in $\mathbb{R}^{2n-\alpha(n)-1}$.

One cannot improve Proposition 4.5 without further specialization, for the following reason: *for any natural number $n > 1$, every smooth n-dimensional manifold X immerses in $\mathbb{R}^{2n-\alpha(n)}$*. This "best-possible" immersion theorem of Cohen [1] is briefly discussed in Remark 9.14.

Proper embeddings are specialized immersions, for which there are improved versions of Propositions 4.3 and 4.5. Specifically, an immersion $X \xrightarrow{f} \mathbb{R}^{2n-p}$ is an *embedding* if it is injective ($f(x) = f(y) \in \mathbb{R}^{2n-p}$ implies $x = y \in X$), and it is *proper* if inverse images of compact sets in $\mathbb{R}^{2n-p}$ are compact in X. For example, any embedding of a compact manifold is necessarily proper.

For any proper embedding $X \xrightarrow{f} \mathbb{R}^{2n-p}$ of a smooth n-dimensional manifold X, for any $\varepsilon > 0$, and for any connected open set $V \subset X$ whose closure $\bar{V}$ is diffeomorphic to a closed disk $D^n \subset \mathbb{R}^n$, let $C_\varepsilon(V)$ denote the set of points $y \in \mathbb{R}^{2n-p}$ such that $|f(x) - y| < \varepsilon$ for some $x \in V$. Since the closure $\overline{C_\varepsilon(V)}$ is compact, the inverse image $f^{-1}(\overline{C_\varepsilon(V)})$ consists of at most finitely many connected components, so that for sufficiently small $\varepsilon > 0$ the set $f^{-1}(\overline{C_\varepsilon(V)})$ contains no points outside the connected component of $\bar{V} \subset X$. In what follows, a *cocoon* $C(V) \subset \mathbb{R}^{2n-p}$ of the image $f(V) \quad \mathbb{R}^{2n-p}$ is any set of the form $C_\varepsilon(V)$, where $\varepsilon > 0$ is chosen in such a way that $f^{-1}(\overline{C_{3\varepsilon}(V)})$ contains no points outside the connected component of $\bar{V} \subset X$.

4.6 Lemma: *Let $E \xrightarrow{\pi} X$ represent the normal bundle ν_f of a proper embedding $X \xrightarrow{f} \mathbb{R}^{2n-p}$ of a smooth n-dimensional manifold X. Then there is a diffeomorphism of some open neighborhood of the zero-section of E onto an open neighborhood $\tilde{E} \subset \mathbb{R}^{2n-p}$ of the image $f(X) \subset \mathbb{R}^{2n-p}$.*

PROOF: Let $x = (x^1, \ldots, x^n)$ be local coordinate functions on some open set in X, so that the restriction of f is given by $f(x) = (f^1(x), \ldots, f^{2n-p}(x))$ for smooth functions $f^1, \ldots, f^{2n-p}$ on an open set $U \subset \mathbb{R}^n$, for which the jacobian matrix $(\partial f^i/\partial x^j) = (f_j^i)$ has rank n. By restriction to a connected open subset $V \subset U$, with closure $\bar{V}$ diffeomorphic to $D^n \subset R^n$, and by a relabeling of coordinates, if necessary, one can suppose that

$$\det \begin{pmatrix} f_1^1 & \cdots & f_1^n \\ \vdots & & \vdots \\ f_n^1 & \cdots & f_n^n \end{pmatrix} \neq 0.$$

It follows for each $x \in V$ and each $y = (y^{n+1}, \ldots, y^{2n-p}) \in \mathbb{R}^{n-p}$ that there is a unique $z = z(x, y) = (z^1, \ldots, z^{2n-p}) \in \mathbb{R}^{2n-p}$ such that

$$\left(\begin{array}{ccc|ccc} f_1^1 & \cdots & f_1^n & f_1^{n+1} & \cdots & f_1^{2-p} \\ \vdots & & \vdots & \vdots & & \vdots \\ f_n^1 & \cdots & f_n^n & f_n^{n+1} & \cdots & f_n^{2n-p} \\ \hline & & & 1 & & 0 \\ & 0 & & & \ddots & \\ & & & 0 & & 1 \end{array}\right) \begin{pmatrix} z^1 \\ \vdots \\ z^n \\ z^{n+1} \\ \vdots \\ z^{2n-p} \end{pmatrix} = \begin{pmatrix} 0 \\ \vdots \\ 0 \\ y^{n+1} \\ \vdots \\ y^{2n-p} \end{pmatrix},$$

hence a well-defined smooth map $V \times \mathbb{R}^{n-p} \xrightarrow{F} \mathbb{R}^{2n-p}$ carrying any point $(x, y) \in V \times \mathbb{R}^{n-p}$ into

$$(f^1(x) + z^1(x, y), \ldots, f^{2n-p}(x) + z^{2n-p}(x, y)) \in \mathbb{R}^{2n-p}.$$

Clearly F is affine in y, so that the restriction of F to $V \times \{0\}$ is f itself, and the restriction of the jacobian determinant of F to $V \times \{0\}$ is

$$\det \begin{pmatrix} f_1^1 & \cdots & f_1^n \\ \vdots & & \vdots \\ f_n^1 & \cdots & f_n^n \end{pmatrix} \neq 0.$$

Since f is an embedding, the inverse function theorem guarantees that the restriction of F to some open neighborhood W of $V \times \{0\}$ is a diffeomorphism. We let $\tilde{E}(V) \subset \mathbb{R}^{2n-p}$ be the (open) intersection $F(W) \cap C(V)$, for any cocoon $C(V) \subset \mathbb{R}^{2n-p}$ of $f(V)$.

The n linear conditions $f_j^1 z^1 + \cdots + f_j^{2n-p} z^{2n-p} = 0$ appearing in the definition of $z(x, y)$ guarantee that for each $(x, y) \in V \times \mathbb{R}^{n-p}$ the vector $z(x, y) \in \mathbb{R}^{2n-p}$ is orthogonal to the tangent space at $f(x)$ of the embedded $f(X)$, in the usual inner product of $\mathbb{R}^{2n-p}$; hence $z(x, y)$ can be identified as a point in the fiber over $x \in V$ of the normal bundle ν_f of the embedding $X \xrightarrow{f} \mathbb{R}^{2n-p}$. Specifically, one can regard $V \times \mathbb{R}^{n-p}$ as a trivialization of the restriction of the normal bundle to $V \subset X$, an open neighborhood of $V \times \{0\}$ being identified with the neighborhood $\tilde{E}(V)$ of $f(V)$. Since $\tilde{E}(V)$ lies in the cocoon $C(V)$, it is clear for any covering $\{V\}$ of X by open sets V of the type described earlier that the union $\bigcup_V \tilde{E}(V)$ is a diffeomorphic image $\tilde{E} \subset \mathbb{R}^{2n-p}$ of an open neighborhood of the zero-section of E, as required.

4.7 Lemma: *Let ν_f be the normal bundle of a proper embedding $X \xrightarrow{f} \mathbb{R}^{2n-p}$ of a smooth manifold X of dimension $n > 0$. Then the $\mathbb{Z}/2$ Euler class vanishes: $e(\nu_f) = 0 \in H^{n-p}(X)$.*

PROOF: By definition, $e(\nu_f) = \sigma^* j^* U_{\nu_f}$ for the $\mathbb{Z}/2$ Thom class $U_{\nu_f} \in H^{n-p}(E, E^*)$, where $E \xrightarrow{\pi} X$ represents ν_f, where $E \xrightarrow{j} E, E^*$ is the inclusion, and where $X \xrightarrow{\sigma} E$ the zero-section. By the excision axiom one may as well replace E by any open neighborhood of the image of σ, which we also denote E, with a corresponding $E^* \subset E$. Then Lemma 4.6 provides the isomorphisms f_0^*, f_1^*, and f^* in the accompanying commutative diagram, and the remain-

$$
\begin{array}{ccccc}
H^{n-p}(E, E^*) & \xrightarrow{j^*} & H^{n-p}(E) & \xrightarrow{\sigma^*} & H^{n-p}(X) \\
\uparrow f_0^* \approx & & \uparrow f_1^* \approx & & \uparrow f^* \approx \\
H^{n-p}(\tilde{E}, \tilde{E}\backslash f(X)) & \xrightarrow{j^*} & H^{n-p}(\tilde{E}) & \xrightarrow{\rho^*} & H^{n-p}(f(X)) \\
\uparrow g_0^* \approx & & \uparrow g_1^* & \nearrow \tau^* & \\
H^{n-p}(\mathbb{R}^{2n-p}, \mathbb{R}^{2n-p}\backslash f(X)) & \xrightarrow{j^*} & H^{n-p}(\mathbb{R}^{2n-p}) & &
\end{array}
$$

ing homomorphisms are induced by the obvious inclusions. The inclusion $\tilde{E} \xrightarrow{g_1} \mathbb{R}^{2n-p}$ induces an excision $\tilde{E}, \tilde{E}\backslash f(X) \xrightarrow{g_0} \mathbb{R}^{2n-p}, \mathbb{R}^{2n-p}\backslash f(X)$, it follows that g_0^* is also an isomorphism. Since $H^{n-p}(\mathbb{R}^{2n-p}) = 0$, it then follows that

$$e(\nu_f) = \sigma^* j^* U_{\nu_f} = f^* \tau^* j^* (g_0^*)^{-1} (f_0^*)^{-1} U_{\nu_f} = 0 \in H^{n-p}(X),$$

as asserted.

4.8 Theorem: *Let $\bar{w}(\tau(X))$ be the dual Stiefel–Whitney class of the tangent bundle $\tau(X)$ of a smooth n-dimensional manifold X, and suppose that there is a proper embedding $X \xrightarrow{f} \mathbb{R}^{2n-p}$; then $\bar{w}_q(\tau(X)) = 0 \in H^q(X)$ for $q \geqq n - p$.*

PROOF: The normal bundle ν_f is an $(n-p)$-plane bundle and one has $\bar{w}(\tau(X)) = w(\nu_f)$, where $w_q(\nu_f) = 0$ for $q > n - p$ as in Lemma 4.2. However, according to Corollary V.2.8 $w_{n-p}(\nu_f)$ is just the $\mathbb{Z}/2$ Euler class $e(\nu_f) \in H^{n-p}(X)$, which vanishes by Lemma 4.7.

4.9 Proposition: *If $n = 2^r$ for $r \geqq 0$, then the real projective space RP^n does not embed in $\mathbb{R}^{2n-1}$.*

PROOF: Since RP^n is closed, a fortiori compact, any embedding is proper. However, $\bar{w}_{n-1}(\tau(RP^n))$ is the nonzero element $e(\gamma_n^1)^{n-1} \in H^{n-1}(RP^n)$ as in

Proposition 4.3, so that Theorem 4.8 forbids an embedding of RP^n into $\mathbb{R}^{2n-1}$.

For any $n > 0$ the number $\alpha(n) > 0$ is described in Definition 4.4.

4.10 Proposition: *For any natural number $n > 0$ there is a smooth closed n-dimensional manifold X which does not embed in $\mathbb{R}^{2n-\alpha(n)}$.*

PROOF: In the proof of Proposition 4.5 it was shown that $\bar{w}_{n-\alpha(n)}(\tau(X)) \neq 0 \in H^{n-\alpha(n)}(X)$ for the closed manifold $X = \prod_{j \in J} RP^{2^j}$. Since any embedding of a closed manifold is necessarily proper, Theorem 4.8 therefore forbids any embedding of X into $\mathbb{R}^{2n-\alpha(n)}$.

5. Stiefel–Whitney Numbers

Let X be any smooth closed n-dimensional manifold, and let $(r_1, \ldots, r_n)$ be any ordered set of natural numbers such that $r_1 + 2r_2 + \cdots + nr_n = n$. We shall use the total Stiefel–Whitney class $w(\tau(X))$ of the tangent bundle $\tau(X)$ to assign an element of $\mathbb{Z}/2$ to each $(r_1, \ldots, r_n)$. One obtains the value $0 \in \mathbb{Z}/2$ for each $(r_1, \ldots, r_n)$ if (and only if) X is the (smooth) boundary $\dot{Y}$ of a smooth compact $(n + 1)$-dimensional manifold Y.

If $1 + w_1(\tau(X)) + \cdots + w_n(\tau(X))$ is the total Stiefel–Whitney class $w(\tau(X)) \in H^*(X)$ of the tangent bundle $\tau(X)$ of the smooth closed n-dimensional manifold X, and if $(r_1, \ldots, r_n)$ is an ordered n-tuple of natural numbers such that $r_1 + 2r_2 + \cdots + nr_n = n$, the product $w_1(\tau(X))^{r_1} \cup \cdots \cup w_n(\tau(X))^{r_n}$ in $H^*(X)$ belongs to $H^n(X)$. The boundary $\dot{X}$ of X is void since X is closed, so that the fundamental $\mathbb{Z}/2$ homology class of Definition 1.4 is an element $\mu_X \in H_n(X)$. Consequently the Kronecker product $H^n(X) \otimes H_n(X) \xrightarrow{\langle , \rangle} \mathbb{Z}/2$ assigns an element of the ground ring $\mathbb{Z}/2$ to the manifold X and the n-tuple $(r_1, \ldots, r_n)$, as follows.

5.1 Definition: Let X be a smooth closed n-dimensional manifold, and let $(r_1, \ldots, r_n)$ be any natural numbers such that $r_1 + 2r_2 + \cdots + nr_n = n$. Then the $(r_1, \ldots, r_n)$th *Stiefel–Whitney number* of X is the element

$$\langle w_1(\tau(X))^{r_1} \cup \cdots \cup w_n(\tau(X))^{r_n}, \mu_X \rangle \in \mathbb{Z}/2.$$

5.2 Proposition: *Suppose that Y is a smooth compact $(n + 1)$-dimensional manifold with boundary $\dot{Y}$, so that $\dot{Y}$ is a smooth closed n-dimensional manifold; then all the Stiefel–Whitney numbers of $\dot{Y}$ vanish.*

PROOF: According to Proposition 1.9 the fundamental $\mathbb{Z}/2$ homology class $\mu_{\dot{Y}} \in H_n(\dot{Y})$ is the image of the fundamental $\mathbb{Z}/2$ class $\mu_Y \in H_{n+1}(Y, \dot{Y})$ under the connecting homomorphism ∂ of the exact homology sequence

$$H_{n+1}(Y) \xrightarrow{j_*} H_{n+1}(Y, \dot{Y}) \xrightarrow{\partial} H_n(\dot{Y}) \xrightarrow{i_*} H_n(Y)$$

for the pair Y, $\dot{Y}$; that is, $\mu_{\dot{Y}} = \partial\mu_Y$. Furthermore, the connecting homomorphism δ of the corresponding exact cohomology sequence

$$H^n(Y) \xrightarrow{i^*} H^n(\dot{Y}) \xrightarrow{\delta} H^{n+1}(Y, \dot{Y}) \xrightarrow{j^*} H^{n+1}(Y)$$

is the adjoint of ∂ with respect to the Kronecker product $\langle\ ,\ \rangle$; that is, $\langle \alpha, \partial v\rangle = \langle \delta\alpha, v\rangle$ for any $\alpha \in H^n(Y)$ and any $v \in H_{n+1}(Y, \dot{Y})$. Consequently the $(r_1, \ldots, r_n)$th Stiefel–Whitney number of $\dot{Y}$ satisfies

$$\langle w_1^{r_1} \cdots w_n^{r_n}, \mu_{\dot{Y}}\rangle = \langle w_1^{r_1} \cdots w_n^{r_n}, \partial\mu_Y\rangle = \langle \delta(w_1^{r_1} \cdots w_n^{r_n}), \mu_Y\rangle,$$

where

$$w_1^{r_1} \cdots w_n^{r_n} = w_1(\tau(\dot{Y}))^{r_1} \cup \cdots \cup w_n(\tau(\dot{Y}))^{r_n};$$

We shall show that $\delta(w_1^{r_1} \cdots w_n^{r_n}) = 0$.

Let $\tau(Y)$ be the tangent bundle of Y itself. Then for the inclusion $\dot{Y} \xrightarrow{i} Y$ of the boundary $\dot{Y}$, the restriction $i^!\tau(Y)$ of $\tau(Y)$ to $\dot{Y}$ is the Whitney sum $\tau(\dot{Y}) \oplus \varepsilon^1$ of the tangent bundle $\tau(\dot{Y})$ and the trivial line bundle ε^1 spanned by the inward-pointing unit vectors normal to $\dot{Y}$, with respect to any riemannian metric on $\tau(Y)$. Since

$$w(\tau(\dot{Y})) = w(\tau(\dot{Y}) \oplus \varepsilon^1) = w(i^!\tau(Y)) = i^*w(\tau(Y))$$

for the total Stiefel–Whitney classes $w(\tau(\dot{Y})) \in H^*(\dot{Y})$ and $w(\tau(Y)) \in H^*(Y)$, it follows that

$$\delta(w_1(\tau(\dot{Y}))^{r_1} \cup \cdots \cup w_n(\tau(\dot{Y}))^{r_n}) = \delta i^*(w_1(\tau(Y))^{r_1} \cup \cdots \cup w_n(\tau(Y))^{r_n}) = 0$$

as desired, since δi^* annihilates $H^n(Y)$ in the exact cohomology sequence of the pair Y, $\dot{Y}$.

5.3 Remark: Proposition 5.2 is due to Pontrjagin [5]. The converse assertion is also true, due to Thom [3, 6–8]. Thus a smooth closed manifold is a boundary if *and only if* all its Stiefel-Whitney numbers vanish. We shall not prove the converse of Proposition 5.2.

The following result depends only on Proposition 5.2.

5.4 Corollary: *If n is even, then the real projective space RP^n is not the boundary of any $(n + 1)$-dimensional manifold.*

Proof: By Proposition 3.1, $w(\tau(RP^n)) = (1 + e(\gamma_n^1))^{n+1}$, with binomial coefficients computed in $\mathbb{Z}/2$, where according to Proposition IV.4.4, $H^*(RP^n)$ is the polynomial ring over $\mathbb{Z}/2$ generated by the $\mathbb{Z}/2$ Euler class $e(\gamma_n^1)$, modulo the relation $e(\gamma_n^1)^{n+1} = 0$. In particular, if n is even, then $\binom{n+1}{n} = 1 \in \mathbb{Z}/2$, so that $w_n(\tau(RP^n))$ is the generator $e(\gamma_n^1)^n \in H^n(RP^n)$. Hence the $(0, \ldots, 0, 1)$th Stiefel–Whitney number of RP^n satisfies $\langle w_n(\tau(RP^n)), \mu_{RP^n} \rangle = 1 \neq 0$, so that Proposition 5.2 prevents RP^n from being a boundary.

5.5 Proposition: *If n is odd, then all Stiefel–Whitney numbers of the real projective space RP^n vanish.*

PROOF: If $n = 2m - 1$ for $m > 0$, then Proposition 3.1 and the technique of Lemma 3.3 give $w(\tau(RP^n)) = (1 + e(\gamma_n^1))^{2m} = (1 + e(\gamma_n^1)^2)^m$; in particular, all the odd Stiefel–Whitney classes of RP^n vanish. However, if $r_1 + 2r_2 + \cdots + nr_n$ is to equal the odd number n, then $r_p \neq 0$ for at least one odd number $p \leqq n$, so that

$$w_1(\tau(RP^n))^{r_1} \cup \cdots \cup w_n(\tau(RP^n))^{r_n} = 0.$$

5.6 Corollary: *If n is odd, then the real projective space RP^n is the boundary of some $(n + 1)$-dimensional manifold.*

PROOF: Remark 5.3 and Proposition 5.5.

5.7 Definition: Two smooth closed n-dimensional manifolds X and X' are *cobordant* if their disjoint union $X + X'$ is the boundary of a smooth compact $(n + 1)$-dimensional manifold Y.

If X is any smooth closed n-dimensional manifold, then $X + X$ is the boundary of $X \times [0, 1]$, so that cobordism is a reflexive relation. Since the disjoint union $X + X'$ is assigned no order cobordism is a symmetric relation. Finally, if $X + X'$ is the boundary of Y and $X' + X''$ is the boundary of Y', then one can identify the common portion X' of Y and Y' to create a smooth compact $(n + 1)$-dimensional manifold Y'' with boundary $X + X''$; hence cobordism is also a transitive relation. In summary, cobordism is an equivalence relation.

Let $[X]$ and $[X']$ denote the cobordism classes of smooth closed manifolds X and X', respectively. If X and X' are of the same dimension, then one easily verifies that the class $[X + X']$ of the disjoint union $X + X'$ depends only on $[X]$ and $[X']$, and for any dimensions one easily verifies that the class $[X \times X']$ of the smooth product $X \times X'$ also depends only on $[X]$ and $[X']$. Hence the set $\mathfrak{N}$ of cobordism classes of smooth closed manifolds admits a *sum* $\mathfrak{N} \times \mathfrak{N} \xrightarrow{+} \mathfrak{N}$ and a *product* $\mathfrak{N} \times \mathfrak{N} \xrightarrow{\cdot} \mathfrak{N}$. Furthermore $\mathfrak{N}$ is graded by dimension: if $\mathfrak{N}_n \subset \mathfrak{N}$ consists of classes represented by smooth closed n-dimensional manifolds, then $\mathfrak{N}$ is the disjoint union of the subsets $\mathfrak{N}_n$, and sums and products satisfy $\mathfrak{N}_n \times \mathfrak{N}_n \xrightarrow{+} \mathfrak{N}_n$ and $\mathfrak{N}_m \times \mathfrak{N}_n \xrightarrow{\cdot} \mathfrak{N}_{m+n}$.

5.8 Lemma: *The set $\mathfrak{N}$ of cobordism classes is a graded commutative ring with respect to the preceding operations*

$$\mathfrak{N}_n \times \mathfrak{N}_n \xrightarrow{+} \mathfrak{N}_n \qquad \text{and} \qquad \mathfrak{N}_m \times \mathfrak{N}_n \xrightarrow{\cdot} \mathfrak{N}_{m+n}.$$

PROOF: This is a trivial verification. In particular, since $X + X$ is the boundary of $X \times [0, 1]$, it follws that $-[X] = [X]$, hence that every element of $\mathfrak{N}$ is of order 2 with respect to addition.

5.9 Definition: The preceding graded commutative ring $\mathfrak{N}$ of cobordism classes of smooth closed manifolds is the *unoriented cobordism ring.*

5.10 Proposition: If $[X] = [X']$ in $\mathfrak{N}_n$ for two smooth closed n-dimensional manifolds X and X', then X and X' have the same Stiefel–Whitney numbers; conversely, if X and X' have the same Stiefel–Whitney numbers, then $[X] = [X']$.

PROOF: The first assertion is a rephrasing of Proposition 5.2, and the second assertion is a rephrasing of Remark 5.3.

There is an alternative way to construct the unoriented cobordism ring $\mathfrak{N}$. Observe that if D^p and D^q are disks of dimensions p and q with boundaries S^{p-1} and S^{q-1}, respectively, then each of the $(p + q - 1)$-dimensional manifolds $D^p \times S^{q-1}$ and $S^{p-1} \times D^q$ has $S^{p-1} \times S^{q-1}$ as its boundary. Suppose that $D^p \times S^{q-1}$ is smoothly embedded in a smooth closed manifold X_0 of dimension $p + q - 1$. One can then remove $D^p \times S^{q-1}$ from X_0, leaving a boundary $S^{p-1} \times S^{q-1}$; since $S^{p-1} \times S^{q-1}$ is also the boundary of $S^{p-1} \times D^q$, one can (smoothly) insert $S^{p-1} \times D^q$ into $X_0 \backslash D^p \times S^{q-1}$, identifying common points on the common boundary $S^{p-1} \times S^{q-1}$, to obtain a new smooth closed manifold X_1, also of dimension $p + q - 1$. Two such manifolds X_0 and X_1 are *surgically related*; in general, surgically related manifolds are not diffeomorphic to each other.

5.11 Definition: Two smooth closed manifolds X_0 and X_q of the same dimension are *surgically equivalent* if and if there are smooth closed manifolds $X_1, \ldots, X_{q-1}$ such that X_{i-1} is surgically related to X_i for each $i = 1, \ldots, q$.

Clearly surgical equivalence is an equivalence relation, and the set of equivalence classes $[X]$ of smooth closed manifolds X forms a commutative semi-ring $\mathfrak{N}'$ with respect to disjoint union and smooth product: $[X] + [X'] = [X + X']$ and $[X] \cdot [X'] = [X \times X']$. The semi-ring $\mathfrak{N}'$ is graded by dimension. We shall show that $\mathfrak{N}'$ is canonically isomorphic to the unoriented cobordism ring $\mathfrak{N}$.

Let $\| \|_p$ and $\| \|_q$ denote the usual euclidean norms on $\mathbb{R}^p$ and $\mathbb{R}^q$. The product $\mathbb{R}^p \times \mathbb{R}^q$ then carries a norm $\| \|$ with $\|(x, y)\| = \max\{\|x\|_p, \|y\|_p\}$, and $D^p \times S^{q-1} \cup S^{p-1} \times D^q$ is the unit sphere in $\mathbb{R}^{p+q}$ with respect to $\| \, \|$. By stretching the unit sphere $D^p \times S^{q-1} \cup S^{p-1} \times D^q$ at the "equator" $S^{p-1} \times S^{q-1}$ it follows that

$$D^p \times S^{q-1} \times \{0\} \cup S^{p-1} \times S^{q-1} \times [0,1] \cup S^{p-1} \times D^q \times \{1\}$$

is the boundary of a $(p+q)$-cell in $\mathbb{R}^{p+q}$, hence homeomorphic to the sphere $S^{p+q-1} \subset \mathbb{R}^{p+q}$.

5.12 Lemma: *Any two surgically related smooth closed manifolds are cobordant; consequently any two surgically equivalent smooth closed manifolds are cobordant.*

PROOF: Suppose that X_0 and X_1 are smooth closed manifolds of dimension $p+q-1$, where one replaces some $D^p \times S^{q-1} \subset X_0$ by $S^{p-1} \times D^q$ to obtain X_1, as before. Then $X_0 \backslash D^p \times S^{q-1}$ and $X_1 \backslash S^{p-1} \times D^q$ are each diffeomorphic to a smooth compact manifold X with boundary $S^{p-1} \times S^{q-1}$. Let Y be the union of the product $X \times [0,1]$ and the $(p+q)$-cell with boundary

$$D^p \times S^{q-1} \times \{0\} \cup S^{p-1} \times S^{q-1} \times [0,1] \cup S^{p-1} \times D^q \times \{1\}$$

described earlier, with the obvious boundary identifications: the two boundaries $D^p \times S^{q-1} \times \{0\}$ are identified with each other, the two boundaries $S^{p-1} \times S^{q-1} \times [0,1]$ are identified with each other, and the two boundaries $S^{p-1} \times D^q \times \{1\}$ are identified with each other. Then Y is a smooth compact $(p+q)$-dimensional manifold with boundary $X_0 \times \{0\} + X_1 \times \{1\}$, as required.

The converse of the second assertion of Lemma 5.12 requires some elementary Morse theory, which we sketch. According to Definition III.6.11 the *differential* of a smooth function $Y \xrightarrow{f} \mathbb{R}$ on a smooth manifold Y is the $C^\infty(Y)$-linear map $\mathscr{E}^*(Y) \xrightarrow{df} C^\infty(Y)$ carrying any smooth vector field $L \in \mathscr{E}^*(Y)$ into the smooth function $Lf \in C^\infty(Y)$. A *critical point* of f is any point $y \in Y$ at which df vanishes; that is, $(Lf)(y) = 0$ for every $L \in \mathscr{E}^*(Y)$.

5.13 Lemma (Regular Interval Lemma): *Let Y be a smooth compact n-dimensional manifold whose boundary is the disjoint union of two smooth closed $(n-1)$-dimensional manifolds X_a and X_b, and let $Y \xrightarrow{f} [a,b] \subset \mathbb{R}$ be a smooth function with no critical points, such that $X_a = f^{-1}(\{a\})$ and $X_b = f^{-1}(\{b\})$. Then there is a diffeomorphism $Y \xrightarrow{g} X_a \times [a,b]$ whose composition with the second projection $X_a \times [a,b] \xrightarrow{\mathrm{pr}_2} [a,b]$ is f itself; a fortiori X_a is diffeomorphic to X_b.*

PROOF: There is a smooth riemannian metric $\mathscr{E}^*(Y) \times \mathscr{E}^*(Y) \xrightarrow{\langle\,,\,\rangle} C^\infty(Y)$ by Proposition III.5.8, which induces an isomorphism $\mathscr{E}(Y) \to \mathscr{E}^*(Y)$ carrying

$df \in \mathscr{E}(Y)$ into a vector field $\operatorname{grad} f \in \mathscr{E}^*(Y)$ such that $\langle M, \operatorname{grad} f\rangle = Mf \in C^\infty(Y)$ for any $M \in \mathscr{E}^*(Y)$. By hypothesis df is nowhere-vanishing; hence grad f is also nowhere-vanishing, so that $\langle \operatorname{grad} f, \operatorname{grad} f\rangle$ is everywhere positive on Y. It follows that there is a unique vector field $L \in \mathscr{E}^*(Y)$ such that $\langle \operatorname{grad} f, \operatorname{grad} f\rangle L = \operatorname{grad} f$, for which one trivially has $Lf = \langle L, \operatorname{grad} f\rangle = 1 \in C^\infty(Y)$. The definition of L also yields the identity $\langle \operatorname{grad} f, \operatorname{grad} f\rangle\langle L, L\rangle = 1$, and since Y is compact, it follows that there are positive constants A and B such that $0 < A \leqq \langle L, L\rangle \leqq B < \infty$ uniformly on Y.

In order to construct $Y \xrightarrow{g} X_a \times [a, b]$ we shall first construct its inverse $X_a \times [a, b] \xrightarrow{h} Y$ by solving ordinary differential equations. In outline, observe that partial differentiation with respect to the coordinate $t \in [a, b]$ provides a vector field $\partial/\partial t$ on $X_a \times [a, b]$. We shall show that there is a uniquely defined h such that $h_*(\partial/\partial t) = L$ on Y and $h(x, a) = x$ for each $x \in X_a$.

For any $c \in [a, b]$ let X_c be the closed set $f^{-1}(\{c\}) \subset Y$. Since Y is compact X_c is also compact, so that it is contained in the union of finitely many open coordinate neighborhoods Y' in Y. As in Proposition I.6.10 one can shrink each of the open coordinate neighborhoods Y' to a smaller open coordinate neighborhood Y'' whose (necessarily compact) closure satisfies $\bar{Y}'' \subset Y'$, in such a way that X_c is also contained in the union of the finitely many open sets Y''; one may as well assume that X_c actually intersects each Y''.

If $(y^1, \ldots, y^n)$ are coordinate functions on one of the former coordinate neighborhoods Y', the restriction of L to Y' is of the form $l^1\, \partial/\partial y^1 + \cdots + l^n\, \partial/\partial y^n$ for smooth functions $l^1, \ldots, l^n$ on Y', as in Lemma III.6.6. Suppose temporarily that c lies in the interior (a, b) of the closed interval $[a, b]$. Then according to the classical existence and uniqueness theorem for ordinary differential equations there are positive constants $\delta > 0$ and $\varepsilon > 0$ and a unique n-tuple $(h^1, \ldots, h^n)$ of smooth real-valued functions on the intersection $X_c \cap \bar{Y}'' \times [c - \delta, c + \varepsilon]$ such that

$$\frac{\partial h^1}{\partial t}(x; t) = l^1(h^1(x; t), \ldots, h^n(x; t)),$$

$$\vdots$$

$$\frac{\partial h^n}{\partial t}(x; t) = l^n(h^1(x; t), \ldots, h^n(x; t)),$$

such that each $x \in X_c \cap \bar{Y}''$ has coordinates $(h^1(x; c), \ldots, h^n(x; c))$, and such that each n-tuple $(h^1(x; t), \ldots, h^n(x; t))$ is the coordinate description of a point in the larger neighborhood Y'. The constants $\delta > 0$ and $\varepsilon > 0$ depend on the geometry (in $\mathbb{R}^n$) of the inclusion $\bar{Y}'' \subset Y'$ and on the constants A and

B described earlier, with $0 < A \leqq \langle L, L\rangle \leqq B < \infty$; however, A and B themselves depend only on the given function $Y \xrightarrow{f} [a, b]$ and the choice of the riemannian metric $\langle\, ,\rangle$, so that δ and ε effectively depend only on the inclusion $\overline{Y''} \subset Y'$. In the special cases $c = a$ or $c = b$ one has $\delta = 0$ or $\varepsilon = 0$, respectively.

One can rewrite the preceding system of differential equations in the form

$$\frac{\partial h^1}{\partial t}\frac{\partial}{\partial y^1} + \cdots + \frac{\partial h^n}{\partial t}\frac{\partial}{\partial y^n} = l^1\frac{\partial}{\partial y^1} + \cdots + l^n\frac{\partial}{\partial y^n},$$

or equally well in the coordinate-free form $h_*(\partial/\partial t) = L$. Since X_c is contained in the union of *finitely* many of the coordinate neighborhoods Y'', the uniqueness of the n-tuples $(h^1, \ldots, h^n)$ implies for $c \in (a, b)$ that there are positive constants $\delta_c > 0$ and $\varepsilon_c > 0$ and a unique map $X_c \times [c - \delta_c, c + \varepsilon_c] \xrightarrow{h} Y$ with $h_*(\partial/\partial t) = L$ over the image of h, and $h(x, c) = x$ for each $x \in X_c$. Since

$$\frac{\partial}{\partial t}(f \circ h) = \left(h_* \frac{\partial}{\partial t}\right) f = Lf = 1,$$

one has $f(h(x, t)) = t$ for each $t \in [c - \delta_c, c + \varepsilon_c]$. Hence for any closed subinterval $[t, t'] \subset [c - \delta_c, c + \varepsilon_c]$ there is a bijection from X_t $(=f^{-1}(\{t\}))$ to $X_{t'}$ $(=f^{-1}(\{t'\}))$ which identifies the X_t-end-point of each integral curve of L with the $X_{t'}$-end-point of the same integral curve; one easily verifies that the preceding bijection is a diffeomorphism from X_t onto $X_{t'}$. In the special cases $c = a$ or $c = b$ one has $\delta_a = 0$ or $\varepsilon_b = 0$, respectively.

To complete the proof of the regular interval lemma it suffices to note that finitely many of the intervals $[c - \delta_c, c + \varepsilon_c]$ cover $[a, b]$, so that there is a unique diffeomorphism $X_a \times [a, b] \xrightarrow{h} Y$ such that $h_*(\partial/\partial t) = L$ on all of Y and $h(x, a) = x$ for $x \in X_a$. It follows as in the local case that $f(h(x, t)) = t$ for any $(x, t) \in X_a \times [a, b]$. Furthermore, if the inverse diffeomorphism $Y \xrightarrow{g} X_a \times [a, b]$ carries $y \in Y$ into $(x, t) \in X_a \times [a, b]$, then $f(y) = t = \mathrm{pr}_2(g(y))$, so that $f = \mathrm{pr}_2 \circ g$ as claimed.

Now let $Y \xrightarrow{f} R$ be a smooth function on a smooth manifold Y of dimension $n = p + q$, and suppose that $y \in Y$ is a critical point of f. If there are coordinate functions $u^1, \ldots, u^p, v^1, \ldots, v^q$ on some neighborhood of y, all vanishing at y, such that

$$f = f(y) - [(u^1)^2 + \cdots + (u^p)^2] + [(v^1)^2 + \cdots + (v^q)^2]$$

in that neighborhood, then y is a *nondegenerate critical point* of f, with *critical value* $f(y) \in \mathbb{R}$. (There is a less restrictive-appearing characterization of nondegenerate critical points; however, for the moment we merely observe that df does indeed vanish at y, so that y is at least a critical point of f in the earlier sense.)

For notational convenience let $u = (u^1, \ldots, u^p)$, $v = (v^1, \ldots, v^q)$, $\|u\|^2 = (u^1)^2 + \cdots + (u^p)^2$, and $\|v\|^2 = (v^1)^2 + \cdots + (v^q)^2$. For any $\varepsilon > 0$ let $D_{2\varepsilon}^{p+q} \subset \mathbb{R}^{p+q}$ be the disk of radius 2ε about $0 \in \mathbb{R}^{p+q}$, consisting of those $(u, v) \in \mathbb{R}^{p+q}$ such that $\|u\|^2 + \|v\|^2 \leqq (2\varepsilon)^2$, and for any $\delta > 0$ let $S_\delta^{p+q-1} \subset \mathbb{R}^{p+q}$ be the sphere of radius δ about $0 \in \mathbb{R}^{p+q}$, consisting of those points $(u, v) \in \mathbb{R}^{p+q}$ such that $\|u\|^2 + \|v\|^2 = \delta^2$. The notations D^p, D^q, S^{p-1}, S^{q-1} without subscripts will denote standard disks and spheres of the indicated dimensions, identified only up to diffeomorphism.

5.14 Lemma: *For any $\varepsilon > 0$ let $f = -\|u\|^2 + \|v\|^2$ on $D_{2\varepsilon}^{p+q}$; then $f^{-1}(\{-\varepsilon^2\})$ is diffeomorphic to $S^{p-1} \times D^q$ and $f^{-1}(\{+\varepsilon^2\})$ is diffeomorphic to $D^p \times S^{q-1}$.*

PROOF: If $0 \leqq \delta \leqq 2\varepsilon$, then the intersection $f^{-1}(\{-\varepsilon^2\}) \cap S_\delta^{p+q-1}$ consists of those $(u, v) \in D_{2\varepsilon}^{p+q}$ such that $-\|u\|^2 + \|v\|^2 = -\varepsilon^2$ and $\|u\|^2 + \|v\|^2 = \delta^2$; that is, $\|u\|^2 = \frac{1}{2}(\delta^2 + \varepsilon^2)$ and $\|v\|^2 = \frac{1}{2}(\delta^2 - \varepsilon^2)$. Clearly $f^{-1}(\{-\varepsilon^2\}) \cap S_\delta^{p+q-1}$ is void for $0 \leqq \delta < \varepsilon$, of the form $S^{p-1} \times \{*\}$ for $\delta = \varepsilon$, and of the form $S^{p-1} \times S^{q-1}$ for $\varepsilon < \delta \leqq 2\varepsilon$, so that $f^{-1}(\{-\varepsilon^2\})$ is diffeomorphic to $S^{p-1} \times D^q$. Similarly $f^{-1}(\{+\varepsilon^2\})$ is diffeomorphic to $D^p \times S^{q-1}$.

5.15 Lemma: *Let Y be a smooth compact n-dimensional manifold whose boundary is the disjoint union of two smooth closed $(n-1)$-dimensional manifolds X_a and X_b, and let $Y \xrightarrow{f} [a, b] \subset \mathbb{R}$ be a smooth function with a single critical point $y \in Y$, which is nondegenerate, whose critical value $c \in \mathbb{R}$ satisfies $a < c < b$, and for which $X_a = f^{-1}(\{a\})$ and $X_b = f^{-1}(\{b\})$. Then X_a is surgically related to X_b.*

PROOF: One has $f = c - \|u\|^2 + \|v\|^2$ in some neighborhood of the critical point y, as in Lemma 5.14, with $(u, v) = (0, 0)$ at y itself. One can choose $\varepsilon > 0$ sufficiently small that $a < c - \varepsilon^2$ and $c + \varepsilon^2 < b$, and also sufficiently small that $D_{2\varepsilon}^{p+q}$ lies in the preceding neighborhood of y. By the regular interval lemma (Lemma 5.13) X_a is diffeomorphic to $X_{c-\varepsilon^2}$ and $X_{c+\varepsilon^2}$ is diffeomorphic to X_b, so that it remains to show that $X_{c-\varepsilon^2}$ is surgically related to $X_{c+\varepsilon^2}$. Let $Y_1 = f^{-1}([c - \varepsilon^2, c + \varepsilon^2]) \cap D_{2\varepsilon}^{p+q}$, and let Y_2 be the closure in Y of the complement $f^{-1}([c-\varepsilon^2, c+\varepsilon^2]) \backslash Y_1$. Then $X_{c-\varepsilon^2} \cap Y_1$ and $X_{c+\varepsilon^2} \cap Y_1$ are diffeomorphic to $S^{p-1} \times D^q$ and $D^p \times S^{q-1}$, respectively, by Lemma 5.14, and since $X_{c-\varepsilon^2} \cap Y_2$ and $X_{c+\varepsilon^2} \cap Y_2$ are mutually diffeomorphic, by the regular interval lemma, it follows that $X_{c-\varepsilon^2}$ is surgically related to $X_{c+\varepsilon^2}$, as required.

5.16 Lemma: *Let Y be a smooth compact n-dimensional manifold whose boundary is the disjoint union of two smooth closed $(n-1)$-dimensional manifolds X_0 and X_1. Then there is a smooth function $Y \xrightarrow{f} [0, 1]$ with $X_0 = f^{-1}(\{0\})$ and $X_1 = f^{-1}(\{1\})$, and with only finitely many critical points, all* *of which are nondegenerate, whose critical levels are mutually distinct real numbers in the open interval* (0, 1).

PROOF: We omit the proof of this classical result, which is Lemma 1 on pages 41–42 of Milnor [9] and Theorem 2.5 on page 9 of Milnor [18]. It also follows from Corollary 6.8 on page 37 of Milnor [13] and from Theorem 1.2 on pages 147–148 of M. W. Hirsch [4].

A function $Y \xrightarrow{f} [0, 1]$ satisfying the conclusion of Lemma 5.16 is an *admissible Morse function.*

5.17 Theorem: *Two smooth closed manifolds X_0 and X_1 of the same dimension are surgically equivalent if and only if they are cobordant.*

PROOF: According to Lemma 5.12 any two surgically equivalent smooth closed manifolds are cobordant. Conversely, if the disjoint union of two smooth closed $(n-1)$-dimensional manifolds X_0 and X_1 is the boundary of a smooth compact n-dimensional manifold Y, Lemma 5.16 provides an admissible Morse function $Y \xrightarrow{f} [0, 1]$ with critical levels $c_1, \ldots, c_r$ satisfying $0 < c_1 < \cdots < c_r < 1$, where r is the number of (nondegenerate) critical points. Let $a_0 = 0$, let $a_q = \frac{1}{2}(c_q + c_{q+1})$ for $q = 1, \ldots, r-1$, let $a_r = 1$, and let $X_{a_q} = f^{-1}(\{a_q\})$ for $q = 0, \ldots, r$. Then Lemma 5.15 asserts that X_{a_q} is surgically related to $X_{a_{q+1}}$ for $q = 0, \ldots, r-1$, so that X_0 is surgically equivalent to X_1, as claimed.

5.18 Corollary: *The semi-ring $\mathfrak{N}'$ of surgical equivalence classes of smooth closed manifolds is canonically isomorphic to the unoriented cobordism ring $\mathfrak{N}$.*

PROOF: Since addition and multiplication in each of $\mathfrak{N}'$ and $\mathfrak{N}$ are induced by disjoint unions and smooth products of smooth closed manifolds, this follows immediately from Theorem 5.17.

5.19 Corollary: *Two smooth closed manifolds represent the same surgical equivalence class if and only if they have the same Stiefel–Whitney numbers.*

PROOF: This follows immediately from Proposition 5.10 and Theorem 5.17.

One can extend Definition 5.1 in an obvious way without altering the preceding results. Let $\mathbb{Z}/2[\sigma_1, \ldots, \sigma_n]$ be the polynomial ring over $\mathbb{Z}/2$ in which each σ_p is assigned degree p, for $p = 1, \ldots, n$; that is, each monomial $\sigma_1^{r_1} \cdots \sigma_n^{r_n} \in \mathbb{Z}/2[\sigma_1, \ldots, \sigma_n]$ is assigned degree $r_1 + 2r_2 + \cdots + nr_n$. A polynomial $p(\sigma_1, \ldots, \sigma_n) \in \mathbb{Z}/2[\sigma_1, \ldots, \sigma_n]$ is *homogeneous of degree n* if it is a sum of monomials of degree n. For any smooth closed n-dimensional manifold X, there is then an element $p(w_1(\tau(X)), \ldots, w_n(\tau(X))) \in H^n(X)$

providing a sum $\langle p(w_1(\tau(X)), \ldots, w_n(\tau(X))), \mu_X \rangle \in \mathbb{Z}/2$ of the Stiefel–Whitney numbers of Definition 5.1, also called a *Stiefel–Whitney number*.

For later convenience, we define particular such Stiefel–Whitney numbers. Let $\mathbb{Z}/2[t_1, \ldots, t_n]$ be the polynomial ring over $\mathbb{Z}/2$ in which each t_p is assigned degree 1, for $p = 1, \ldots, n$. According to the fundamental property of elementary symmetric polynomials, there is a unique $s_n(\sigma_1, \ldots, \sigma_n) \in \mathbb{Z}/2[\sigma_1, \ldots, \sigma_p]$, which becomes $t_1^n + \cdots + t_n^n \in \mathbb{Z}/2[t_1, \ldots, t_n]$ when one replaces $\sigma_1, \ldots, \sigma_n$ by the elementary symmetric polynomials $t_1 + \cdots + t_n, \ldots, t_1 \cdots t_n$; clearly $s_n(\sigma_1, \ldots, \sigma_n)$ is homogeneous of degree n in the sense of the preceding paragraph.

The following definition will be used in Remark 9.23.

5.20 Definition: For any $n > 0$ and any smooth closed n-dimensional manifold X, the number

$$\langle s_n(w_1(\tau(X)), \ldots, w_n(\tau(X))), \mu_X \rangle \in \mathbb{Z}/2$$

is the *basic Stiefel–Whitney number* $s_n(X)$ of X.

6. Stiefel–Whitney Genera

The set $\mathscr{U}$ of diffeomorphism classes of smooth, not necessarily orientable, closed manifolds X is a commutative semi-ring with respect to disjoint unions and smooth products. The subset $\mathscr{I} \subset \mathscr{U}$ of sums of two copies of any $X \in \mathscr{U}$ is an ideal in $\mathscr{U}$, in the obvious sense, and there is an equivalence relation $\sim$ in $\mathscr{U}$ with $X_0 \sim X_1$ if and only if there are elements $X_2 + X_2$ and $X_3 + X_3$ in $\mathscr{I}$ with $X_0 + X_2 + X_2 = X_1 + X_3 + X_3$. The algebraic operations in $\mathscr{U}$ induce corresponding operations in the quotient $\mathscr{U}/\mathscr{I}$, and if $[X] \in \mathscr{U}/\mathscr{I}$ is the equivalence class of $X \in \mathscr{U}$, then the relation $[X] + [X] = 0 \in \mathscr{U}/\mathscr{I}$ implies that $\mathscr{U}/\mathscr{I}$ is a commutative *ring* $\hat{\mathscr{U}}$ in which every element is of order two.

The quotient of $\mathscr{U}$ by the ideal represented by smooth closed boundaries is precisely the unoriented cobordism ring $\mathfrak{N}$ of Definition 5.9: the natural epimorphism $\mathscr{U} \to \mathfrak{N}$ factors out boundaries. The purpose of this section is to introduce certain other homomorphisms $\mathscr{U} \to \mathbb{Z}/2$ which will serve as models for later constructions.

For convenience we use the same notation X to denote either a smooth closed manifold in $\mathscr{U}$ or its equivalence class $[X] \in \hat{\mathscr{U}}$. Since every element of $\hat{\mathscr{U}}$ is of order two, $\hat{\mathscr{U}}$ can be regarded as a (graded) $\mathbb{Z}/2$-algebra.

For any n-dimensional $X \in \mathscr{U}$ the $\mathbb{Z}/2$ cohomology module $H^q(X)$ vanishes for $q > n$, so that $H^{**}(X) = H^*(X)$. Let $\tau(X)$ be the tangent bundle of X, and for any formal power series $f(t) \in \mathbb{Z}/2[[t]]$ with leading term $1 \in \mathbb{Z}/2$ let $u_f(\tau(X)) \in H^*(X)$ be the resulting multiplicative $\mathbb{Z}/2$ class, as in Definition V.1.4. Since X is closed, in the usual sense that its boundary $\dot{X}$ is empty, the fundamental $\mathbb{Z}/2$ homology class of Definition 1.4 is an element $\mu_X \in H_n(X)$. The Kronecker product $\langle u_f(\tau(X)), \mu_X\rangle \in \mathbb{Z}/2$, in which the homogeneous element μ_X of degree n annihilates all cohomology classes except in degree n, clearly depends only on X as an element of $\mathcal{U}$.

6.1 Definition: For any formal power series $f(t) \in \mathbb{Z}/2[[t]]$ with leading term $1 \in \mathbb{Z}/2$, and for any $X \in \mathcal{U}$, the Kronecker product $\langle u_f(\tau(X)), \mu_X\rangle \in \mathbb{Z}/2$ is the *f-genus* $G(f)(X)$ of X.

6.2 Proposition: *For any formal power series $f(t) \in \mathbb{Z}/2[[t]]$ with leading term $1 \in \mathbb{Z}/2$, there is a semi-ring homomorphism $\mathcal{U} \xrightarrow{G(f)} \mathbb{Z}/2$ carrying each $X \in \mathcal{U}$ into the f-genus $G(f)(X) \in \mathbb{Z}/2$.*

PROOF: Since addition in $\mathcal{U}$ is induced by disjoint union of manifolds, the additivity of $G(f)$ is trivial. For any $X_1 \in \mathcal{U}$ and $X_2 \in \mathcal{U}$ the tangent bundle of the product $X_1 \times X_2 \in \mathcal{U}$ is given by $\tau(X_1 \times X_2) = \mathrm{pr}_1^!\tau(X_1) \oplus \mathrm{pr}_2^!\tau(X_2) = \tau(X_1) + \tau(X_2)$, for the projections $X_1 \times X_2 \xrightarrow{\mathrm{pr}_1} X_1$ and $X_1 \times X_2 \xrightarrow{\mathrm{pr}_2} X_2$, so that the cross-product version of the Whitney product formula (Proposition V.2.11) gives

$$\begin{aligned} u_f(\tau(X_1 \times X_2)) &= u_f(\tau(X_1) + \tau(X_2)) \\ &= u_f(\tau(X_1)) \times u_f(\tau(X_2)) \in H^*(X_1 \times X_2). \end{aligned}$$

The characterization of fundamental $\mathbb{Z}/2$ homology classes appearing in Corollary 1.5 immediately implies that the fundamental $\mathbb{Z}/2$ homology class $\mu_{X_1 \times X_2} \in H_*(X_1 \times X_2)$ of $X_1 \times X_2$ is the cross product $\mu_{X_1} \times \mu_{X_2} \in H_*(X_1 \times X_2)$ of the fundamental $\mathbb{Z}/2$ homology classes $\mu_{X_1} \in H_*(X_1)$ and $\mu_{X_2} \in H_*(X_2)$. Since μ_{X_1} and μ_{X_2} are homogeneous, their degrees being $\dim X_1$ and $\dim X_2$, the classical relation between cohomology cross-products and homology cross-products implies

$$\begin{aligned} G(f)(X_1 \times X_2) &= \langle u_f(\tau(X_1 \times X_2)), \mu_{X_1 \times X_2}\rangle \\ &= \langle u_f(\tau(X_1)) \times u_f(\tau(X_2)), \mu_{X_1} \times \mu_{X_2}\rangle \\ &= \langle u_f(\tau(X_1)), \mu_{X_1}\rangle\langle u_f(\tau(X_2)), \mu_{X_2}\rangle \\ &= G(f)(X_1) \cdot G(f)(X_2), \end{aligned}$$

as desired. (See page 217 of Dold [8], e.g., for the cross-product relation; there are no $\pm$ signs in the present computation because the coefficient ring is $\mathbb{Z}/2$.)

One can equally well apply the term *f-genus* to the $\mathbb{Z}/2$-algebra homomorphism $\hat{\mathscr{U}} \xrightarrow{G(f)} \mathbb{Z}/2$ induced by Proposition 6.2.

Any formal power series $f(t) \in \mathbb{Z}/2[[t]]$ with leading term $1 \in \mathbb{Z}/2$ induces a multiplicative sequence $\sum_{n \geq 0} P_n(u_1, \ldots, u_n) \in \mathbb{Z}/2[[u_1, u_2, \ldots]]$ as in Definition V.2.9. Furthermore, if X is of dimension n, then Proposition V.2.10 guarantees that $G(f)(X) = \langle P_n(w_1(\tau(X)), \ldots, w_n(\tau(X))), \mu_X \rangle$. Thus $G(f)(x)$ is a $\mathbb{Z}/2$-linear combination of the Stiefel–Whitney numbers of X; it is therefore reasonable to call $G(f)$ a *Stiefel–Whitney genus*, as in the title of this section. According to Proposition 5.2 this remark implies that $G(f)(X) = 0$ whenever X is the boundary $\dot{Y}$ of a smooth compact $(n+1)$-dimensional manifold Y. Consequently $\mathscr{U} \xrightarrow{G(f)} \mathbb{Z}/2$ annihilates the kernel of the natural epimorphism $\mathscr{U} \to \mathfrak{N}$ onto the unoriented cobordism ring $\mathfrak{N}$, so that $G(f)$ can equally well be regarded as a $\mathbb{Z}/2$-algebra homomorphism $\mathfrak{N} \to \mathbb{Z}/2$, which we shall also denote $G(f)$. However, there are many nonzero elements of $\mathfrak{N}$ which lie in the intersection of the kernels of all $\mathbb{Z}/2$-algebra homomorphisms $\mathfrak{N} \to \mathbb{Z}/2$, as we shall learn in Remark 9.33.

The simplest Stiefel–Whitney genus $\mathscr{U} \xrightarrow{G(f)} \mathbb{Z}/2$ is induced by the polynomial $f(t) = 1 + t \in \mathbb{Z}/2[t]$ $(\subset \mathbb{Z}/2[[t]])$. In this case the corresponding multiplicative sequence $\sum_{n \geq 0} P_n(u_1, \ldots, u_n)$ satisfies $P_n(u_1, \ldots, u_n) = u_n$ for each $n > 0$, as we observed following Definition V.2.9. Hence if X is a smooth closed n-dimensional manifold one has $u_{f,n}(\tau(X)) = w_n(\tau(X)) \in H^n(X)$, and by Corollary V.2.8 $w_n(\tau(X))$ is the $\mathbb{Z}/2$ Euler class $e(\tau(X))$ of $\tau(X)$. Consequently

$$\begin{aligned} G(f)(X) &= \langle u_f(\tau(X)), \mu_X \rangle = \langle u_{f,n}(\tau(X)), \mu_X \rangle \\ &= \langle w_n(\tau(X)), \mu_X \rangle = \langle e(\tau(X)), \mu_X \rangle \in \mathbb{Z}/2. \end{aligned}$$

6.3 Definition: For any smooth closed manifold $X \in \mathscr{U}$ the $\mathbb{Z}/2$ *Euler characteristic* $\chi_2(X)$ is the element $\langle e(\tau(X)), \mu_X \rangle \in \mathbb{Z}/2$.

Since we have just observed that the $\mathbb{Z}/2$ Euler characteristic is merely a specialized Stiefel–Whitney genus, Proposition 6.2 guarantees that it too can be regarded as a $\mathbb{Z}/2$-algebra homomorphism $\hat{\mathscr{U}} \xrightarrow{\chi_2} \mathbb{Z}/2$ or as a $\mathbb{Z}/2$-algebra homomorphism $\mathfrak{N} \xrightarrow{\chi_2} \mathbb{Z}/2$; the kernel of the latter homomorphism is geometrically characterized in Remark 9.25.

7. $\mathbb{Z}/2$ Thom Forms

For any smooth closed manifold $X \in \mathscr{U}$ let $H^*(X) \xrightarrow{D_P} H_*(X)$ be the $\mathbb{Z}/2$ Poincaré duality map $\cap \mu_X$ of Corollary 2.4 and let $H^*(X) \otimes H^*(X) \xrightarrow{(\,,\,)_P} \mathbb{Z}/2$ be the corresponding bilinear symmetric $\mathbb{Z}/2$ *Poincaré form*, with

$(\alpha, \beta)_{\mathrm{P}} = \langle \alpha, D_{\mathrm{P}}\beta \rangle = \langle \alpha \cup \beta, \mu_X \rangle \in \mathbb{Z}/2$ for any $(\alpha, \beta) \in H^*(X) \times H^*(X)$; since D_{P} is an isomorphism $(,)_{\mathrm{P}}$ is nondegenerate. We shall construct a bilinear symmetric $\mathbb{Z}/2$ form $H_*(X) \otimes H_*(X) \xrightarrow{(\,,\,)_{\mathrm{T}}} \mathbb{Z}/2$ which is dual to $(\,,\,)_{\mathrm{P}}$, in the obvious sense, hence a specific inverse $H_*(X) \xrightarrow{D_{\mathrm{T}}} H^*(X)$ of the $\mathbb{Z}/2$ Poincaré duality map. The form $(\,,\,)_{\mathrm{T}}$ is an element $j^*T_X \in H^n(X \times X)$, where n is the dimension of X; its behavior will provide the major result of the next section. In this section itself we shall furthermore show that if $X \xrightarrow{\Delta} X \times X$ is the diagonal map, then $\Delta^* j^* T_X$ is the $\mathbb{Z}/2$ Euler class $e(\tau(X)) \in H^n(X)$ of the tangent bundle $\tau(X)$. As in the rest of the chapter, all coefficients lie in $\mathbb{Z}/2$.

Let $E \xrightarrow{\pi} X$ represent the tangent bundle $\tau(X)$ of a smooth closed manifold $X \in \mathscr{U}$. One can identify X with the image $\sigma(X) \subset E$ of the zero section $X \xrightarrow{\sigma} E$, and since one can also identify X with the image $\Delta(X) \subset X \times X$ of the diagonal embedding $X \xrightarrow{\Delta} X \times X$ there is a canonical diffeomorphism $\sigma(X) \to \Delta(X)$. One can extend $\sigma(X) \to \Delta(X)$ in many ways to a diffeomorphism of open neighborhoods of $\sigma(X) \subset E$ and $\Delta(X) \subset X \times X$, as in the following lemma.

Recall that two maps $\tilde{E} \xrightarrow{F_0} X \times X$ and $\tilde{E} \xrightarrow{F_1} X \times X$ are homotopic relative to a subset $\sigma(X) \subset E$ if they are the restrictions to $\tilde{E} \times \{0\}$ and $\tilde{E} \times \{1\}$, respectively, of a map $\tilde{E} \times [0,1] \xrightarrow{F} X \times X$, such that $F(x,t) \in X \times X$ is independent of $t \in [0,1]$ whenever $x \in \sigma(X)$. In the following lemma $X \times X \xrightarrow{\mathrm{pr}_1} X$ and $X \times X \xrightarrow{\mathrm{pr}_2} X$ are the first and second projections of the product $X \times X$.

7.1 Lemma: *Let $E \xrightarrow{\pi} X$ represent the tangent bundle $\tau(X)$ of a smooth closed manifold $X \in \mathscr{U}$. Then there are embeddings $\tilde{E} \xrightarrow{F_0} X \times X$ and $\tilde{E} \xrightarrow{F_1} X \times X$ of some open neighborhood $\tilde{E} \subset E$ of the image $\sigma(X) \subset E$ of the zero-section $X \xrightarrow{\sigma} E$, which restrict to the diffeomorphism $\sigma(X) \to \Delta(X)$, and such that each of the compositions*

$$X \xrightarrow{\sigma} \tilde{E} \xrightarrow{F_0} X \times X \xrightarrow{\mathrm{pr}_1} X$$

and

$$X \xrightarrow{\sigma} \tilde{E} \xrightarrow{F_1} X \times X \xrightarrow{\mathrm{pr}_2} X$$

is the identity on X. Furthermore, there is a homotopy F from F_0 to F_1, relative to $\tau(X) \subset E$, which induces a homotopy from the restriction

$$\tilde{E} \backslash \sigma(X) \xrightarrow{F_0} X \times X \backslash \Delta(X)$$

to the restriction

$$\tilde{E} \backslash \sigma(X) \xrightarrow{F_1} X \times X \backslash \Delta(X).$$

PROOF: The initial stage of the construction can be found on pages 108–109 of Bishop and Crittenden [1], on pages 32–40 of Helgason [1], or on pages 32–40 of Helgason [2]. One introduces an *affine connection* in $\tau(X)$, such as the *Levi–Civita connection* associated to a riemannian metric on $\tau(X)$. This provides an *exponential map* $\tilde{E}_x \xrightarrow{\exp_x} X$ for each $x \in X$, defined on an open neighborhood $\tilde{E}_X \subset E_X$ of $0 \in E_X$, with the following properties: each $\exp_x$ is a diffeomorphism such that $\exp_x 0 = x \in X$, and there is an open neighborhood $\tilde{E} \subset E$ of $\sigma(X) \subset E$ such that each $\exp_x$ is the restriction to $\tilde{E}_x \subset \tilde{E}$ of a map $\tilde{E} \to X$. One can therefore define $\tilde{E} \times [0,1] \xrightarrow{F} X \times X$, as required, by setting

$$F(e,t) = (\exp_{\pi(e)}(te), \exp_{\pi(e)}(-(1-t)e)).$$

For any $X \in \mathscr{U}$ the $\mathbb{Z}/2$ cohomology module $H^*(X \times X, X \times X\backslash\Delta(X))$ is an $H^*(X)$-module with respect to the composition

$$\begin{aligned} &H^*(X) \otimes H^*(X \times X, X \times X\backslash\Delta(X)) \\ &\xrightarrow{\mathrm{pr}_1^* \otimes \mathrm{id}} H^*(X \times X) \otimes H^*(X \times X, X \times X\backslash\Delta(X)) \\ &\xrightarrow{\ \cup\ } H^*(X \times X, X \times X\backslash\Delta(X)) \end{aligned}$$

carrying any $\alpha \otimes T \in H^*(X) \otimes H^*(X \times X, X \times X\backslash\Delta(X))$ into $(\alpha \times 1) \cup T \in H^*(X \times X, X \times X\backslash\Delta(X))$. The $\mathbb{Z}/2$ cohomology module $H^*(E, E^*)$ is also an $H^*(X)$-module, as usual, with respect to the product

$$H^*(X) \otimes H^*(E, E^*) \xrightarrow{\pi^* \otimes \mathrm{id}} H^*(E) \otimes H^*(E, E^*) \xrightarrow{\ \cup\ } H^*(E, E^*)$$

carrying $\alpha \otimes U \in H^*(X) \otimes H^*(E, E^*)$ into $\pi^*\alpha \cup U \in H^*(E, E^*)$. For any $x \in X$ let

$$X, X\backslash\{x\} = \{x\} \times X, \{x\} \times X\backslash\Delta(X) \xrightarrow{l_x} X \times X, X \times X\backslash\Delta(X)$$

be the *left inclusion*, and let $E_x, E_x^* \xrightarrow{j_x} E, E^*$ be the inclusion of the fiber pair over x, where $E \xrightarrow{\pi} X$ represents the tangent bundle $\tau(X)$ as usual.

7.2 Lemma: *For each $X \in \mathscr{U}$ and each $x \in X$ there is an $H^*(X)$-module isomorphism G_l and a $\mathbb{Z}/2$-module isomorphism $G_{l,x}$ such that the diagram*

$$\begin{array}{ccc} H^*(X \times X, X \times X\backslash\Delta(X)) & \xrightarrow[\approx]{G_l} & H^*(E, E^*) \\ \downarrow{\scriptstyle l_x^*} & & \downarrow{\scriptstyle j_x^*} \\ H^*(X, X\backslash\{x\}) & \xrightarrow[\approx]{G_{l,x}} & H^*(E_x, E_x^*) \end{array}$$

of $\mathbb{Z}/2$-module homomorphisms commutes.

PROOF: Let $\tilde{D} \subset X \times X$ be the image $F_0(\tilde{E})$ of the embedding F_0 of Lemma 7.1, define a neighborhood $Y \subset X$ of $x \in X$ by setting $\{x\} \times Y = (\{x\} \times X) \cap \tilde{D}$, and observe that the inclusion l_x induces a commutative diagram

$$\begin{array}{ccc}
H^*(X \times X, X \times X\backslash\Delta(X)) & \xrightarrow{\approx} & H^*(\tilde{D}, \tilde{D}\backslash\Delta(X)) \\
\downarrow{\scriptstyle l_x^*} & & \downarrow{\scriptstyle l_x^*} \\
H^*(\{x\} \times X, \{x\} \times X\backslash\Delta(X)) & \xrightarrow{\approx} & H^*(\{x\} \times Y, \{x\} \times Y\backslash\Delta(X)) \\
\downarrow{\scriptstyle =} & & \downarrow{\scriptstyle =} \\
H^*(X, X\backslash\{x\}) & \xrightarrow{\approx} & H^*(Y, Y\backslash\{x\})
\end{array}$$

with horizontal excision isomorphisms. If $\tilde{E}_x = \tilde{E} \cap E_x$, $\tilde{E}_x^* = \tilde{E} \cap E_x^*$, and $\tilde{E}^* = \tilde{E} \cap E^*$, the inclusion $E_x, E_x^* \xrightarrow{j_x} E, E^*$ also induces a commutative diagram

$$\begin{array}{ccc}
H^*(E, E^*) & \xrightarrow{\approx} & H^*(\tilde{E}, \tilde{E}^*) \\
\downarrow{\scriptstyle j_x^*} & & \downarrow{\scriptstyle j_x^*} \\
H^*(E_x, E_x^*) & \xrightarrow{\approx} & H^*(\tilde{E}_x, \tilde{E}_x^*)
\end{array}$$

with horizontal excision isomorphisms. Since the composition

$$X \to \tilde{E} \xrightarrow{F_0} X \times X \xrightarrow{\mathrm{pr}_1} X$$

is the identity map, F_0 induces a commutative diagram

$$\begin{array}{ccc}
\tilde{E}_x, \tilde{E}_x^* & \xrightarrow{F_{0,x}} & \{x\} \times Y, \{x\} \times Y\backslash\Delta(X) = Y, Y\backslash\{x\} \\
\downarrow{\scriptstyle j_x} & & \downarrow{\scriptstyle l_x} \\
\tilde{E}, \tilde{E}^* & \xrightarrow{F_0} & \tilde{D}, \tilde{D}\backslash\Delta(X)
\end{array}$$

with vertical inclusions, hence a commutative diagram

$$\begin{array}{ccc}
H^*(\tilde{D}, \tilde{D}\backslash\Delta(X)) & \underset{\approx}{\xrightarrow{F_0^*}} & H^*(\tilde{E}, \tilde{E}^*) \\
\downarrow{\scriptstyle l_x^*} & & \downarrow{\scriptstyle j_x^*} \\
H^*(Y, Y\backslash\{x\}) & \underset{\approx}{\xrightarrow{F_{0,x}^*}} & H^*(\tilde{E}_x, \tilde{E}_x^*)
\end{array}$$

with isomorphisms F_0^* and $F_{0,x}^*$. To complete the proof one defines G_l and $G_{l,x}$ to be the obvious compositions of excision isomorphisms with F_0^* and $F_{0,x}^*$, respectively. The proof that G_l is an $H^*(X)$-module isomorphism is a direct verification.

One can interchange the two factors $X \times X$ in Lemma 7.2, in which case $H^*(X \times X, X \times X \backslash \Delta(X))$ is an $H^*(X)$-module with respect to the composition carrying

$$\beta \otimes T \in H^*(X) \otimes H^*(X \times X, X \times X \backslash \Delta(X))$$

into

$$(1 \times \beta) \cup T \in H^*(X \times X, X \times X \backslash \Delta(X));$$

for any $x \in X$ the *right inclusion is*

$$X, X \backslash \{x\} = X \times \{x\}, X \times \{x\} \backslash \Delta(X) \xrightarrow{r_x} X \times X, X \times X \backslash \Delta(X).$$

As in Lemma 7.2 one then has an $H^*(X)$-module isomorphism G_r and $\mathbb{Z}/2$-module isomorphisms $G_{r,x}$ such that the diagrams

$$\begin{array}{ccc} H^*(X \times X, X \times X \backslash \Delta(X)) & \xrightarrow[\approx]{G_r} & H^*(E, E^*) \\ \downarrow r_x^* & & \downarrow j_x^* \\ H^*(X, X \backslash \{x\}) & \xrightarrow[\approx]{G_{r,x}} & H^*(E_x, E_x^*) \end{array}$$

of $\mathbb{Z}/2$-module homomorphisms commute.

In the following lemma we temporarily ignore the two $H^*(X)$-module structures of $H^*(X \times X, X \times X \backslash \Delta(X))$.

7.3 Lemma: *The isomorphisms G_l and G_r from $H^*(X \times X, X \times X \backslash \Delta(X))$ to $H^*(E, E^*)$ are the same $\mathbb{Z}/2$-module isomorphisms.*

PROOF: G_l is defined up to excision isomorphisms as the $\mathbb{Z}/2$-module isomorphism $H^*(\tilde{D}, \tilde{D} \backslash \Delta(X)) \xrightarrow{F_0^*} H^*(\tilde{E}, \tilde{E}^*)$, and G_r is defined up to the same excision isomorphisms as the $\mathbb{Z}/2$-module isomorphism $H^*(\tilde{D}, \tilde{D} \backslash \Delta(X)) \xrightarrow{F_1^*} H^*(\tilde{E}, \tilde{E}^*)$. By Lemma 7.1 there is a homotopy F from $\tilde{E} \xrightarrow{F_0} \tilde{D} \subset X \times X$ to $\tilde{E} \xrightarrow{F_1} \tilde{D} \subset X \times X$ relative to $\sigma(X) \subset \tilde{E}$, which also induces a homotopy from the restriction

$$\tilde{E} \backslash \sigma(X) \xrightarrow{F_0} \tilde{D} \backslash \Delta(X) \subset X \times X \backslash \Delta(X)$$

to the restriction

$$\tilde{E} \backslash \sigma(X) \xrightarrow{F_1} \tilde{D} \backslash \Delta(X) \subset X \times X \backslash \Delta(X).$$

Hence $F_0^* = F_1^*$, so that $G_l = G_r$.

7.4 Lemma: *For any $\beta \in H^*(X)$ and any $T \in H^*(X \times X, X \times X\backslash\Delta(X))$ one has $T \cup (\beta \times 1) = T \cup (1 \times \beta) \in H^*(X \times X, X \times X\backslash\Delta(X))$; hence*

$$j^*T \cup (\beta \times 1) = j^*T \cup (1 \times \beta) \in H^*(X \times X)$$

for the inclusion $X \times X \xrightarrow{j} X \times X, X \times X\backslash\Delta(X)$.

PROOF: Since the $\mathbb{Z}/2$-algebra $H^*(X \times X, X \times X\backslash\Delta(X))$ is commutative, it suffices to show that $(\beta \times 1) \cup T = (1 \times \beta) \cup T$. However, Lemma 7.3 gives

$$\begin{aligned} G_l((\beta \times 1) \cup T) &= \pi^*\beta \cup G_l T \\ &= \pi^*\beta \cup G_r T = G_r((1 \times \beta) \cup T) \end{aligned}$$

for the common $\mathbb{Z}/2$-module isomorphism $G_l = G_r$.

Lemmas 7.3 and 7.4 together imply that G_l and G_r are the same $H^*(X)$-module isomorphism, which we henceforth denote

$$H^*(X \times X, X \times X\backslash\Delta(X)) \xrightarrow{G} H^*(E, E^*).$$

7.5 Definition: For any smooth closed n-dimensional manifold $X \in \mathscr{U}$ the *diagonal $\mathbb{Z}/2$ Thom class* $T_X \in H^n(X \times X, X \times X\backslash\Delta(X))$ is the inverse image $G^{-1}U_{\tau(X)}$ of the $\mathbb{Z}/2$ Thom class $U_{\tau(X)} \in H^n(E, E^*)$ of the tangent bundle $\tau(X)$; the $\mathbb{Z}/2$ *Thom form* of $X \in \mathscr{U}$ is the image $j^*T_X \in H^n(X \times X)$ of T_X under the inclusion-induced homomorphism $H^n(X \times X, X \times X\backslash\Delta(X)) \xrightarrow{j^*} H^n(X \times X)$.

There are useful alternative characterizations of the diagonal $\mathbb{Z}/2$ Thom class T_X. Recall from Lemma 1.1 and Corollary 1.5 that for each $x \in X$ the $\mathbb{Z}/2$ homology module $H_n(X, X\backslash\{x\})$ is free on the single generator $j_{x,*}\mu_X$, where $X \xrightarrow{j_x} X, X\backslash\{x\}$ is an inclusion and $\mu_X \in H_n(X)$ is the fundamental $\mathbb{Z}/2$ homology class. (The boundary $\dot{X}$ of any $X \in \mathscr{U}$ is empty, so that $H_n(X, \dot{X}) = H_n(X)$.) Hence the $\mathbb{Z}/2$ cohomology module $H^n(X, X\backslash\{x\})$ is also free on a single generator, which we denote ω_x.

7.6 Proposition: *For any smooth closed n-dimensional manifold $X \in \mathscr{U}$ and any $x \in X$, let*

$$X, X\backslash\{x\} = \{x\} \times X, \{x\} \times X\backslash\Delta(X) \xrightarrow{l_x} X \times X, X \times X\backslash\Delta(X)$$

and

$$X, X\backslash\{x\} = X \times \{x\}, X \times \{x\}\backslash\Delta(X) \xrightarrow{r_x} X \times X, X \times X\backslash\Delta(X)$$

be left and right inclusions, respectively. Then the diagonal $\mathbb{Z}/2$ Thom class $T_X \in H^n(X \times X, X \times X\backslash\Delta(X))$ is uniquely characterized by the property that $l_x^*T_X = \omega_x \in H^n(X, X\backslash\{x\})$ for each $x \in X$; similarly, T_X is also uniquely characterized by the property that $r_x^*T_X = \omega_x \in H^n(X, X\backslash\{x\})$ for each $x \in X$.

PROOF: To prove the first assertion it suffices to reexamine the diagram

$$\begin{array}{ccc} H^*(X \times X, X \times X\backslash\Delta(X)) & \xrightarrow[\approx]{G} & H^*(E, E^*) \\ \Big\downarrow l_x^* & & \Big\downarrow j_x^* \\ H^*(X, X\backslash\{x\}) & \xrightarrow[\approx]{G_{1,x}} & H^*(E_x, E_x^*) \end{array}$$

of Lemma 7.2, where $GT_X = U_{\tau(X)}$, and to recall from Definition IV.1.4 that the $\mathbb{Z}/2$ Thom class $U_{\tau(X)} \in H^n(E, E^*)$ is the unique class such that $j_x^* U_{\tau(X)}$ is the generator of $H^*(E_x, E_x^*)$ for each $x \in X$. The proof of the second assertion is similar.

The following result is a partial rephrasing of Proposition 7.6.

7.7 Lemma: *Let $\mu_X \in H_n(X)$ be the fundamental $\mathbb{Z}/2$ homology class of a smooth closed n-dimensional manifold $X \in \mathscr{U}$, with $\mathbb{Z}/2$ Thom form $j^*T_X \in H^n(X \times X)$; then $\langle j^*T_X, 1 \times \mu_X\rangle = 1 \in \mathbb{Z}/2$ and similarly $\langle j^*T_X, \mu_X \times 1\rangle = 1 \in \mathbb{Z}/2$.*

PROOF: One starts with an inclusion diagram

$$\begin{array}{ccc} X = \{x\} \times X & \xrightarrow{j_x} & \{x\} \times X, \{x\} \times X\backslash\Delta(X) = X, X\backslash\{x\} \\ \Big\downarrow l_x & & \Big\downarrow l_x \\ X \times X & \xrightarrow{j} & X \times X, X \times X\backslash\Delta(X). \end{array}$$

By applying $H^*(\ )$ and $H_*(\ )$ and computing Kronecker products, it follows that

$$\begin{aligned} \langle j^*T_X, 1 \times \mu_X\rangle &= \langle j^*T_X, l_{x,*}\mu_X\rangle = \langle l_x^* j^* T_X, \mu_X\rangle \\ &= \langle j_x^* l_x^* T_X, \mu_X\rangle = \langle l_x^* T_X, (j_x)_*\mu_X\rangle \\ &= \langle \omega_x, (j_x)_*\mu_X\rangle = 1 \end{aligned}$$

for the unique generators $\omega_x \in H^n(X, X\backslash\{x\}$ and $(j_x)_*\mu_X \in H_n(X, X\backslash\{x\})$. A similar proof yields the second assertion.

Here is the main result of this section.

7.8 Proposition: *For any smooth closed n-dimensional manifold $X \in \mathscr{U}$, let $j^*T_X \in H^n(X \times X)$ be the $\mathbb{Z}/2$ Thom form of Definition 7.5, and let $H^*(X) \xrightarrow{D_P} H_*(X)$ be the Poincaré duality isomorphism $\cap\mu_X$ of Corollary 2.4. Then $\langle j^*T_X, a \times D_P\beta\rangle = \langle \beta, a\rangle \in \mathbb{Z}/2$ for any $a \in H_*(X)$ and any $\beta \in H^*(X)$, and $\langle j^*T_X, D_P\alpha \times b\rangle = \langle \alpha, b\rangle \in \mathbb{Z}/2$ for any $\alpha \in H^*(X)$ and $b \in H_*(X)$.*

PROOF: Observe that if a and β are homogeneous of different degrees then $\langle j^*T_X, a \times D_P\beta\rangle$ and $\langle \beta, a\rangle$ both vanish; hence one may as well assume that a and β are homogeneous of the same degree, in which case the cap product $\beta \cap a$ is just $\langle \beta, a\rangle 1 \in H_0(X)$. Since the coefficient ring is $\mathbb{Z}/2$, cup products commute, and one also has the classical identity $(\theta \times \varphi) \cap (\mu \times \nu) = (\theta \cap \mu) \times (\varphi \cap \nu)$ relating cap and cross products (as in Spanier [4, p. 255], for example). Consequently

$$\begin{aligned}
&\langle j^*T_X, a \times D_P\beta\rangle \\
&\quad = \langle j^*T_X, (1 \cap a) \times (\beta \cap \mu_X)\rangle = \langle j^*T_X, (1 \times \beta) \cap (a \times \mu_X)\rangle \\
&\quad = \langle j^*T_X \cup (1 \times \beta), a \times \mu_X\rangle = \langle j^*T_X \cup (\beta \times 1), a \times \mu_X\rangle \\
&\quad = \langle j^*T_X, (\beta \times 1) \cap (a \times \mu_X)\rangle = \langle j^*T_X, (\beta \cap a) \times (1 \cap \mu_X)\rangle \\
&\quad = \langle j^*T_X, \langle \beta, a\rangle 1 \times \mu_X\rangle = \langle \beta, a\rangle\langle j^*T_X, 1 \times \mu_X\rangle = \langle \beta, a\rangle,
\end{aligned}$$

as claimed, by Lemmas 7.4 and 7.7. Similarly, if α and b are homogeneous of the same degree one has

$$\begin{aligned}
&\langle j^*T_X, D_P\alpha \times b\rangle \\
&\quad = \langle j^*T_X, (\alpha \cap \mu_X) \times (1 \cap b)\rangle = \langle j^*T_X, (\alpha \times 1) \cap (\mu_X \times b)\rangle \\
&\quad = \langle j^*T_X \cup (\alpha \times 1), \mu_X \times b\rangle = \langle j^*T_X \cup (1 \times \alpha), \mu_X \times b\rangle \\
&\quad = \langle j^*T_X, (1 \times \alpha) \cap (\mu_X \times b)\rangle = \langle j^*T_X, (1 \cap \mu_X) \times (\alpha \cap b)\rangle \\
&\quad = \langle j^*T_X, \mu_X \times \langle \alpha, b\rangle 1\rangle = \langle \alpha, b\rangle\langle j^*T_X, \mu_X \times 1\rangle = \langle \alpha, b\rangle.
\end{aligned}$$

Observe that the preceding proof does not use the property that the $\mathbb{Z}/2$-module homomorphism $H^*(X) \xrightarrow{D_P} H_*(X)$ is actually an isomorphism.

Recall that the $\mathbb{Z}/2$ Poincaré form is a symmetric bilinear form

$$H^*(X) \times H^*(X) \xrightarrow{(\,,\,)_P} Z/2$$

defined by setting $(\alpha, \beta)_P = \langle \alpha, D_P\beta\rangle = \langle \alpha \cup \beta, \mu_X\rangle$ for any $(\alpha, \beta) \in H^*(X) \times H^*(X)$, and observe that the $\mathbb{Z}/2$ Thom form $j^*T_X \in H^*(X \times X)$ of Definition 7.5 can be regarded as a bilinear form

$$H_*(X) \times H_*(X) \xrightarrow{(\,,\,)_T} \mathbb{Z}/2$$

by setting $(a, b)_T = \langle j^*T_X, a \times b\rangle$ for any $(a, b) \in H_*(X) \times H_*(X)$. Since $H^*(X)$ is canonically isomorphic to $\mathrm{Hom}_{\mathbb{Z}/2}(H_*(X), \mathbb{Z}/2)$, the $\mathbb{Z}/2$ Thom form induces a unique $\mathbb{Z}/2$-module homomorphism $H_*(X) \xrightarrow{D_T} H^*(X)$ such that $\langle D_Ta, b\rangle = (a, b)_T$ for any $(a, b) \in H_*(X) \times H_*(X)$.

7.9 Corollary: *For any $X \in \mathscr{U}$, the compositions D_PD_T and D_TD_P are the identity isomorphisms on $H_*(X)$ and $H^*(X)$, respectively; that is, $D_T = D_P^{-1}$, so that the $\mathbb{Z}/2$ Poincaré form $(\,,\,)_P$ and $\mathbb{Z}/2$ Thom form $(\,,\,)_T$ are inverse non-degenerate bilinear forms.*

PROOF: For any $(\beta, a) \in H^*(X) \times H_*(X)$ one has $\langle \beta, D_{\mathrm{P}} D_{\mathrm{T}} a \rangle = (\beta, D_{\mathrm{T}} a)_{\mathrm{P}} = (D_{\mathrm{T}} a, \beta)_{\mathrm{P}} = \langle D_{\mathrm{T}} a, D_{\mathrm{P}} \beta \rangle = (a, D_{\mathrm{P}} \beta)_{\mathrm{T}} = \langle j^* T_X, a \times D_{\mathrm{P}} \beta \rangle = \langle \beta, a \rangle$ by Proposition 7.8, so that $D_{\mathrm{P}} D_{\mathrm{T}}$ is an isomorphism. Similarly, for any $(\alpha, b) \in H^*(X) \times H_*(X)$, one has $\langle D_{\mathrm{T}} D_{\mathrm{P}} \alpha, b \rangle = (D_{\mathrm{P}} \alpha, b)_{\mathrm{T}} = \langle j^* T_X, D_{\mathrm{P}} \alpha \times b \rangle = \langle \alpha, b \rangle$ by Proposition 7.8, so that $D_{\mathrm{T}} D_{\mathrm{P}}$ is an isomorphism.

For later convenience we exhibit the duality between $(\ ,\ )_{\mathrm{P}}$ and $(\ ,\ )_{\mathrm{T}}$ in terms of D_{P} and D_{T}, as it occurs in the preceding proof:

$$\langle D_{\mathrm{T}} a, D_{\mathrm{P}} \beta \rangle = \langle \beta, a \rangle \in \mathbb{Z}/2 \qquad \textit{for any} \quad (\beta, a) \in H^*(X) \times H_*(X).$$

Here is one of the most useful properties of $\mathbb{Z}/2$ Thom forms.

7.10 Proposition: *Let $X \in \mathscr{U}$ be any smooth closed n-dimensional manifold with $\mathbb{Z}/2$ Thom form $j^* T_X \in H^n(X \times X)$, and let $X \xrightarrow{\Delta} X \times X$ be the diagonal map. Then the $\mathbb{Z}/2$ Euler class of the tangent bundle $\tau(X)$ is given by $e(\tau(X)) = \Delta^*(j^* T_X) \in H^n(X)$.*

PROOF: Let $E \xrightarrow{\pi} X$ represent $\tau(X)$. Then the vertical arrows on the left-hand side of the clearly commutative diagram

$$\begin{array}{ccccc}
H^n(X \times X, X \times X \backslash \Delta(X)) & \xrightarrow{j^*} & H^n(X \times X) & \searrow^{\Delta^*} & \\
\downarrow \approx & & \downarrow & & \\
H^n(\tilde{D}, \tilde{D} \backslash \Delta(X)) & \xrightarrow{j^*} & H^n(\tilde{D}) & \xrightarrow{\Delta^*} & H^n(X) \\
\downarrow \approx\ F_0^* & & \downarrow \approx & & \| \\
H^n(\tilde{E}, \tilde{E} \backslash \sigma(X)) & \xrightarrow{j^*} & H^n(\tilde{E}) & \xrightarrow{\sigma^*} & H^n(X) \\
\uparrow \approx & & \uparrow & \nearrow^{\sigma^*} & \\
H^n(E, E^*) & \xrightarrow{j^*} & H^n(E) & &
\end{array}$$

are precisely the isomorphisms used to construct the isomorphism G_l $(= G)$ of Lemma 7.2; hence one has a commutative diagram

$$\begin{array}{ccccc}
H^n(X \times X, X \times X \backslash \Delta(X)) & \xrightarrow{j^*} & H^n(X \times X) & \searrow^{\Delta^*} & \\
\downarrow \approx\ G & & & & H^n(X). \\
H^n(E, E^*) & \xrightarrow{j^*} & H^n(E) & \nearrow^{\sigma^*} &
\end{array}$$

Since the $\mathbb{Z}/2$ Euler class of $\tau(X)$ is given in Definition IV.3.1 by setting $e(\tau(X)) = \sigma^* j^* U_{\tau(X)}$ for the $\mathbb{Z}/2$ Thom class $U_{\tau(X)} \in H^n(E, E^*)$, it follows that

$$j^* \Delta^* T_X = \sigma^* j^* G T_X = \sigma^* j^* U_{\tau(X)} = e(\tau(X)),$$

as claimed.

8. The Thom–Wu Theorem

We shall show for any formal power series $f(t) \in \mathbb{Z}/2[[t]]$ with leading term $1 \in \mathbb{Z}/2$ that the multiplicative $\mathbb{Z}/2$ class $u_f(\tau(X)) \in H^*(X)\ (\subset H^{**}(X))$ of the tangent bundle $\tau(X)$ of any smooth closed manifold $X \in \mathscr{U}$ depends only on the $\mathbb{Z}/2$ homology and cohomology of X itself. Hence $u_f(\tau(X))$ is homotopy invariant (up to canonical isomorphisms); in particular, $u_f(\tau(X))$ is independent of the smooth structure assigned to X. The proof uses the Steenrod square Sq of Remark V.7.14, which is not part of elementary algebraic topology; nevertheless, the result is too important to omit.

8.1 Lemma: *For any smooth closed n-dimensional manifold $X \in \mathscr{U}$ let $j^*T_X \in H^n(X \times X)$ be the $\mathbb{Z}/2$ Thom form, and let $w \in H^*(X)$ be the total Stiefel–Whitney class $w(\tau(X))$ of the tangent bundle $\tau(X)$; then*

$$j^*T_X \cup (w \times 1) = j^*T_X \cup (1 \times w) = \mathrm{Sq}\, j^*T_X \in H^*(X \times X),$$

for the Steenrod square Sq *of Remark* V.7.14.

Proof: Let $E \xrightarrow{\pi} X$ represent $\tau(X)$ and let $U_{\tau(X)} \in H^n(E, E^*)$ be the usual $\mathbb{Z}/2$ Thom class. By Remark V.7.18 one then has an identity $\pi^*w \cup U_{\tau(X)} = \mathrm{Sq}\, U_{\tau(X)} \in H^*(E, E^*)$ to which one applies the $\mathbb{Z}/2$-module isomorphism

$$H^*(E, E^*) \xrightarrow{G_l^{-1}} H^*(X \times X, X \times X\backslash\Delta(X))$$

of Lemma 7.2 to obtain

$$(w \times 1) \cup T_X = \mathrm{Sq}\, T_X \in H^*(X \times X, X \times X\backslash\Delta(X)).$$

Since Sq is natural, and since $H^*(X \times X)$ is commutative with respect to cup product, one can then apply the inclusion-induced homomorphism $H^*(X \times X, X \times X\backslash\Delta(X)) \xrightarrow{j^*} H^*(X \times X)$ and Lemma 7.4 to complete the proof.

For any $X \in \mathscr{U}$ the Steenrod square Sq induces a transpose $\mathbb{Z}/2$-module homomorphism $H_*(X) \xrightarrow{\mathrm{Sq}_X} H_*(X)$, defined by the requirement that $\langle \beta, \mathrm{Sq}_X a\rangle = \langle \mathrm{Sq}\,\beta, a\rangle$ for any $a \in H_*(X)$ and any $\beta \in H^*(X)$.

8.2 Lemma: *For any $X \in \mathscr{U}$ and any elements $a \in H_*(X)$, $b \in H_*(X)$, and $T \in H^*(X \times X)$ one has*

$$\langle \mathrm{Sq}\, T, a \times b\rangle = \langle T, \mathrm{Sq}_X a \times \mathrm{Sq}_X b\rangle \in \mathbb{Z}/2.$$

Proof: Since T is a $\mathbb{Z}/2$ linear combination of cross products $\alpha \times \beta \in H^*(X \times X)$, and since $\mathrm{Sq}(\alpha \times \beta) = \mathrm{Sq}\,\alpha \times \mathrm{Sq}\,\beta$ by Remark V.7.16, it suffices to observe that

$$\begin{aligned}\langle \mathrm{Sq}(\alpha \times \beta), a \times b\rangle &= \langle \mathrm{Sq}\,\alpha \times \mathrm{Sq}\,\beta, a \times b\rangle = \langle \mathrm{Sq}\,\alpha, a\rangle\langle \mathrm{Sq}\,\beta, b\rangle \\ &= \langle \alpha, \mathrm{Sq}_X a\rangle\langle \beta, \mathrm{Sq}_X b\rangle = \langle \alpha \times \beta, \mathrm{Sq}_X a \times \mathrm{Sq}_X b\rangle.\end{aligned}$$

In the following result $H_*(X) \xrightarrow{D_T} H^*(X)$ is the inverse of the $\mathbb{Z}/2$ Poincaré duality isomorphism $H^*(X) \xrightarrow{D_P = \cap \mu_X} H_*(X)$, as in Corollary 7.9.

8.3 Proposition (Thom and Wu): *For any smooth closed manifold* $X \in \mathcal{U}$ *the total Stiefel–Whitney class* $w(\tau(X))$ *of the tangent bundle* $\tau(X)$ *satisfies the identity* $w(\tau(X)) = \operatorname{Sq} D_T \operatorname{Sq}_X \mu_X \in H^*(X)$ *for the fundamental* $\mathbb{Z}/2$ *class* $\mu_X \in H_*(X)$.

PROOF: Set $w(\tau(X)) = w \in H^*(X)$ as before. Then for any $a \in H_*(X)$ one has

$$\begin{aligned}
\langle w, a\rangle &= \langle D_T D_P w, a\rangle \\
&= (D_P w, a)_T = \langle j^* T_X, D_P w \times a\rangle = \langle j^* T_X, (w \cap \mu_X) \times (1 \cap a)\rangle \\
&= \langle j^* T_X, (w \times 1) \cap (\mu_X \times a) = \langle j^* T_X \cup (w \times 1), \mu_X \times a\rangle \\
&= \langle \operatorname{Sq} j^* T_X, \mu_X \times a\rangle = \langle j^* T_X, \operatorname{Sq}_X \mu_X \times \operatorname{Sq}_X a\rangle \\
&= (\operatorname{Sq}_X \mu_X, \operatorname{Sq}_X a)_T = \langle D_T \operatorname{Sq}_X \mu_X, \operatorname{Sq}_X a\rangle \\
&= \langle \operatorname{Sq} D_T \operatorname{Sq}_X \mu_X, a\rangle \in \mathbb{Z}/2,
\end{aligned}$$

by Lemmas 8.1 and 8.2.

Proposition 8.3 is a variant of the original result of Thom [2] and Wu [3]. Another formulation is given in Corollary 8.5.

8.4 Theorem (Thom and Wu): *For any smooth closed manifold* $X \in \mathcal{U}$ *and any formal power series* $f(t) \in \mathbb{Z}/2[[t]]$ *with leading term* $1 \in \mathbb{Z}/2$, *the* $\mathbb{Z}/2$ *multiplicative class* $u_f(\tau(X)) \in H^*(X)$ *depends only on the algebraic structure of* $H_*(X)$ *and* $H^*(X)$; *in particular,* $u_f(\tau(X))$ *is independent of the smooth structure assigned to* X.

PROOF: Since the fundamental $\mathbb{Z}/2$ class $\mu_X \in H_n(X)$ and the operators Sq_X, D, Sq depend only on algebraic properties of $H_*(X)$ and $H^*(X)$, the identity $w(\tau(X)) = \operatorname{Sq} D_T \operatorname{Sq}_X \mu_X$ of Proposition 8.3 implies the result for the total Stiefel–Whitney class, which is the special case $f(t) = 1 + t$. However, Proposition V.2.10 provides polynomial computations for *any* multiplicative $\mathbb{Z}/2$ class $u_f(\tau(X))$ in terms of $w(\tau(X))$, which are also independent of $\tau(X)$.

Since $H_*(X)$ and $H^*(X)$ depend only on the homotopy type of $X \in \mathcal{U}$, the preceding result can be rephrased as follows: *For any* $X \in \mathcal{U}$ *the* $\mathbb{Z}/2$ *multiplicative classes* $u_f(\tau(X))$ *depend only on the homotopy type of* X.

Here are some more corollaries of Proposition 8.3.

8.5 Corollary (Wu): *For any smooth closed manifold* $X \in \mathcal{U}$ *the Wu class* $\operatorname{Wu}(\tau(X))$ *of the tangent bundle* $\tau(X)$ *is the unique class* $v \in H^*(X)$ *such that* $\langle v \cup \alpha, \mu_X\rangle = \langle \operatorname{Sq} \alpha, \mu_X\rangle$ *for every* $\alpha \in H^*(X)$.

PROOF: Since $\langle v \cup \alpha, \mu_X \rangle = (v, \alpha)_P$ for the nondegenerate $\mathbb{Z}/2$ Poincaré form $(\, , \,)_P$, uniqueness is clear. It remains to observe that

$$\begin{aligned}\langle \mathrm{Wu}(\tau(X)) \cup \alpha, \mu_X \rangle &= \langle \mathrm{Sq}^{-1} w(\tau(X)) \cup \alpha, \mu_X \rangle \\ &= (D_T \, \mathrm{Sq}_X \, \mu_X, \alpha)_P = \langle D_T \, \mathrm{Sq}_X \, \mu_X, D_P \alpha \rangle \\ &= \langle \alpha, \mathrm{Sq}_X \, \mu_X \rangle = \langle \mathrm{Sq} \, \alpha, \mu_X \rangle\end{aligned}$$

by Remark V.7.20 and the reformulation of Corollary 7.9.

8.6 Corollary: *If* $X \in \mathscr{U}$ *is of dimension* n*, then* $\mathrm{Wu}(\tau(X)) \in H^*(X)$ *vanishes in all dimensions* p *such that* $2p > n$.

PROOF: If $a \in H_p(X)$, then $D_T a \in H^{n-p}(X)$ and

$$\langle \mathrm{Wu}(\tau(X)), a \rangle = \langle D_T \, \mathrm{Sq}_X \, \mu_X, a \rangle = \langle D_T a, \mathrm{Sq}_X \, \mu_X \rangle = \langle \mathrm{Sq} \, D_T a, \mu_X \rangle.$$

By the dimension axiom for Steenrod squares (Remark V.7.14) $\mathrm{Sq}^q \, D_T \, a = 0 \in H^{n-p+q}(X)$ for $q > n - p$. Hence, if $p > n - p$ one has $\mathrm{Sq}^p \, D_T a = 0 \in H^n(X)$, so that $\langle \mathrm{Wu}(\tau(X)), a \rangle = 0$, for every $a \in H_p(X)$.

8.7 Corollary: *For any* $X \in \mathscr{U}$ *of dimension* $n > 0$ *one has* $\langle \mathrm{Wu}(\tau(X)), \mu_X \rangle = 0 \in \mathbb{Z}/2$.

PROOF: Since $2n > n$, this is a consequence of Corollary 8.6.

9. Remarks and Exercises

9.1 Remark: The $\mathbb{Z}/2$ fundamental class $\mu_X \in H_n(X, \dot{X}; \mathbb{Z}/2)$ of Definition 1.4 was defined only for *smooth* n-dimensional compact manifolds X with boundary $\dot{X}$ because the given proof of its existence required results (Propositions I.8.4 and I.9.7) which were established only in the triangulable case. However, with additional effort one can extend Definition 1.4 to n-dimensional compact *topological* manifolds.

If an n-dimensional compact topological manifold X with boundary $\dot{X}$ is *oriented* in the sense described in the next volume, then for *any* commutative ring Λ with unit there is a unique fundamental class $\mu_X \in H_n(X, \dot{X}; \Lambda)$; the case $\Lambda = \mathbb{Z}$ suffices. We shall obtain this result for the smooth case in Volume 2 as an oriented analog of Proposition 1.3. As before, smoothness is not really required; proofs of the topological case can be found in Dold [8, pp. 259–267], Massey [6, pp. 200–205], Milnor and Stasheff [1, pp. 273–274], Spanier [4, pp. 299–306], and (for closed topological manifolds) in Samelson [1].

9.2 Remark: An alternative proof of the identity $\partial\mu_X = \mu_{\dot{X}}$ of Proposition 1.9 is given in Spanier [4, p. 304].

9.3 Remark: Here is an alternative proof that the $\mathbb{Z}/2$ Poincaré duality map $H^*(X) \xrightarrow{D_P = \cap\mu_X} H_*(X)$ of Corollary 2.4 is an isomorphism. The fact that D_P is an isomorphism is not used in the proof of Proposition 7.8; furthermore, the Thom form $j^*T_X \in H^*(X \times X)$ and induced homomorphism $H_*(X) \xrightarrow{D_T} H^*(X)$ were constructed independently of D_P. However, according to Corollary 7.9 the compositions $D_T D_P$ and $D_P D_T$ are identity isomorphisms on $H^*(X)$ and $H_*(X)$, respectively; hence D_P is an isomorphism with inverse D_T.

9.4 Remark: If X is an *oriented* compact smooth manifold with boundary $\dot{X}$, then one can replace the coefficient ring $\mathbb{Z}/2$ of Theorem 2.3 by any commutative ring Λ with unit to obtain an *oriented* Poincaré–Lefschetz duality theorem. We shall do so in the next volume, using the obvious analog of Lemma 2.2; a different algebraic format for the proof is given on pages 1–17 of Browder [1].

Poincaré–Lefschetz duality is also valid for *topological* manifolds, using the coefficient ring $\mathbb{Z}/2$ except in the oriented case. One of the standard proofs is similar to the proof of Theorem 2.3; it can be found in Dold [8, pp. 291–298], Greenberg [1, pp. 162–189], Milnor and Stasheff [1, pp. 276–280], and Spanier [4, pp. 296–297.]. There are also proofs of the same nature in Borel [3] and in Griffiths [1], using Alexander–Spanier and Čech cohomology, respectively, rather than singular cohomology.

Given a smooth closed *oriented* manifold X, one can establish Poincaré duality for arbitrary coefficients by a procedure analogous to that of Remark 9.3; the details appear in Milnor [3, pp. 51–52] and Milnor and Stasheff [1, pp. 127–128]. A similar technique applies to closed *topological* manifolds; the details appear in Spanier [3] and in Samelson [1].

Finally, there are older proofs of the Poincaré–Lefschetz duality theorem for triangulable manifolds (a fortiori for smooth manifolds) in Mayer [2], Lefschetz [1, pp. 188–204], and Maunder [1, pp. 170–199]. A survey of Poincaré–Lefschetz duality and related topics appears in Dold [7], and a 1965 bibliography of proofs of Poincaré duality appears in Samelson [1].

9.5 Remark: Stiefel–Whitney ($\mathbb{Z}/2$ cohomology) classes of tangent bundles of smooth manifolds were first constructed in Stiefel [1] and in Whitney [2] in 1935, using obstruction theory, as noted in Remark V.7.1. Whitney presented the details of his original construction in Whitney [3, 4, 6], and some related work appeared in Rohlin [3], also using obstruction theory. In

Pontrjagin [1] classifying space techniques were applied exclusively to tangent bundles to extend the results of Stiefel [1]; the same techniques were applied in the detailed exposition of Pontrjagin [5].

Chern [1] and Pontrjagin [2] used integral formulas to express Stiefel–Whitney classes of tangent bundles. Classifying space techniques like those of Pontrjagin were then developed in Chern [2] for tangent bundles; they were extended to vector bundles in general in Chern [3] and in Wu [2, 5].

9.6 Remark: The general computation of Stiefel–Whitney classes via Steenrod squares (Exercise V.7.17) appeared in Thom [1], which concerns arbitrary real vector bundles. However, the method was instantly applied to tangent bundles in Thom [2], Wu [3], and Thom [4].

There are other methods for computing Stiefel–Whitney classes of tangent bundles of smooth manifolds. For example, Bucur and Lascu [1] compute Stiefel–Whitney classes of smooth closed manifolds by a method originally applied in B. Segre [1] to algebraic varieties. The alternate computations of Nash [1] and of Teleman [1–3] form part of the next remark.

9.7 Remark: We already know from Remark III.13.34 that Thom [4] shows that the fiber homotopy type of the tangent bundle $\tau(X)$ of a smooth closed manifold X is independent of the smooth structure assigned to X; a fortiori the same conclusion applies to the J-equivalence class of $\tau(X)$. (These results were subsequently strengthened in Atiyah [1] and in Benlian and Wagoner [1].) We also know from Remark V.7.10 that Thom [4] shows that Stiefel–Whitney classes of vector bundles depend only on the J-equivalence classes of the bundles. Consequently for any smooth (closed) manifold X the Stiefel–Whitney class $w(\tau(X))$ is independent of the smooth structure assigned to X. (The same result was established in Theorem 8.4 by methods of Thom [2] and Wu [3].)

This is perhaps the best place to recall from the remainder of Remark V.7.10 that one can most easily explain the preceding result by assigning Stiefel–Whitney classes directly to topological manifolds X in a way which produces $w(\tau(X))$ whenever X happens to be smooth. The first such construction is that of Nash [1]. A later construction begins with the microbundles of Milnor [11, 14], reinterpreted as tangent topological $\mathbb{R}^n$ bundles in Kister [1, 2], to which one assigns total Stiefel–Whitney classes as in Teleman [1, 2, 3], for example; the latter Stiefel–Whitney classes agree with those of Nash.

9.8 Remark: The combinatorial construction on page 342 of Stiefel [1] and a later sketch in Whitney [5] assign $\mathbb{Z}/2$ *homology* classes (rather than $\mathbb{Z}/2$ *cohomology* classes) to smooth manifolds, using barycentric subdivisions of given triangulations. These constructions were revitalized by Cheeger [1], Sullivan [1], Halperin and Toledo [1], and Latour [1], with the observation that the $\mathbb{Z}/2$ homology classes so assigned to a smooth closed manifold X are the $\mathbb{Z}/2$ Poincaré duals of the Stiefel–Whitney classes $w_i(\tau(X))$ of the tangent bundle $\tau(X)$. Halperin and Toledo [1] provided the first complete proof of the duality assertion; later proofs were given by L. R. Taylor [1] and Blanton and McCrory [1]. Goldstein and Turner [1] give a variant of Whitney's construction which is explicitly independent of barycentric subdivision, Banchoff [1] gives a visually appealing interpretation of the construction, and McCrory [1] and Porter [1] give an intepretation which is based upon singularities of mappings of the manifold X into euclidean spaces. Related constructions can be found in Akin [1] and in Goldstein and Turner [2].

9.9 Remark: The preceding remark suggests that Stiefel–Whitney *cohomology* classes of tangent bundles also lead a life of their own, without the machinery of vector bundles in general. In fact, an axiomatic characterization of Stiefel–Whitney classes of tangent bundles is given in Blanton and Schweitzer [1], and another axiomatic characterization of such classes is given in Stong [7]. The value of the Blanton–Schweitzer axioms is demonstrated in L. R. Taylor [1] and in Blanton and McCrory [1], where they are used to prove the duality theorem mentioned in the preceding remark. The key to Blanton and Schweitzer [1] lies in Remark III.13.44.

9.10 Remark: For any smooth manifold X the classical *Whitney duality theorem* in $H^*(X; \mathbb{Z}/2)$ is merely the Whitney product formula for the Whitney sum $\tau(X) \oplus \nu_f$ of the tangent bundle $\tau(X)$ and the normal bundle ν_f of any immersion $X \xrightarrow{f} \mathbb{R}^{2n-k}$; the result was first announced without proof in Whitney [6], as already noted in Remark V.7.8. An analogous duality theorem in $H_*(X; \mathbb{Z}/2)$ was established in Halperin and Toledo [2], helping to verify that the $\mathbb{Z}/2$ homology classes of Remark 9.8 are Poincaré duals of the corresponding Stiefel–Whitney (cohomology) classes; the homology duality is visually interpreted in Banchoff and McCrory [1]. Finally, the suggestion formulated in Remark 9.9 is further dramatized by a *direct* combinatorial construction in Banchoff and McCrory [2] of the dual Stiefel–Whitney (cohomology) classes $\bar{w}_i(\tau(X))$ of any triangulated manifold X.

9.11 Exercise: Use the formula

$$\tau(G^m(\mathbb{R}^{m+n})) \oplus (\gamma_n^m \otimes \gamma_n^m) = (m+n)\gamma_n^m$$

of Exercise III.13.27 to compute the total Stiefel–Whitney class $w(\tau(G^n(\mathbb{R}^{m+n}))$ $\in H^*(G_m(\mathbb{R}^{m+n}))$ of the tangent bundle $\tau(G^m(\mathbb{R}^{m+n}))$ of the real Grassmann manifold $G^m(\mathbb{R}^{m+n})$. [*Hint*: Do Exercise V.7.9 first.]

9.12 Exercise: For any smooth closed manifold X the tangent bundle $\tau(X \times X)$ of the product $X \times X$ is trivially the sum $\tau(X) + \tau(X)$. For any riemannian metric $\langle\, ,\rangle$ on $\tau(X)$ one can therefore define a riemannian metric on $\tau(X \times X)$ by setting $\langle e + e', f + f'\rangle = \langle e, f\rangle + \langle e', f'\rangle$ for elements e, e', f, f' in the total space of $\tau(X)$ such that $\pi e = \pi f$ and $\pi e' = \pi f'$, where π is the projection onto X. Hence one can define the normal bundle ν_Δ of the diagonal map $X \xrightarrow{\Delta} X \times X$ by requiring $\tau(X) \oplus \nu_\Delta = \Delta^! \tau(X \times X)$ as in Definition 4.1. Show that ν_Δ is $\tau(X)$ itself.

9.13 Remark: Some general immersion theorems were briefly described in Remark I.10.13, and a classical nonimmersion theorem was proved in Proposition 4.5. Here is a very general necessary and sufficient immersion criterion.

If $X \xrightarrow{f} \mathbb{R}^{2n-p}$ is any immersion of a smooth n-dimensional manifold X, then the normal bundle ν_f is of rank $n - p$ and $\tau(X) \oplus \nu_f = \varepsilon^{2n-p}$ over X, as in Definition 4.1. Conversely, according to M. W. Hirsch [1], if there is any real vector bundle ξ of rank $n - p$ over X such that $\tau(X) \oplus \xi = \varepsilon^{2n-p}$, then there is an immersion $X \xrightarrow{f} \mathbb{R}^{2n-p}$ whenever X is compact.

The "easy" Whitney immersion theorem of Remark I.10.13 trivially provides an immersion $X \xrightarrow{h} \mathbb{R}^{2n+1}$ for any smooth n-dimensional manifold X, and according to another result of M. W. Hirsch [1] there is a smooth homotopy $X \times [0, 1] \to \mathbb{R}^{2n+1}$ relating any two such immersions, h_0 and h_1, such that each $X \times \{t\} \to \mathbb{R}^{2n+1}$ is also an immersion. It follows that the classifying maps $X \to G^{n+1}(\mathbb{R}^\infty)$ for the normal bundles ν_{h_0} and ν_{h_1} are homotopic, hence that $\nu_{h_0} = \nu_{h_1}$; that is, the normal bundle ν_h is uniquely defined by X itself. (According to Kervaire [2] this result was already known for *embeddings* $X \xrightarrow{h} \mathbb{R}^{2n+1}$. In fact, for $n > 1$ Wu [8] had proved an even stronger result: any two smooth *embeddings* $X \xrightarrow{h_0} \mathbb{R}^{2n+1}$ and $X \xrightarrow{h_1} \mathbb{R}^{2n+1}$ of a smooth n-dimensional manifold X are *isotopic* in the sense that there is a smooth map $X \times [0, 1] \xrightarrow{h} \mathbb{R}^{2n+1}$ such that each $X \xrightarrow{h_t} \mathbb{R}^{2n+1}$ is itself a smooth embedding.)

One can always weaken any two immersions $X \xrightarrow{f} \mathbb{R}^{2n-p}$ and $X \xrightarrow{g} \mathbb{R}^{2n-q}$ with normal bundles ν_f and ν_g, using inclusion maps $\mathbb{R}^{2n-p} \to \mathbb{R}^{2n+1}$ and $\mathbb{R}^{2n-q} \to \mathbb{R}^{2n+1}$ to obtain immersions $X \xrightarrow{f} \mathbb{R}^{2n-p} \to \mathbb{R}^{2n+1}$ and $X \xrightarrow{g} \mathbb{R}^{2n-q} \to \mathbb{R}^{2n+1}$ with normal bundles $\nu_f \oplus \varepsilon^{p+1}$ and $\nu_g \oplus \varepsilon^{q+1}$, respectively. According to the preceding paragraph one then has $\nu_f \oplus \varepsilon^{p+1} = \nu_g \oplus \varepsilon^{q+1}$; that is, in the terminology of Remark III.13.36, *any* two normal bundles of X determine the same stable equivalence class, the *stable normal bundle of* X.

By combining the preceding results it follows that the question of a "best-possible" immersion $X \xrightarrow{f} \mathbb{R}^{2n-p}$ of a given smooth n-dimensional manifold X is equivalent to the determination of the *geometric dimension* of the stable normal bundle of X, as in Remark III.13.40. Specifically, if the geometric dimension of the stable normal bundle of X is $n - p$ or less then X immerses in $\mathbb{R}^{2n-p}$, at least when X is compact.

9.14 Remark: The "best-possible" immersion conjecture of Remark I.10.13 was proved by Cohen [1], using results of R. L. W. Brown [2, 3] and Brown and Peterson [4, 5]: any smooth n-dimensional compact manifold X admits a smooth immersion $X \to \mathbb{R}^{2n-\alpha(n)}$, where $\alpha(n)$ is the number of 1's in the dyadic expansion of the dimension n. According to Proposition 4.5 one can do no better.

Here is an outline of the proof. Let $BO(m)$ denote the classifying space $G^m(\mathbb{R}^\infty)$ for real m-plane bundles, as in Remark III.13.4, and let BO denote the classifying space $\lim_m BO(m)$ for stable real vector bundles, as in Remark III.13.36. Since the $\mathbb{Z}/2$ cohomology rings $H^*(BO(m))$ are polynomial rings with one generator $w_1(\gamma^m), \ldots, w_m(\gamma^m)$ in each degree $1, \ldots, m$, the identities $w_i(\gamma^m \oplus \varepsilon^1) = w_i(\gamma^m)$ and the naturality of Stiefel–Whitney classes imply that the $\mathbb{Z}/2$ cohomology ring $H^*(BO)$ is a polynomial ring with one generator w_i in each degree $i > 0$.

Now let X be any smooth n-dimensional compact manifold, and let ν_X be its stable normal bundle; as in Remark III.13.36, there is a stable homotopy classifying map $X \xrightarrow{f_X} BO$ for ν_X. According to Remark 9.13, one must show that the geometric dimension of ν_X is at most $n - \alpha(n)$; that is, one must factor f_X in the form $X \to BO(n - \alpha(n)) \to BO$, up to homotopy.

(According to the result of Wu [8] and M. W. Hirsch [1] cited in Remark 9.13, one can regard ν_X more concretely as an $(n + 1)$-plane bundle ν_h whose classifying map $X \to BO(n + 1)$ is to be factored in the form $X \to BO(n - \alpha(n)) \to BO(n + 1)$, up to homotopy. However, the stable classifying space BO is more convenient than $BO(n + 1)$.

Let $I_n \subset H^*(BO)$ be the ideal of all polynomials in $w_1, w_2, \ldots$ lying in the kernel of all the homomorphisms $H^*(BO) \xrightarrow{f_X^*} H^*(X)$ induced by the stable homotopy classifying maps f_X, for all smooth n-dimensional manifolds X. According to Brown and Peterson [4, 5] there is a topological space BO/I_n and a universal map $BO/I_n \xrightarrow{j_n} BO$ with the following two properties: (i) each f_X can be factored in the form $X \to BO/I_n \xrightarrow{j_n} BO$, and (ii) the inclusion $I_n \subset H^*(BO)$ induces a short exact sequence

$$O \longrightarrow I_n \longrightarrow H^*(BO) \xrightarrow{j_n^*} H^*(BO/I_n) \longrightarrow O.$$

Briefly, BO/I_n is itself a classifying space for stable normal bundles of smooth n-dimensional manifolds, and its $\mathbb{Z}/2$ cohomology ring is as small as possible. Brown and Peterson also obtain a weakened form $MO/I_n \to MO(n - \alpha(n)) \to MO$ of a possible factorization $BO/I_n \to BO(n - \alpha(n)) \to BO$ of j_n.

Recall from Remark I.10.13 that R. L. W. Brown [2, 3] showed that every smooth closed n-dimensional manifold is cobordant to a manifold that immerses in $\mathbb{R}^{2n-\alpha(n)}$. Cohen [1] uses this result to construct a space X_n and maps f_n and g_n for which the compositions $X_n \xrightarrow{f_n} BO(n - \alpha(n)) \to BO$ and $X_n \xrightarrow{g_n} BO/I_n \xrightarrow{j_n} BO$ are homotopic maps from X_n to BO; the corresponding weakened compositions $TX_n \xrightarrow{Tf_n} MO(n - \alpha(n)) \to MO$ and $TX_n \xrightarrow{Tg_n} MO/I_n \xrightarrow{Tj_n} MO$ are then homotopic maps from TX_n to MO. Cohen also constructs a map σ_n such that $MO/I_n \xrightarrow{\sigma_n} TX_n \xrightarrow{Tg_n} MO/I_n$ is homotopic to the identity and

$$TX_n \xrightarrow{Tg_n} MO/I_n \xrightarrow{\sigma_n} TX_n \xrightarrow{Tf_n} MO(n - \alpha(n))$$

is homotopic to Tf_n. Consequently the composition $MO/I_n \xrightarrow{Tg_n \circ \sigma_n} MO(n - \alpha(n)) \to MO$ is a new weak version of a possible factorization of j_n, homotopic to the version of the preceding paragraph; moreover, Cohen strengthens the new version to an actual factorization $BO/I_n \to BO(n - \alpha(n)) \to BO$ of j_n. It follows that the stable homotopy classifying map f_X of the stable normal bundle ν_X of any smooth compact n-dimensional manifold $X \in \mathscr{M}$ can be factored in the form $X \to BO/I_n \to BO(n - \alpha(n)) \to BO$, and hence that ν_X is of geometric dimension at most $n - \alpha(n)$; consequently X immerses in $\mathbb{R}^{2n-\alpha(n)}$ as desired, by Remark 9.13.

9.15 Remark: Theorem 4.8 first appeared in its present form in Chern and and Spanier [2], following unpublished work of Hopf. The result can also be found in Milnor [3, pp. 43–44], Husemoller [1, pp. 261–262], and Milnor and Stasheff [1, p. 120].

There is an important partial converse in Haefliger and Hirsch [1], as follows. Let X be a smooth closed *k-connected* n-dimensional manifold with $0 \leqq 2k < n - 4$; then X embeds in $\mathbb{R}^{2n-k-1}$ if and only if $\bar{w}_{n-k-1}(\tau(X)) = 0 \in H^{n-k-1}(X)$. (A manifold is *k-connected* whenever it is connected and the homotopy groups $\pi_1(X), \ldots, \pi_k(X)$ vanish.) Thus, in this special case, the embedding criterion of Theorem 4.8 is both necessary and sufficient.

The case $k = 0$ of the Haefliger–Hirsch theorem will be used in the next remark. Let X be *any* smooth closed manifold of dimension $n > 4$; then X embeds in $\mathbb{R}^{2n-1}$ if and only if $\bar{w}_{n-1}(\tau(X)) = 0 \in H^{n-1}(X)$. Observe that the connectedness condition is entirely deleted; one simply applies the $k = 0$ version of the Haefliger–Hirsch theorem separately to each connected component of X.

9.16 Remark: Some general embedding theorems were briefly suggested in Remark I.10.11, and a classical nonembedding theorem was proved in Proposition 4.10. This is an appropriate place for more details about the strongest known general embedding theorem.

According to M. W. Hirsch [2], any smooth *open* n-dimensional manifold whatsoever embeds in $\mathbb{R}^{2n-1}$. Accordingly we henceforth consider only smooth *closed* n-dimensional manifolds, for $n = 1, 2, 3, \ldots$.

The "hard" Whitney embedding $X \to \mathbb{R}^{2n}$ is itself best-possible for smooth closed manifolds of dimensions $n = 1$ and $n = 2$; for according to Proposition 4.9 the real projective spaces RP^1 and RP^2 do *not* embed in $\mathbb{R}^1$ and $\mathbb{R}^3$, respectively. On the other hand, every *orientable* closed manifold of dimension $n = 2$ is one of the familiar orientable surfaces of some genus p, which visibly embed in $\mathbb{R}^3$.

According to Wall [1] every smooth 3-dimensional manifold X embeds in $\mathbb{R}^5$.

The "hard" Whitney embedding $X \to \mathbb{R}^{2n}$ is best-possible for smooth closed manifolds X of dimension $n = 4$; for according to Proposition 4.9 the projective space RP^4 does not embed in $\mathbb{R}^7$. However, there is a conjecture that every smooth closed *orientable* 4-dimensional manifold does embed in $\mathbb{R}^7$, and many special cases of this conjecture have been verified in M. W. Hirsch [3], Watabe [1, 2], and Boéchat and Haefliger [1, 2], for example.

At the end of the preceding remark we learned that a smooth closed manifold X of dimension $n > 4$ embeds in $\mathbb{R}^{2n-1}$ if and only if the dual Stiefel–Whitney class $\overline{w}_{n-1}(\tau(X)) \in H^{n-1}(X)$ vanishes. However, according to Massey [1, 3] and Massey and Peterson [1] one has $\overline{w}_{n-1}(\tau(X)) = 0$ for all orientable manifolds of any dimension $n > 1$ and for all nonorientable manifolds whose dimension n is *not* of the form 2^r. It follows that every smooth closed *orientable* manifold of dimension $n > 4$ embeds in $\mathbb{R}^{2n-1}$; the same result is true for smooth closed nonorientable manifolds of dimension $n \neq 2^r$. Incidentally, the proviso $n \neq 2^r$ is necessary in the nonorientable case; for according to Proposition 4.9 the projective space RP^n does not embed in $\mathbb{R}^{2n-1}$ for $n = 2^r$.

The preceding results justify the conclusion of Remark I.10.11 that, except for easily identifiable exceptions, every smooth n-dimensional manifold can be embedded in $\mathbb{R}^{2n-1}$. The case of orientable 4-dimensional manifolds is probably not an exception to the general rule; however, at the moment it is not known whether every smooth closed orientable 4-dimensional manifold embeds in $\mathbb{R}^7$.

Portions of the preceding results were obtained independently by other authors. S. P. Novikov [1] showed that every simply connected smooth manifold of odd dimension $n > 6$ can be embedded in $\mathbb{R}^{2n-1}$. Wu [9] showed that every smooth closed *orientable* manifold of dimension $n > 4$ can be embedded in $\mathbb{R}^{2n-1}$. Rohlin [6] showed that every smooth *nonorientable* 3-dimensional manifold embeds in $\mathbb{R}^5$.

9.17 Remark: In view of the preceding abundance of smooth embeddings $X \xrightarrow{f} \mathbb{R}^{2n-1}$ of smooth n-dimensional manifolds X, it is of interest to examine the normal bundles ν_f of such embeddings. According to Massey [2], ν_f has a nonvanishing section if and only if $w_2(\nu_f) \cup w_{n-2}(\nu_f) = 0 \in H^n(X; \mathbb{Z}/2)$; since $w(\nu_f) = \bar{w}(\tau(X))$, as in Lemma 4.2, the latter condition is just $\bar{w}_2(\tau(X)) \cup \bar{w}_{n-2}(\tau(X)) = 0 \in H^n(X; \mathbb{Z}/2)$.

9.18 Remark: The "best-possible" embedding conjecture of Remark I.10.11 states that every smooth n-dimensional manifold X embeds in $\mathbb{R}^{2n-\alpha(n)+1}$. One justification for the conjecture is the following result of R. L. W. Brown [2, 3]: If X is closed, then it is cobordant to a smooth closed n-dimensional manifold which *does* embed in $\mathbb{R}^{2n-\alpha(n)+1}$. There is a related result in R. L. W. Brown [4]: a necessary and sufficient condition for X to be cobordant to a smooth closed n-dimensional manifold which embeds in S^{n+k} is that all Stiefel–Whitney numbers involving $\bar{w}_i(\tau(X))$ for $i \geqq k$ vanish.

According to M. W. Hirsch [1], if a smooth closed n-dimensional manifold X admits a smooth embedding $X \xrightarrow{f} \mathbb{R}^{2n-k+1}$ for which the normal bundle ν_f has a nowhere-vanishing section, then there is an immersion $X \to \mathbb{R}^{2n-k}$. The condition on ν_f is necessary since Mahowald and Peterson [1] construct a smooth closed n-dimensional manifold X with a smooth embedding $X \xrightarrow{f} \mathbb{R}^{2n-k+1}$, for which there is no immersion $X \to \mathbb{R}^{2n-k}$ whatsoever; since $k > \alpha(n)$ in this example, there is no contradiction with the "best-possible" embedding conjecture.

9.19 Remark: Immersions and embeddings of real projective spaces RP^n are of special interest, partly because the nonimmersion and nonembedding techniques of Propositions 4.3 and 4.9 can easily be extended to other dimensions. However, the following Exercise does *not* in general represent best-possible results.

9.20 Exercise: Show that if $n = 2^r + q$ for $q \geqq 0$, then RP^n does not immerse in $\mathbb{R}^{2^{r+1}-2}$ and does not embed in $\mathbb{R}^{2^{r+1}-1}$.

9.21 Remark: Many best-possible immersions and embeddings are known for real projective spaces RP^n, and there is now a large literature on the subject. Most of the results known through 1979 are catalogued in Berrick [1]; there are also earlier catalogs and large bibliographies in Gitler [1] and in James [1, 2].

The problem of enumerating immersions and embeddings (up to isotopy) of real projective spaces is considered in Larmore and Rigdon [1, 2] and in Yasui [2, 3], for example.

9.22 Remark: Best-possible immersions and embeddings of complex projective spaces CP^n are also of interest, although most of the known results involve techniques which will be introduced only in later chapters of this work. For example, for any $n > 0$ Atiyah and Hirzebruch [1] show that the complex projective space CP^n, of real dimension $2n$, does not embed in $\mathbb{R}^{4n-2\alpha(n)}$. Sanderson and Schwarzenberger [1] use this nonembedding theorem to show for certain values of n that CP^n does *not* immerse in $\mathbb{R}^{4n-2\alpha(n)-1}$, and Sigrist and Suter [1] find necessary conditions for which CP^n *does* immerse in $\mathbb{R}^{4n-2\alpha(n)-1}$. As of 1977 all known nonimmersion results for CP^n were consequences of a general technique of Davis and Mahowald [2]; these results depend on the geometric dimensions of Whitney sums $m\gamma_n^1$ of the canonical complex line bundle γ_n^1 over CP^n, and on the Atiyah–Todd number $i(m, n)$, briefly mentioned in Remark III.13.42. (According to Remark III.13.29 the corresponding immersion problem for real projective spaces RP^n is equivalent to finding the geometric dimensions of Whitney sums $m\gamma_n^1$ of the canonical real line bundle γ_n^1 over RP^n, for all $m > 0$ and $n > 0$.)

Yasui [1] enumerates certain embeddings of complex projective spaces CP^n (up to isotopy). Oproiu [2] generalizes some of the nonembedding results for real projective spaces RP^n $(=G^1(\mathbb{R}^{n+1}))$ to the particular Grassmann manifolds $G^2(\mathbb{R}^{n+2})$ and $G^3(\mathbb{R}^{n+3})$, and Opriou [3] contains further such generalizations; immersions of Grassmann manifolds are considered in Hiller and Stong [1]. Kobayashi [1] considers certain immersions and embeddings of lens spaces.

9.23 Remark: Following some initial work of Rohlin [1, 2, 4], the unoriented cobordism ring $\mathfrak{N}$ was completely computed in Thom [3, 6–8]. Specifically, $\mathfrak{N}$ is a graded polynomial algebra $\mathbb{Z}/2[x_2, x_4, x_5, x_6, x_8, \ldots]$ with one generator x_n in each degree n not of the form $2^q - 1$. Furthermore, each generator x_n can be represented by *any* smooth closed n-dimensional manifold X_n such that the basic Stiefel–Whitney number of Definition 5.20 satisfies $s_n(X_n) = 1 \in \mathbb{Z}/2$. Specific manifolds representing the generators x_5 and x_{2m} for all $m > 0$ were constructed in Thom [8], and manifolds representing the remaining generators of $\mathfrak{N}$ were constructed in Dold [3]. Alternative sets of manifolds X_n representing the generators x_n of $\mathfrak{N}$ are given in Milnor [17], in Stong [6], and in Royster [1], for example.

Complete expositions of Thom's computation of the unoriented cobordism ring $\mathfrak{N}$ are given in Liulevicius [1, 3] and Stong [2]. There are also surveys containing some of the computation in Milnor [10], Rohlin [5], Poénaru [1], and Gray [1].

Quillen [1, 2] applied *formal groups* to compute the unoriented cobordism ring $\mathfrak{N}$ via Steenrod squares, and Brown and Peterson [3] used Steenrod squares in an entirely different way to compute $\mathfrak{N}$. Quillen's approach is given in detail in Bröcker and tom Dieck [1], and surveys of formal groups and their applications to the computation of $\mathfrak{N}$ appear in Schochet [1], Buhštaber *et al.* [1], and Karoubi [1].

9.24 Remark: Unoriented cobordism is a relatively weak equivalence relation, so that one expects each class in $\mathfrak{N}$ to be represented by at least one manifold with special properties. For example, it has already been noted that according to R. L. W. Brown [2, 3] each class in $\mathfrak{N}$ contains at least one representative satisfying the "best-possible" immersion conjecture and at least one representative satisfying the "best-possible" embedding conjecture. According to Stong [4], each class of positive degree in $\mathfrak{N}$ can also be represented by a fibration over the real projective plane RP^2.

9.25 Remark: Certain classes in $\mathfrak{N}$ contain representatives with further special properties, especially those classes in the kernel of the $\mathbb{Z}/2$ Euler characteristic $\mathfrak{N} \xrightarrow{\chi_2} \mathbb{Z}/2$ of Definition 6.3. According to Conner and Floyd [1], an unoriented cobordism class lies in the kernel of χ_2 if and only if it can be represented by the total space of a fiber bundle over S^1 with structure group $\mathbb{Z}/2$. According to Stong and Winkelnkemper [1], such classes are characterized by the property that there is a representative which admits a locally free action by the product group $S^1 \times S^1$. According to Iberkleid [1], such classes are also characterized by the property that they can be represented by at least one smooth closed manifold X whose tangent bundle $\tau(X)$ splits into a Whitney sum of real line bundles. This implies that *every* smooth closed manifold is cobordant to a smooth closed manifold X such that the Whitney sum $\tau(X) \oplus \varepsilon^1$ splits into a Whitney sum of real line bundles; furthermore, according to Stong [5], if $2 \leqq 2k < n$, then *every* smooth closed n-dimensional manifold is cobordant to a smooth closed manifold X such that $\tau(X) = \xi \oplus \eta$ for a $2k$-plane bundle ξ and an $(n-2k)$-plane bundle η. Finally, according to R. L. W. Brown [1], every even degree cobordism class in the kernel of $\mathfrak{N} \xrightarrow{\chi_2} \mathbb{Z}/2$ can be represented by a fibration over the 2-sphere S^2.

9.26 Remark: One can obtain results similar to those of the preceding remark by imposing restrictions on other Stiefel–Whitney numbers than the one $\langle w_n(\tau(X)), \mu_X \rangle$ which defines the $\mathbb{Z}/2$ Euler characteristic. For example, the result of R. L. W. Brown [1] continues as follows: If X is a smooth closed manifold of odd dimension n, then X is cobordant to a fibration over S^2 whenever $\langle w_2(\tau(X)) \cup w_{n-2}(\tau(X)), \mu_X \rangle = 0$.

According to Milnor [17] the unoriented cobordism class of a given manifold X contains a complex manifold if and only if all Stiefel–Whitney numbers constructed from at least one odd-dimensional Stiefel–Whitney class $w_p(\tau(X))$ vanish; this happens if and only if the unoriented cobordism class of X also contains the square $Y \times Y$ of a smooth closed manifold Y.

Finally, Yoshida [3] shows that if X is a smooth closed n-dimensional manifold such that all Stiefel–Whitney numbers other than those constructed from $w_1(\tau(X)), \ldots, w_{n-k}(\tau(X))$ vanish, for some $k \leqq 6$, then X is cobordant to a smooth closed manifold with at least k linearly independent vector fields. (The special case $k = 1$ is an instant corollary of the result of Conner and Floyd [1] described in Remark 9.25.)

9.27 Remark: In some cases the vanishing of certain Stiefel–Whitney *numbers* of a smooth closed manifold X implies that X is cobordant to a manifold for which related Stiefel–Whitney *classes* vanish. Here are three such results.

(i) Recall from Remark V.7.28 that a (smooth closed) *spin manifold* X is characterized by the conditions $w_1(\tau(X)) = 0$ and $w_2(\tau(X)) = 0$. Suppose that one knows only that X is a smooth closed manifold such that every Stiefel–Whitney number containing one of the factors $w_1(\tau(X))$ or $w_2(\tau(X))$ vanishes. Then, according to Anderson, Brown, and Peterson [1, 2], X is cobordant to a spin manifold.

(ii) Milnor [17], already cited in Remark 9.26, has the following further consequence. If X is a smooth closed manifold such that every Stiefel–Whitney number containing any odd factor $w_{2p+1}(\tau(X))$ vanishes, then X is cobordant to a smooth closed manifold X' such that $w_{2p+1}(\tau(X')) = 0$ for every $p \geqq 0$.

(iii) Let X be a smooth closed n-dimensional manifold, let $i_1, \ldots, i_s$ satisfy $2i_1 \geqq n + 1, \ldots, 2i_s \geqq n + 1$, and suppose that every Stiefel–Whitney number containing one of the factors $w_{i_1}(\tau(X)), \ldots, w_{i_s}(\tau(X))$ vanishes. Then X is cobordant to a smooth closed manifold X' such that $w_{i_1}(\tau(X')) = 0, \ldots, w_{i_s}(\tau(X')) = 0$. This result was established under the more stringent conditions $2i_1 > n + 1, \ldots, 2i_s > n + 1$ by Reed [1], and the same version is in Wall [3, p. 17]; the present version is due to Papastavridis [1, 3].

9.28 Exercise: Show for any $n > 0$ that the complex projective space CP^n has the same Stiefel–Whitney numbers as the square $RP^n \times RP^n$ of the real projective space RP^n, hence that CP^n is cobordant to $RP^n \times RP^n$. Given this result, construct a specific cobordism from CP^n to $RP^n \times RP^n$. (The second part of this problem is nontrivial; it is solved in Stong [3].)

9.29 Exercise: According to Corollary 5.6, for each $n > 0$ the real projective space RP^{2n-1} is the boundary of a smooth compact $2n$-dimensional manifold X_{2n}; find such manifolds X_{2n}. (See Husemoller [1, pp. 262–263] for example.)

9.30 Remark: The definition of surgical equivalence and the proof of Theorem 5.17, that surgical equivalence is cobordism, appeared independently in Milnor [9] and in Wallace [1–4]. Wallace uses the terminology "spherical modification" in place of "surgery."

9.31 Remark: The existence of the admissible Morse functions required to complete the proof of Theorem 5.17 is easily established in Milnor [13, 18] and in M. W. Hirsch [4], as already indicated in Lemma 5.16. Other general accounts of Morse theory are given in Pitcher [1] and Morse and Cairns [1], for example.

Admissible Morse functions are also easily constructed for any smooth manifold X by means of the following technique, in Morse [1]. Given any embedding $X \to \mathbb{R}^m$, there is a dense set of linear functionals $\mathbb{R}^m \to \mathbb{R}$ whose restrictions to X are smooth functions all of whose critical points are nondegenerate. It remains only to separate critical values, as in Smale [1, 2].

9.32 Remark: Given a smooth manifold X, it is of interest to construct an admissible Morse function $X \xrightarrow{f} \mathbb{R}$ with as few critical points as possible. According to M. W. Hirsch [2], if X is not closed, then there is a nonconstant smooth function $X \xrightarrow{f} \mathbb{R}$ with no critical points whatsoever; hence the question centers on closed manifolds.

According to Reeb [1] and Milnor [1], any smooth closed n-dimensional manifold X with only two critical points is *homeomorphic* (but not necessarily diffeomorphic) to the sphere S^n. A related set of conditions, involving an entire family of Morse functions, is used in Rayner [1] to characterize those manifolds X which are *diffeomorphic* to S^n.

One easily constructs an admissible Morse function $RP^2 \xrightarrow{f} \mathbb{R}$ with only three critical points. Eells and Kuiper [1, 2] and Banchoff and Takens [1] study other manifolds with the same property.

Specific admissible Morse functions are constructed for Grassmann manifolds in Hangan [1] and in Alexander [1], for lens spaces in Vrănceanu [1], and for other special cases in Vrănceanu [2] and Masuda [1], for example.

Aside from the intrinsic interest of constructing admissible Morse functions with a minimal number of critical points on a given smooth manifold X, one is interested in relating the latter number to the Ljusternik–Schnirelmann category of X (Remark I.10.2), a relation which is studied in Threlfall [1] and Takens [1], for example. A similar question for cobordism classes of smooth closed manifolds is studied in Mielke [1–4].

9.33 Remark: Since any $\mathbb{Z}/2$-algebra homomorphism $\mathfrak{N} \xrightarrow{\varphi} \mathbb{Z}/2$ whatsoever can only assume one of the values $0 \in \mathbb{Z}/2$ or $1 \in \mathbb{Z}/2$ on each of the generators $x_2 \in \mathfrak{N}$ and $x_4 \in \mathfrak{N}$, for example, it follows that the element $(x_2)^3 x_4 + x_2(x_4)^2 \in \mathfrak{N}$ of degree 10 lies in the kernel of every such φ. Consequently any manifold $X \in \mathscr{U}$ representing the element $(x_2)^3 x_4 + x_2(x_4)^2 \in \mathfrak{N}$ lies in the kernel of *every* Stiefel–Whitney genus $\mathscr{U} \xrightarrow{G(f)} \mathbb{Z}/2$. Thus one cannot compute all Stiefel–Whitney numbers in terms of Stiefel–Whitney genera: there are smooth closed manifolds that are not boundaries all of whose Stiefel–Whitney genera vanish.

It is probably not difficult to compute the ideal $\bigcap_f \ker G(f) \subset \mathfrak{N}$ of those unoriented cobordism classes lying in the kernel of every Stiefel–Whitney genus $\mathfrak{N} \xrightarrow{G(f)} \mathbb{Z}/2$, or to compute the quotient $\mathfrak{N}/\bigcap_f \ker G(f)$. However, the geometric significance of such computations is not clear to the author.

9.34 Exercise: Observe that the structure theorem $\mathfrak{N} = \mathbb{Z}/2[x_2, x_4, x_5, x_6, x_8, \ldots]$ of Remark 9.23 implies that every smooth closed 3-dimensional manifold is the boundary of a smooth compact 4-dimensional manifold, a result first announced in Rohlin [1]. It follows from Proposition 5.2 that all the Stiefel–Whitney numbers of any smooth closed 3-dimensional manifold vanish. Prove the latter statement directly, without using Remark 9.23.

9.35 Exercise: Show that the $\mathbb{Z}/2$ Thom form $j^* T_{RP^n} \in H^n(RP^n \times RP^n)$ of any real projective space RP^n is the sum $\sum_{p+q=n} e(\gamma_n^1)^p \times e(\gamma_n^1)^q$, for the generator $e(\gamma_n^1) \in H^1(RP^n)$.

9.36 Exercise: Show that the $\mathbb{Z}/2$ Thom form $j^* T_{S^n} \in H^n(S^n \times S^n)$ of any sphere S^n is the sum $1 \times \omega(S^n) + \omega(S^n) \times 1$, for the generator $\omega(S^n) \in H^n(S^n)$.

9.37 Remark: The conclusion of Theorem 8.4 can be reformulated as follows: *For any smooth closed n-dimensional manifold X the Stiefel–Whitney classes* $w_1(\tau(X)), \ldots, w_n(\tau(X))$ *are independent of the smooth structure assigned to X*. This result was also formulated in Remark V.7.10, with indications of the alternative proof given in Thom [4].

9.38 Remark: According to Remark V.7.27 there are *no* "universal" polynomial relations $P(w_1(\tau(X)), \ldots, w_n(\tau(X))) = 0 \in H^p(X)$ of weighted degree p which are valid for every smooth closed manifold X of dimension $n \geqq 2p$; this was proved independently in E. H. Brown [2] and in Stong [1]. However, the total Wu class $\mathrm{Wu}(\tau(X)) = 1 + \mathrm{Wu}_1(\tau(X)) + \cdots + \mathrm{Wu}_n(\tau(X))$ of the tangent bundle $\tau(X)$ of an n-dimensional manifold X is of the form $\sum_{p=0}^{n} P_p(w_1(\tau(X)), \ldots, w_p(\tau(X))$, $\sum_{p\geqq 0} P_p(u_1, \ldots, u_p)$ being the multiplicative sequence associated to the "Wu series" of Exercises V.7.21 and V.7.22, where each $P_p(u_1, \ldots, u_p)$ is a nontrivial polynomial of weighted degree p. Since $\mathrm{Wu}_p(\tau(X)) = 0$ whenever $2p > n$, by Corollary 8.6, the *Wu polynomials* $P_p(u_1, \ldots, u_p)$ *do* provide nontrivial relations $P_p(w_1(\tau(X)), \ldots, w_p(\tau(X))) = 0 \in H^p(X)$ which are valid for every smooth closed manifold X of given dimension $n < 2p$. It follows that the ideal generated by the Wu polynomials $P_p(u_1, \ldots, u_p)$ such that $2p > n$ is contained in the ideal of *all* polynomials P such that $P(w_1(\tau(X)), \ldots, w_n(\tau(X))) = 0$ for every smooth closed manifold X of dimension n. The latter ideal was completely determined in Brown and Peterson [1, 2], as indicated in Remark V.7.27.

9.39 Remark: For any given $n > 0$ let $P(u_1, \ldots, u_n)$ be a polynomial over $\mathbb{Z}/2$ of weighted degree n such that $P(w_1(\tau(X)), \ldots, w_n(\tau(X))) = 0 \in H^n(X)$ for every smooth closed manifold X of dimension n. According to Dold [2], in this special case the polynomial $P(u_1, \ldots, u_n)$ then lies in the ideal generated by the Wu polynomials $P_p(u_1, \ldots, u_p)$ for $2p > n$. This result does not generalize to polynomials of weighted degree less than n.

9.40 Remark: Let $X \xrightarrow{f} Y$ be any map of smooth closed manifolds X, Y of dimensions m and n, respectively, and let $H_*(X) \xrightarrow{f_*} H_*(Y)$ be the induced $\mathbb{Z}/2$ *homology* homomorphism. Since X and Y are closed, one also has Poincaré duality isomorphisms $H^*(X) \xrightarrow{D_{\mathrm{P}}} H_*(X)$ and $H^*(Y) \xrightarrow{D_{\mathrm{P}}} H_*(Y)$, as in Corollary 2.4 and Remark 9.3, with specific inverses $H_*(X) \xrightarrow{D_{\mathrm{T}}} H^*(X)$ and $H_*(Y) \xrightarrow{D_{\mathrm{T}}} H^*(Y)$, as in Corollary 7.9. The composition

$$H^*(X) \xrightarrow{D_{\mathrm{P}}} H_*(X) \xrightarrow{f_*} H_*(Y) \xrightarrow{D_{\mathrm{T}} = D_{\mathrm{P}}^{-1}} H^*(Y)$$

is the *Gysin homomorphism* $H^*(X) \xrightarrow{f_{\mathrm{H}}} H^*(Y)$. Since f_{H} proceeds in the reverse direction (= *Umkehr*) of the usual cohomology homomorphism $H^*(Y) \xrightarrow{f^*} H^*(X)$, the Gysin homomorphism is an example of an *Umkehr homomorphism*; more general *Umkehr* homomorphisms will be constructed in the third volume of this work, along with an explanation of the notation f_{H} and a proof of a general result which includes the following special case.

Recall from Remark V.7.14 that the Steenrod square Sq is *natural* in the sense that

$$\begin{array}{ccc} H^*(Y) & \xrightarrow{\mathrm{Sq}} & H^*(Y) \\ \downarrow{\scriptstyle f^*} & & \downarrow{\scriptstyle f^*} \\ H^*(X) & \xrightarrow{\mathrm{Sq}} & H^*(X) \end{array}$$

commutes for any map $X \xrightarrow{f} Y$, and that Sq is *multiplicative* in the sense that it satisfies the Cartan formula $\mathrm{Sq}(\alpha \cup \beta) = \mathrm{Sq}\,\alpha \cup \mathrm{Sq}\,\beta$. It is reasonable to ask whether Sq is in some sense natural with respect to the Gysin homomorphism $H^*(X) \xrightarrow{f_{\mathrm{H}}} H^*(Y)$ induced by any map $X \xrightarrow{f} Y$.

The answer is affirmative only if one introduces a correction consisting of cup products by the Wu classes $\mathrm{Wu}(\tau(X)) \in H^*(X)$ and $\mathrm{Wu}(\tau(Y)) \in H^*(Y)$ of the tangent bundles $\tau(X)$ and $\tau(Y)$. Specifically, the $\mathbb{Z}/2$ *Riemann–Roch theorem* asserts that

$$\begin{array}{ccccc} H^*(X) & \xrightarrow{\mathrm{Sq}} & H^*(X) & \xrightarrow{\cup \mathrm{Wu}(\tau(X))} & H^*(X) \\ \downarrow{\scriptstyle f_{\mathrm{H}}} & & & & \downarrow{\scriptstyle f_{\mathrm{H}}} \\ H^*(Y) & \xrightarrow{\mathrm{Sq}} & H^*(Y) & \xrightarrow{\cup \mathrm{Wu}(\tau(Y))} & H^*(Y) \end{array}$$

commutes. This result underscores the importance of Wu classes.

The preceding $\mathbb{Z}/2$ Riemann–Roch theorem is a slightly specialized version of the main result of Atiyah and Hirzebruch [2], later generalized by Spanier [2]. However, it is also a special case of a much broader *generalized Riemann–Roch theorem*, which will be proved in Volume 3. In the most useful special case the Steenrod square $H^*(X; \mathbb{Z}/2) \xrightarrow{\mathrm{Sq}} H^*(X; \mathbb{Z}/2)$ will be replaced by the *Chern character* $K^{\oplus}(X) \xrightarrow{\mathrm{ch}(X)} H^*(X; \mathbb{Q})$ and the correction factor $\mathrm{Wu}(\tau(X)) \in H^*(X; \mathbb{Z}/2)$ will be replaced by the *Todd class* $\mathrm{td}(\tau(X)) \in H^*(X; \mathbb{Q})$, along with corresponding changes for Y.

9.41 Exercise: Let X be a smooth compact n-dimensional manifold with boundary $\dot{X}$ and $\mathbb{Z}/2$ fundamental class $\mu_X \in H_n(X, \dot{X})$. Use *covariant* Mayer–Vietoris functors on $\mathscr{Q}(\mathring{X})$ to show for each $q \in \mathbb{Z}$ that the cap product $H^{n-q}(X, \dot{X}) \xrightarrow{\cap \mu_X} H_q(X)$ is a $\mathbb{Z}/2$-module isomorphism. (This is also a $\mathbb{Z}/2$ Poincaré–Lefschetz duality theorem.)

References

The notation MR **55** 12555 indicates that the given work is reviewed in *Mathematical Reviews*, Vol. **55**, review number 11255; page numbers are used in lieu of review numbers in Vols. **1–19** of *Mathematical Reviews*. The notation MR **20** 3539 = S 384-3 indicates that review MR **20** 3539 is reprinted in Steenrod [7, p. 384, review number 3 on that page]. Steenrod [7] contains reviews of papers in algebraic and differential topology, topological groups, and homological algebra for the years 1940–1967, corresponding to Vols. 1–34 of *Mathematical Reviews*. The final italic numbers in each entry identify pages on which the given reference is cited.

Accascina, G.

[1] A variant of Segal's construction of classifying spaces. *Atti Accad. Naz. Lincei Rend. Cl. Sci. Fis. Mat. Natur.* (*8*) **57** (1974), 606–610. MR **55** 11255. *103.*

Adams, J. F.

[1] On the nonexistence of elements of Hopf invariant one. *Bull. Amer. Math. Soc.* **64** (1958), 279–282. MR **20** 3539 = S 384-3. *247.*

[2] On the non-existence of elements of Hopf invariant one. *Ann. of Math.* (2) **72** (1960), 20–104. MR **25** 4530 = S 384-4. *247.*

[3] On formulae of Thom and Wu. *Proc. London Math. Soc.* (3) **11** (1961), 741–752. MR **25** 2613 = S 718-1. *245.*

[4] Vector fields on spheres. *Ann. of Math.* (2) **75** (1962), 602–632. MR **25** 2614 = S 739-1. *248.*

[5] "Algebraic Topology—A Student's Guide." Cambridge Univ. Press, London and New York, 1972. MR **56** 3824. *314.*

Adams, J. F., and Atiyah, M. F.

[1] *K*-theory and the Hopf invariant. *Quart. J. Math. Oxford Ser.* (*2*) **17** (1966), 31–38. MR **33** 6618 = S 388-3. *247.*

Adams, J. F., and Walker, A. G.

[1] On complex Stiefel manifolds. *Proc. Cambridge Philos. Soc.* **61** (1965), 81–103. MR **30** 1516 = S 739-4. *189.*

Adem, J.

[1] The iteration of the Steenrod squares in algebraic topology. *Proc. Nat. Acad. Sci. U.S.A.* **38** (1952), 720–726. MR **14** 306 = S 460-4. *244.*

[2] The relations on Steenrod powers of cohomology classes. "Algebraic Geometry and Topology: A Symposium in Honor of S. Lefschetz," pp. 191–238. Princeton Univ. Press, Princeton, New Jersey, 1957. MR **19** 50 = S 462-1. *244.*

Adyan, S. I.

[1] Algorithmic unsolvability of problems of recognition of certain properties of groups. (Russian.) *Dokl. Akad. Nauk SSSR* (*N.S.*) **103** (1955), 533–535. MR **18** 455. *52.*

Ahlfors, L. V., and Sario, L.

[1] "Riemann Surfaces." Princeton Univ. Press, Princeton, New Jersey, 1960. MR **22** 5729. *53.*

Akin, E.

[1] Stiefel-Whitney homology classes and bordism. *Trans. Amer. Math. Soc.* **205** (1975), 341–359. MR **50** 11288. *290.*

[2] *K*-theory doesn't exist. *J. Pure Appl. Algebra* **12** (1978), 177–179. MR **57** 13945. *101.*

Alexander, J. C.

[1] Morse functions on Grassmannians. *Illinois J. Math.* **15** (1971), 672–681. MR **44** 4777. *299.*

Allaud, G.

[1] Concerning universal fibrations and a theorem of E. Fadell. *Duke Math. J.* **37** (1970), 213–224. MR **43** 2704. *103.*

Anderson, D. W., Brown, E. H., Jr., and Peterson, F. P.

[1] Spin cobordism. *Bull. Amer. Math. Soc.* **72** (1966), 256–260. MR **32** 8349 = S 796-4. *298.*

[2] The structure of the spin cobordism ring. *Ann. of Math.* (2) **86** (1967), 271–298. MR **36** 2160. *298.*

Arnold, J. E., Jr.

[1] Local to global theorems in the theory of Hurewicz fibrations. *Trans. Amer. Math. Soc.* **164** (1972), 179–188. MR **45** 4415. *99.*

Artin, E., and Braun, H.

[1] Vorlesungen über algebraische Topologie. *Math. Sem. Univ. Hamburg, Hamburg* (1964). MR **29** 2791 = S 18-1. *4.*
[2] "Introduction to Algebraic Topology." Merrill Publ., Columbus, Ohio, 1969. MR **40** 888. *4, 42, 212.*

Atiyah, M. F.

[1] Thom complexes. *Proc. London Math. Soc.* (*3*) **11** (1961), 291–310. MR **24** A1727 = S 323-2. *189, 212, 289.*
[2] "*K*-Theory." Benjamin, New York, 1967. MR **36** 7130. *178, 182.*

Atiyah, M. F., and Hirzebruch, F.

[1] Quelques théorèmes de non-plongement pour les variétés différentiables. *Bull. Soc. Math. France* **87** (1959), 383–396. MR **22** 5055 = 764-5. *296.*
[2] Cohomologie-Operationen und charakteristische Klassen. *Math. Z.* **77** (1961), 149–187. MR **27** 6285 = S 718-2. *245, 302.*
[3] Bott periodicity and the parallelizability of the spheres. *Proc. Cambridge Philos. Soc.* **57** (1961), 223–226. MR **23** A3578 = S 323-1. *184, 246.*

Atiyah, M. F., and Rees, E.

[1] Vector bundles on projective 3-space. *Invent. Math.* **35** (1976), 131–153. MR **54** 7870. *186.*

Atiyah, M. F., and Todd, J. A.

[1] On complex Stiefel manifolds. *Proc. Cambridge Philos. Soc.* **56** (1960), 342–353. MR **24** A2392 = S 737-4. *189.*

Auslander, L., and MacKenzie, R. E.

[1] On the topology of tangent bundles. *Proc. Amer. Math. Soc.* **10** (1959), 627–632. MR **21** 6574 = S 690-4. *178.*
[2] "Introduction to Differentiable Manifolds." McGraw-Hill, New York, 1963. MR **28** 4462 = S 661-1. [2nd ed., 1977. MR **57** 10717.] *50, 51, 97, 178.*

Banchoff, T. F.

[1] Stiefel–Whitney homology classes and singularities of projections for polyhedral manifolds. "Differential Geometry" [*Proc. Symp. Pure Math.* **27**, Part 1 (1973)], pp. 333–347. American Mathematical Society, Providence, Rhode Island, 1975. MR **51** 14090. *290.*

Banchoff, T. F., and McCrory, C.

[1] Whitney duality and singularities of projections. "Geometry and Topology" [*Proc. Latin Amer. School of Math., 3rd* (1976)], Lecture Notes in Mathematics Vol. 597, pp. 68–81. Springer-Verlag, Berlin and New York, 1977. MR **56** 6678. *290.*

[2] A combinatorial formula for normal Stiefel–Whitney classes. *Proc. Amer. Math. Soc.* **76** (1979), 171–177. MR **80h** 57031. *290.*

Banchoff, T. F., and Takens, F.

[1] Height functions on surfaces with three critical points. *Illinois J. Math.* **19** (1975), 325–335. MR **51** 14079. *299.*

Barratt, M. G., and Mahowald, M. E.

[1] The metastable homotopy of $O(n)$. *Bull. Amer. Math. Soc.* **70** (1964), 758–760. MR **31** 6229 = S 1168-4. *188, 247.*

Bartle, R. G.

[1] "The Elements of Real Analysis," 2nd ed. Wiley, New York, 1976. MR **52** 14179. *30.*

Bendersky, M.

[1] Characteristic classes of n-manifolds immersing in $\mathbb{R}^{n+k}$. *Math. Scand.* **31** (1972), 293–300. MR **48** 3058. *246.*

[2] "Generalized Cohomology and K-Theory," Matematisk Institut, Lecture Notes Series No. 33, 1971/1972. Aarhus Univ., Aarhus (1972). MR **49** 3892. *49.*

Benlian, R., and Wagoner, J.

[1] Type d'homotopie fibré et réduction structurale des fibrés vectoriels. *C. R. Acad. Sci. Paris Ser. A-B* **265** (1967), A207–A209. MR **36** 4576. *187, 189, 212, 289.*

Berrick, A. J.

[1] Projective space immersions, bilinear maps and stable homotopy groups of spheres. *Proc. Topol. Symp., Seigen, 1979* Lecture Notes in Mathematics, Vol. 788, pp. 1–22. Springer-Verlag, Berlin and New York, 1980. MR **82b** 57015. *185, 295.*

Bernstein, I.

[1] On the Lusternik–Schnirelmann category of Grassmannians. *Math. Proc. Cambridge Philos. Soc.* **79** (1976), 129–134. MR **53** 4047. *48.*

Bing, R. H.

[1] An alternative proof that 3-manifolds can be triangulated. *Ann. of Math.* (2) **69** (1959), 37–65. MR **20** 7269 = S 153-2. *53.*

Birkhoff, G.

[1] Lie groups simply isomorphic with no linear group. *Bull. Amer. Math. Soc.* **42** (1936), 883–888. *91.*

Bishop, R. L., and Crittenden, R. J.

[1] "Geometry of Manifolds." Academic Press, New York, 1964. MR **29** 6401. *100, 278.*

Blanchard, A.

[1] La cohomologie réelle d'un espace fibré à fibre kählérienne. *C. R. Acad. Sci. Paris* **239** (1954), 1342–1343. MR **16** 737 = S 278-5. *104.*

Blanton, J. D., and McCrory, C.

[1] An axiomatic proof of Stiefel's conjecture. *Proc. Amer. Math. Soc.* **77** (1979), 409–414. MR **80k** 55052. *290.*

Blanton, J. D., and Schweitzer, P. A.

[1] Axioms for characteristic classes of manifolds. "Differential Geometry" [*Proc. Symp. Pure Math.* **27**, Part 1 (1973)], pp. 349–356. American Mathematical Society, Providence, Rhode Island, 1975. MR **51** 11534. *290.*

Boéchat, J., and Haefliger, A.

[1] Plongements différentiables des variétés de dimension 4 dans $\mathbb{R}^7$. *C. R. Acad. Sci. Paris Sér. A-B* **266** (1968), A1226–A1228. MR **37** 5882. *294.*

[2] Plongements différentiables des variétés orientées de dimension 4 dans $\mathbb{R}^7$. "Essays on Topology and Related Topics," pp. 156–166. Springer-Verlag, Berlin and New York, 1970. MR **42** 5273. *294.*

Bognar, M.

[1] A finite 2-dimensional CW complex which cannot be triangulated. *Acta Math. Acad. Sci. Hungar.* **29** (1977), 107–112. MR **55** 9106. *49.*

Boone, W. W.

[1] Certain simple, unsolvable problems of group theory. IV. *Nederl. Akad. Wetensch. Proc. Ser. A* **58** = *Indag. Math.* **17** (1955), 571–577. MR **20** 5230. *52.*

[2] Certain simple, unsolvable problems of group theory. V, VI. *Nederl. Akad. Wetensch. Proc. Ser. A* **60** = *Indag. Math.* **19** (1957), 22–27, 227–232. MR **20** 5231. *52.*

[3] The word problem. *Ann. of Math.* **70** (1959), 207–265. MR **31** 3485. *52.*

Boone, W. W., Haken, W., and Poénaru, V.

[1] On recursively unsolvable problems in topology and their classification. "Contributions to Mathematical Logic" (*Colloq., Hannover, 1966*), pp. 37–74. North-Holland Publ., Amsterdam, 1968. MR **41** 7695. *52.*

Boothby, W. M.

[1] "An Introduction to Differentiable Manifolds and Riemannian Geometry." Academic Press, New York, 1975. MR **54** 13956. *50.*

Bordoni, M.

[1] Sulle classi caratteristiche di un fibrato di sfere. (English summary.) *Atti Accad. Naz. Lincei Rend. Cl. Sci. Fis. Mat. Natur.* (8) **55** (1973), 404–414 (1974). MR **51** 6839. *242.*

Borel, A.

[1] Sur la cohomologie des espaces fibrés principaux et des espaces homogènes de groupes de Lie compacts. *Ann. of Math.* (2) **57** (1953), 115–207. MR **14** 490 = S 310-4. *213.*
[2] La cohomologie mod 2 de certains espaces homogènes. *Comment. Math. Helv.* **27** (1953), 165–197. MR **15** 244 = S 312-1. *245.*
[3] The Poincaré duality in generalized manifolds. *Michigan Math. J.* **4** (1957), 227–239. MR **20** 4842 = S 642-2. *288.*

Borel, A., and Hirzebruch, F.

[1] Characteristic classes and homogeneous spaces. I. *Amer. J. Math.* **80** (1958), 458–538. MR **21** 1586 = S 1147-5. *97, 185, 239, 241.*
[2] Characteristic classes and homogeneous spaces. II. *Amer. J. Math.* **81** (1959), 315–382. MR **22** 988 = S 1148-1. *247.*

Borsuk, K.

[1] Sur les prolongements des transformations continues. *Fund. Math.* **28** (1937), 99–110. *98.*
[2] On the Lusternik–Schnirelmann category in the theory of shape. *Fund. Math.* **99** (1978), 35–42. MR **57** 4158. *48.*

Bott, R.

[1] On symmetric products and the Steenrod squares. *Ann. of Math.* (2) **57** (1953), 579–590. MR **15** 54 = S 454-2. *243.*

Bott, R., and Milnor, J.

[1] On the parallelizability of the spheres. *Bull. Amer. Math. Soc.* **64** (1958), 87–89. MR **21** 1590 = S 737-2. *184, 247.*

Bott, R., and Tu, L. W.

[1] "Differential Forms in Algebraic Topology." Springer-Verlag. Berlin and New York, 1982. *56, 212.*

Bourgin, D. G.

[1] The paracompactness of the weak simplicial complex. *Proc. Nat. Acad. Sci. U.S.A.* **38** (1952), 305–313. MR **14** 70 = S 554-5. *50.*

Braemer, J. M.

[1] Classes caractéristiques des fibrés vectoriels. *Publ. Dép. Math.* (*Lyon*) **6** (1969), fasc. 2, 119–129. MR **41** 2699. *241.*

Britton, J. L.

[1] The word problem for groups. *Proc. London Math. Soc.* (*3*) **8** (1958), 493–506. MR **23** A2326. *52.*

Bröcker, T., and tom Dieck, T.

[1] "Kobordismentheorie." Lecture Notes in Mathematics, Vol. 178. Springer-Verlag, Berlin and New York, 1970. MR **43** 1202. *297.*

Browder, W.

[1] "Surgery on Simply-Connected Manifolds." Springer-Verlag, Berlin and New York, 1972. MR **50** 11272. *288.*

Browder, W., and Livesay, G. R.

[1] Fixed point free involutions on homotopy spheres. *Bull. Amer. Math. Soc.* **73** (1967), 242–245. MR **34** 6781 = S 1116-7. *55.*

Brown, E. H.

[1] Twisted tensor products. I. *Ann. of Math.* (2) **69** (1959), 223–246. MR **21** 4423 = S 283-2. *104.*

[2] Nonexistence of low dimension relations between Stiefel–Whitney classes. *Trans. Amer. Math. Soc.* **104** (1962), 374–382. MR **25** 5521 = S 719-2. *246, 301.*

Brown, E. H., and Peterson, F. P.

[1] Relations between Stiefel–Whitney classes of manifolds. *Bull. Amer. Math. Soc.* **69** (1963), 228–230. MR **26** 4369 = S 719-5. *245, 246, 301.*

[2] Relations among characteristic classes. I. *Topology Suppl.* **1** (1964), 39–52. MR **29** 629 = S 720-1. *246, 301.*

[3] Computation of the unoriented cobordism ring. *Proc. Amer. Math. Soc.* **55** (1976), 191–192. MR **52** 15507. *297.*

[4] On immersions of n-manifolds. *Adv. in Math.* **24** (1977), 74–77. MR **55** 9119. *52, 292.*

[5] A universal space for normal bundles of n-manifolds. *Comment. Math. Helv.* **54** (1979), 405–430. MR **80k** 57055. *52, 292.*

Brown, R.

[1] Two examples in homotopy theory. *Proc. Cambridge Philos. Soc.* **62** (1966), 575–576. MR **34** 5085 = S 1011-4. *99.*

Brown, R. F.

[1] Path fields on manifolds. *Trans. Amer. Math. Soc.* **118** (1965), 180–191. MR **30** 3482 = S 633-1. *180, 242.*

Brown, R. F., and Fadell, E.

[1] Nonsingular path fields on compact topological manifolds. *Proc. Amer. Math. Soc.* **16** (1965), 1342–1349. MR **32** 1709 = S 633-3. *180, 242.*

Brown, R. L. W.

[1] Cobordism and bundles over spheres. *Michigan Math. J.* **16** (1969), 315–320. MR **40** 3558. *297, 298.*
[2] Imbeddings, immersions, and cobordism of differentiable manifolds. *Bull. Amer. Math. Soc.* **76** (1970), 763–766. MR **41** 4559. *50, 52, 292, 293, 295, 297.*
[3] Immersions and embeddings up to cobordism. *Canad. J. Math.* **23** (1971), 1102–1115. MR **45** 6017. *50, 52, 292, 293, 295, 297.*
[4] Stiefel–Whitney numbers and maps cobordant to embeddings. *Proc. Amer. Math. Soc.* **48** (1975), 245–250. MR **50** 11289. *295.*

Bucur, I., and Lascu, Al.

[1] Les classes de Segre d'une variété différentiable. *Rev. Math. Pures Appl.* **4** (1959), 661–664. MR **23** A1386 = S 717-6. *289.*

Buhštaber, V. M., Miščenko, A. S., and Novikov, S. P.

[1] Formal groups and their rôle in the apparatus of algebraic topology. (Russian.) *Uspehi Mat. Nauk* **26** (1971), 131–154. MR **56** 3862. *297.*

Burali-Forti, C.

[1] "Introduction à la géométrie différentielle suivant la méthode de H. Grassmann." Gauthier-Villars, Paris, 1897. *56.*

Cairns, S. S.

[1] On the triangulation of regular loci. *Ann. of Math.* **35** (1934), 579–587. *52.*
[2] Triangulation of the manifold of class 1. *Bull. Amer. Math. Soc.* **41** (1935), 549–552. *52.*
[3] Triangulated manifolds and differentiable manifolds. "Lectures in Topology," pp. 143–157. Univ. of Michigan Press, Ann Arbor, Michigan, 1941. MR **3** 133 = S 566-7. *52.*
[4] The triangulation problem and its rôle in analysis. *Bull. Amer. Math. Soc.* **52** (1946), 545–571. MR **8** 166 = S 666-1. *52.*
[5] A simple triangulation method for smooth manifolds. *Bull. Amer. Math. Soc.* **67** (1961), 389–390. MR **26** 6978 = S 566-8. *52.*
[6] "Introductory Topology." Ronald Press, New York, 1961. MR **22** 9964 = S 14-1. *49, 51.*

Cappell, S. E., and Shaneson, J. L.

[1] Some new four-manifolds. *Ann. of Math.* (2) **104** (1976), 61–72. MR **54** 6167. *55.*

Cartan, H.

[1] Une théorie axiomatique des carrés de Steenrod. *C. R. Acad. Sci. Paris* **230** (1950), 425–427. MR **12** 42 = S 452-1. *243.*
[2] Sur l'itération des opérations de Steenrod. *Comment. Math. Helv.* **29** (1955), 40–58. MR **16** 847 = S 461-4. *244.*
[3] Espaces fibrés et homotopie. *Sém. Henri Cartan Ecole Normale Supérieure, 1949/1950*; 18 exposés par A. Blanchard, A. Borel, H. Cartan, J.-P. Serre, and W.-T. Wu, 2nd ed. Secrétariat mathématique, Paris, 1956. MR **18** 409 = S 28-4. *97, 243.*
[4] Théorie des fibrés principaux. *Sém. Henri Cartan Ecole Normale Supérieure, 1956/1957.* Secrétariat mathématique, Paris, 1958. MR **22** 1896c = S 37-3. *102.*
[5] Invariant de Hopf et opérations cohomologiques secondaires. *Sém. Henri Cartan Ecole Normale Supérieure, 1958/1959.* Secrétariat mathématique, Paris, 1959. MR **23** A1367 = S 42-1. *247.*

Cartier, P.

[1] Structure topologique des groupes de Lie généraux. *Sém. "Sophus Lie" de l'Ecole Normale Supérieure* (22-1)–(22-20). Secrétariat mathématique, Paris, 1955. MR **17** 384. *86, 104.*

Cavanaugh, J. M.

[1] Algebraic vector bundles over the *l*-hole torus. *Illinois J. Math.* **19** (1975), 336–343. MR **51** 14044. *188.*

Chan, W. M., and Hoffman, P.

[1] Vector bundles over suspensions. *Canad. Math. Bull.* **17** (1974), 483–484. MR **51** 9057. *189.*

Cheeger, J.

[1] A combinatorial formula for Stiefel–Whitney classes. "Topology of Manifolds" [*Proc. Inst., Univ. of Georgia (1969)*], pp. 470–471. Markham, Chicago, Illinois, 1970. *290.*

Cheeger, J., and Kister, J. M.

[1] Counting topological manifolds. *Topology* **9** (1970), 149–151. MR **41** 1055. *53.*

Chern, S. S.

[1] Integral formulas for the characteristic classes of sphere bundles. *Proc. Nat. Acad. Sci. U.S.A.* **30** (1944), 269–273. MR **6** 106 = S 713-1. *289.*
[2] On the characteristic ring of a differentiable manifold. *Acad. Sinica Sci. Record* **2** (1947), 1–5. MR **9** 297 = S 713-3. *179, 239, 240, 241, 289.*
[3] On the multiplication in the characteristic ring of a sphere bundle. *Ann. of Math. (2)* **49** (1948), 362–372. MR **9** 456 = S 713-4. *179, 239, 240, 241, 289.*

Chern, S. S., and Spanier, E.

[1] The homology structure of sphere bundles. *Proc. Nat. Acad. Sci. U.S.A.* **36** (1950), 248–255. MR **12** 42 = S 293-5. *104, 212.*

[2] A theorem on orientable surfaces in four-dimensional space. *Comment. Math. Helv.* **25** (1951), 205–209. MR **13** 492 = S 762-8. *293.*

Chern, S. S., and Sun, Y.-F.

[1] The imbedding theorem for fibre bundles. *Trans. Amer. Math. Soc.* **67** (1949), 286–303. MR **11** 378 = S 302-3. *102.*

Chevalley, C.

[1] "Theory of Lie Groups. I." Princeton Univ. Press, Princeton, New Jersey, 1946. MR **7** 412 = S 868-1. *180.*

Cockroft, W. H.

[1] On the Thom isomorphism theorem. *Proc. Cambridge Philos. Soc.* **58** (1962), 206–208. MR **26** 3066 = S 295-4. *212.*

Coelho, M.

[1] Théorie simpliciale de la catégorie de Lusternik–Schnirelmann. *C. R. Acad. Sci. Paris Sér. A-B* **281** (1975), A199–A202. MR **52** 6702. *48.*

Cohen, R.

[1] The immersion conjecture for differentiable manifolds. *Ann. of Math.* (to be published). *52, 262, 292, 293.*

Connell, E. H.

[1] Characteristic classes. *Illinois J. Math.* **14** (1970), 496–521. MR **41** 4571. *47, 49, 104.*

Conner, P. E., and Floyd, E. E.

[1] Fibring within a cobordism class. *Michigan Math. J.* **12** (1965), 33–47. MR **31** 4038 = S 790-1. *297, 298.*

Cooke, G. E., and Finney, R. L.

[1] "Homology of Cell Complexes" (based on lectures by Norman E. Steenrod). Princeton Univ. Press, Princeton, New Jersey, and Univ. of Tokyo Press, Tokyo, 1967. MR **36** 2142. *48, 49.*

Curtis, M. L.

[1] The covering homotopy theorem. *Proc. Amer. Math. Soc.* **7** (1956), 682–684. MR **18** 60 = S 260-3. *99.*

Curtis, M. L., and Lashof, R.

[1] Homotopy equivalence of fiber bundles. *Proc. Amer. Math. Soc.* **9** (1958), 178–182. MR **20** 4838 = S 305-2. *103.*

Davis, D. M., and Mahowald, M. E.

[1] The geometric dimension of some vector bundles over projective spaces. *Trans. Amer. Math. Soc.* **205** (1975), 295–315. MR **51** 9058. *189.*
[2] Immersions of complex projective spaces and the generalized vector field problem. *Proc. London Math. Soc.* (*3*) **35** (1977), 333–344. MR **56** 3839. *296.*

de Carvalho, C. A. A.

[1] Classes de Smith associées à un espace fibré. Classes caractéristiques. *C. R. Acad. Sci. Paris* **247** (1958), 1947–1950. MR **21** 5192 = S 314-6. *239.*

de Rham, G.

[1] "Variétés Différentiables." Hermann, Paris, 1955. MR **16** 957. *50.*

de Sapio, R.

[1] Differential structures on a product of spheres. *Comment. Math. Helv.* **44** (1969), 61–69. MR **39** 4857. *54.*
[2] Differential structures on a product of spheres. II. *Ann. of Math.* (*2*) **89** (1969), 305–313. MR **39** 7611. *54.*

Dediu, M.

[1] Campi di vettori tangenti sullo spazio lenticolare $L^7(3)$. (French summary.) *Atti Accad. Naz. Lincei Rend. Cl. Sci. Fis. Mat. Natur.* (*8*) **58** (1975), 14–17. MR **54** 6154. *185.*
[2] Tre campi di vettori tangenti indipendenti sugli spazi lenticolari di dimensione $4n + 3$. (French summary.) *Atti Accad. Naz. Lincei Rend. Cl. Sci. Fis. Mat. Natur.* (*8*) **58** (1975), 174–178. MR **54** 6155. *185.*
[3] Campi di vettori tangenti sugli spazi lenticolari di dimensione tre. (French summary.) *Atti Accad. Peloritana Pericolanti Cl. Sci. Fis. Natur.* **54** (1974), 329–334. MR **80a** 57001. *185.*

Deligne, P., and Sullivan, D.

[1] Fibrés vectoriels complexes à groupe structural discret. (English summary.) *C. R. Acad. Sci. Paris Sér. A-B* **281** (1975), A1081–A1083. MR **53** 1587. *185.*

Delzant, A.

[1] Définition des classes de Stiefel–Whitney d'un module quadratique sur un corps de caractéristique différente de 2. *C. R. Acad. Sci. Paris* **255** (1962), 1366–1368. MR **26** 175 = S 1319-2. *248.*

Derwent, J.

[1] On the covering homotopy theorem. *Nederl. Akad. Wetensch. Proc. Ser. A* **62** [*Indag. Math.* **21**] (1959), 275–279. MR **21** 6580 = S 260-1. *99.*

[2] Inverses for fiber spaces. *Proc. Amer. Math. Soc.* **19** (1968), 1491–1494. MR **38** 715. *181.*

[3] A note on numerable covers. *Proc. Amer. Math. Soc.* **19** (1968), 1130–1132. MR **38** 2778. *98.*

Dieudonné, J.

[1] Une généralisation des espaces compacts. *J. Math. Pure Appl.* **23** (1944), 65–76. MR **7** 134. *30.*

Dold, A.

[1] Über faserweise Homotopieäquivalenz von Faserräumen. *Math. Z.* **62** (1955), 111–136. MR **17** 519 = S 267-1. *103.*

[2] Vollständigkeit der Wuschen-Relationen zwischen den Stiefel–Whitneyschen Zahlen differenzierbarer Mannigfaltigkeiten. *Math. Z.* **65** (1956), 200–206. MR **18** 143 = S 716-1. *301.*

[3] Erzeugende der Thomschen Algebra $\mathfrak{N}$. *Math. Z.* **65** (1956), 25–35. MR **18** 60 = S 785-3. *296.*

[4] Relations between ordinary and extraordinary cohomology. "Colloquium on Algebraic Topology," pp. 2–9. Matematisk Institut, Aarhus Univ., Aarhus, 1962. MR **26** 3565 = S 16-5. [Reprinted on pp. 167–177 of Adams [5].] *104.*

[5] Partitions of unity in the theory of fibrations. *Ann. of Math.* (2) **78** (1963), 223–255. MR **27** 5264 = S 262-3. *98, 102, 103.*

[6] "On General Cohomology," Chapters 1–9. Aarhus Univ., Aarhus, 1968. MR **40** 8045. *104.*

[7] (Co-)homology properties of topological manifolds. *Conf. Topol. Manifolds, 1967* pp. 47–57. Prindle, Weber, and Schmidt, Boston, Massachusetts, 1968. MR **40** 2087. *288.*

[8] "Lectures on Algebraic Topology." Springer-Verlag. New York and Berlin, 1972. MR **54** 3685. [2nd ed., 1980. MR **82c** 55001.] *4, 20, 49, 244, 256, 257, 258, 275, 287, 288.*

Dold, A., and Lashof, R.

[1] Principal quasi-fibrations and fibre homotopy equivalence of bundles. *Illinois J. Math.* **3** (1959), 285–305. MR **21** 331 = S 306-1. *102, 103.*

Dowker, C. H.

[1] Topology of metric complexes. *Amer. J. Math.* **74** (1952), 555–577. MR **13** 965 = S 554-2. *18, 49, 158.*

Draper, J. A.

[1] Homotopy epimorphisms and Lusternik–Schnirelmann category. *Proc. Amer. Math. Soc.* **50** (1975), 471–476. MR **51** 4225. *48.*

Dugundji, J.

[1] Note on CW polytopes. *Portugal. Math.* **11** (1952), 7–10. MR **14** 74 = S 554-6. *50.*
[2] "Topology." Allyn and Bacon, Boston, Massachusetts, 1966. MR **33** 1824. *30, 39, 82, 129.*

Dupont, J. L.

[1] "*K*-Theory," Lecture Notes Series No. 11. Matematisk Institut, Aarhus Univ., Aarhus, 1968. MR **43** 3324. *178.*
[2] On homotopy invariance of the tangent bundle. I, II. *Math. Scand.* **26** (1970), 5–13, 200–220. MR **42** 8516. *187, 212.*
[3] *K*-theory obstructions to the existence of vector fields. *Acta Math.* **133** (1974), 67–80. MR **54** 13929. *184.*

Eckmann, B.

[1] Zur Homotopietheorie gefaserter Räume. *Comment. Math. Helv.* **14** (1942), 141–192. MR **3** 317 = S 258-2. *97.*

Eells, J., Jr.

[1] Fibre bundles. "Global Analysis and its Applications," Vol. I, pp. 53–82. International Atomic Energy Agency, Vienna, 1974. MR **55** 11253. *97.*

Eells, J., Jr., and Kuiper, N. H.

[1] Closed manifolds which admit nondegenerate functions with three critical points. *Nederl. Akad. Wetensch. Proc. Ser. A* **64** = *Indag. Math.* **23** (1961), 411–417. MR **25** 2612 = S 836-3. *299.*
[2] Manifolds which are like projective planes. *Inst. Hautes Etudes Sci. Publ. Math.* **14** (1962), 5–46. MR **26** 3075 = S 838-2. *299.*
[3] An invariant for certain smooth manifolds. *Ann. Mat. Pura Appl.* (4) **60** (1962), 93–110. MR **27** 6280 = S 832-2. *54.*

Ehresmann, C.

[1] Sur la topologie de certains espaces homogènes. *Ann. of Math.* **35** (1934), 396–443. *240.*
[2] Sur la topologie de certaines variétés algébriques réelles. *J. Math. Pures Appl.* **104** [= (9) **16**] (1937), 69–100. *240.*
[3] Espaces fibrés associés. *C. R. Acad. Sci. Paris* **213** (1941), 762–764. MR **5** 148 = S 265-1. *97, 100.*
[4] Espaces fibrés de structures comparables. *C. R. Acad. Sci. Paris* **214** (1942), 144–147. MR **4** 146 = S 265-2. *97, 100.*
[5] Sur les espaces fibrés associés à une variété différentiable. *C. R. Acad. Sci. Paris* **216** (1943), 628–630. MR **5** 214 = S 690-1. *179.*

Ehresmann, C., and Feldbau, J.

[1] Sur les propriétés d'homotopie des espaces fibrés. *C. R. Acad. Sci. Paris* **212** (1941), 945–948. MR **3** 58 = S 264-5. *97, 100.*

Eilenberg, S.

[1] Algebraic topology. "Lectures on Modern Mathematics," Vol. I, pp. 98–114. Wiley, New York, 1963. MR **31** 2719 = S 26-8. *248.*

Eilenberg, S., and Steenrod, N. E.

[1] Axiomatic approach to homology theory. *Proc. Nat. Acad. Sci. U.S.A.* **31** (1945), 117–120. MR **6** 279 = S 191-5. *56.*

[2] "Foundations of Algebraic Topology." Princeton Univ. Press, Princeton, New Jersey, 1952. MR **14** 398 = S 8-2. *4.*

Ellis, H. G.

[1] Directional differentiation in the plane and tangent vectors on C^r manifolds. *Amer. Math. Monthly* **82** (1975), 641–645. MR **51** 14139. *180.*

Fadell, E.

[1] On fiber spaces. *Trans. Amer. Math. Soc.* **90** (1959), 1–14. MR **21** 330 = S 260-7. *103.*

[2] On fiber homotopy equivalence. *Duke Math. J.* **26** (1959), 699–706. MR **22** 233 = S 261-1. *103.*

[3] The equivalence of fiber spaces and bundles. *Bull. Amer. Math. Soc.* **66** (1960), 50–53. MR **22** 2992 = S 261-2. *103.*

[4] Generalized normal bundles for locally-flat imbeddings. *Trans. Amer. Math. Soc.* **114** (1965). 488–513. MR **31** 4037 = S 633-2. *180, 242.*

Feder, S.

[1] The reduced symmetric product of projective spaces and the generalized Whitney theorem. *Illinois J. Math.* **16** (1972), 323–329. MR **45** 6025. *212.*

Fintushel, R. A., and Stern, R. J.

[1] An exotic free involution on S^4. *Ann. of Math.* (2) **113** (1981), 357–365. *54.*

Flanders, H.

[1] Development of an extended exterior differential calculus. *Trans. Amer. Math. Soc.* **75** (1953), 311–326. MR **15** 161. *180.*

Fossum, R.

[1] Vector bundles over spheres are algebraic. *Invent. Math.* **8** (1969), 222–225. MR **40** 3537. *188.*

Fox, R. H.

[1] On the Lusternik–Schnirelmann category. *Ann. of Math.* (2) **42** (1941), 333–370. MR **2** 320 = S 250-4. *48.*

[2] Topological invariants of the Lusternik–Schnirelmann type. "Lectures in Topology," pp. 293–295. Univ. of Michigan Press, Ann Arbor, Michigan, 1941. MR **3** 135 = S 250-5. *48.*

[3] On fibre spaces. I. *Bull. Amer. Math. Soc.* **49** (1943), 555–557. MR **5** 48 = S 258-4. *97.*

[4] On fibre spaces. II. *Bull. Amer. Math. Soc.* **49** (1943), 733–735. MR **5** 104 = S 259-1. *98.*

Fuchs, M.

[1] A modified Dold–Lashof construction that does classify H-principal fibrations. *Math. Ann.* **192** (1971), 328–340. MR **45** 1167. *102, 103.*

Ganea, T.

[1] Sur quelques invariants numériques du type d'homotopie. *Cahiers Topologie Géom. Différentielle* **9** (1967), 181–241. MR **37** 5870. *48.*

[2] Lusternik–Schnirelmann category and strong category. *Illinois J. Math.* **11** (1967), 417–427. MR **37** 4814. *48.*

[3] Some problems on numerical homotopy invariants. *Sym. Algebraic Topol.* Lecture Notes in Math., Vol. 249, pp. 23–30. Springer-Verlag, Berlin and New York, 1971. MR **49** 3910. *48.*

Gel'fand, I. M. and Fuks, D. B.

[1] Classifying spaces for principal bundles with Hausdorff bases. (Russian.) *Dokl. Akad. Nauk SSSR* **181** (1968), 515–518. [*English transl.*: *Sov. Math.Dokl.* **9** (1968), 851–854.] MR **38** 716. *102.*

Giambelli, G. Z.

[1] Risoluzione del problema degli spazi secanti. *Mem. Reale Accad. Sci. Torino* (2) **52** (1902), 171–211. *239.*

Ginsburg, M.

[1] On the homology of fiber spaces. *Proc. Amer. Math. Soc.* **15** (1964), 423–431. MR **28** 5444 = S 286-4. *104.*

Gitler, S.

[1] Immersion and embedding of manifolds. "Algebraic Topology" [*Proc. Symp. Pure Math.* **22** (1970)], pp. 87–96. American Mathematical Society, Providence, Rhode Island, 1971. MR **47** 4275. *50, 185, 295.*

Glover, H. H., Homer, W. D., and Strong, R.

[1] Splitting the tangent bundles of projective spaces. *Indiana Univ. Math. J.* **31** (1982), 161–166. *188.*

Goldstein, R. Z., and Turner, E. C.

[1] A formula for Stiefel–Whitney homology classes. *Proc. Amer. Math. Soc.* **58** (1976), 339–342. MR **54** 3724. *290.*
[2] Stiefel–Whitney homology classes of quasi-regular cell complexes. *Proc. Amer. Math. Soc.* **64** (1977), 157–162. MR **57** 7617. *290.*

Gordon, R.

[1] Vector fields on spheres. *Sém. Heidelberg–Saarbrücken–Strasbourg sur la K-théorie (1967/1968)*, Lecture Notes in Mathematics, Vol. 136., pp. 254–264. Springer-Verlag, Berlin and New York, 1970. MR **42** 2501. *248.*

Gottlieb, D. H.

[1] The total space of universal fibrations. *Pacific J. Math.* **46** (1973), 415–417. MR **48** 9717. *102.*

Grassmann, H.

[1] "Die lineale Ausdehnungslehre, ein neuer Zweig der Mathematik, dargestellt und durch Anwendung auf die übrigen Zweige der Mathematik, wie auf die Statik, Mechanik, die Lehre vom Magnetismus und die Kristallonomie erläutert." Wigand, Leipzig, 1844. [Reprinted in "Gesammelte Werke," Bd. I, 1. Teil, Teubner, Leipzig, 1894.] *55.*
[2] "Die Ausdehnungslehre, vollständig und in strenger Form bearbeitet." Verlag von Th. Chr. Fr. Enslin, Berlin, 1862. [Reprinted in "Gesammelte Werke," Bd. I, 2. Teil, Teubner, Leipzig, 1896.] *55.*

Gray, B.

[1] "Homotopy Theory—An Introduction to Algebraic Topology." Academic Press, New York, 1975. MR **53** 6528. *21*, *48*, *49*, *212*, *243*, *297.*

Greenberg, M. J.

[1] "Lectures on Algebraic Topology." Benjamin, New York, 1967. MR **35** 6137. [Revised edition: Addison-Wesley, Reading, Massachusetts, 1981.] *4*, *212*, *213*, *214*, *288.*

Greene, R. E., and Wu, H.

[1] Embedding of open Riemannian manifolds by harmonic functions. *Ann. Inst. Fourier (Grenoble)* **25** (1975), 215–235. MR **52** 3583. *50.*

Griffiths, H. B.

[1] Locally trivial homology theories, and the Poincaré duality theorem. *Bull. Amer. Math. Soc.* **64** (1958), 367–370. MR **20** 4843 = S 642-3. *288.*

Griffiths, P., and Harris, J.

[1] "Principles of Algebraic Geometry." Wiley, New York, 1978. MR **80b** 14001. *214*, *240.*

Grothendieck, A.

[1] La théorie des classes de Chern. *Bull. Soc. Math. France* **86** (1958), 137–154. MR **22** 6818. *213, 241.*

Guillemin, V., and Pollack, A.

[1] "Differential Topology." Prentice-Hall, Englewood Cliffs, New Jersey, 1974. MR **50** 1276. *50.*

Gysin, W.

[1] Zur Homologietheorie der Abbildungen und Faserungen von Mannigfaltigkeiten. *Comment. Math. Helv.* **14** (1942), 61–122. MR **3** 317 = S 293-2. *104, 212.*

Haefliger, A., and Hirsch, M. W.

[1] On the existence and classification of differentiable embeddings. *Topology* **2** (1963), 129–135. MR **26** 6981 = S 767-1. *293.*

Halperin, S., and Toledo, D.

[1] Stiefel–Whitney homology classes. *Ann. of Math.* (*2*) **96** (1972), 511–525. MR **47** 1072. *290.*

[2] The product formula for Stiefel–Whitney homology classes. *Proc. Amer. Math. Soc.* **48** (1975), 239–244. MR **51** 1836. *290.*

Hangan, T.

[1] A Morse function on Grassmann manifolds. *J. Differential Geom.* **2** (1968), 363–367. MR **39** 6357. *299.*

Hardie, K. A.

[1] A note on fibrations and category. *Michigan Math. J.* **17** (1970), 351–352. MR **44** 1025. *48, 104.*

[2] On $\mathrm{cat}^i X$. *J. London Math. Soc.* (2) **3** (1971), 91–92. MR **45** 2703. *48.*

[3] A note on the category of the telescope. *Canad. Math. Bull.* **20** (1977), 107. MR **56** 6650. *48.*

Held, R. P., and Sjerve, D.

[1] Sur certains complexes de Thom et le J-homomorphisme. *C. R. Acad. Sci. Paris Sér. A-B* **275** (1972), A1293–A1295. MR **47** 5871. *212.*

[2] A propos du type d'homotopie stable des espaces projectifs tronqués. *C. R. Acad. Paris. Sér. A-B* **276** (1973), A1665–A1668. MR **48** 5069. *212.*

[3] On the stable homotopy type of Thom complexes. *Canad. J. Math.* **25** (1973), 1285–1294. MR **49** 3911. *212.*

Helgason, S.

[1] "Differential Geometry and Symmetric Spaces." Academic Press, New York, 1962. MR **26** 2986. *278*.
[2] "Differential Geometry, Lie Groups, and Symmetric Spaces." Academic Press, New York, 1978. MR **80k** 53081. *278*.

Heller, A.

[1] Singular homology in fibre bundles. *Ann. of Math.* (*2*) **55** (1952), 232–249. MR **13** 967 = S 302-5. *102*.

Hill, R. O., Jr.

[1] On the geometric dimension of stable real vector bundles. *Bol. Soc. Mat. Mexicana* (*2*) **17** (1972), 42–58. MR **49** 3935. *189*.

Hiller, H., and Stong, R. E.

[1] Immersion dimensions for real Grassmannians. *Math. Ann.* **255** (1981), 361–367. *296*.

Hilton, P. J.

[1] "An Introduction to Homotopy Theory," Cambridge Tracts in Mathematics and Mathematical Physics No. 43. Cambridge Univ. Press, London and New York, 1953. MR **15** 52 = S 336-1. *49*.

Hilton, P. J., and Wylie, S.

[1] "Homology Theory: An Introduction to Algebraic Topology." Cambridge Univ. Press., London and New York, 1960. MR **22** 5963 = S 13-3. *4*, *49*.

Hirsch, G.

[1] Un isomorphisme attaché aux structures fibrées. *C. R. Acad. Sci. Paris* **227** (1948), 1328–1330. MR **10** 558 = S 281-3. *104*.
[2] Sur les groupes d'homologie des espaces fibrés. *Bull. Soc. Math. Belg.* **1** (1947–1948) (1949), 24–33. MR **11** 194 = S 281-4. *104*.
[3] L'anneau de cohomologie d'un espace fibré et les classes caractéristiques. *C. R. Acad. Sci. Paris* **229** (1949), 1297–1299. MR **11** 379 = S 281-5. *104*.
[4] Sur la structure multiplicative de l'anneau de cohomologie d'un espace fibré. *C. R. Acad. Sci. Paris* **230** (1950), 46–48. MR **11** 379 = S 281-6. *104*.
[5] Homology invariants and fibre bundles. *Proc. Internat. Congr. Math., Cambridge, Massachusetts, 1950* Vol. 2, pp. 383–389. American Mathematical Society, Providence, Rhode Island, 1952. MR **13** 486 = S 281-7. *104*.
[6] La structure homologique des espaces fibrés et leur classification. *Congr. Nat. Sci., 3rd, Bruxelles*, Vol. 2, pp. 59–63. Fédération belge des sociétés scientifiques, Bruxelles, 1950. MR **17** 882 = S 266-3. *104*.

[7] Quelques relations entre l'homologie dans les espaces fibrés et les classes caractéristiques relatives à un groupe de structure. *Colloq. topolog. (espaces fibrés), Bruxelles, 1950* pp. 123–136. Georges Thone, Liège; Masson, Paris, 1951. MR **13** 56 = S 310-2. *241.*
[8] Sur les groupes d'homologie des espaces fibrés. *Bull. Soc. Math. Belg.* **6** (1953–1954), 79–96. MR **16** 1142 = S 281-8. *104.*

Hirsch, M. W.

[1] Immersions of manifolds. *Trans. Amer. Math. Soc.* **93** (1959), 242–276. MR **22** 9980 = S 750-4. *52, 291, 292, 295.*
[2] On imbedding differentiable manifolds in euclidean space. *Ann. of Math.* (2) **73** (1961), 566–571. MR **23** A2223 = S 765-2. *294, 299.*
[3] On embedding 4-manifolds in $\mathbb{R}^7$. *Proc. Cambridge Philos. Soc.* **61** (1965), 657–658. MR **32** 462 = S 611-2. *294.*
[4] "Differential Topology." Springer-Verlag, Berlin and New York, 1976. MR **56** 6669. *50, 159, 178, 273, 299.*

Hirsch, M. W., and Milnor, J.

[1] Some curious involutions of spheres. *Bull. Amer. Math. Soc.* **70** (1964), 372–377. MR **31** 751 = S 1115-6. *54.*

Hirzebruch, F.

[1] On Steenrod's reduced powers, the index of inertia, and the Todd genus. *Proc. Nat. Acad. Sci. U.S.A.* **39** (1953), 951–956. MR **16** 159 = S 724-1. *243.*
[2] "Neue topologische Methoden in der algebraischen Geometrie." Springer-Verlag, Berlin and New York, 1956. MR **18** 509 = S 656-2. *241.*
[3] "Topological Methods in Algebraic Geometry," 3rd enlarged ed. Springer-Verlag, Berlin and New York, 1966. MR **34** 2573. *241, 243.*

Hochschild, G.

[1] "The Structure of Lie Groups." Holden-Day, San Francisco, California, 1965. MR **34** 7696 = S 873-4. *86, 91, 104.*

Hodge, W. V. D.

[1] The base for algebraic varieties of a given dimension on a Grassmannian variety. *J. London Math. Soc.* **16** (1941), 245–255. MR **3** 304. *240.*
[2] The intersection formulae for a Grassmannian variety. *J. London Math. Soc.* **17** (1942), 48–64. MR **4** 52. *240.*
[3] Tangent sphere-bundles and canonical models of algebraic varieties. *J. London Math. Soc.* **27** (1952), 152–159. MR **13** 771. *239.*

Hodge, W. V. D., and Pedoe, D.

[1] "Methods of Algebraic Geometry," Vol. II. Cambridge Univ. Press, London and New York, 1952. MR **13** 972. [Reprinted, 1968.] *240.*

Holmann, H.

[1] "Vorlesung über Faserbündel." Aschendorffsche Verlagsbuchhandlung, Münster, 1962. MR **28** 5446 = S 16-6. *97.*

Hoo, C. S.

[1] The weak weak category of a space. *Canad. Math. Bull.* **14** (1971), 49–51. MR **45** 2705. *48.*

Horrocks, G.

[1] On extending vector bundles over projective space. *Quart. J. Math. Oxford Ser.* (2) **17** (1966), 14–18. MR **33** 4062. *185.*

Horvath, R.

[1] Normal bundles of topological manifolds. *Indian J. Math.* **12** (1970), 117–124. MR **46** 8225. *212.*

Hsiang, W.-C.

[1] On Wu's formula of Steenrod squares on Stiefel–Whitney classes. *Bol. Soc. Mat. Mexicana* (2) **8** (1963), 20–25. MR **29** 609 = S 317-2. *245.*

Hsiang, W.-C., and Szczarba, R. H.

[1] On the tangent bundle of a Grassmann manifold. *Amer. J. Math.* **86** (1964), 698–704. MR **30** 2523 = S 691-1. *185.*

Hu, S.-T.

[1] On generalising the notion of fibre spaces to include the fibre bundles. *Proc. Amer. Math. Soc.* **1** (1950), 756–762. MR **12** 435 = S 259-3. *99.*
[2] "Homotopy Theory." Academic Press, New York, 1959. MR **21** 5186 = S 336-4. *97.*
[3] "Elements of General Topology." Holden-Day, San Francisco, California, 1964. MR **31** 1643. *49, 56.*
[4] "Homology Theory: A First Course in Algebraic Topology." Holden-Day, San Francisco, California, 1966. MR **36** 875. *4, 213.*
[5] "Cohomology Theory." Markham Publ., Chicago, Illinois, 1968. MR **38** 2765. *4, 42.*
[6] "Differentiable Manifolds." Holt, New York, 1969. MR **39** 6343. *50.*

Huebsch, W.

[1] On the covering homotopy theorem. *Ann. of Math.* (2) **61** (1955), 555–563. MR **19** 974 = S 259-8. *99.*
[2] Covering homotopy. *Duke Math. J.* **23** (1956), 281–291. MR **19** 974 = S 259-9. *99.*

Hurewicz, W.

[1] On the concept of fiber space. *Proc. Nat. Acad. Sci. U.S.A.* **41** (1955), 956–961. MR **17** 519 = S 260-2. *99.*

Hurewicz, W., and Steenrod, N. E.

[1] Homotopy relations in fibre spaces. *Proc. Nat. Acad. Sci. U.S.A.* **27** (1941), 60–64. MR **2** 323 = S 258-1. *97, 98.*

Hurewicz, W., and Wallman, H.

[1] "Dimension Theory." Princeton Univ. Press, Princeton, New Jersey, 1941. MR **3** 312 = S 1-1. *48, 51.*

Husch, L. S.

[1] Hurewicz fibrations need not be locally trivial. *Proc. Amer. Math. Soc.* **61** (1976), 155–156. MR **56** 6656. *99.*

Husemoller, D.

[1] "Fibre Bundles." McGraw-Hill, New York, 1966. MR **37** 4821. [2nd ed., Springer-Verlag, Berlin and New York, 1975. MR **51** 6805.] *97, 104, 178, 182, 188, 214, 248, 293, 299.*

Iberkleid, W.

[1] Splitting the tangent bundle. *Trans. Amer. Math. Soc.* **191** (1974), 53–59. MR **50** 1269. *297.*

Ishimoto, H.

[1] A note on homotopy invariance of tangent bundles. *Nagoya Math. J.* **31** (1968), 247–250. MR **36** 4577. *187.*

Iwasawa, K.

[1] On some types of topological groups. *Ann. of Math.* (2) **50** (1949), 507–558. MR **10** 679 = S 910-1. *86, 104.*

Jaber, H., and Alkutibi, S. S.

[1] A note on Hurewicz fibrations. *Bull. College Sci.* (*Baghdad*) **16** (1975), 153–157. MR **52** 15454. *99.*

James, I. M.

[1] Euclidean models of projective spaces. *Bull. London Math. Soc.* **3** (1971), 257–276. MR **45** 7729. *185, 295.*

[2] Two problems studied by Heinz Hopf. "Lectures on Algebraic and Differential Topology, 1971," Lecture Notes in Mathematics, Vol. 279. Springer-Verlag, Berlin and New York, 1972. MR **51** 1840. *295.*

[3] On category, in the sense of Lusternik-Schnirelmann. *Topology* **17** (1978), 331–348. MR **80i** 55001. *48.*

Jupp, P. E.

[1] Classification of certain 6-manifolds. *Proc. Cambridge Philos. Soc.* **73** (1973), 293–300. MR **47** 2626. *53.*

Kahn, D. W.

[1] "Introduction to Global Analysis." Academic Press, New York, 1980. MR **81i** 58001. *178.*

Kaminker, J.

[1] The tangent bundle of an H-manifold. *Proc. Amer. Math. Soc.* **41** (1973), 305–308. MR **47** 7732. *187.*

Kandelaki, T. K.

[1] The equivalence of the categories of vector bundles and projective modules of finite type over a Banach algebra. (Russian. Georgian and English summaries.) *Sakharth. SSR Mecn. Akad. Moambe* **83** (1976), 33–36. MR **55** 13423. *180.*

Karoubi, M.

[1] Cobordisme et groupes formels (d'après D. Quillen et tom Dieck). *Sém. Bourbaki, 1971/1972* Exp. No. 408, pp. 141–165. Lecture Notes in Mathematics, Vol. 317. Springer-Verlag, Berlin and New York, 1973. MR **54** 1263. *297.*

[2] "K-Theory: An Introduction." Springer-Verlag, Berlin and New York, 1978. MR **58** 7605. *178, 248.*

Kawakubo, K.

[1] Smooth structures on $S^p \times S^q$. *Proc. Japan Acad.* **45** (1969), 215–218. MR **40** 2107. *54.*

[2] Smooth structures on $S^p \times S^q$. *Osaka J. Math.* **6** (1969), 165–196. MR **41** 6228. *54.*

Kelley, J. L.

[1] "General Topology." Van Nostrand, New York, 1955. MR **16** 1136. *39, 82.*

Kelley, J. L., and Pitcher, E.

[1] Exact homomorphism sequences in homology theory. *Ann. of Math.* (2) **48** (1947), 682–709. MR **9** 52 = S 274-6. *56.*

Kervaire, M. A.

[1] Courbure intégrale généralisée et homotopie. *Math. Ann.* **131** (1956), 219–252. MR **19** 160 = S 834-1. *184*.
[2] Sur le fibré normal à une variété plongée dans l'espace euclidien. *Bull. Soc. Math. France* **87** (1959), 397–401. MR **22** 5054 = S 764-4. *291*.
[3] A manifold which does not admit any differentiable structure. *Comment. Math. Helv.* **34** (1960), 257–270. MR **25** 2608 = S 574-3. *54*.

Kervaire, M. A., and Milnor, J.

[1] Groups of homotopy spheres. I. *Ann. of Math.* (2) **77** (1963), 504–537. MR **26** 5584 = S 832-3. *54*.

Kirby, R., and Siebenmann, L. C.

[1] On the triangulation of manifolds and the Hauptvermutung. *Bull. Amer. Math. Soc.* **75** (1969), 742–749. MR **39** 3500. [Reprinted on pp. 299–306 of Kirby and Siebenmann [2].] *53*.
[2] "Foundational Essays on Topological Manifolds, Smoothings, and Triangulations." Princeton Univ. Press, Princeton, New Jersey, and Univ. of Tokyo Press, Tokyo, 1977. MR **58** 31082. *53*, *54*.

Kister, J. M.

[1] Microbundles are fibre bundles. *Bull. Amer. Math. Soc.* **69** (1963), 854–857. MR **27** 6283 = S 581-2. *180*, *289*.
[2] Microbundles are fibre bundles. *Ann. of Math.* (2) **80** (1964), 190–199. MR **31** 5216 = S 581-3. *180*, *289*.
[3] Inverses of Euclidean bundles. *Michigan Math. J.* **14** (1967), 349–352. MR **35** 3695. *181*.

Kleiman, S. L.

[1] Problem 15. Rigorous foundation of Schubert's enumerative calculus. "Mathematical Developments Arising from Hilbert Problems" [*Proc. Symp. Pure Math.* **28**, Part 2 (1974)], pp. 445–482. American Mathematical Society, Providence, Rhode Island, 1976. MR **55** 2946. *240*.

Kleiman, S. L., and Laksov, D.

[1] Schubert calculus. *Amer. Math. Monthly* **79** (1972), 1061–1082. MR **48** 2152. *56*, *240*.

Kobayashi, S., and Nomizu, K.

[1] "Foundations of Differential Geometry." Vol. I. Wiley (Interscience), New York, 1963. MR **27** 2945. *100*.

Kobayashi, T.

[1] Immersions and embeddings of lens spaces. *Hiroshima Math. J.* **2** (1972), 345–352. MR **48** 7300. *296*.

Kobayashi, T., Maki, H., and Yoshida, T.

[1] Remarks on extendible vector bundles over lens spaces and real projective spaces. *Hiroshima Math. J.* **5** (1975), 487–497. MR **52** 6716. *186*.

Koszul, J. L.

[1] "Lectures on Fibre Bundles and Differential Geometry." Tata Institute of Fundamental Research, Bombay, 1960. MR **42** 3698. *100*.

Kudo, T.

[1] Homological properties of fibre bundles. *J. Inst. Polytech. Osaka City Univ. Ser. A. Math.* **1** (1950), 101–114. MR **13** 56 = S 275-2. *104*.
[2] Homological structure of fibre bundles. *J. Inst. Polytech. Osaka City Univ. Ser A. Math.* **2** (1952), 101–140. MR **14** 1111 = S 275-3. *104*.

Kuiper, N. H.

[1] The quotient space of CP(2) by complex conjugation is the 4-sphere. *Math. Ann.* **208** (1974), 175–177. MR **49** 11541. *55*.

Kuiper, N. H., and Lashof, R. K.

[1] Microbundles and bundles. I. Elementary theory. *Invent. Math.* **1** (1966), 1–17. MR **35** 7339. *181*.

Kurogi, T.

[1] A remark on microbundle. *Tamkang J. Math.* **2** (1971), 95–100. MR **46** 9982. *181*.

Lam, K. Y.

[1] Cup product in projective spaces. *Proc. Amer. Math. Soc.* **24** (1970), 832–833. MR **41** 4576. *213*.
[2] Fiber homotopic trivial bundles over complex projective spaces. *Proc. Amer. Math. Soc.* **33** (1972), 211–212. MR **45** 2731. *189*.
[3] Sectioning vector bundles over real projective spaces. *Quart. J. Math. Oxford. Ser (2)* **23** (1972), 97–106. MR **45** 6024. *185*.
[4] A formula for the tangent bundle of flag manifolds and related manifolds. *Trans. Amer. Math. Soc.* **213** (1975), 305–314. MR **55** 4196. *185*.

Lang, S.

[1] "Differentiable Manifolds." Addison-Wesley, Reading, Massachusetts, 1972. MR **55** 4241. *50*, *178*.

Lanteri, A.

[1] Un atlante differenziabile delle grassmaniane reali e questioni di orientabilità. *Ist. Lombardo Accad. Sci. Lett. Rend. A* **109** (1975), 337–347. MR **54** 6174. *56*.

Larmore, L. L., and Rigdon, R. D.

[1] Enumerating immersions and embeddings of projective spaces. *Pacific J. Math.* **64** (1976), 471–492. MR **55** 1359. *296.*
[2] Enumerating normal bundles of immersions and embeddings of projective spaces. *Pacific J. Math.* **70** (1977), 207–220. MR **57** 10712. *296.*

Lashof, R.

[1] The tangent bundle of a topological manifold. *Amer. Math. Monthly* **79** (1972), 1090–1096. MR **47** 2605. *178, 181.*

Lashof, R., and Rothenberg, M.

[1] Microbundles and smoothing. *Topology* **3** (1965), 357–388. MR **31** 752 = S 571-1. *181.*
[2] Triangulation of manifolds. I, II. *Bull. Amer. Math. Soc.* **75** (1969), 750–757. MR **40** 895. *53.*

Latour, F.

[1] Variétés géométriques et résolutions. I. Classes caractéristiques. *Ann. Sci. Ecole Norm. Sup.* (*4*) **10** (1977), 1–72. MR **57** 17661. *290.*

Lees, J. A.

[1] "Notes on Bundle Theory. Lectures 1974," Lecture Notes Series No. 42. Matematisk Institut, Aarhus Univ., Aarhus, 1974. MR **50** 11235. *97.*

Lefschetz, S.

[1] "Introduction to Topology." Princeton Univ. Press, Princeton, New Jersey, 1949. MR **11** 193 = S 7-1. *288.*

Leray, J.

[1] L'anneau d'homologie d'une représentation. *C.R. Acad. Sci. Paris* **222** (1946), 1366–1368. MR **8** 49 = S 274-1. *104.*
[2] Structure de l'anneau d'homologie d'une représentation. *C.R. Acad. Sci. Paris* **222** (1946), 1419–1422. MR **8** 49 = S 274-2. *104.*
[3] Propriétés de l'anneau d'homologie de la projection d'un espace fibré sur sa base. *C.R. Acad. Sci. Paris* **223** (1946), 395–397. MR **8** 166 = S 274-3. *104.*
[4] Sur l'anneau d'homologie de l'espace homogène, quotient d'un groupe clos par un sous-groupe abélien, connexe, maximum. *C.R. Acad. Sci. Paris* **223** (1946), 412–415. MR **8** 166 = S 274-4. *104.*
[5] L'homologie d'un espace fibré dont la fibre est connexe. *J. Math. Pures Appl.* (*9*) **29** (1950), 169–213. MR **12** 521 = S 277-1. *104.*

Liulevicius, A.

[1] A proof of Thom's theorem. *Comment. Math. Helv.* **37** (1962/1963), 121–131. MR **26** 3058 = S 788-3. *296.*

[2] "Characteristic Classes and Cobordism," Part I. Matematisk Institut, Aarhus Univ., Aarhus, 1967. MR **41** 2700. *97*, *102*, *240*.

[3] "On Characteristic Classes." Matematisk Institut, Aarhus Univ., Aarhus, 1968. MR **41** 1065. *97*, *296*.

Livesay, G. R.

[1] Fixed point free involutions on the 3-sphere. *Ann. of Math.* (2) **72** (1960), 603–611. MR **22** 7131 = S 1114-5. *54*.

Livesay, G. R., and Thomas, C. B.

[1] Involutions on homotopy spheres. *Proc. Conf. Transformat. Groups, New Orleans, Louisiana, 1967*, pp.143–147. Springer-Verlag, Berlin and New York, 1968. MR **42** 1147. *55*.

Ljusternik, L. A.

[1] Topology of Functional Spaces and the Calculus of Variations in the Large. (Russian. English summary.) *Trav. Inst. Math. Stekloff* **19** (1947). MR **9** 596 = S 813-4. [*English transl.*: The topology of the calculus of variations in the large. "Translations of Mathematical Monographs," Vol. **16**. American Mathematical Society, Providence, Rhode Island, 1966. MR **36** 906.] *48*.

Ljusternik, L., and Schnirelmann, L.

[1] "Topological Methods in Problems of the Calculus of Variations." (Russian.) Research Institute for Math. and Mech., Moscow, 1930. [A French translation is given under the transliteration Lusternik and Schnirelmann [1].] *48*.

Lønsted, K.

[1] Vector bundles over finite CW-complexes are algebraic. *Proc. Amer. Math. Soc.* **38** (1973), 27–31. MR **47** 5862. *180*, *188*.

López de Medrano, S.

[1] "Involutions on Manifolds." Springer-Verlag, Berlin and New York, 1971. MR **45** 7747. *55*.

Luft, E.

[1] Covering of manifolds with open cells. *Illinois J. Math.* **13** (1969), 321–326. MR **39** 3496. *48*.

[2] Covering of 2-dimensional manifolds with open cells. *Arch. Math.* (*Basel*) **22** (1971), 536–544. MR **46** 895. *48*.

Lundell, A. T., and Weingram, S.

[1] "The Topology of CW Complexes." Van Nostrand Reinhold, New York, 1969. *21*, *37*, *49*, *50*.

Lusternik, L., and Schnirelmann, L.

[1] Méthodes topologiques dans les problèmes variationels. "Actualités Scientifiques et Industrielles," Vol. 188. Hermann, Paris, 1934. [This is a French translation of Ljusternik and Schnirelmann [1].] *48, 328.*

Lusztig, G.

[1] Construction of a universal bundle over arbitrary polyhedra. (Romanian.) *Stud. Cerc. Mat.* **18** (1966), 1215–1219. MR **38** 3869. *102.*

Mahammed, N., Piccinini, R., and Suter, U.

[1] "Some Applications of Topological K-Theory." North-Holland Publ., Amsterdam, 1980. MR **82f** 55009. *248.*

Mahowald, M.

[1] On the metastable homotopy of $O(n)$. *Proc. Amer. Math. Soc.* **19** (1968), 639–641. MR **37** 918. *188.*

Mahowald, M., and Peterson, F. P.

[1] Secondary cohomology operations on the Thom class. *Topology* **2** (1963), 367–377. MR **28** 612 = S 758-3. *295.*

Maki, H.

[1] Extendible vector bundles over lens spaces mod 3. *Osaka J. Math.* **7** (1970), 397–407. MR **43** 2728. *186.*

Mal'cev, A.

[1] On the theory of Lie groups in the large. (Russian). *Rec. Math.* [*Mat. Sbornik*] *N.S.* **16** (**58**) (1945), 163–190. MR **7** 115. *103.*

Marchiafava, S. and Romani, G.

[1] Classi caratteristiche dei fibrati quaternionali generalizzati. (English summary.) *Atti Accad. Naz. Lincei Rend. Cl. Sci. Fis. Mat. Natur.* (*8*) **56** (1974), 899–906. MR **53** 6557. *247.*

[2] Sui fibrati con struttura quaternionale generalizzata. *Ann. Mat. Pura Appl.* (4) **107** (1975), 131–157. MR **53** 6558. *247.*

[3] Ancora sulle classi di Stiefel-Whitney dei fibrati quaternionali generalizzati. (English summary.) *Atti Accad. Naz. Lincei Rend. Cl. Sci. Fis. Mat. Natur.* (*8*) **61** (1976), 438–447. MR **58** 13064. *247.*

Markov, A.

[1] The insolubility of the problem of homeomorphy. *Dokl. Akad. Nauk SSSR* **121** (1958), 218–220. (Russian.) MR **20** 4260 = S 592-1. *52.*

[2] Unsolvability of certain problems in topology. *Dokl. Akad. Nauk SSSR* **123** (1958), 978–980. (Russian.) MR **21** 2224 = S 592-2. *52.*

[3] Insolubility of the problem of homeomorphy. *Proc. Internat. Congr. Math., 1958* pp. 300–306. (Russian.) Cambridge Univ. Press, London and New York, 1960. MR **22** 5962 = S 592-3. *52.*

Massey, W. S.

[1] On the Stiefel-Whitney classes of a manifold. *Amer. J. Math.* **82** (1960), 92–102. MR **22** 1918 = S 716-5. *246, 294.*

[2] Normal vector fields on manifolds. *Proc. Amer. Math. Soc.* **12** (1961), 33–40. MR **23** A2222 = S 764-2. *295.*

[3] On the Stiefel–Whitney classes of a manifold. II. *Proc. Amer. Math. Soc.* **13** (1962), 938–942. MR **25** 5522 = S 717-1. *246, 294.*

[4] "Algebraic Topology: An Introduction." Harcourt, New York, 1967. MR **35** 2271. [Reprinted 1977. MR **56** 6638.] *52, 53.*

[5] The quotient space of the complex projective plane under conjugation is a 4-sphere. *Geomet. Dedicata* **2** (1973), 371–374. MR **49** 6262. *55.*

[6] "Singular Homology Theory." Springer-Verlag, Berlin and New York, 1980. MR **81g** 55002. *4, 49, 213, 287.*

Massey, W. S., and Peterson, F. P.

[1] On the dual Steifel–Whitney classes of a manifold. *Bol Soc. Mat. Mexicana (2)* **8** (1963), 1–13. MR **29** 628 = S 717-2. *246, 294.*

Masuda, K.

[1] Morse functions on some algebraic varieties. *Kōdai Math. Sem. Rep.* **26** (1974/1975), 216–229. MR **54** 6168. *299.*

Maunder, C. R. F.

[1] "Algebraic Topology." Van Nostrand Reinhold, New York, 1970. *21, 49, 213, 288.*

Maurin, K.

[1] "Calculus of Variations and Classical Field Theory. I," Lecture Notes Series No. 34. Matematisk Institut, Aarhus Univ., Aarhus, 1972. MR **58** 7670. *48.*

May, J. P.

[1] Classifying spaces and fibrations. *Mem. Amer. Math. Soc.* **1** (1975). MR **51** 6806. *49, 97, 102.*

Mayer, W.

[1] Über abstrakte Topologie. *Monatsch. Math. Phys.* **36** (1929), 1–42. *56.*

[2] Duality theorems. *Fund. Math.* **35** (1948), 188–202. MR **6** 280 = S 202-3. *288.*

McCrory, C.

[1] Euler singularities and homology operations. "Differential Geometry" [*Proc. Symp. Pure Math.* **27**, Part 1 (1973)], pp. 371–380. American Mathematical Society, Providence, Rhode Island, 1975. MR **51** 14089. *290.*

Metzler, W.

[1] Beispiele zu Unterteilungsfragen bei CW- und Simplizial-Komplexen. *Arch. Math.* (*Basel*) **18** (1967), 513–519. MR **36** 4550. *49.*

Micali, A. and Revoy, Ph.

[1] Modules quadratiques. *Univ. Sci. Tech. Languedoc*; *U.E.R. Math., Montpellier, Cahiers Math.* **10** (1977). MR **58** 10974. *248.*

Mielke, M. V.

[1] Cobordism properties of manifolds of small category. *Proc. Amer. Math. Soc.* **21** (1969), 332–334. MR **38** 5232. *300.*
[2] Rearrangement of spherical modifications. *Pacific J. Math.* **28** (1969), 143–150. MR **39** 963. *300.*
[3] Spherical modifications and coverings by cells. *Duke Math. J.* **36** (1969), 49–53. MR **39** 4858. *300.*
[4] Spherical modifications and the strong category of manifolds. *J. Austral. Math. Soc.* **9** (1969), 449–454. MR **39** 7626. *300.*

Milgram, R. J.

[1] The bar construction and abelian H-spaces. *Illinois J. Math.* **11** (1967), 242–250. MR **34** 8404 = S 309-3. *102.*

Milnor, J.

[1] On manifolds homeomorphic to the 7-sphere. *Ann. of Math.* (*2*) **64** (1956), 399–405. MR **18** 498 = S 828-1. *54, 299.*
[2] Construction of universal bundles. II. *Ann. of Math.* (*2*) **63** (1956), 430–436. MR **17** 1120 = S 304-5. *102, 103.*
[3] "Lectures on Characteristic Classes, 1957," Mimeographed notes. Princeton University, Princeton, New Jersey, 1958. *104, 212, 240, 243, 288, 293.*
[4] Some consequences of a theorem of Bott. *Ann. of Math.* (*2*) **68** (1958), 444–449. MR **21** 1591 = S 737-3. *184, 247.*
[5] On the existence of a connection with curvature zero. *Comment. Math. Helv.* **32** (1958), 215-228. MR **20** 2020 = S 699-3. *185.*
[6] Differentiable structures on spheres. *Amer J. Math.* **81** (1959), 962–972. MR **22** 990 = S 828-5. *54.*
[7] Sommes de variétés différentiables et structures différentiables des sphères. *Bull. Soc. Math. France* **87** (1959), 439–444. MR **22** 8518 = S 829-1. *54.*
[8] On spaces having the homotopy type of a CW-complex. *Trans. Amer. Math. Soc.* **90** (1959), 272–280. MR **20** 6700 = S 491-3. *21, 49.*

[9] A procedure for killing homotopy groups of differentiable manifolds. "Differential Geometry" [*Proc. Symp. Pure Math.* **3** (1960)], pp. 39–55. American Mathematical Society, Providence, Rhode Island, 1961. MR **24** A556 = S 823-1. *273, 299.*

[10] A survey of cobordism theory. *Enseignement Math.* (*2*) **8** (1962), 16–23. MR **27** 2989 = S 670-2. *297.*

[11] Topological manifolds and smooth manifolds. *Proc. Internat. Congr. Math., Stockholm, 1962* pp. 132–138. Inst. Mittag-Leffler, Djursholm, 1963. MR **28** 4533a = S 580-4. *180, 289.*

[12] Spin structures on manifolds. *Enseignement Math.* (*2*) **9** (1963), 198–203. MR **28** 622 = S 792-3. *247.*

[13] "Morse Theory," Annals of Mathematics Studies No. 51. Princeton Univ. Press, Princeton, New Jersey, 1963. MR **29** 634 = S 662-2. *273, 299.*

[14] Microbundles. I. *Topology* **3**, *Suppl. 1* (1964), 53–80. MR **28** 4533b = S 580-5. *180, 289.*

[15] "Topology from the Differentiable Viewpoint." Univ. Press of Virginia, Charlottesville, Virginia, 1965. MR **37** 2239. *50.*

[16] Remarks concerning spin manifolds. "Differential and Combinatorial Topology" (A Symposium in Honor of Marston Morse), pp. 55–62. Princeton Univ. Press, Princeton, New Jersey, 1965. MR **31** 5208 = S 792-4. *54, 247.*

[17] On the Stiefel–Whitney numbers of complex manifolds and of spin manifolds. *Topology* **3** (1965), 223–230. MR **31** 5207 = S 793-5. *296, 298.*

[18] "Lectures on the *h*-Cobordism Theorem." Princeton Univ. Press, Princeton, New Jersey, 1965. MR **32** 8352 = S 665-2. *273, 299.*

[19] Algebraic *K*-theory and quadratic forms. *Invent. Math.* **9** (1970), 318–344. MR **41** 5465. *248.*

Milnor, J., and Stasheff, J. D.

[1] "Characteristic Classes," Annals of Mathematics Studies No. 76. Princeton Univ. Press, Princeton, New Jersey and Univ. of Tokyo Press, Tokyo, 1974. MR **55** 13428. *37, 56, 104, 178, 212, 240, 243, 287, 288, 293.*

Miyazaki, H.

[1] The paracompactness of CW-complexes. *Tôhoku Math. J.* (2) **4** (1952), 309–313. MR **14** 894 = S 554-7. *50.*

Moise, E.

[1] Affine structures in 3-manifolds. V. The triangulation theorem and the Hauptvermutung. *Ann. of Math.* (*2*) **56** (1952), 96–114. MR **14** 72 = S 152-3. *53.*

Moore, N.

[1] Algebraic vector bundles over the 2-sphere. *Invent. Math.* **14** (1971), 167–172. MR **45** 3403. *188.*

Moran, D. A.

[1] Minimal cell coverings of some sphere bundles. *Comment. Math. Univ. Carolinae* **14** (1973), 647–650. MR **48** 9716. *48, 104.*

[2] Minimal cell coverings of sphere bundles over spheres. *Comment. Math. Univ. Carolinae* **16** (1975), 147–149. MR **51** 1834. *48, 104.*

[3] Cell coverings and residual sets of closed manifolds. *Illinois J. Math.* **20** (1976), 516–518. MR **53** 9220. *48.*

[4] Corrigendum et addendum ad: "Minimal cell coverings of sphere bundles over spheres." *Comment. Math. Univ. Carolinae* **20** (1979), 189–190. MR **80m** 55003. *48.*

Morrow, J. A.

[1] The tangent bundle of the long line. *Proc. Amer. Math. Soc.* **23** (1969), 458. MR **39** 7622. *180.*

Morse, M.

[1] The existence of non-degenerate functions on a compact differentiable m-manifold M. *Ann. Mat. Pura Appl.* (4) **49** (1960), 117–128. MR **22** 11399 = S 809-3. *299.*

Morse, M., and Cairns, S. S.

[1] "Critical Point Theory in Global Analysis and Differential Topology: An Introduction." Academic Press, New York, 1969. MR **39** 6358. *299.*

Mosher, R. E., and Tangora, M. C.

[1] "Cohomology Operations and Applications in Homotopy Theory." Harper, New York, 1968. MR **37** 2223. *243.*

Mostow, G. D.

[1] Self-adjoint groups. *Ann. of Math.* (2) **62** (1955), 44–55. MR **16** 1088. *86, 104.*

Munkres, J. R.

[1] "Elementary Differential Topology." Annals of Math. Studies No. 54. Princeton Univ. Press, Princeton, New Jersey, 1963. MR **29** 623 = S 661-2 [Rev. ed., 1966. MR **33** 6637 = S 662-1.] *50, 52.*

Myers, R.

[1] Free involutions on lens spaces. *Topology* **20** (1981), 313–318. *55.*

Myers, S. B.

[1] Algebras of differentiable functions. *Proc. Amer. Math. Soc.* **5** (1954), 917–922. MR **16** 491. *180.*

Nachbin, L.

[1] Algebras of finite differential order and the operational calculus. *Ann. of Math.* (2) **70** (1959), 413–437. MR **21** 7444. *180.*

Nagata, J.-i.

[1] "Modern Dimension Theory." Wiley (Interscience), New York, 1965. MR **34** 8380 = S 1-2. *48, 51.*

Nash, J.

[1] A path space and the Stiefel–Whitney classes. *Proc. Nat. Acad. Sci. U.S.A.* **41** (1955), 320–321. MR **17** 80 = S 715-2. *180, 242, 289.*

Newns, W. F., and Walker, A. G.

[1] Tangent planes to a differentiable manifold. *J. London Math. Soc.* **31** (1956), 400–407. MR **18** 821. *180.*

Novikov, P. S.

[1] On the algorithmic unsolvability of the word problem in group theory. (Russian.) *Trudy Mat. Inst. Steklov* No. 44, (1955). MR **17** 706. [*English transl.*: *Amer. Math. Soc. Transl.* (2) **9** (1958), 1–122. MR **19** 1158.] *52.*

Novikov, S. P.

[1] Imbedding of simply connected manifolds into Euclidean space. (Russian.) *Dokl. Akad. Nauk SSSR* **138** (1961), 775–778. MR **24** A2969 = S 766-1. *294.*

[2] Differentiable sphere bundles. (Russian.) *Izv. Akad. Nauk SSSR Ser. Mat.* **29** (1965), 71–96. MR **30** 4266 = S 848-2. [*English transl.*: see Seven papers on algebra, algebraic geometry, and algebraic topology. *Amer. Math. Soc. Transl.* (*2*) **63** (1967). MR **36** 2.] *179.*

Okonek, C., Schneider, M., and Spindler, H.

[1] Vector bundles on complex projective spaces. *Progr. Math.* **3**, 1980. *186.*

Ono, Y.

[1] Notes on the Lusternik–Schnirelmann category and H-spaces. *Bull. Kyoto Univ. Educ. Ser. B.* (34) (1969), 1–4. MR **39** 4843. *48.*

[2] On the mod p category and the fibrations. *Bull. Kyoto Univ. Educ. Ser. B* No. 40 (1972), 5–9. MR **46** 2667. *48, 104.*

Oproiu, V.

[1] Some multiplication formulas in the cohomology ring of the Grassmann manifolds. *Proc. Inst. Math. Iasi, 1974* pp. 67–74. Editura Acad. R.S.R., Bucharest, 1976. MR **56** 1335. *241.*

[2] Some non-embedding theorems for the Grassmann manifolds $G_{2,n}$ and $G_{3,n}$. *Proc. Edinburgh Math. Soc.* (*2*) **20** (1976/1977), 177–185. MR **56** 3870. *296.*

[3] Some results concerning the non-embedding codimension of Grassmann manifolds in Euclidean spaces. *Rev. Roumaine Math. Pures Appl.* **26** (1981), 275–286. *296.*

Orlick, P., and Rourke, C. P.

[1] Free involutions on homotopy $(4k+3)$-spheres. *Bull. Amer. Math. Soc.* **74** (1968), 949–953. MR **37** 4819. *55.*

Osborn, H.

[1] Intrinsic characterizations of tangent spaces. *Proc. Amer. Math. Soc.* **16** (1965), 591–594. MR **31** 3963. *180.*
[2] Modules of differentials, I. *Math. Ann.* **170** (1967), 221–244. MR **35** 4839. *180.*
[3] Derivations of C^∞ functions. *Math. Z.* **101** (1967), 265–268. MR **36** 802. *180, 184.*
[4] Modules of differentials, II. *Math. Ann.* **175** (1968), 146–158. MR **36** 3268. *180.*
[5] Derivations of commutative algebras. *Illinois J. Math.* **13** (1969), 137–144. MR **38** 4466. *180.*
[6] Base spaces. *Indiana Univ. Math. J.* **28** (1979), 745–755. MR **81i** 55017. *47, 49, 104.*

Osborne, R. P., and Stern, J. L.

[1] Covering manifolds with cells. *Pacific J. Math.* **30** (1969), 201–207. MR **39** 6327.

Palais, R. S.

[1] Manifolds of sections of fiber bundles and the calculus of variations. "Nonlinear Functional Analysis" [*Proc. Symp. Pure Math.* **18**, Part 1, (1968)], pp. 195–205. American Mathematical Society, Providence, Rhode Island, 1970. MR **42** 6854. *48, 104.*

Papastavridis, S.

[1] Killing characteristic classes by surgery. *Proc. Amer. Math. Soc.* **63** (1977), 353–358. MR **55** 13442. *298.*
[2] Imbeddings, immersions, and characteristic classes of differentiable manifolds. *Proc. Amer. Math. Soc.* **69** (1978), 177–180. MR **58** 7649. *246.*
[3] On killing characteristic classes by cobordism. *Manuscripta Math.* **29** (1979), 85–92. MR **80j** 57023. *298.*

Papy, G.

[1] Sur la définition intrinsèque des vecteurs tangents. *C.R. Acad. Sci. Paris* **241** (1955), 19–20. MR **16** 1152. *180.*
[2] Sur la définition intrinsèque des vecteurs tangents à une variété de classe C^r lorsque $1 \leqq r < \infty$. *C.R. Acad. Sci. Paris* **242** (1956), 1573–1575. MR **17** 892. *180.*

Patterson, R. R.

[1] The Hasse invariant of a vector bundle. *Trans. Amer. Math. Soc.* **150** (1970), 425–443. MR **42** 3790. *248.*

Piccinini, R. A.

[1] CW-Complexes, Homology Theory. "Queen's Papers in Pure and Applied Mathematics," No. 34. Queen's Univ., Kingston, Ontario, 1973. MR **50** 14731. *49.*

Pieri, M.

[1] Sul problema degli spazi secanti. Nota 2ª. Rendiconti della Reale Istituto Lombardo (2) **27** (1894). 258–273. *239.*

Pieri, M., and Zeuthen, H. G.

[1] Géométrie énumérative. Encyclopédie des sciences mathématiques **3**, Part 1, 260–331. B. G. Teubner, Leipzig, 1915. *240.*

Pitcher, E.

[1] Inequalities of critical point theory. *Bull. Amer. Math. Soc.* **64** (1958), 1–30. MR **20** 2648 = S 808-1. *299.*

Poénaru, V.

[1] On the geometry of differentiable manifolds. "Studies in Modern Topology," M.A.A. Studies in Mathematics, Vol. 5, pp. 165–207. Mathematical Association of America, Prentice-Hall, Englewood Cliffs, New Jersey, 1968. MR **36** 5957. *297.*

Pontrjagin, L. S.

[1] Characteristic cycles on manifolds (Russian). *C.R. Dokl. Acad. Sci. URSS (N.S.)* **35** (1942), 34–37. MR **4** 147 = S 721-5. *179, 239, 240, 289.*

[2] On some topologic invariants of Riemannian manifolds (Russian). *C.R. Dokl. Acad. URSS (N.S.)* **43** (1944), 91–94. MR **6** 182 = S 721-6. *289.*

[3] Characteristic cycles. (Russian.) *C.R. Dokl. Acad. Sci. URSS (N.S.)* **47** (1945), 242–245. MR **7** 138 = S 302-2. *179.*

[4] Classification of some skew products. (Russian.) *C.R. Dokl. Acad. Sci. URSS (N.S.)* **47** (1945), 322–325. MR **7** 138 = S 302-1. *179.*

[5] Characteristic cycles on differentiable manifolds. (Russian.) *Mat. Sb.* **21** (63) (1947), 233–284. MR **9** 243 = S 722-2. [*English transl.*: *Amer. Math. Soc. Transl.* No. 32 (1950). MR **12** 350 = S 723-1.] *179, 239, 240, 266, 289.*

Porter, R. D.

[1] Characteristic classes and singularities of mappings. "Differential Geometry" [*Proc. Symp. Pure Math.* **27**, Part 1 (1973)], pp. 397–402. American Mathematical Society, Providence, Rhode Island, 1975. MR **55** 9122. *290.*

[2] "Introduction to Fibre Bundles," Lecture Notes in Pure and Applied Math., Vol. 31, Dekker, New York, 1977. MR **56** 9526. *97, 102.*

Postnikov, M. M.

[1] On paracompactness of cell-like polyhedra. (Russian.) *Uspehi Mat. Nauk* **20** (1965), 226–230. MR **41** 9240. *50.*

Price, J. F.

[1] "Lie Groups and Compact Groups." London Math. Soc. Lecture Notes Series, Vol. 25. Cambridge Univ. Press, London and New York, 1977. MR **56** 8743. *91.*

Price, T. M.

[1] Towards classifying all manifolds. *Math. Chronicle* **7** (1978), 1–47. MR **81i** 57008. *53.*

Quillen, D.

[1] On the formal group laws of unoriented and complex cobordism theory. *Bull. Amer. Math. Soc.* **75** (1969), 1293–1298. MR **40** 6565. *297.*

[2] Elementary proofs of some results of cobordism theory using Steenrod operations. *Advances in Math.* **7** (1971), 29–56. MR **44** 7566. *297.*

Rabin, M. O.

[1] Recursive unsolvability of group theoretic problems. *Ann. of Math.* (2) **67** (1958), 172–194. MR **22** 1611 = S 592-4. *52.*

Radó, T.

[1] Über den Begriff der Riemannschen Fläche. *Acta Litt. Sci. Szeged* **2** (1925), 101–121. *53.*

Raymond, F.

[1] Local triviality for Hurewicz fiberings of manifolds. *Topology* **3** (1965), 43–57. MR **28** 2554 = S 262-4. *99.*

Rayner, C. B.

[1] Immersions for which the Gauss map is nowhere singular. *J. London Math. Soc.* **43** (1968), 679–680. MR **37** 6949. *299.*

Reeb, G.

[1] Sur certaines propriétés topologiques des variétés feuilletées. *Actualités Sci. Ind. No. 1183* = *Publ. Inst. Math. Univ. Strasbourg* **11**, pp. 91–154, 157–158. Hermann, Paris, 1952. MR **14** 1113 = S 849-2. *299.*

Reed, J.

[1] Killing cohomology classes by surgery. *Proc. Adv. Study Inst. Algebraic Topol. Aarhus* Vol. II, pp. 446–454. Matematisk Institut, Aarhus Univ., Aarhus, 1970. MR **42** 6846. *298.*

Rees, E.

[1] Some rank two bundles on $P_n C$ whose Chern classes vanish. "Variétés Analytiques Compactes" (*Colloq., Nice*, 1977), Lecture Notes in Math., Vol. 683, pp. 25–28. Springer-Verlag, Berlin, and New York, 1978. MR **80f** 55016. *186.*

Reidemeister, K.

[1] Homotopieringe und Linsenräume. *Abh. Math. Sem. Univ. Hamburg* **11** (1935), 102–109. *55.*

Rohlin, V. A.

[1] A three-dimensional manifold is the boundary of a four-dimensional one. (Russian.) *Dokl. Akad. Nauk SSSR (N.S.)* **81** (1951), 355–357. MR **14** 72 = S 783-2. *296, 300.*
[2] New results in the theory of four-dimensional manifolds. (Russian.) *Dokl. Akad. Nauk SSSR (N.S.)* **84** (1952), 221–224. MR **14** 573 = S 783-3. *296.*
[3] Intrinsic definition of Pontryagin's characteristic cycles. (Russian.) *Dokl. Akad. Nauk SSSR (N.S.)* **84** (1952), 449–452. MR **14** 306 = S 723-6. *288.*
[4] Intrinsic homologies (Russian). *Dokl. Akad. Nauk SSSR (N.S.)* **89** (1953), 789–792. MR **15** 53 = S 783-4. *296.*
[5] Theory of intrinsic homologies (Russian). *Uspehi Mat. Nauk* **14** (1959), 3–20. MR **22** 11402 = S 668-3. [*English transl.*: Intrinsic homology theory. *Amer. Math. Soc. Transl.* (*2*) **30** (1963), 255–271. MR **27** 5252 = S 668-4.] *297.*
[6] The embedding of non-orientable three-manifolds into five-dimensional Euclidean space. (Russian.) *Dokl. Akad. Nauk SSSR* **160** (1965), 549–551. [*English transl.*: *Sov. Math. Dokl.* **6** (1965), 153–156.] MR **32** 1719 = S 770-3. *295.*

Rohlin, V. A., and Fuks, D. B.

[1] "A Beginning Course in Topology: Geometrical Chapters." Nauka, Moscow, 1977. (Russian.) MR **58** 31080. *49, 50, 97.*

Royden, H. L.

[1] "Real Analysis." Macmillan, New York, 1963. MR **27** 1540. [2nd ed., 1968.] *30.*

Royster, D. C.

[1] Aspherical generators of unoriented cobordism. *Proc. Amer. Math. Soc.* **66** (1977), 131–137. MR **56** 6680. *296.*

Samelson, H.

[1] On Poincaré duality. *J. Analyse Math.* **14** (1965), 323–336. MR **31** 5203 = S 203-4. *287, 288.*

Sanchez Giralda, T.

[1] A way of introducing the tangent space at a point of a differentiable manifold. (Spanish.) *Rev. Mat. Hisp.-Amer.* (*4*) **33** (1973), 54–58. MR **50** 5835. *180.*

Sanderson, B. J., and Schwarzenberger, R. L. E.

[1] Non-immersion theorems for differentiable manifolds. *Proc. Cambridge Philos. Soc.* **59** (1963), 319–322. MR **26** 5589 = S 758-1. *296.*

Schochet, C.

[1] "Cobordism from an Algebraic Point of View," Lecture Notes Series No. 29. Matematisk Institut, Aarhus Univ., Aarhus, 1971. MR **45** 6020. *297.*

Schön, R.

[1] Fibrations over a CWh-base. *Proc. Amer. Math. Soc.* **62** (1976), 165–166. MR **55** 4165. *212.*

Schubert, H. C. H.

[1] "Kalkül der abzählenden Geometrie." Teubner, Leipzig, 1879. [Reprinted by Springer-Verlag, Berlin and New York, 1979]. *239.*

[2] Lösung des Charakteristiken-Problems für lineare Räume beliebiger Dimension. [Written in 1886 and presented in 1889.] Mitteil. Math. Ges. Hamburg **1** (1889), 134–155. *239, 240.*

Schultz, R.

[1] Smooth structures on $S^p \times S^q$. *Ann. of Math.* (2) **90** (1969), 187–198. MR **40** 3560. *54.*

[2] On the inertia group of a product of spheres. *Trans. Amer. Math. Soc.* **156** (1971), 137–153. MR **43** 1209. *54.*

[3] Some recent results on topological manifolds. *Amer. Math. Monthly* **78** (1971), 941–952. MR **45** 6015. *53.*

[4] A generalization of Thom classes and characteristic classes to nonspherical fibrations. *Canad. J. Math.* **26** (1974), 138–144. MR **49** 3934. *212.*

Schwartz, M.-H.

[1] Fibre Spaces (Spanish). *Rev. Mat. Elem.* **5** (1961), supl. MR **30** 583 = S 24-7. *97.*

Schwarzenberger, R. L. E.

[1] Vector bundles on algebraic surfaces. *Proc. London Math. Soc.* (3) **11** (1961), 601–622. MR **25** 1160. *186.*

[2] Vector bundles on the projective plane. *Proc. London Math. Soc.* (3) **11** (1961), 623–640. MR **25** 1161. *186.*

[3] Extendible vector bundles over real projective space. *Quart. J. Math. Oxford Ser.* (2) **17** (1966). 19–21. MR **33** 3310 = S 300-3. *186.*

[4] "Topics in Differential Topology," Publications of the Ramanujan Institute, No. 3. Ramanujan Inst., Univ. of Madras, Madras, 1972. MR **53** 9236. *186.*

Segal, G.

[1] Classifying spaces and spectral sequences. *Inst. Hautes Etudes Sci. Publ. Math.* No. 34 (1968), 105–112. MR **38** 718. *102, 103.*

Segal, G. B., and Stretch, C. T.

[1] Characteristic classes for permutation representations. *Math. Proc. Cambridge Philos. Soc.* **90** (1981), 265–278. *248.*

Segre, B.

[1] Nuovi metodi e resultati nella geometria sulle varietà algebriche. *Ann. Mat. Pura Appl.* (4) **35** (1953), 1–127. MR **15** 822. *289.*

Segre, C.

[1] Mehrdimensionale Räume. Article III.C.7 (1921), *Encyklopädie der Math. Wiss.* **3**, 769–972. B.G. Teubner, Leipzig, 1921-1928. *56.*

Seifert, H., and Threlfall, W.

[1] "Lehrbuch der Topologie." Teubner, Leipzig, 1934. [*English transl.*: "A Textbook of Topology." Academic Press, New York, 1980. MR **82b** 55001.] *51, 52.*

Serre, J.-P.

[1] Homologie singulière des espaces fibrés. I. La suite spectrale. *C.R. Acad. Sci. Paris* **231** (1950), 1408–1410. MR **12** 520 = S 277-2. *99, 104.*
[2] Homologie singulière des espaces fibrés. Applications. *Ann. of Math.* (2) **54** (1951), 425–505. MR **13** 574 = S 359-1. *104.*
[3] Géométrie algébrique et géométrie analytique. *Ann. Inst. Fourier* **6** (1956), 1–42. MR **18** 511. *186.*
[4] Modules projectifs et espaces fibrés à fibre vectorielle. *Séminaire P. Dubreil, 1957/1958* Fasc. **2**, Exposé 23. Secrétariat mathématique, Paris, 1958. S 298-3. *179.*

Severi, F.

[1] I fondamenti della geometria numerativa. Annali di Mat. Pura ed Applicada (4) **19** (1940), 153–242. *240.*
[2] "Grundlagen der abzählende Geometrie." Wolfenbütteler Verlaganstalt, Wolfenbüttel, 1948. *240.*

Shiraiwa, K.

[1] A note on tangent bundles. *Nagoya Math. J.* **29** (1967), 259–267. MR **35** 1042. *187.*
[2] A note on tangential equivalences. *Nagoya Math. J.* **33** (1968), 53–56. MR **38** 1694. *187.*

Siegel, J.

[1] On a space between BH and B_∞. *Pacific J Math.* **60** (1975), 235–246. MR **53** 6556. *103.*

Sigrist, F., and Suter, U.

[1] On immersions $CP^n \looparrowright \mathbb{R}^{4n-2\alpha(n)}$. "Algebraic Topology" (*Proc. Conf., Vancouver*, 1977), Lecture Notes in Mathematics, Vol. 673, pp. 106–115. Springer-Verlag, Berlin and New York, 1978. MR **80g** 57042. *296.*

Sikorski, R.

[1] Differential modules. *Colloq. Math.* **24** (1971/1972), 45–79. MR **58** 2845. *180.*

Singhof, W.

[1] On the Lusternik–Schnirelmann category of Lie groups. *Math. Z.* **145** (1975), 111–116. MR **52** 11897. *48.*
[2] On the Lusternik–Schnirelmann category of Lie groups. II. *Math. Z.* **151** (1976), 143–148. MR **54** 13902. *48.*

Sjerve, D.

[1] Geometric dimension of vector bundles over lens spaces. *Trans. Amer. Math. Soc.* **134** (1968), 545–557. MR **38** 1695. *189.*

Smale, S.

[1] The generalized Poincaré conjecture in higher dimensions. *Bull. Amer. Math. Soc.* **66** (1960), 373–375. MR **23** A2220 = S 830-2. *299.*
[2] Generalized Poincaré's conjecture in dimensions greater than four. *Ann. of Math.* (2) **74** (1961), 391–406. MR **25** 580 = S 830-5. *299.*

Spanier, E.

[1] Homology theory of fiber bundles. *Proc. Internat. Congr. Math., Cambridge, Massachusetts, 1950* Vol. 2, pp. 390–396. American Mathematical Society, Providence, Rhode Island, 1952. MR **13** 486 = S 278-1. *104.*
[2] A formula of Atiyah and Hirzebruch. *Math. Z.* **80** (1962), 154–162. MR **26** 779 = S 719-1. *302.*
[3] Duality in topological manifolds. *Colloq. Topol. Bruxelles, 1964* pp. 91–111. Librairie Univ. Louvain, 1966. MR **36** 3363. *288.*
[4] "Algebraic Topology." McGraw-Hill, New York, 1966. MR **35** 1007. *4, 42, 49, 99, 104, 213, 243, 244, 256, 257, 258, 283, 287, 288.*

Staples, E. B.

[1] A short and elementary proof that a product of spheres is parallelizable if one of them is odd. *Proc. Amer. Math. Soc.* **18** (1967), 570–571. MR **36** 2165. *184.*

Stasheff, J. D.

[1] A classification theorem for fibre spaces. *Topology* **2** (1963), 239–246. MR **27** 4235 = S 307-2. *103, 212.*
[2] *H*-spaces and classifying spaces: foundations and recent developments. "Algebraic Topology" [*Proc. Symp. Pure Math.* **22** (1970)], pp. 247–272. American Mathematical Society, Providence, Rhode Island, 1971. MR **47** 9612. *103.*
[3] Parallel transport and classification of fibrations. *Conf. Topolog. Methods Algebraic Topol., 1973* Lecture Notes in Mathematics, Vol. 428, pp. 1–17. Springer-Verlag, Berlin, and New York, 1974. MR **51** 11501. *103.*

Steenrod, N. E.

[1] Topological methods for the construction of tensor functions. *Ann. of Math.* (2) **43** (1942), 116–131. MR **3** 144 = S 689-2. *179.*
[2] The classification of sphere bundles. *Ann. of Math.* (2) **45** (1944), 294–311. MR **5** 214 = S 301-6. *97, 179.*
[3] Products of cocycles and extensions of mappings. *Ann. of Math.* (2) **48** (1947), 290–320. MR **9** 154 = S 451-9. *243.*
[4] "The Topology of Fibre Bundles." Princeton Univ. Press, Princeton, New Jersey, 1951. MR **12** 522 = S 7-7. *97, 99, 101, 102, 238.*
[5] Reduced powers of cohomology classes. *Ann. of Math.* (2) **56** (1952), 47–67. MR **13** 966 = S 452-3. *243.*

[6] A convenient category of topological spaces. *Michigan Math. J.* **14** (1967), 133–152. MR **35** 970. *48.*

[7] "Reviews of Papers in Algebraic and Differential Topology, Topological Groups, and Homological Algebra." American Mathematical Society, Providence, Rhode Island, 1968. MR **38** 3853. *303.*

[8] Milgram's classifying space of a topological group. *Topology* **7** (1968), 349–368. MR **38** 1675. *102.*

[9] Cohomology operations, and obstructions to extending continuous functions. *Adv. in Math.* **8** (1972), 371–416. MR **45** 7705. *243.*

Steenrod, N. E., and Epstein, D. B. A.

[1] "Cohomology Operations," Annals of Math. Studies No. 50. Princeton Univ. Press, Princeton, New Jersey, 1962. MR **26** 3056 = S 338-1. *243.*

Steenrod, N. E., and Whitehead, J. H. C.

[1] Vector fields on the n-sphere. *Proc. Nat. Acad. Sci. U.S.A.* **37** (1951), 58–63. MR **12** 847 = S 736-3. *247.*

Stern, R. J.

[1] Stability implies normal and disc bundles. *Bull. Amer. Math. Soc.* **79** (1973), 133–135. MR **46** 9978. *239, 242.*

[2] On topological and piecewise linear vector fields. *Topology* **14** (1975), 257–269. MR **52** 15460. *239, 242.*

Stiefel, E.

[1] Richtungsfelder und Fernparallelismus in Mannigfaltigkeiten. *Comment. Math. Helv.* **8** (1935/1936), 305–353. *179, 184, 238, 239, 288, 289.*

[2] Über Richtungsfelder in den projektiven Räumen und einen Satz aus der reellen Algebra. *Comment. Math. Helv.* **13** (1941), 201–218. MR **3** 61 = S 735-4. *247.*

Stone, A. H.

[1] Paracompactness and product spaces. *Bull. Amer. Math. Soc.* **54** (1948), 977–982. MR **10** 204. *39, 50, 114.*

Stong, R. E.

[1] Relations among Stiefel–Whitney classes. *Proc. Amer. Math. Soc.* **15** (1964), 151–153. MR **28** 2542 = S 720-4. *246, 301.*

[2] "Notes on Cobordism Theory." Princeton Univ. Press, Princeton, New Jersey and Univ. of Tokyo Press, Tokyo, 1968. MR **40** 2108. *296.*

[3] A cobordism. *Proc. Amer. Math. Soc.* **35** (1972), 584–586. MR **46** 4554. *299.*

[4] On fibering of cobordism classes. *Trans. Amer. Math. Soc.* **178** (1973), 431–447. MR **47** 4282. *297.*

[5] Subbundles of the tangent bundle. *Trans. Amer. Math. Soc.* **200** (1974), 185–197. MR **50** 8573. *297.*

[6] Induced cobordism theories—an example. *Michigan Math. J.* **23** (1976), 303–307. MR **56** 6683. *296.*

[7] Stiefel–Whitney classes of manifolds. *Pacific J. Math.* **68** (1977), 271–276. MR **56** 6672. *290.*

Stong, R. E., and Winkelnkemper, H. E.

[1] Locally free actions and Stiefel–Whitney numbers. II. *Proc. Amer. Math. Soc.* **66** (1977), 367–371. MR **56** 13235. *297.*

Sullivan, D.

[1] Combinatorial invariants of analytic spaces. *Proc. Liverpool Singularities Symp. I (1969/1970)*, Lecture Notes in Mathematics, Vol. 192, pp. 165–168. Springer-Verlag, Berlin, and New York, 1971. MR **43** 4063. *290.*

[2] Inside and outside manifolds. *Proc. Internat. Congr. Math. Vancouver, 1974* Vol. 1, pp. 201–207. Canadian Mathematical Congress, Montreal, 1975. MR **54** 13915. *53.*

Sutherland, W. A.

[1] Fibre homotopy equivalence and vector fields. *Proc. London Math. Soc.* (*3*) **15** (1965), 543–556. MR **31** 763 = S 734-5. *189.*

[2] Vanishing of Stiefel–Whitney classes. *Proc. Amer. Math. Soc.* **31** (1972), 637. MR **45** 1168. *246.*

Švarc, A. S.

[1] The genus of a fiber space.(Russian.) *Dokl. Akad. Nauk SSSR N.S.* **119** (1958), 219–222. MR **21** 1598 = S 268-4. *104.*

[2] The genus of a fibered space. (Russian.) *Trudy Moskov. Mat. Obšč.* **10** (1961), 217–272. MR **27** 4233 = S 269-4. *104.*

[3] The genus of a fibre space.(Russian.) *Trudy Moskov. Mat. Obšč.* **11** (1962), 99–126. MR **27** 1963 = S 270-1. *104.*

Swan, R. G.

[1] Vector bundles and projective modules. *Trans. Amer. Math. Soc.* **105** (1962), 264–277. MR **26** 785 = S 298-4. *179.*

Switzer, R. M.

[1] "Algebraic Topology—Homotopy and Homology." Springer-Verlag, Berlin and New York, 1975. MR **52** 6695. *21, 49, 97.*

Szigeti, F.

[1] On covering homotopy theorems. *Ann. Univ. Sci. Budapest Eötvös Sect. Math.* **13** (1970), 89–91. MR **47** 1048. *99.*

Ta, M.

[1] On the homology groups of real projective spaces. (Romanian; French summary.) *Stud. Cerc. Mat.* **28** (1976), 485–488. MR **55** 4137. *213.*

Takens, F.

[1] The minimal number of critical points of a function on a compact manifold and the Lusternik–Schnirelman category. *Invent. Math.* **6** (1968), 197–244. MR **38** 5235. *48, 300.*
[2] The Lusternik–Schnirelman categories of a product space. *Compositio Math.* **22** (1970), 175–180. MR **44** 7549. *48.*

Takizawa, S.

[1] On characteristic classes of sphere bundles. (Japanese.) *Sûgaku* **8** (1956/1957), 229–245. MR **22** 5045 = S 25-6. *239.*

Tamura, I.

[1] Classification des variétés différentiables, $(n-1)$-connexes, sans torsion, de dimension $2n+1$. *Sém Henri Cartan, 1962/1963* Exp. 16-19. Secrétariat mathématique, Paris, 1964. MR **28** 3453 = S 839-3. *53.*
[2] Classification of $(n-1)$-connected differentiable manifolds of dimension $2n+1$ (Japanese). *Sûgaku* **16** (1964/1965), 69–80. MR **37** 924. *53.*
[3] On the classification of sufficiently connected manifolds. *J. Math. Soc. Japan* **20** (1968), 371–389. MR **37** 5892. *53.*

Tango, H.

[1] An example of an indecomposable vector bundle of rank $n-1$ on $\mathbb{P}^n$. *J. Math. Kyoto Univ.* **16** (1976), 137–141. MR **53** 5593. *188.*

Taniguchi, H.

[1] Some examples of differentiable sphere bundles. *J Fac. Sci. Univ. Tokyo Sect. I* **16** (1969), 477–495. MR **41** 9281. *179.*

Taylor, L. E.

[1] The tangent space to a C^k manifold. *Bull. Amer. Math. Soc.* **79** (1973), 746. MR **47** 5894. *180.*

Taylor, L. R.

[1] Stiefel–Whitney homology classes. *Quart. J. Math. Oxford Ser.* (*2*) **28** (1977), 381–387. MR **58** 24286. *290.*

Teleman, N.

[1] Fiber bundle with involution and characteristic classes. *Atti Accad. Naz. Lincei Rend. Cl. Sci. Fis. Mat. Natur.* (*8*) **54** (1973), 49–56. MR **50** 3223. *242, 289.*

[2] A variant construction of Stiefel–Whitney classes of a topological manifold. *Atti Accad. Naz. Lincei Rend. Cl. Sci. Fis. Mat. Natur.* (*8*) **54** (1973), 426–433. MR **50** 11259. *242, 289.*

[3] Characteristic classes of fiber bundle with involution. *Ann. Mat. Pura Appl.* (*4*) **101** (1974), 65–90. MR **50** 8547. *242, 289.*

Thom, R.

[1] Classes caractéristiques et *i*-carrés. *C. R. Acad. Sci. Paris* **230** (1950), 427–429. MR **12** 42 = S 714-1. *104, 211, 212, 239, 243, 289.*

[2] Variétés plongées et *i*-carrés. *C. R. Acad. Sci. Paris* **230** (1950), 507–508. MR **12** 42 = S 714-2. *286, 289.*

[3] Quelques propriétés des variétés-bords. *Colloq. Topol. Strasbourg, 1951* No. V. La Bibliothèque Nationale et Univ. de Strasbourg, 1952. MR **14** 492 = S 784-1. *266, 296.*

[4] Espaces fibrés en sphères et carrés de Steenrod. *Ann. Sci. Ecole Norm. Sup.* (*3*) **69** (1952), 109–182. MR **14** 1004 = S 714-6. *103, 187, 211, 212, 239, 241, 242, 243, 289, 300.*

[5] Une théorie intrinsèque des puissances de Steenrod. *Colloq. Topol. Strasbourg, 1951* No. VII. La Bibliothèque Nationale et Univ. de Strasbourg, 1952. MR **14** 491 = S 453-1. *243.*

[6] Variétés différentiables cobordantes. *C. R. Acad. Sci. Paris* **236** (1953), 1733–1735. MR **14** 1112 = S 784-3. *266, 296.*

[7] Variétés différentiables cobordantes. Géométrie différentielle. *Colloq. Internat. C.N.R.S., Strasbourg, 1953* pp. 143–149. Centre National de la Recherche Scientifique, Paris, 1953. MR **15** 643 = S 785-1. *266, 296.*

[8] Quelques propriétés globales des variétés différentiables. *Comment. Math. Helv.* **28** (1954), 17–86. MR **15** 890 = S 785-2. *266, 296.*

Thomas, E.

[1] On tensor products of *n*-plane bundles. *Arch. Math.* **10** (1959), 174–179. MR **21** 5959 = S 315-2. *241.*

[2] Characteristic classes and differentiable manifolds. "Classi Caratteristiche e Questioni Connesse," pp. 113–187. Centro Internazionale Matematico Estivo, 1966. Edizioni Cremonese, Rome, 1967. MR **36** 5959. *238.*

Thomas, T. Y.

[1] Imbedding theormes in differential geometry. *Bull. Amer. Math. Soc.* **45** (1939), 841–850. MR **1** 88 = S 762-3. *50.*

[2] Lecture notes on Whitney's theory of the imbedding of differentiable manifolds in Euclidean Space. *Revista Ci. Lima* **46** (1944), 29–60. MR **5** 274 = S 762-4. *50.*

Threlfall, W.

[1] Stationäre Punkte auf geschlossenen Mannigfaltigkeiten. *Jber. Deutsch. Math. Verein.* **51** (1941), 14–33. MR **3** 62 = S 805-3. *300.*

Tietze, H.

[1] Über die topologischen Invarianten mehrdimensionaler Mannigfaltigkeiten. *Mh. Math. Phys.* **19** (1908), 1–118. *55.*

Todd, J. A.

[1] On Giambelli's formulae for incidences of linear spaces. *J. London Math. Soc.* **6** (1931), 209–216. *239.*

tom Dieck, T.

[1] Klassifikation numerierbarer Bündel. *Arch. Math.* (*Basel*) **17** (1966), 395–399. MR **34** 6776 = S 309-2. *102.*
[2] Partitions of unity in homotopy theory. *Compositio Math.* **23** (1971), 159–167. MR **45** 2702. *103.*
[3] On the homotopy type of classifying spaces. *Manuscripta Math.* **11** (1974), 41–49. MR **50** 3222. *103.*

van Kampen, E. R.

[1] Komplexe in euklidische Räumen. *Abh. Math. Sem. Hamburg* **9** (1932), 72–78, 152–153. *51.*

Varadarajan, K.

[1] On fibrations and category. *Math. Z.* **88** (1965), 267–273. MR **31** 5199 = S 263-3. *104.*

Vastersavendts, M.-L.

[1] Sur les groupes d'homologie de certains espaces fibrés. *Acad. Roy. Belg. Bull. Cl. Sci.* (*5*) **50** (1964), 624–636. MR **30** 4265 = S 287-1. *104.*

Vazquez, R.

[1] Generalized characteristic classes and Steenrod squares in the Gysin sequence of a space fibred by spheres. (Spanish.) *Rev. Un. Mat. Argentina* **19** (1960), 207–216. MR **24** A1131 = S 295-1. *242.*
[2] On the cohomology ring modulo 2 of a spherically fibered space. (Spanish.) *Bol. Soc. Mat. Mexicana* (*2*) **6** (1961), 1–4. MR **24** A2391 = S 295-2. *242.*

Vick, J. W.

[1] "Homology Theory." Academic Press, New York, 1973. MR **51** 11475. *4.*

Vietoris, L.

[1] Über die Homologiegruppen der Vereinigung zweier Komplexe. *Monatsch. Math. Phys.* **37** (1930), 159–162. *56.*

Vogt, R. M.

[1] Convenient categories of topological spaces for algebraic topology. *Proc. Adv. Study Inst. Algebraic Topol. Aarhus, 1970* Vol. III, pp. 649–656. Matematisk Institut, Aarhus Univ., Aarhus, 1970. MR **49** 11475. *48.*
[2] Convenient categories of topological spaces for homotopy theory. *Arch. Math.* (*Basel*) **22** (1971), 545–555. MR **45** 9323. *48.*

Vrănceanu, G.

[1] Sopra le funzioni di Morse degli spazi lenticolari. (French summary.) *Atti Accad. Naz. Lincei Rend. Cl. Sci. Fis. Mat. Natur.* (*8*) **51** (1971), 497–499. MR **48** 9743. *299.*
[2] Fonctions de Morse sur les espaces exotiques. (English summary.) *Ann. Mat. Pura Appl.* (*4*) **98** (1974), 33–46. MR **49** 9856. *299.*

Wall, C. T. C.

[1] All 3-manifolds imbed in 5-space. *Bull. Amer. Math. Soc.* **71** (1965), 564–567. MR **30** 5324 = S 770-2. *294.*
[2] Classification problems in differential topology. VI. Classification of $(s-1)$-connected $(2s+1)$-manifolds. *Topology* **6** (1967), 273–296. MR **35** 7343. *53.*
[3] "Surgery on Compact Manifolds." Academic Press, New York, 1970. MR **55** 4217. *298.*

Wallace, A. H.

[1] Modifications and cobounding manifolds. *Canad. J. Math.* **12** (1960), 503–528. MR **23** A2887 = S 822-1. *299.*
[2] Modifications and cobounding manifolds. II. *J. Math. Mech.* **10** (1961), 773–809. MR **24** A555 = S 822-2. *299.*
[3] Modifications and cobounding manifolds. III. *J. Math. Mech.* **11** (1962), 979–990. MR **26** 3068 = S 822-3. *299.*
[4] Modifications and cobounding manifolds. IV. *J. Math. Mech.* **12** (1963), 445–484. MR **27** 4246 = S 822-4. *299.*
[5] "Differential Topology: First Steps." Benjamin, New York, 1968. MR **36** 7150. [2nd printing, 1977, MR **55** 9098.] *50.*
[6] "Algebraic Topology: Homology and Cohomology." Benjamin, New York, 1970. MR **43** 4023. *4.*

Wang, H.-C.

[1] The homology groups of the fibre bundles over a sphere. *Duke Math. J.* **16** (1949), 33–38. MR **10** 468 = S 293-3. *104.*

Watabe, T.

[1] On imbedding closed 4-dimensional manifolds in Euclidean space. *Sci. Rep. Niigata Univ. Ser. A* No. 3 (1966), 9–13. MR **34** 8422 = S 768-6. *294.*
[2] $SO(r)$-cobordism and embedding of 4-manifolds. *Osaka J. Math.* **4** (1967), 133–140. MR **36** 2163. *294.*

Whitehead, G. W.

[1] "Elements of Homotopy Theory." Springer-Verlag, Berlin and New York, 1978. MR **80b** 55001. *48, 49, 97, 238.*

Whitehead, J. H. C.

[1] On C^1-complexes. *Ann. of Math.* (*2*) **41** (1940), 809–824. MR **2** 73 = S 566-6. *52.*
[2] On incidence matrices, nuclei and homotopy types. *Ann. of Math.* (2) **42** (1941), 1197–1239. MR **3** 142. *55.*

[3] Combinatorial homotopy. I. *Bull. Amer. Math. Soc.* **55** (1949), 213–245. MR **11** 48 = S 489-5. *17, 49.*

[4] A certain exact sequence. *Ann. of Math.* (*2*) **52** (1950), 51–110. MR **12** 43 = S 358-1. *21, 49.*

[5] Manifolds with transverse fields in euclidean space. *Ann. of Math.* (*2*) **73** (1961), 154–212. MR **23** A2225 = S 573-1. *52.*

Whitney, H.

[1] Differentiable manifolds in Euclidean space. *Proc. Nat. Acad. Sci. U.S.A.* **21** (1935), 462–464. *50, 51.*

[2] Sphere spaces. *Proc. Nat. Acad. Sci. U.S.A.* **21** (1935), 464–468. *97, 179, 238, 239, 288.*

[3] Differentiable manifolds. *Ann. of Math.* **37** (1936), 645–680. *50, 51, 288.*

[4] Topological properties of differentiable manifolds. *Bull. Amer. Math. Soc.* **43** (1937), 785–805. *97, 179, 288.*

[5] On the theory of sphere-bundles. *Proc. Nat. Acad. Sci. U.S.A.* **26** (1940), 148–153. MR **1** 220 = S 296-1. *97, 179, 289.*

[6] On the topology of differentiable manifolds. "Lectures in Topology," pp. 101–141. Univ. of Michigan Press, Ann Arbor, Michigan, 1941. MR **3** 133 = S 712-2. *97, 179, 212, 241, 288, 290.*

[7] The self-intersections of a smooth n-manifold in $2n$-space. *Ann. of Math.* (*2*) **45** (1944), 220–246. MR **5** 273 = S 753-3. *28, 50.*

[8] The singularities of a smooth n-manifold in $(2n - 1)$-space. *Ann. of Math.* (*2*) **45** (1944), 247–293. MR **5** 274 = S 753-4. *51.*

[9] Complexes of manifolds. *Proc. Nat. Acad. Sci. U.S.A.* **33** (1947), 10–11. MR **8** 398 = S 754-1. *50.*

[10] "Geometric Integration Theory." Princeton Univ. Press, Princeton, New Jersey, 1957. MR **19** 309 = S 659-1. *52.*

Wilkens, D. L.

[1] Closed $(s - 1)$-connected $(2s + 1)$-manifolds, $s = 3, 7$. *Bull. London Math. Soc.* **4** (1972), 27–31. MR **46** 6378. *53.*

Wilson, W. S.

[1] A new relation on the Stiefel–Whitney classes of spin manifolds. *Illinois J. Math.* **17** (1973), 115–127. MR **46** 9981. *247.*

Woodward, L. M.

[1] Vector fields on spheres and a generalization. *Quart. J. Math. Oxford Ser.* (2) **24** (1973), 357–366. MR **48** 5093. *248.*

Wu, W. -T.

[1] On the product of sphere bundles and the duality theorem modulo two. *Ann. of Math.* (*2*) **49** (1948), 641–653. MR **10** 203 = S 309-4. *241.*

[2] Les classes caractéristiques d'un espace fibré. "Espaces fibrés et homotopie: Séminaire Henri Cartan, 1949/1950," Exposés 17 and 18, 2nd ed. Secrétariat mathématique, Paris, 1956. MR **18** 409 = S 28-4. *239, 289.*

[3] Classes caractéristiques et i-carrés d'une variété. *C. R. Acad. Sci. Paris* **230** (1950), 508–511. MR **12** 42 = S 714-3. *239, 245, 286, 289.*

[4] Les i-carrés dans une variété grassmannienne. *C. R. Acad. Sci. Paris* **230** (1950), 918–920. MR **12** 42 = S 309-5. *245.*

[5] Sur les classes caractéristiques des structures fibrées sphériques. *Actualités Sci. Ind. No. 1183 = Publ. Inst. Math. Univ. Strasbourg* **11**, pp. 5–89, 155–156. Hermann, Paris, 1952. MR **14** 1112 = S 311-2. *179, 239, 289.*

[6] Sur les puissances de Steenrod. *Colloq. Topol. Strasbourg, 1951* No. 9. La Bibliothèque Nationale, Univ. Strasbourg (1952). MR **14** 491 = S 453-2. *243.*

[7] On squares in Grassmannian manifolds. *Acta Sci. Sinica* **2** (1953), 91–115. MR **16** 611 = S 310-1. *245.*

[8] On the isotopy of C^r-manifolds of dimension n in euclidean $(2n + 1)$-space. *Sci. Record (N.S.)* **2** (1958), 271–275. MR **21** 3027 = S 763-2. *291, 292.*

[9] On the imbedding of orientable manifolds in a euclidean space. *Sci. Sinica* **12** (1963), 25–33. MR **26** 5586 = S 766-3. *294.*

[10] "A Theory of Imbedding, Immersion, and Isotopy of Polytopes in a Euclidean Space." Science Press, Peking, 1965. MR **35** 6146.[Reprinted 1974.] *51.*

Yasui, T.

[1] The reduced symmetric product of a complex projective space and the embedding problem. *Hiroshima Math. J.* **1** (1971), 27–40. MR **47** 2624. *296.*

[2] Note on the enumeration of embeddings of real projective spaces. *Hiroshima Math. J.* **3** (1973), 409–418. MR **49** 3982. *296.*

[3] Note on the enumeration of embeddings of real projective spaces. II. *Hiroshima Math. J.* **6** (1976), 221–225. MR **55** 9121. *296.*

Yoshida, T.

[1] On the vector bundles $m\xi_n$ over real projective spaces. *J. Sci. Hiroshima Univ. Ser. A-I Math.* **32** (1968), 5–16. MR **38** 6622. *185.*

[2] Corrections to: "On the vector bundles $m\xi_n$ over real projective spaces." *Hiroshima Math. J.* **2** (1972), 231–232. MR **47** 9617. *185.*

[3] A note on vector fields up to bordism. *Hiroshima Math. J.* **8** (1978), 63–69. MR **57** 4196. *298.*

Zeuthen, H. G.

[1] "Lehrbuch der abzählende Methoden der Geometrie." Teubner, Leipzig, 1914. *240*

Glossary of Notation

Roman alphabet

$A \in GL(n, \mathbb{C})$	any element of the complex general linear group	87
$A^* \in GL(n, \mathbb{C})$	the adjoint of $A \in GL(n, \mathbb{C})$	87
$\lvert A \rvert \in GL(n, \mathbb{C})$	the modulus of $A \in GL(n, \mathbb{C})$	89
$\mathscr{A}\mathscr{b}^{\oplus}$	the category of $\mathbb{Z}$-graded abelian groups	45
BG	classifying space for a topological group G	101
B^I	barycenter of a simplex I	9
$\mathscr{B}$	category of base spaces	2, 5ff, 7
$\mathbb{C}$	complex field, in its usual topology	
$\mathbb{C}^n$	standard n-dimensional complex vector space, in its usual topology	34
CP^n	complex projective space $(= G^1(\mathbb{C}^{n+1}))$	23, 34
CP^∞	complex projective space $(= G^1(\mathbb{C}^\infty))$	23, 34
$C(V) \subset \mathbb{R}^{2n-p}$	cocoon of $V \subset X$	262
CW complex		20
CW space		19ff, 20
CW structure		20
CW structure of projective spaces		22, 23
$\mathbb{C}^\infty$	the complex vector space $\lim_n \mathbb{C}^n$ in the weak topology	34
$C^0(X)$	the ring of continuous functions $X \to \mathbb{R}$	116
$C^0(X)$-modules $\mathscr{F}$		117, 118, 119
$C^0(X)$-modules $\mathscr{F}^*$		119

$C^\infty(X)$	the ring of smooth functions $X \to \mathbb{R}$	119
$C^\infty(X)$-module $\mathscr{E}(X)$ of smooth differentials on X		134, 135, 183
$C^\infty(X)$-module $\mathscr{E}^*(X)$ of smooth vector fields on X		127, 129, 130, 131
$C^\infty(X)$-module $\mathscr{F}$ of smooth sections $X \to E$		122
$C^\infty(X)$-module $\mathscr{F}^*$		123
$C^\infty(X)$-module $\mathscr{F}^{**}$		133
$\mathscr{C}(G, F)$	category of families of fibers, with respect to a group action $G \times F \to F$	61
D_P	Poincaré duality map $\cap\, \mu_X$	250, 258, 276
D_T	inverse Poincaré duality map D_P^{-1}	277, 283
$D^n \subset \mathbb{R}^n$	the closed unit n-disk	19
$dg \in \mathscr{E}(X)$	the differential of $g \in C^\infty(X)$	134, 183, 184
E	the total space of a family $E \xrightarrow{\pi} X$ of fibers (especially a coordinate bundle $E \xrightarrow{\pi} X$)	57, 61, 67
EG	the total space of a universal G-bundle γ_G	102
E_x	the fiber $\pi^{-1}(\{x\})$ over $x \in X$ of a fiber bundle represented by $E \xrightarrow{\pi} X$	67
$e^A \in GL(n, \mathbb{C})$	the exponential of $A \in \mathrm{End}\, \mathbb{C}^n$	87
$\mathscr{E}(X)$	the $C^\infty(X)$-module of smooth differentials on X	134, 135, 183
$\mathscr{E}^*(X)$	the $C^\infty(X)$-module of smooth vector fields on X	127, 129, 130, 131
$e(\xi) \in H^m(X; \mathbb{Z}/2)$	the $\mathbb{Z}/2$ Euler class of a real m-plane bundle ξ over $X \in \mathscr{B}$	196
F	the fiber of any family of fibers (especially a coordinate bundle), also applied to fibre bundles (and especially to fiber bundles)	3, 57, 60
$\mathscr{F}$	a $C^0(X)$-module of (continuous) sections $X \to E$	117, 118, 119
$\mathscr{F}$	a $C^\infty(X)$-module of smooth sections $X \to E$	122
$\mathscr{F}^*$	the first conjugate of $\mathscr{F}$	118, 119, 123
$\mathscr{F}^{**}$	the second conjugate of $\mathscr{F}$	133
f-genus	a homomorphism $\mathscr{U} \xrightarrow{G(f)} \mathbb{Z}/2$ or $\mathfrak{N} \xrightarrow{G(f)} \mathbb{Z}/2$ assigned to a formal power series $f(t) \in \mathbb{Z}/2[[t]]$	274ff, 275, 276
$f^!\xi$	pullback of ξ along $X \xrightarrow{f} X'$	3, 63
G	structure group	3, 58, 60
$G \times F \to F$	action of a transformation group G (structure group) on a fiber F	2, 60, 65

G-related isomorphisms		60
G-related set of homomorphisms		60
$G(f)$	Stiefel–Whitney f-genus	274ff, 275, 276
G/K	homogeneous space	83
$GL(n, \mathbb{C})$	complex general linear group	59, 87
$GL(m, \mathbb{R})$	real general linear group	1, 58
$GL^+(m, \mathbb{R})$	the component of the identity in $GL(m, \mathbb{R})$	59, 90
$G^m(\mathbb{C}^{n+m})$	complex Grassmann manifold	34
$G^m(\mathbb{C}^\infty)$	complex Grassmann manifold	34
$G^m(\mathbb{R}^{m+n})$	real Grassmann manifold	34
$G^m(\mathbb{R}^\infty)$	real Grassmann manifold	34
$\tilde{G}^m(\mathbb{R}^m)$	the total space $O(\gamma^m)$ of $o(\gamma^m)$	166
$G^1(\mathbb{C}^{n+1})$	complex projective space CP^n	34
$G^1(\mathbb{C}^\infty)$	complex projective space CP^∞	34
$G^1(\mathbb{R}^{n+1})$	real projective space RP^n	34
$G^1(\mathbb{R}^\infty)$	real projective space RP^∞	34
$g^!\xi$	pullback of ξ along $X \xrightarrow{g} X'$	63
$H^*(G^m(\mathbb{R}^{m+n}); \mathbb{Z}/2)$		227ff, 230
$H^*(G^m(\mathbb{R}^\infty); \mathbb{Z}/2)$		227ff, 229
$H^*(\mathbb{R}P^n; \mathbb{Z}/2)$		199ff, 201
$H^*(\mathbb{R}P^\infty; \mathbb{Z}/2)$		199ff, 200
I	a two-sided ideal in $\bigotimes V$ with $\bigotimes V/I = \bigwedge V$	35
$I_1, \ldots, I_q$ and I	simplexes in a simplicial complex K, which also constitute the vertices of the barycentric subdivision K'	9
$i(m, n)$	Atiyah–Todd number	189, 296
J	a complex structure in a real vector bundle	173
j^*T_X	a Thom form $(\ ,\ )_T$	277, 281, 282, 283
K	an abstract simplicial complex	8
K'	the first barycentric subdivision of K	9
K_0	the vertex set of K	8
$\tilde{K}$	a family of functions $K \to [0, 1]$	9
$\tilde{K}_0$	a family of functions $K_0 \to [0, 1]$	8
$\|K\| \subset \tilde{K}_0$	the simplicial space of K	8
$\|K'\| \subset \tilde{K}$	the simplicial space of K'	9
$\|K\|_m (= \|K\|)$	metric simplicial space of K	17, 18
$\|K\|_w (= \|K\|)$	weak simplicial space of K	17, 18
$\|K\|^*$	the metric telescope of $\|K\|$	12
K-theory		170
$\mathscr{K}$	Steenrod's convenient category of compactly generated spaces	48
$\lim_n X_n$ (in the weak topology)		19, 20
$\ln A \in GL(n, \mathbb{C})$	the logarithm of a positive element $A \in GL(n, \mathbb{C})$	87
l_x	left inclusion	278, 281

m-plane bundle	any real or complex vector bundle of rank m	1, 106, 170
$\mathscr{M}$	the category of smooth manifolds and smooth maps	2, 23ff, 25
$\mathfrak{M}$	a category of modules over a fixed commutative ring	41
$\mathfrak{M}_R^{\oplus}$	the category of $\mathbb{Z}$-graded modules over the commutative ring R	45
$\mathbb{N}$	the natural numbers $\{0, 1, 2, \ldots\}$	
n-cell (for $n = 0, 1, 2, \ldots$)		19
n-disk D^n (for $n = 1, 2, 3, \ldots$)		19
n-plane bundle	any real or complex vector bundle of rank n	106, 170
$(n-1)$-sphere S^{n-1} (for $n = 1, 2, 3, \ldots$)		19
nth type of topological space (for $n = 0, 1, 2, \ldots$)		6
n-skeleton (for $n = 0, 1, 2, \ldots$)		20
n-sphere S^n (for $n = 0, 1, 2, \ldots$)		19
$\mathfrak{N}$	the unoriented cobordism ring	250, 268, 273
$\mathfrak{N}'$	the unoriented surgical equivalence ring	268, 273
$O(m) \subset GL(m, \mathbb{R})$	the orthogonal group, usually acting on $\mathbb{R}^m$ or on S^{m-1}	3, 59, 81, 86, 90
$O^4(m) \subset GL^+(m, \mathbb{R})$	the rotation group, usually acting on $\mathbb{R}^m$ or on S^{m-1}	59, 81, 86, 90
$O(1) \subset GL(1, \mathbb{R})$	the orthogonal group (acting on S^0)	160
$O(\xi)$	the total space of the unique coordinate bundle representing an orientation bundle $o(\xi)$ [in particular, $O(\gamma_n^m)$ is denoted $\tilde{G}^m(\mathbb{R}^{m+n})$ and $O(\gamma^m)$ is denoted $\tilde{G}^m(\mathbb{R}^\infty)$]	162, 163
$o(\xi)$	the orientation bundle of a real vector bundle ξ over $X \in \mathscr{B}$	160, 161, 162, 163
$\mathcal{O}(V)$	the category of open sets in $V \subset X$ for $X \in \mathscr{B}$	43
$\mathcal{O}(X)$	the category of open sets in $X \in \mathscr{B}$	41
$(\, , \,)_{\mathrm{P}}$	Poincaré form	276, 277, 283, 284
P_ξ	the projective bundle of a real vector bundle ξ, or the total space of the projective bundle	3, 202
p-simplex (for $p = 0, 1, 2, \ldots$)		8
q-dimensional abstract simplicial complex K (for $q = 0, 1, 2, \ldots$)		9
q-dimensional simplicial space $\lvert K \rvert$ (for $q = 0, 1, 2, \ldots$)		9
$\mathscr{Q}_s(V) \subset \mathcal{O}(V)$	a specific family of open sets in $V \subset X$	43
$\mathscr{Q}(\mathring{X})$	a category of open sets in the interior $\mathring{X}$ of a manifold X	46, 252ff, 256ff
$\mathbb{R}$	real field, in its usual topology	

$\mathbb{R}^n$	standard n-dimensional real vector space, in its usual topology	34
RP^n	real projective space $(=G^1(\mathbb{R}^{n+1}))$	22, 23, 34
RP^∞	real projective space $(=G^1(\mathbb{R}^\infty))$	22, 23, 34
$\mathbb{R}^\infty$	the real vector space $\lim_n \mathbb{R}^n$ in the weak topology	22
r_x	right inclusion	280, 281
Sq	the Steenrod square	243ff, 285ff
S^{n-1}	the $(n-1)$-sphere (for $n = 1, 2, 3, \ldots$)	19
S^n	the n-sphere (for $n = 0, 1, 2, \ldots$)	19
S^0	the 0-sphere (consisting of two points)	19, 162, 167
$T_X \in H^m(X \times X, X \times X \backslash \Delta(X))$	the diagonal Thom class of $X \in \mathcal{M}$	281ff
$(,)_T$	Thom form $j^* T_X$	276ff, 277, 283, 284
$U(n) \subset GL(n, \mathbb{C})$	the unitary group (usually acting on $\mathbb{C}^n$ or on S^{2n-1})	59, 81, 86, 89
$U_\xi \in H^n(E, E^*; \mathbb{Z}/2)$	$\mathbb{Z}/2$ Thom class of a real n-plane bundle ξ	3, 191, 192ff, 194ff, 211
$u_f(\xi)$	multiplicative $\mathbb{Z}/2$ class of a real vector bundle ξ with respect to a formal power series $f(t) \in \mathbb{Z}/2[[t]]$	2, 3, 215ff, 216
$u_{1/f}(\xi)$	dual of a $\mathbb{Z}/2$ multiplicative class $u_f(\xi)$	237, 238
$\mathcal{U}$	the semi-ring of smooth closed unoriented manifolds	274
$\mathcal{U}$	the category of spaces homotopy equivalent to spaces in Steenrod's category $\mathcal{K}$	49
V	any of the real or complex vector spaces $\mathbb{R}^{m+n}$, $\mathbb{R}^\infty$, $\mathbb{C}^{m+n}$, $\mathbb{C}^\infty$ (used to generate the tensor algebra $\bigotimes V$ and the exterior algebra $\bigwedge V = \bigotimes V/I$	34
$V_\xi \in H^n(P_{\xi\oplus 1}, P_\xi; \mathbb{Z}/2)$	$\mathbb{Z}/2$ Thom class of a real n-plane bundle ξ	209ff, 211
$W_\xi \in H^n(P_{\xi\oplus 1}, P^*_{\xi\oplus 1}; \mathbb{Z}/2)$	$\mathbb{Z}/2$ Thom class of a real n-plane bundle ξ	211
$w_q(\xi)$	the qth Stiefel–Whitney class of a real vector bundle ξ	4, 216
$w(\xi)$	the total Stiefel–Whitney class of a real vector bundle ξ	2, 4, 215, 216
$\bar{w}(\xi)$	the dual Stiefel–Whitney class of a real vector bundle ξ	2, 237
$\mathcal{W}$	the category of spaces homotopy equivalent to metric simplicial spaces	2, 13, 21
$\mathcal{W}_0$	the category of spaces homotopy equivalent to countable metric simplicial spaces	14

$\dot{X}$	the boundary of a manifold X	24
$\mathring{X}$	the interior of a manifold X	25
$\dot{Y}$	the boundary of a manifold Y	24
$\mathring{Y}$	the interior of a manifold Y	25
$\mathbb{Z}$	the ring of integers	
$\mathbb{Z}/2$	the field of integers modulo 2	2

Greek alphabet

$\alpha(n)$	number of 1's in the dyadic expansion of n	2, 50, 52, 246, 261, 292
$\Gamma(X)$	real line bundle group over $X \in \mathscr{B}$	206ff, 207, 208, 213, 214
(Γ, Φ), (Γ', Φ'), $(\Gamma^{\oplus}, \Phi^{\oplus})$, $(\Gamma^{\oplus}, \Phi^{\oplus})$, ...	morphisms of transformation groups	70, 71, 72, 81, 109 110, 111, 115, 161, 162, 172, 173, 176, 177, 178
γ^m	universal complex m-plane bundle	170
γ^m	universal real m-plane bundle	3, 149
$\tilde{\gamma}^m$	universal oriented m-plane bundle $\pi(\gamma^m)^!\gamma^m$	164
γ_n^m	canonical complex m-plane bundle	170
γ_n^m	canonical real m-plane bundle	144
$\tilde{\gamma}_n^m$	canonical oriented m-plane bundle $\pi(\gamma_n^m)^!\gamma_n^m$	166
Δ	diagonal map $X \to X \times X$	110
$\Delta(X)$	image of a diagonal map Δ	277
ε^m, ε^n, ...	trivial real or complex vector bundles of ranks m, n, ...	112
ζ, ζ', ...	complex vector bundles (usually)	170
$\zeta_{\mathbb{R}}$	realification of a complex vector bundle ζ	172, 173
η, η', ... (and also ξ, ξ', ...)	real vector bundles	106, 107
$\Theta^{\lrcorner}$	the linear map $\bigwedge^m V \to V$ induced by a linear functional Θ on $\bigwedge^{m-1} V$	35
λ, λ', ...	real or complex line bundles	106, 202, 204, 206
λ_ξ	splitting bundle (over P_ξ) of a real vector bundle ξ	3, 202
μ, μ', ...	real or complex line bundles (such as λ, λ', ...)	
$\mu_X \in H_n(X, \dot{X}; \mathbb{Z}/2)$	$\mathbb{Z}/2$ fundamental class of a compact manifold X with boundary $\dot{X}$	249, 253
$\cap\mu_x$	cap product by μ_X	256, 258, 302
ν_f	normal bundle of an immersion $X \xrightarrow{f} \mathbb{R}^{2n-p}$	260
ξ	arbitrary fiber bundle	3, 63, 64
ξ, ξ', ... (and also η, η', ...)	real vector bundles	1, 106, 107
$\xi_{\mathbf{P}}$	principal G-bundle	99

$\xi \mid U$	restriction of ξ to a subspace U, given as the pullback $i^!\xi$ along the inclusion i of U	63
π	the projection $E \to X$ of a family of fibers (especially the projection of a coordinate bundle)	1, 57, 61, 64, 67
$\pi(\gamma^m)^!\gamma^m$	the universal oriented m-plane bundle $\tilde{\gamma}^m$	164
$\pi(\gamma_n^m)^!\gamma_n^m$	a canonical oriented m-plane bundle $\tilde{\gamma}_n^m$	166
σ	section $X \to E$ of a coordinate bundle $E \xrightarrow{\pi} X$	3, 82
$\sigma(X) \subset E$	image of the zero section $X \xrightarrow{\sigma} E$	3, 277
$\sigma_\xi(t)$	the characteristic polynomial of a real vector bundle ξ	220, 222
τ	zero-section $X \to P_{\xi \oplus 1}$ for a real vector bundle ξ	210
$\tau(RP^n)$	tangent bundle of RP^n	138
$\tau(X)$	tangent bundle of any smooth manifold $X \in \mathscr{M}$	2, 4, 130
Φ_ξ	$\mathbb{Z}/2$ Thom isomorphism of a real vector bundle ξ	199
χ_2	the $\mathbb{Z}/2$ Euler characteristic $\mathscr{U} \to \mathbb{Z}/2$	276, 297
Ψ_i	local trivialization $E \mid U_i \to U_i \times F$	57, 67
ψ_i^j	transition function $U_i \cap U_j \to G$	58, 67

Index

D

E

F

M

U

V

W

Z

Pure and Applied Mathematics

A Series of Monographs and Textbooks

RECENT TITLES

CARL L. DEVITO. Functional Analysis

MICHIEL HAZEWINKEL. Formal Groups and Applications

SIGURDUR HELGASON. Differential Geometry, Lie Groups, and Symmetric Spaces

ROBERT B. BURCKEL. An Introduction to Classical Complex Analysis: Volume 1

JOSEPH J. ROTMAN. An Introduction to Homological Algebra

C. TRUESDELL AND R. G. MUNCASTER. Fundamentals of Maxwell's Kinetic Theory of a Simple Monatomic Gas: Treated as a Branch of Rational Mechanics

BARRY SIMON. Functional Integration and Quantum Physics

GRZEGORZ ROZENBERG AND ARTO SALOMAA. The Mathematical Theory of L Systems.

DAVID KINDERLEHRER and GUIDO STAMPACCHIA. An Introduction to Variational Inequalities and Their Applications.

H. SEIFERT AND W. THRELFALL. A Textbook of Topology; H. SEIFERT. Topology of 3-Dimensional Fibered Spaces

LOUIS HALLE ROWEN. Polynominal Identities in Ring Theory

DONALD W. KAHN. Introduction to Global Analysis

DRAGOS M. CVETKOVIC, MICHAEL DOOB, AND HORST SACHS. Spectra of Graphs

ROBERT M. YOUNG. An Introduction to Nonharmonic Fourier Series

MICHAEL C. IRWIN. Smooth Dynamical Systems

JOHN B. GARNETT. Bounded Analytic Functions

EDUARD PRUGOVEČKI. Quantum Mechanics in Hilbert Space, Second Edition

M. SCOTT OSBORNE AND GARTH WARNER. The Theory of Eisenstein Systems

K. A. ZHEVLAKOV, A. M. SLIN'KO, I. P. SHESTAKOV, AND A. I. SHIRSHOV. Translated by HARRY SMITH. Rings That Are Nearly Associative

JEAN DIEUDONNÉ. A Panorama of Pure Mathematics; Translated by I. Macdonald

JOSEPH G. ROSENSTEIN. Linear Orderings

AVRAHAM FEINTUCH AND RICHARD SAEKS. System Theory: A Hilbert Space Approach

ULF GRENANDER. Mathematical Experiments on the Computer

HOWARD OSBORN. Vector Bundles: Volume 1, Foundations and Stiefel-Whitney Classes

IN PREPARATION

ROBERT B. BURCKEL. An Introduction to Classical Complex Analysis: Volume 2

RICHARD V. KADISON AND JOHN R. RINGROSE. Fundamentals of the Theory of Operator Algebras, Volume 1, Volume 2

BARRETT O'NEILL. Semi-Riemannian Geometry: With Applications to Relativity

EDWARD B. MANOUKIAN. Renormalization

E. J. MCSHANE. Unified Integration

A. P. MORSE. A Theory of Sets, Revised and Enlarged Edition
BHASKARA RAO AND BHASKARA RAO. Theory of Charges